FRACTALS, DIFFUSION, AND RELAXATION IN DISORDERED COMPLEX SYSTEMS

A SPECIAL VOLUME OF ADVANCES IN CHEMICAL PHYSICS
VOLUME 133

PART A

FRACTALS, DIFFUSION, AND RELAXATION IN DISORDERED COMPLEX SYSTEMS

ADVANCES IN CHEMICAL PHYSICS

VOLUME 133

PART A

Edited By

WILLIAM T. COFFEY AND YURI P. KALMYKOV

Series Editor

STUART A. RICE

Department of Chemistry
and
The James Franck Institute
The University of Chicago
Chicago, Illinois

WILEY-
INTERSCIENCE
AN INTERSCIENCE PUBLICATION
JOHN WILEY & SONS, INC.

For general information on our other products and services or for technical support, please contact our Customer Care Department within the United States at (800) 762-2974, outside the United States at (317) 572-3993 or fax (317) 572-4002.

Wiley also publishes its books in a variety of electronic formats. Some content that appears in print may not be available in electronic formats. For more information about Wiley products, visit our web site at www.wiley.com.

Library of Congress Catalog Number: 58-9935

ISBN-13 978-0-470-04607-4 (Set)
ISBN-10 0-470-04607-4 (Set)
ISBN-13 978-0-471-72507-7 (Part A)
ISBN-10 0-471-72507-2 (Part A)
ISBN-13 978-0-471-72508-4 (Part B)
ISBN-10 0-471-72508-0 (Part B)

Printed in the United States of America

10 9 8 7 6 5 4 3 2 1

CONTRIBUTORS TO VOLUME 133

ELI BARKAI, Department of Chemistry and Biochemistry, Notre Dame University, Notre Dame, Indiana 46566, USA; and Department of Physics, Bar Ilan University, Ramat Gan 52900, Israel

ALEXANDER BRODIN, Experimentalphysik II, Universität Bayreuth, D 95440 Bayreuth, Germany

THOMAS BLOCHOWICZ, Institute fur Festkörperphysik, Technische Universität Darmstadt, D 64289 Darmstadt, Germany

SIMONE CAPACCIOLI, Dipartimento di Fisica and INFM, Università di Pisa, I-56127, Pisa, Italy; and CNR-INFM Center "SOFT: Complex Dynamics in Structured Systems," Università di Roma "La Sapienza," I-00185 Roma, Italy

RICCARDO CASALINI, Naval Research Laboratory, Washington, DC 20375, USA; and Chemistry Department, George Mason University, Fairfax, Virginia 20030, USA

ALEKSEI V. CHECHKIN, Institute for Theoretical Physics, National Science Center, Kharkov Institute of Physics and Technology, Kharkov 61108, Ukraine

WILLIAM T. COFFEY, Department of Electronic and Electrical Engineering, School of Engineering, Trinity College, Dublin 2 Ireland

YURI FELDMAN, Department of Applied Physics, The Hebrew University of Jerusalem, Jerusalem 91904, Israel

VSEVOLD Y. GONCHAR, Institute for Theoretical Physics, National Science Center, Kharkov Institute of Physics and Technology, Kharkov 61108, Ukraine

PAOLO GRIGOLINI, Department of Physics, University of North Texas, Denton, Texas, 76203 USA; and Department of Physics, University of Pisa, Pisa, Italy

YURI P. KALMYKOV, Laboratoire de Mathématiques et Physique des Systèmes, Universite de Perpignan, 66860 Perpignan Cedex, France

JOSEPH KLAFTER, School of Chemistry, Tel Aviv University, Tel Aviv 69978, Israel

FRIEDRICH KREMER, Universität Leipzig, Fakultat für Physik und Geowissenschaften, 04103 Leipzig, Germany

MASARU KUNO, Department of Chemistry and Biochemistry, Notre Dame University, Notre Dame, Indiana 46566, USA; and Department of Physics, Bar Ilan University, Ramat Gan 52900, Israel

GENNADY MARGOLIN, Department of Chemistry and Biochemistry, Notre Dame University, Notre Dame, Indiana 46556, USA

RALF METZLER, NORDITA–Nordic Institute for Theoretical Physics, DK-2100 Copenhagen Danish Denmark

KIA L. NGAI Naval Research Laboratory, Washington, DC 20375, USA

VITALY V. NOVIKOV, Odessa National Polytechnical University, 65044 Odessa, Ukraine

MARIAN PALUCH, Institute of Physics, Silesian University, 40-007 Katowice, Poland

NOÉLLE POTTIER, Matière et Systèmes Complexes, UMR 7057 CNRS and Université Paris 7—Denis Diderot, 75251 Paris Cedex 05, France

VLADIMIR PROTASENKO, Department of Chemistry and Biochemistry, Notre Dame University, Notre Dame, Indiana 46566, USA; and Department of Physics, Bar Ilan University, Ramat Gan 52900, Israel

ALEXANDER PUZENKO, Department of Applied Physics, The Hebrew University of Jerusalem, Jerusalem 91904, Israel

C. M. ROLAND, Naval Research Laboratory, Washington, DC 20375, USA

ERNST A. RÖSSLER, Experimentalphysik II, Universität Bayreuth, D 95440 Bayreuth, Germany

YAROSLAV RYABOV, Department of Applied Physics, The Hebrew University of Jerusalem, Jerusalem 91904, Israel. Maryland Center of Biomolecular Structure and Organization, University of Maryland, College Park, Maryland 20742-3360, USA

ANATOLI SERGHEI, Fakultat fur Physik und Geowissenschaften, Universität Leipzig, 04103 Leipzig, Germany

SERGEY V. TITOV, Institute of Radio Engineering and Electronics of the Russian Academy of Seciences, Fryazino, Moscow Region, 141190, Russian Federation

BRUCE J. WEST, Mathematical & Information Sciences Directorate, U.S. Army Research Office, Research Triangle Park, North Carolina 27709, USA

INTRODUCTION

Few of us can any longer keep up with the flood of scientific literature, even in specialized subfields. Any attempt to do more and be broadly educated with respect to a large domain of science has the appearance of tilting at windmills. Yet the synthesis of ideas drawn from different subjects into new, powerful, general concepts is as valuable as ever, and the desire to remain educated persists in all scientists. This series, *Advances in Chemical Physics*, is devoted to helping the reader obtain general information about a wide variety of topics in chemical physics, a field that we interpret very broadly. Our intent is to have experts present comprehensive analyses of subjects of interest and to encourage the expression of individual points of view. We hope that this approach to the presentation of an overview of a subject will both stimulate new research and serve as a personalized learning text for beginners in a field.

STUART A. RICE

PREFACE

Fractals, Diffusion, and Relaxation in Disordered Complex Systems, which is the subject of the present anthology, may be said to have evolved in two stages: (1) in the course of conversations with Stuart Rice during a remarkably pleasant lunch at the University of Chicago following the Indianapolis meeting of the American Physical Society in March 2002 and (2) following the Royal Irish Academy Conference on *Diffusion and Relaxation in Disordered Fractal Systems* held in Dublin in September 2002 [1]. During each of these meetings, the necessity of reviewing the progress both experimental and theoretical which has been made in our understanding of physical systems with relaxation differing substantially from exponential behavior was recognized. Furthermore, it was considered that the *Advances in Chemical Physics*, in line with its stated aspirations and with its wide circulation, would provide an ideal means of attaining this goal.

For the best part of three centuries the fractional calculus constituted a subject area mainly of interest to mathematicians. Indeed many great mathematicians such as Leibniz, L'Hôpital, Euler, Fourier, Abel, Liouville, Weierstrass, Riemann, Letnikov, Wiener, Lévy, and Hardy, to name but a few, have contributed to its development (for a historical survey see Ref. 2). In contrast, applications of fractional calculus in other branches of science have appeared only sporadically—for example, the application to the propagation of disturbances on transmission lines in the context of Heaviside's operational calculus and Kohlrausch's stretched exponential decay law [2,14]. However, the situation radically changed toward the end of the last century following the appearance of the famous books of Benoit Mandelbrot on fractals [3]. Thus, over the past few decades, the fractional calculus has no longer been restricted to the realm of pure mathematics and probability theory [2,4]. Indeed many scientists have discovered that the behavior of a variety of complex systems (such as glasses, liquid crystals, polymers, proteins, biopolymers, living organisms, or even ecosystems) may be successfully described by fractional calculus; thus it appears that complex systems governed by fractional differential equations play a dominant role in both the exact and life sciences [5]. In particular in the context of applications in physics and chemistry, the fractional calculus allows one to describe complex systems exhibiting anomalous relaxation behavior in much the same way as the normal relaxation of simple systems [6]. Examples include charge transport in amorphous semiconductors, the spread of contaminants in underground water, relaxation in polymer systems, and tracer dynamics in both polymer networks and arrays of convection rolls, and so on [6]. In general, the diffusion and relaxation processes in such complex systems no longer follow Gaussian statistics so that the temporal evolution of these systems deviates from the corresponding standard laws (where the mean-square displacement of a particle is proportional to the time between observations) for normal diffusion such as exhibited by classical Brownian particles. Furthermore, following the development for complex systems of higher-order experimental resolutions or via a combination of different probe techniques, the deviations from the classical diffusion and relaxation laws have become ever more apparent. Thus the ever larger data windows that are becoming accessible bring ever more refinement to the experimental data [5], with the result that fractional diffusion and kinetic equations have become extremely powerful tools for the description of anomalous relaxation and diffusion processes in such systems. In the present anthology we have tried to present a comprehensive account of the present state of the subject. It is obvious, however, that we cannot survey completely such an enormous area of modern research, and inevitably many important topics will have been omitted. In order to remedy this defect, we remark that the interested reader can find additional information concerning anomalous diffusion and relaxation and applications of fractional calculus in physics, chemistry, biology, radio engineering, and so on, in various review articles and books, a selection of which is given in Refs. 5–23.

Roughly speaking, the contents of the two-volume anthology may be divided into four experimental and seven theoretical chapters that may be described as follows.

Chapter 1, "Dielectric Relaxation Phenomena in Complex Materials," by Y. Feldman, A. Puzenko, and Y. Ryabov, concerns dielectric spectroscopy studies of the structure, dynamics, and macroscopic behavior of materials, which may broadly be described by the generic term *complex systems.* Complex systems constitute an almost universal class of materials including associated liquids, polymers, biomolecules, colloids, porous materials, doped ferroelectric crystals, and so on. These systems are characterized by a new "*mesoscopic*" length scale, intermediate between molecular and macroscopic. The mesoscopic structures of complex systems typically arise from fluctuations or competing interactions and exhibit a rich variety of static and dynamic behavior. This growing field is interdisciplinary; it complements solid-state and statistical physics, and it overlaps considerably with chemistry, chemical engineering, materials science, and even biology. A common theme in complex systems is that while such materials are *disordered* on the molecular scale and *homogeneous* on the macroscopic scale, they usually possess a *certain degree of order* on a *intermediate*, or *mesocopic*, scale due to the delicate balance of interaction and thermal effects. The authors demonstrate how dielectric spectroscopy studies of complex systems can be applied to determine both their structures and dynamics, how they both arise, and how both may influence the macroscopic behavior.

The *glass transition* is an unsolved problem of condensed mater physics. This question is addressed in chapter 2 by T. Blochowicz, A. Brodin, and E. Rössler, entitled "Evolution of the Dynamic Susceptibility in Supercooled Liquids and Glasses." The emergence of the mode coupling theory of the glass transition has prompted the compilation of a large body of information on the glass transition phenomenon as well as on the glassy state that is reviewed in this contribution. Thus this chapter focuses on describing the evolution of the dynamic susceptibility; that is, its characteristic changes while supercooling a molecular liquid. The authors provide information on the relevant molecular dynamics, and a comparison between experiment and theory is given. The phenomenon is essentially addressed from an experimental point of view, by simultaneously discussing the results from three different probe techniques, namely quasi-elastic light scattering, dielectric spectroscopy, and nuclear magnetic resonance spectroscopy. The application of each of the three methods allows one to investigate the dynamics in the 0- to 1-THz frequency range. The crossover from liquid dynamics at the highest temperatures to glassy dynamics at moderate temperatures as well as the crossover to solid-state behavior at the lowest temperatures near the glass transition temperature, is described in detail. In addition, some remarks on the evolution of the susceptibility down to cryogenic temperatures are given. The lesson to be drawn from this contribution is that an understanding of the dynamics of disordered systems can only be achieved by joint application of the various techniques covering a large frequency range.

In many complex systems such as glasses, polymers, and proteins, temporal evolutions differ as we have seen from the conventional exponential decay laws (and are often much slower). Very slowly relaxing systems remain out of equilibrium over very long times, and they display aging effects so that the time scale of response and correlation functions increases with the age of the system (i.e., the time elapsed since its preparation): Older systems relax more slowly than younger ones. Chapter 3 by N. Pottier, entitled "Slow Relaxation, Anomalous Diffusion, and Aging in Equilibrated or Nonequilibrated Environments," describes recent developments in the physics of slowly relaxing out of equilibrium systems. Questions specifically related to out-of-equilibrium dynamics, such as (1) aging effects and (2) their description by means of an effective temperature, are discussed in the framework of a simple model. A system well adapted to the analysis of these concepts is a diffusing particle in contact with an environment, which is either itself in equilibrium (thermal bath) or out of equilibrium (aging medium). In an aging environment, the diffusing particle acts as a thermometer: Independent measurements, at the same age of the medium, of the particle mobility and mean-square displacement yield the effective temperature of the medium.

Time-dependent fluctuations in the spectra of individual molecules appear in many single-molecule experiments. Since the dynamics of a single molecule is typically strongly coupled to the dynamics of the local environment of that molecule, it is not unusual that the time trace of the intensity of a single molecule should exhibit stochastic behavior. It is frequently assumed that the process of photon emission is stationary and ergodic. In contrast, the correlation function of single nanocrystals (or quantum dots) is nonstationary and nonergodic; thus these systems exhibit statistical behaviour very different from other single emitting objects. In this context, G. Margolin, V. Protasenko, M. Kuno, and E. Barkai in Chapter 4, entitled "Power-Law Blinking Quantum Dots: Stochastic and Physical Models." discuss simple models that may explain the nonergodic behaviour of nanocrystals. The authors use a stochastic model to discuss statistical properties of blinking nanocrystals and to illustrate the concept of non ergodicity and aging. They study intensity correlation functions and discuss ensemble average correlation functions for both capped and uncapped nanocrystals. Different modes of aging appear; that is, a nonvanishing dependence of the correlation functions on the age of the system exists, and this dependence has different functional forms in each of the two cases. The authors also discuss nonergodicity of intensity fluctuations for capped nanocrystals, comparing trajectory (time) and ensemble intensity mean values and correlation functions. They analyze experimental data and show that due to weak ergodicity breaking, the time-averaged intensity of blinking dots is a random variable even for long measurement times. The distribution of the time-averaged intensity is not centered around the ensemble-averaged intensity; instead the authors find very large fluctuations, in good agreement with the predictions of stochastic theory.

The main purpose of Chapter 5 by P. Grigolini, "The Continuous-Time Random Walk Versus the Generalized Master Equation," is to show that the interpretation of certain experimental results concerning the spectroscopy of blinking quantum dots and single molecules requires new theoretical methods. He argues that traditional methods of statistical mechanics, based on either the quantum or the classical Liouville equation—and thus based on densities—must be replaced by the continuous-time random walk model introduced by Montroll and Weiss in 1965. To justify this change, the author reviews the recent work done in deriving Lévy anomalous diffusion from a Liouville equation formalism. He demonstrates that this method, which is satisfactory for Poisson statistics, cannot reproduce the numerical and experimental results in the non-Poisson case. Using the continuous-time random walk formalism, the author determines the generalized master equations that should arise from the Liouville method and also proves that such equation are characterized by aging. He shows that, in spite of making a given generalized master equation totally equivalent to the continuous-time random walk picture, an external field perturbing the generalized master equations yields effects distinctly different from those obtained by applying the same external field in the continuous-time random walk picture. Here there is no need for the reader to know *a priori* the projection approach to the generalized master equations, and the fundamentals of continuous-time random walk calculus needed are included in the chapter. Thus this chapter aims at being an elementary introduction to these techniques and thus will be accessible to both researchers and graduate and undergraduate students with no special knowledge of the formalism.

In Chapter 6, entitled "Fractal Physiology, Complexity, and the Fractional Calculus," B. J. West concentrates on describing the new area of medicine called fractal physiology and focuses on the complexity of the human body and the characterization of that complexity through fractal measures. It is demonstrated that not only various anatomical structures within the human body—such as the convoluted surface of the brain, the lining of the bowel, neural networks, and placenta—are fractal, but also the output of many other dynamical physiological systems. For example, the time series for the interbeat intervals of the heart, interbreath intervals, and interstride intervals have all been shown to be fractal or multifractal statistical processes. Consequently, the fractal dimension turns out to be a significantly better indicator of health than more traditional measures, such as heart rate, breathing rate, and average gait. The observation that human physiology is fractal was first made by the author and his collaborators in the 1980s, based on the analysis of the data sets mentioned above. Subsequently, it was determined that the appropriate methodology for describing the dynamics of fractal time series is the fractional calculus, using either the fractional Langevin equation or the fractional diffusion equation, both of which are discussed in a biomedical context. The general goal of this chapter is to understand how complex phenomena in human physiology can be faithfully described using dynamical models involving fractional stochastic differential equations.

Now various structures—for example, aggregates of particles in colloids, certain binary solutions, polymers, composites, and so on—can be conceived as fractal. Materials with a fractal structure belong to a wide class of inhomogeneous media and may exhibit properties differing from those of uniform matter, like crystals, ordinary composites, or homogeneous fluids. Thus in Chapter 7 by V. Novikov, entitled *"Physical Properties of Fractal Structures,"* hierarchical structure models are applied to study the dielectric, conductive, and elastic properties of inhomogeneous media with a chaotic, fractal structure. The power of the fractional calculus is demonstrated using as example the derivation of certain known patterns of anomalous, nonexponential dielectric relaxation of an inhomogeneous medium in the time domain. It is explicitly assumed that the fractional derivative is related to the dimensionality of a temporal fractal ensemble (in the sense that the relaxation times are distributed over a self-similar fractal system). The proposed fractal model of inhomogeneous media exhibiting nonexponential relaxation behavior is constructed by selecting groups of hierarchically subordinated ensembles (subclusters, clusters, superclusters, etc.) from the entire statistical set available. Different relaxation functions are derived assuming that the actual (physical) ensemble of relaxation times is confined between the upper and lower limits of self-similarity. It is predicted that at times shorter than the relaxation time at the lowest (primitive) self-similarity level, the relaxation should be of a classical, Debye-like type, irrespective of the pattern of nonclassical relaxation at longer times. The material described in this chapter can be used in the analysis of the frequency dependence of the dielectric permittivity, the conductivity, and the elastic parameters of various materials. Providing both a critical evaluation of characterization methods and a quantitative description of composition-dependent properties, the material surveyed is of particular interest to researchers in materials and polymer science.

Chapter 8 by W. T. Coffey, Y. P. Kalmykov, and S. V. Titov, entitled "Fractional Rotational Diffusion and Anomalous Dielectric Relaxation in Dipole Systems," provides an introduction to the theory of fractional rotational Brownian motion and microscopic models for dielectric relaxation in disordered systems. The authors indicate how anomalous relaxation has its origins in anomalous diffusion and that a physical explanation of anomalous diffusion may be given via the continuous time random walk model. It is demonstrated how this model may be used to justify the fractional diffusion equation. In particular, the Debye theory of dielectric relaxation of an assembly of polar molecules is reformulated using a fractional noninertial Fokker–Planck equation for the purpose of extending that theory to explain anomalous dielectric relaxation. Thus, the authors show how the Debye rotational diffusion model of dielectric relaxation of polar molecules (which may be described in microscopic fashion as the diffusion limit of a discrete time random walk on the surface of the unit sphere) may be extended via the continuous-time random walk to yield the empirical Cole–Cole, Cole–Davidson, and Havriliak–Negami equations of anomalous dielectric relaxation from a microscopic model based on a kinetic equation, just as the Debye model. These kinetic equations are obtained from a generalization of the noninertial Fokker–Planck equation of conventional Brownian motion to fractional kinetics governed by the Cole–Cole, Cole–Davidson, and Havriliak–Negami relaxation mechanisms. As particular examples, approximate solutions of the fractional diffusion equation are derived for anomalous noninertial rotational diffusion in various potentials. It is shown that a knowledge of the effective relaxation times for *normal* rotational diffusion is sufficient to predict accurately the *anomalous* dielectric relaxation behavior of the system for all time scales of interest. Furthermore, the inertia-corrected Debye model of rotational Brownian motion of polar molecules is generalized to fractional dynamics (anomalous diffusion) using the fractional Klein–Kramers equation. The result can be considered as a generalization of the solution for the normal Brownian motion in a periodic cosine potential to fractional dynamics (giving rise to anomalous diffusion) and also represents a generalization of Fröhlich's model of relaxation over a potential barrier.

Chapter 9 by A.V. Chechkin, V.Y. Gonchar, J. Klafter, and R. Metzler, entitled "Fundamentals of Lévy Flight Processes," reviews recent developments in the fractional dynamics of Lévy flights under the influence of an external force field and for non trivial boundary conditions—in particular, first passage time problems. The Lévy flights are formulated in terms of a space-fractional Fokker–Planck equation, in which the usual Laplacian is replaced by the Riesz–Weyl fractional operator. The authors discuss the intriguing behavior of this type of random process in external fields; for example, for potentials of harmonic or softer types, the variance diverges and the stationary solution has the same Lévy index as the external noise. In contrast, for steeper than harmonic potentials, the solution leaves the basin of attraction of Lévy stable densities, and multimodal structures appear. The first passage time problem of Lévy flights exhibits a universal character in the sense that the force-free first passage time density exhibits Sparre Andersen universality. This is discussed in detail, and it is compared to the problem of first arrival in Lévy flights. The authors also address the question of the validity of Lévy flights as a description of a physical system due to their diverging variance—for example, arguing that for a massive particle, dissipative nonlinearities may lead to a finite variance.

Now on decreasing temperature or increasing pressure a noncrystallizing liquid will *vitrify*; that is, the structural relaxation time, τ_α, becomes so long that the system cannot attain an equilibrium configuration in the time available. Such theories as exist, including the well-known free volume and configurational entropy models, explain the glass transition by invoking a single quantity governing τ_α. Thus the dispersion of the structural relaxation is either so not addressed at all or else derived merely as afterthought and so is independent of τ_α. Thus, in these models the time dependence of the relaxation process bears no fundamental relation to τ_α and its dynamical properties. In Chapter 10 by K. Ngai, R. Casalini, S. Capaccioli, M. Paluch, and C. M. Roland, entitled "Dispersion of the Structural Relaxation and the Vitrification of Liquids," the authors show from disparate experimental data that the dispersion (i.e., time dependence of the relaxation time or distribution of relaxation times) of the structural relaxation originating from many-molecule dynamics is a fundamental parameter governing τ_α and so controls its various properties. Large bodies of experimental data are presented or cited in order to support this conclusion in a convincing fashion. It appears that without considering dispersion as a fundamental physical entity at the outset of any theory of vitrification, many general experimental features of the molecular dynamics of supercooled liquids will remain unexplained.

Glass-forming systems have been studied for decades using a variety of experimental tools measuring microscopic or macroscopic physical quantities. Thus the conjecture that the glass transition has an inherent length scale has led to numerous studies on confined glassy dynamics. In this context, thin polymer films are of special interest. Chapter 11 by F. Kremer and A. Serghei, entitled "Molecular Dynamics in Thin Polymer films," contributes to this discussion. The authors address from an experimental point of view many interesting topics such as ensuring both reproducible preparation and reproducible measurements of thin polymer films, the influence of the molecular architecture of polymers on their dynamics in thin layers, the effect of confinement in thin polymer films giving rise to novel dynamic modes, methods for the determination of the glass transition temperature, and so on.

The Guest Editors and authors are very grateful to the Series Editior, Stuart A. Rice, for the opportunity to produce this anthology. We would like to thank Dr. Sergey V. Titov and Ms Christine Moore for their excellent help in the preparation of the manuscripts. We would also like to thank Dr. David Burns and Dr. Michael Milligan of USAF, EOARD London for facilitating *Window on Science visits* to the United States during the course of which this project was conceived, as well as Professor Werner Blau of Trinity College Dublin for financial support from the HEA Ireland PRTLI Nanomaterials project and the Trinity College Dublin Trust.

November 2005

WILLIAM T. COFFEY
YURI P. KALMYKOV

Dublin and Perpignan

References

1. Proceedings available as special issue (Diffusion and Relaxation in Disordered Complex Systems) of *Journal of Molecular Liquids* **114**, No. 1–3 (2004), guest editor W. T. Coffey.

2. K. B. Oldham and J. Spanier, *The Fractional Calculus*, Academic Press, New York, London, 1974.

3. B. B. Mandelbrot, *Fractals: Form, Chance and Dimension*, Freeman, San Francisco, 1982; *The Fractal Geometry of Nature*, Freeman, San Francisco, 1982.

4. K. S. Miller and B. Ross, *An Introduction to the Fractional Calculus and Fractional Differential Equations*, Wiley, New York, 1993.

5. I. Sokolov, J. Klafter, and A. Blumen, Fractional Kinetics. *Physics Today*, Nov. 2002, p. 48.

6. R. Metzler and J. Klafter, The random walk's guide to anomalous diffusion: A fractional dynamics approach. *Phys. Rep.* **339**, 1 (2000).

7. C. J. F. Böttcher and P. Bordewijk, *Theory of Electric Polarization*, Vol. 2, Elsevier, Amsterdam, 1973.

8. J.-P. Bouchaud and A. Georges, Anomalous diffusion in disordered media: Statistical mechanisms, models and physical applications. *Phys. Rep.* **195**, 127 (1990).

9. N. G. McCrum, B. E. Read, and G. Williams, *Anelastic and Dielectric Effects in Polymeric Solids* Dover, New York, 1991.

10. M. F. Shlesinger, G. M. Zaslavsky, and J. Klafter, Strange Kinetics, *Nature* **363**, 31 (1993).

11. R. Richert and Blumen, eds., *Disorder Effects on Relaxation Processes*, Springer-Verlag, Berlin, 1994.

12. A. Bunde and S. Havlin, eds., *Fractals in Disordered Systems*, Springer-Verlag, Berlin, 1996.

13. A. K. Jonscher, *Universal Relaxation Law*, Chelsea Dielectric Press, London, 1996.

14. W. Paul and J. Baschnagel, *Stochastic Processes from Physics to Finance*, Springer-Verlag, Berlin, 1999.

15. R. Hilfer, ed., *Applications of Fractional Calculus in Physics*, World Scientific, River Edge, N J, 2000.

16. E. Donth, *The Glass Transition: Relaxation Dynamics in Liquids and Disordered Materials*, Springer-Verlag, Berlin, 2000.

17. G. M. Zaslavsky, Chaos, fractional kinetics and anomalous transport. *Phys. Rep.* **371**, 461 (2002).

18. F. Kremer and A. Schönhals, eds., *Broadband Dielectric Spectroscopy*, Springer, Berlin, 2002.

19. A. A. Potapov, *Fractals in Radiophysics and Electromagnetic Detections*, Logos, Moscow, 2002.

20. B. J. West, M. Bologna, and Grigonlini, *Physics of Fractal Operators*, Springer, New York, 2003.

21. W. T. Coffey, Y. P. Kalmykov, and J. T. Waldron, *The Langevin Equation*, 2nd ed, World Scientific, Singapore, 2004.

22. R. Metzler and J. Klafter, The restaurant at the end of the random walk: Recent development in the description of anomalous transport by fractional dynamics. *J. Phys. A: Math Gen* 37, 1505 (2004)

23. J. Klafter and I. Sokolov, Anomalous diffusion spreads its wings. *Physics Today*, Aug. 2005 p. 29.

CONTENTS PART A

CONTENTS PART B

CHAPTER 1

DIELECTRIC RELAXATION PHENOMENA IN COMPLEX MATERIALS

YURI FELDMAN, ALEXANDER PUZENKO, and YAROSLAV RYABOV[1]

*Department of Applied Physics, The Hebrew University of Jerusalem,
Jerusalem 91904, Israel*

CONTENTS

[1]*Present Address:* Center for Biomolecular Structure and Organization, University of Maryland, College Park, Margland 20742-3360,USA.

Fractals, Diffusion, and Relaxation in Disordered Complex Systems: A Special Volume of Advances in Chemical Physics, Volume 133, Part A, edited by William T. Coffey and Yuri P. Kalmykov. Series editor Stuart A Rice.

1

I. INTRODUCTION

Recent years have witnessed extensive research in soft condensed matter physics in order to investigate the structure, dynamics, and macroscopic behavior of *complex systems* (CS). CS are a very broad and general class of materials that are typically noncrystalline. Polymers, biopolymers, colloid systems (emulsions and microemulsions), biological cells, porous materials, and liquid crystals can all be regarded as CS. All of these systems exhibit a common feature: the new *mesoscopic* length scale, intermediate between molecular and macroscopic. The dynamic processes occurring in CS include different length and time scales. Both fast and ultra-slow molecular rearrangements take place within the microscopic, mesoscopic, and macroscopic organization of the systems. A common theme in CS is that while the materials are disordered at the molecular scale and homogeneous at the macroscopic scale, they usually possess a certain amount of order at an intermediate, so-called mesoscopic, scale due to a delicate balance of internal interactions and thermal effects. A simple exponential relaxation law and the classical model of Brownian diffusion cannot adequately describe the relaxation phenomena and kinetics in such materials. This kind of nonexponential relaxation behavior and anomalous diffusion phenomena is today called "strange kinetics" [1,2]. Generally, the complete characterization of these relaxation behaviors requires the use of a variety of techniques in order to span the relevant ranges in frequency. In this approach, *dielectric spectroscopy* (DS) has its own advantages. The modern DS technique may overlap extremely wide frequency (10^{-6}–10^{12} Hz), temperature ($-170\,°C$ to $+500\,°C$), and pressure ranges [3–5]. DS is especially sensitive to intermolecular interactions and is able to monitor cooperative processes at the molecular level. Therefore, this method is more appropriate than any other to monitor such different scales of molecular motions. It provides a link between the investigation of the properties of the individual constituents of the complex material via molecular spectroscopy and the characterization of its bulk properties.

This chapter concentrates on the results of DS study of the structure, dynamics, and macroscopic behavior of complex materials. First, we present an introduction to the basic concepts of dielectric polarization in static and time-dependent fields, before the dielectric spectroscopy technique itself is reviewed for both frequency and time domains. This part has three sections, namely, broadband dielectric spectroscopy, time-domain dielectric spectroscopy, and a section where different aspects of data treatment and fitting routines are discussed in detail. Then, some examples of dielectric responses observed in various disordered materials are presented. Finally, we will consider the experimental evidence of non-Debye dielectric responses in several complex disordered systems such as microemulsions, porous glasses, porous silicon, H-bonding liquids, aqueous solutions of polymers, and composite materials.

In the writing of this chapter we have not sought to cover every aspect of the dielectric relaxation of complex materials. Rather, our aim has been to demonstrate the usefulness of dielectric spectroscopy for such systems, using its application to selected examples as illustrations.

II.   DIELECTRIC POLARIZATION, BASIC PRINCIPLES

A.   Dielectric Polarization in Static Electric Fields

A dielectric sample, when placed in an external electric field E, acquires a nonzero *macroscopic dipole moment* indicating that the dielectric is polarized under the influence of the field. The *polarization* P of the sample, or dipole density, can be presented in a very simple way:

$$P = \frac{\langle M \rangle}{V} \tag{1}$$

where $\langle M \rangle$ is the macroscopic dipole moment of the whole sample volume V, which is comprised of permanent microdipoles (i.e., coupled pairs of opposite charges) as well as dipoles that are not coupled pairs of microcharges within the electroneutral dielectric sample. The brackets $\langle \rangle$ denote ensemble average. In the linear approximation the macroscopic polarization of the dielectric sample is proportional to the strength of the applied external electric field E [6]:

$$P_i = \varepsilon_0 \chi_{ik} E_k \qquad (2)$$

where χ_{ik} is *the tensor of the dielectric susceptibility* of the material and $\varepsilon_0 = 8.854 \times 10^{-12}$ [F·m^{-1}] is the dielectric permittivity of the vacuum. If the dielectric is isotropic and uniform, χ is a scalar and Eq. (2) will be reduced to the more simple form:

$$P = \varepsilon_0 \chi E \qquad (3)$$

According to the macroscopic Maxwell approach, matter is treated as a continuum, and the field in the matter in this case is the direct result of the electric displacement (electric induction) vector D, which is the electric field corrected for polarization [7]:

$$D = \varepsilon_0 E + P \qquad (4)$$

For an uniform isotropic dielectric medium, the vectors D, E, P have the same direction, and the susceptibility is coordinate-independent; therefore

$$D = \varepsilon_0 (1 + \chi) E = \varepsilon_0 \varepsilon E \qquad (5)$$

where $\varepsilon = 1 + \chi$ is the relative *dielectric permittivity*. Traditionally, it is also called the dielectric constant, because in the linear regime it is independent of the field strength. However, it can be a function of many other variables. For example, for time variable fields it is dependent on the frequency of the applied electric field, sample temperature, sample density (or pressure applied to the sample), sample chemical composition, and so on.

1. Types of Polarization

For uniform isotropic systems and static electric fields, from (3)–(5) we have

$$P = \varepsilon_0 (\varepsilon - 1) E \qquad (6)$$

The applied electric field gives rise to a dipole density through the following mechanisms:

Deformation polarization: It can be divided into two independent types:

Electron polarization—the displacement of nuclei and electrons in the atom under the influence of an external electric field. Because electrons are very light, they have a rapid response to the field changes; they may even follow the field at optical frequencies.

Atomic polarization—the displacement of atoms or atom groups in the molecule under the influence of an external electric field.

Orientation polarization: The electric field tends to direct the permanent dipoles. The rotation is counteracted by the thermal motion of the molecules. Therefore, the orientation polarization is strongly dependent on the frequency of the applied electric field and on the temperature.

Ionic polarization: In an ionic lattice, the positive ions are displaced in the direction of an applied field while the negative ions are displaced in the opposite direction, giving a resultant dipole moment to the whole body. The ionic polarization demonstrates only weak temperature dependence and is determined mostly by the nature of the interface where the ions can accumulate. Many cooperative processes in heterogeneous systems are connected with ionic polarization.

To investigate the dependence of the polarization on molecular quantities, it is convenient to assume the polarization P to be divided into two parts: the induced polarization P_α, caused by translation effects, and the dipole polarization P_μ, caused by the orientation of the permanent dipoles. Note that in ionic polarization the transport of charge carriers and their trapping can also create induced polarization.

$$P_\alpha + P_\mu = \varepsilon_0(\varepsilon - 1)E \qquad (7)$$

We can now define two major groups of dielectrics: polar and nonpolar. A *polar dielectric* is one in which the individual molecules possess a permanent dipole moment even in the absence of any applied field; that is, the center of positive charge is displaced from the center of negative charge. A *nonpolar dielectric* is one where the molecules possess no dipole moment unless they are subjected to an electric field. The mixture of these two types of dielectrics is common for complex liquids, and the most interesting dielectric processes occur at their phase borders or liquid–liquid interfaces.

Due to the long range of the dipolar forces, an accurate calculation of the interaction of a particular dipole with all other dipoles of a specimen would be very complicated. However, a good approximation can be made by considering that the dipoles beyond a certain distance, say some radius a, can be replaced by a continuous medium having the macroscopic dielectric properties. Thus, the dipole, whose interaction with the rest of the specimen is being calculated, may be considered as surrounded by a sphere of radius a containing a discrete number of dipoles. In order that this approximation should be valid, the dielectric properties of the whole region within the sphere should be equal to those of a macroscopic specimen; that is, it should contain a sufficient number of molecules to render fluctuations very small [7,8]. This approach can be successfully used also for the calculation of dielectric properties of ionic self-assembled liquids. Here the system can be considered to be mono-dispersed, consisting of spherical polar water droplets dispersed in a nonpolar medium [9].

Inside the sphere where the interactions take place, the use of statistical mechanics is required. To represent a dielectric with dielectric constant ε, consisting of polarizable molecules with a permanent dipole moment, Fröhlich [6] introduced a continuum with dielectric constant ε_∞ in which point dipoles with a moment $\boldsymbol{\mu}_d$ are embedded. In this model, $\boldsymbol{\mu}_d$ has the same nonelectrostatic interactions with the other point dipoles as the molecule had, while the polarizability of the molecules can be imagined to be smeared out to form a continuum with dielectric constant ε_∞ [7].

In this case, the induced polarization is equal to the polarization of the continuum with $\varepsilon = \varepsilon_\infty$, so that one can write

$$P_\alpha = \varepsilon_0(\varepsilon_\infty - 1)E \tag{8}$$

The orientation polarization is given by the dipole density due to the dipoles $\boldsymbol{\mu}_d$. If we consider a sphere with volume V comprised of dipoles, one can write

$$P_\mu = \frac{1}{V}\langle M_d \rangle \tag{9}$$

where $M_d = \sum_{i=1}^{N} (\boldsymbol{\mu}_d)_i$ is the average component in the direction of the field of the moment due to the dipoles in the sphere.

In order to describe the correlations between the orientations (and also between the positions) of the ith molecule and its neighbors, Kirkwood introduced a correlation factor g, which may be written $g = \sum_{j=1}^{N}\langle \cos\theta_{ij}\rangle$, where θ_{ij} denotes the angle between the orientation of the ith and the jth dipole [7]. An approximate expression for the Kirkwood correlation factor can be derived by taking only nearest-neighbor interactions into account. It reads as follows:

$$g = 1 + z\langle \cos\theta_{ij}\rangle \tag{10}$$

Here the sphere is shrunk to contain only the ith molecule and its z nearest neighbors. Correlation factor g will differ from 1 when $\langle \cos\theta_{ij}\rangle \neq 0$ that is, when there is a correlation between the orientations of neighboring molecules. When the molecules tend to direct themselves with parallel dipole moments, $\langle \cos \theta_{ij} \rangle$ will be positive and $g > 1$. When the molecules prefer to arrange themselves with antiparallel dipoles, then $g < 1$. Both cases are observed experimentally [6–8]. If there is no specific correlation, then $g = 1$. If the correlations are not negligible, detailed information about the molecular interactions is required for the calculations of g.

For experimental estimation of the correlation factor g the Kirkwood–Fröhlich equation [7]

$$g\mu_d^2 = \frac{\varepsilon_0 9 k_B T V}{N} \frac{(\varepsilon - \varepsilon_\infty)(2\varepsilon + \varepsilon_\infty)}{\varepsilon(\varepsilon_\infty + 2)^2} \tag{11}$$

is used, which gives the relation between the dielectric constant ε and the dielectric constant of induced polarization ε_∞. Here $k_B = 1.381 \times 10^{-23}$ [J·K^{-1}] is the Boltzmann constant and T is the absolute temperature. The correlation factor is extremely useful in understanding the short-range molecular mobility and interactions in self-assembled systems [10].

B. Dielectric Polarization in Time-Dependent Electric Fields

When an external field is applied to a dielectric, polarization of the material reaches its equilibrium value, not instantaneously but rather over a period of time. By analogy, when the field is suddenly removed, the polarization decay caused by thermal motion follows the same law as the relaxation or decay function of dielectric polarization $\phi(t)$:

$$\phi(t) = \frac{P(t)}{P(0)} \tag{12}$$

where $P(t)$ is a time-dependent polarization vector. The relationship for the dielectric displacement vector $D(t)$ for time-dependent fields may be written as follows [6,11]:

$$D(t) = \varepsilon_0 \left[\varepsilon_\infty E(t) + \int_{-\infty}^{t} \dot{\Phi}(t') E(t - t') dt' \right] \tag{13}$$

In (13) $D(t) = \varepsilon_0 E(t) + P(t)$, and $\Phi(t)$ is the dielectric response function $\Phi(t) = (\varepsilon_s - \varepsilon_\infty)[1 - \phi(t)]$, where ε_s and ε_∞ are the low- and high-frequency limits of the dielectric permittivity, respectively. The complex dielectric permittivity $\varepsilon^*(\omega)$ (where ω is the angular frequency) is connected with the relaxation function by a very simple relationship [6,11]:

$$\frac{\varepsilon^*(\omega) - \varepsilon_\infty}{\varepsilon_s - \varepsilon_\infty} = \hat{L}\left[-\frac{d}{dt}\phi(t)\right] \tag{14}$$

where $\hat{L}$ is the Laplace transform operator, which is defined for an arbitrary time-dependent function $f(t)$ as

$$\hat{L}[f(t)] \equiv F(\omega) = \int_0^\infty e^{-pt} f(t)\, dt \tag{15}$$

$$p = x + i\omega, \quad \text{where} \quad x \to 0 \quad \text{and } i \text{ is an imaginary unit}$$

Relation (14) gives equivalent information on dielectric relaxation properties of the sample being tested both in frequency and in time domain. Therefore the dielectric response might be measured experimentally as a function of either frequency or time, providing data in the form of a dielectric spectrum $\varepsilon^*(\omega)$ or the macroscopic relaxation function $\phi(t)$.

For example, when a macroscopic relaxation function obeys the simple exponential law

$$\phi(t) = \exp(-t/\tau_m) \tag{16}$$

where τ_m represents the characteristic relaxation time, the well-known Debye formula for the frequency-dependent dielectric permittivity can be obtained by substitution of (16) into (15) [6–8,11]:

$$\frac{\varepsilon^*(\omega) - \varepsilon_\infty}{\varepsilon_s - \varepsilon_\infty} = \frac{1}{1 + i\omega\tau_m} \tag{17}$$

For many of the systems being studied, the relationship above does not sufficiently describe the experimental results. The Debye conjecture is simple and elegant. It enables us to understand the nature of dielectric dispersion. However, for most of the systems being studied, the relationship above does not sufficiently describe the experimental results. The experimental data are better described by nonexponential relaxation laws. This necessitates empirical relationships, which formally take into account the distribution of relaxation times.

1. Dielectric Response in Frequency and Time Domains

In the most general sense, non-Debye dielectric behavior can be described in terms of a continuous distribution of relaxation times, $G(\tau)$ [11].

This implies that the complex dielectric permittivity can be presented as follows:

$$\frac{\varepsilon^*(\omega) - \varepsilon_\infty}{\varepsilon_s - \varepsilon_\infty} = \int\limits_0^\infty \frac{G(\tau)}{1 + i\omega\tau} d\tau \qquad (18)$$

where the distribution function $G(\tau)$ satisfies the normalization condition

$$\int\limits_0^\infty G(\tau) d\tau = 1 \qquad (19)$$

The corresponding expression for the decay function is

$$\phi(t) = \int\limits_0^\infty G(\tau) \exp\left(-\frac{t}{\tau}\right) d\tau \qquad (20)$$

By virtue of the relationship (14) between frequency and time representation it must be clearly understood that the $G(\tau)$ calculation does not provide in itself anything more than another way of describing the dynamic behavior of dielectrics in time domain [12]. Moreover, such a calculation is a mathematically ill-posed problem [13,14], which leads to additional mathematical difficulties.

The majority of cases of non-Debye dielectric spectrum have been described by the so-called Havriliak–Negami (HN) relationship [8,11,15]:

$$\varepsilon^*(\omega) = \varepsilon_\infty + \frac{\varepsilon_s - \varepsilon_\infty}{[1 + (i\omega\tau_m)^\alpha]^\beta}, \qquad 0 \le \alpha, \ \beta \le 1 \qquad (21)$$

Here α and β are empirical exponents. The specific case $\alpha = 1$, $\beta = 1$ gives the Debye relaxation law; the case $\beta = 1$, $\alpha \ne 1$ corresponds to the so-called Cole–Cole (CC) equation [16]; and the case $\alpha = 1$, $\beta \ne 1$ corresponds to the Cole–Davidson (CD) formula [17]. The high- and low-frequency asymptotes of relaxation processes are usually assigned to Jonscher's power-law wings $(i\omega)^{(n-1)}$ and $(i\omega)^m$ ($0 < n, m \le 1$ are Jonscher stretch parameters) [18,19]. Notice that the real part $\varepsilon'(\omega)$ of the complex dielectric permittivity is proportional to the imaginary part $\sigma''(\omega)$ of the complex ac-conductivity $\sigma^*(\omega)$, $\varepsilon'(\omega) \propto -\sigma''(\omega)/\omega$, and the dielectric losses $\varepsilon''(\omega)$ are proportional to the real part $\sigma'(\omega)$ of the ac-conductivity, $\varepsilon''(\omega) \propto \sigma'(\omega)/\omega$. The asymptotic power law for $\sigma^*(\omega)$ has been termed *universal* due to its appearance in many types of disordered systems [20,21]. Progress has been made recently in understanding the physical significance of the empirical parameters α, β and the exponents of the Jonscher wings [22–26].

An alternative approach to DS study is to examine the dynamic molecular properties of a substance directly in the time domain. In the linear response approximation, the fluctuations of polarization caused by thermal motion are the same as for the macroscopic rearrangements induced by the electric field [27,28]. Thus, one can equate the relaxation function $\phi(t)$ and the macroscopic dipole correlation function (DCF) $\Psi(t)$ as follows:

$$\phi(t) \cong \Psi(t) = \frac{\langle M(0)M(t)\rangle}{\langle M(0)M(0)\rangle} \tag{22}$$

where $M(t)$ is the macroscopic fluctuating dipole moment of the sample volume unit, which is equal to the vector sum of all the molecular dipoles. The laws governing the DCF are directly related to the structural and kinetic properties of the sample and characterize the macroscopic properties of the system under study. Thus, the experimental function $\Phi(t)$ and hence $\phi(t)$ or $\Psi(t)$ can be used to obtain information on the dynamic properties of the dielectric under investigation.

The dielectric relaxation of many complex systems deviates from the classical exponential Debye pattern (16) and can be described by the Kohlrausch–Williams–Watts (KWW) law or the *stretched exponential law* [29,30]

$$\phi(t) = \exp\left\{-\left(\frac{t}{\tau_m}\right)^\nu\right\} \tag{23}$$

with a characteristic relaxation time τ_m and empirical exponent $0 < \nu \leq 1$. The KWW decay function can be considered as a generalization of Eq. (16) becoming Debye's law when $\nu = 1$. Another common experimental observation of DCF is the asymptotic power law [18,19]

$$\phi(t) = A\left(\frac{t}{\tau_1}\right)^{-\mu}, \quad t \geq \tau_1 \tag{24}$$

with an amplitude A, an exponent $\mu > 0$, and a characteristic time τ_1 which is associated with the effective relaxation time of the microscopic structural unit. This relaxation power law is sometimes referred to in the literature as describing anomalous diffusion when the mean square displacement does not obey the linear dependency $\langle R^2 \rangle \sim t$. Instead, it is proportional to some power of time $\langle R^2 \rangle \sim t^\gamma$ $(0 < \gamma < 2)$ [31–33]. In this case, the parameter τ_1 is an effective relaxation time required for the charge carrier displacement on the minimal structural unit size. A number of approaches exist in order to describe such kinetic processes: Fokker–Planck equation [34], propagator representation [35,36], different models of dc- and ac-conductivity [20,25], and so on.

In the frequency domain, Jonscher's power-law wings, when evaluated by ac-conductivity measurements, sometimes reveal a dual transport mechanism with different characteristic times. In particular, they treat anomalous diffusion as a random walk in fractal geometry [31] or as a thermally activated hopping transport mechanism [37].

An example of a phenomenological decay function that has different short- and long-time asymptotic forms (with different characteristic times) can be presented as follows [38,39]:

$$\phi(t) = A\left(\frac{t}{\tau_1}\right)^{-\mu} \exp\left\{-\left(\frac{t}{\tau_m}\right)^{\nu}\right\} \qquad (25)$$

This function is the product of the KWW and power-law dependencies. The relaxation law (25) in time domain and the HN law (21) in the frequency domain are rather generalized representations that lead to the known dielectric relaxation laws. The fact that these functions have power-law asymptotes has inspired numerous attempts to establish a relationship between their various parameters [40,41]. In this regard, the exact relationship between the parameters of (25) and the HN law (21) should be a consequence of the Laplace transform according to (14) [11,12]. However, there is currently no concrete proof that this is indeed so. Thus, the relationship between the parameters of equations (21) and (25) seems to be valid only asymptotically.

In summary, we must say that unfortunately there is as yet no generally acknowledged opinion about the origin of the non-Debye dielectric response. However, there exist a significant number of different models which have been elaborated to describe non-Debye relaxation in some particular cases. In general these models can be separated into three main classes:

a. The models comprising the first class are based on the idea of a relaxation time distribution and regard non-Debye relaxation as a cumulative effect arising from the combination of a large number of microscopic relaxation events obeying the appropriate distribution function. These models—for example, the concentration fluctuations model [42], the mesoscopic mean-field theory for supercooled liquids [43], or the recent model for ac-conduction in disordered media [20]—are derived and closely connected to the microscopic background of the relaxation process. However, they cannot answer the question concerning the origin of the very elegant empirical equations of (21) or (25).

b. The second type of model is based on the idea of a Debye type relaxation equation combined with derivatives of noninteger order (e.g., Refs. 22, 26, 44, and 45). This approach is immediately able to reproduce all known empirical expressions for non-Debye relaxation. However, they are rather formal models and they do not provide a link to the microscopic relaxations.

c. In a certain sense the third class of models provides the bridge between the two previous cases. On the one hand, they are based on the microscopic relaxation properties, but, on the other hand, reproduce the known empirical expressions for non-Debye relaxation. The most famous and definitely most elaborate example of such a description is the application of the continuous random walk theory to the anomalous transport problem (see the very detailed review of this problem in Ref. 31).

Later, we will discuss in detail two examples of such models: the model of relaxation peak broadening, which describes a relaxation of the Cole–Cole type [46], and the model of coordination spheres for relaxations of the KWW type [47].

C. Relaxation Kinetics

It has already been mentioned that the properties of a dielectric sample are a function of many experimentally controlled parameters. In this regard, the main issue is the temperature dependence of the characteristic relaxation times—that is, relaxation kinetics. Historically, the term "kinetics" was introduced in the field of Chemistry for the temperature dependence of chemical reaction rates. The simplest model, which describes the dependence of reaction rate k on temperature T, is the so-called Arrhenius law [48]:

$$k = k_0 \exp\left(-\frac{E_a}{k_B T}\right) \tag{26}$$

where E_a is the activation energy and k_0 is the preexponential factor corresponding to the fastest reaction rate at the limit $T \to \infty$. In his original paper [48], Arrhenius deduced this kinetic law from transition state theory. The basic idea behind (26) addressed the single-particle transition process between two states separated by the potential barrier of height E_a.

The next development of the chemical reaction rate theory was provided by Eyring [49–51], who suggested a more advanced model

$$k = \frac{k_B T}{\hbar} \exp\left(\frac{\Delta S}{k_B} - \frac{\Delta H}{k_B T}\right) \tag{27}$$

where ΔS is activation entropy, ΔH is activation enthalpy, and $\hbar = 6.626 \times 10^{-34}$ [J·s] is Planck's constant. As in the case of (26), the Eyring law (27) is based on the idea of a transition state. However, in contrast to the Arrhenius model (26), the Eyring equation (27) is based on more accurate evaluations of the equilibrium reaction rate constant, producing the extra factor proportional to temperature.

The models (26) and (27), used to explain the kinetics of chemical reaction rates, were also found to be very useful for other applications. Taking into account the relationship $\tau \sim 1/k$, these equations can describe the temperature dependence of the relaxation time τ for dielectric or mechanical relaxations provided by the transition between the initial and final states separated by an energy barrier.

The relaxation kinetics of the Arrhenius and Eyring types were found for an extremely wide class of systems in different aggregative states [7,52–54]. Nevertheless, in many cases, these laws cannot explain the experimentally observed temperature dependences of relaxation rates. Thus, to describe the relaxation kinetics, especially for amorphous and glass-forming substances [55–59], many authors have used the Vogel–Fulcher–Tammann (VFT) law:

$$\ln\left(\frac{\tau}{\tau_0}\right) = \frac{E_{VFT}}{k_B(T - T_{VFT})} \tag{28}$$

where T_{VFT} is the VFT temperature and E_{VFT} is the VFT energy. This model was first proposed in 1921 by Vogel [60]. Shortly afterwards it was independently discovered by Fulcher [61] and then utilized by Tammann and Hesse [62] to describe their viscosimetric experiments. It is currently widely held that VFT relaxation kinetics has found its explanation in the framework of the Adam and Gibbs model [63]. This model is based on the Kauzmann concept [64,65], which states that the configurational entropy is supposed to disappear for an amorphous substance at temperature T_k. Thus, the coincidence between the experimental data and VFT law is usually interpreted as a sign of cooperative behavior in a disordered glass-like state.

An alternative explanation of the VFT model (28) is based on the free volume concept introduced by Fox and Floury [66–68] to describe the relaxation kinetics of polystyrene. The main idea behind this approach is that the probability of movement of a polymer molecule segment is related to the free volume availability in a system. Later, Doolittle [69] and Turnbull and Cohen [70] applied the concept of free volume to a wider class of disordered solids. They suggested a similar relationship

$$\ln\left(\frac{\tau}{\tau_0}\right) = \frac{v_0}{v_f} \tag{29}$$

where v_0 is the volume of a molecule (a mobile unit) and v_f is the free volume per molecule (per mobile unit). Thus, if the free volume grows linearly with temperature $v_f \sim T - T_k$, the VFT law (28) is immediately obtained.

Later, the VFT kinetic model was generalized by Bendler and Shlesinger [71]. Starting from the assumption that the relaxation of an amorphous solid is caused by some mobile defects, they deduced the relationship between τ and T in the form

$$\ln\left(\frac{\tau}{\tau_0}\right) = \frac{B}{(T - T_k)^{3/2}} \qquad (30)$$

where B is a constant dependent on the defect concentration and the characteristic correlation length of the defects space distribution [71]. Model (30) is not as popular as the VFT law, however it has been found to be useful for some substances [72,73].

Another type of kinetics pattern currently under discussion is related to the so-called Mode-Coupling Theory (MCT) developed by Götze and Sjögren [74]. In the MCT the cooperative relaxation process in supercooled liquids and amorphous solids is considered to be a critical phenomenon. The model predicts a dependence of relaxation time on temperature for such substances in the form

$$\tau \sim (T - T_c)^{-\gamma} \qquad (31)$$

where T_c is a critical temperature and γ is the critical MCT exponent. Relation (31) was introduced for the first time by Bengtzelius et al. [75] to discuss the temperature dependence of viscosity for methyl-cyclopene and later was utilized for a number of other systems [76,77].

Beside monotonous relaxation kinetics, which is usually treated using one of the above models, there is experimental evidence for nonmonotonous relaxation kinetics [78]. Some of these experimental examples can be described by the model

$$\ln\left(\frac{\tau}{\tau_0}\right) = \frac{E_a}{k_B T} + C \exp\left(-\frac{E_b}{k_B T}\right) \qquad (32)$$

which can be applied to the situation when dielectric relaxation occurs within a confined geometry. Now the parameters of (32) are the "confinement factor," C (a small value of C denotes weak confinement), and two characteristic activation energies of relaxation process: E_a, which is the activation energy of the process in the absence of confinement, and E_b, which characterizes the temperature dependency of free volume in a confined geometry [78]. In contrast to the VFT equation, which is based on the assumption of linear growth of sample free volume with temperature, Eq. (32) implies that due to the confinement the free volume shrinks an increase within temperature.

III. BASIC PRINCIPLES OF DIELECTRIC SPECTROSCOPY AND DATA ANALYSES

The DS method occupies a special place among the numerous modern methods used for physical and chemical analysis of materials because it enables investigation of relaxation processes in an extremely wide range of characteristic times (10^5–10^{-13} s). Although the method does not possess the selectivity of NMR or ESR, it offers important and sometimes unique information about the dynamic and structural properties of substances. DS is especially sensitive to intermolecular interactions, and therefore cooperative processes may be monitored. It provides a link between the properties of the bulk and individual constituents of a complex material (see Fig. 1).

However, despite its long history of development, this method is not so widely used because the large frequency range (10^{-6}–10^{12} Hz), overlapped by discrete frequency domain methods, requires a complex and expensive equipment. Also, for various reasons, not all frequency ranges have been equally available for measurement. Thus, investigations of samples with variable properties over time (for example, nonstable emulsions or biological systems) have been difficult to conduct. In addition, low-frequency measurements of conductive systems were strongly limited due to electrode polarization. All these reasons mentioned above mean that reliable information

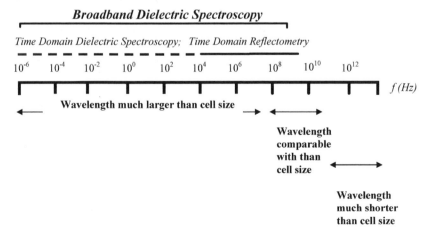

Figure 1.   The frequency band of dielectric spectroscopy.

on dielectric characteristics of a substance can only be obtained in limited frequency ranges. As a result the investigator has only part of the dielectric spectrum at his disposal to determine the relaxation parameters.

The successful development of the time-domain dielectric spectroscopy method (generally called time-domain spectroscopy, TDS) [79–86] and broadband dielectric spectroscopy (BDS) [3,87–90] have radically changed the attitude towards DS, making it an effective tool for investigation of solids and liquids on the macroscopic, mesoscopic, and, to some extent, microscopic levels.

A. Basic Principles of the BDS Methods

As mentioned previously, the complex dielectric permittivity $\varepsilon^*(\omega)$ can be measured by DS in the extremely broad frequency range 10^{-6}–10^{12} Hz (see Fig. 1). However, no single technique can characterize materials over all frequencies. Each frequency band and loss regime requires a different method. In addition to the intrinsic properties of dielectrics, their aggregate state, and dielectric permittivity and losses, the extrinsic quantities of the measurement tools must be taken into account. In this respect, most dielectric measurement methods and sample cells fall into three broad classes [3,4,91]:

(a) Those in which the dielectric properties are measured by means of impedance, Z, or admittance, Y, where the sample in a measuring cell is treated as a parallel or series circuit of an ideal (parallel plate or cylindrical) capacitor and an active resistor. All these methods may be called Lumped-Impedance Methods and are largely used at low frequencies (LF) (10^{-6}–10^7 Hz) and in the radio-frequency (RF) range of the spectrum up to 1 GHz. The electromagnetic (EM) wavelength in these methods is much larger than the sample cell size (see Figure 1). To cover the frequency range 10^{-6}–10^7 Hz, dielectric analyzers that consist of a Fourier correlation analysis in combination with dielectric converters or impedance analyzers (10–10^7 Hz) are used [3,4,90]. At higher frequencies (10^6–10^9 Hz) RF-reflectometry or spectrum analyzers are employed [3,92].

(b) Those in which the dielectric interacts with traveling and standing electromagnetic waves and can be called "Wave Methods" (10^9–10^{11} Hz) [3,4,8]. In this frequency range, network analyzers as well as waveguide and cavity techniques can be applied. The wavelength in these methods is comparable to the sample cell size (see Fig. 1).

(c) Those (10^{10}–10^{12} Hz) in which the wavelengths are much shorter than the sample cell size (see Fig. 1). In these cases, quasi-optical setups like interferometers or oversized cavity resonators are applied [3,4,93]. At sufficiently high frequencies, quasi-optical methods essentially become optical methods.

The LF measurements (a) are provided by means of impedance/admittance analyzers or automatic bridges. Another possibility is to use a frequency response analyzer. In lumped-impedance measurements for a capacitor, filled with a sample, the complex dielectric permittivity is defined as [3]

$$\varepsilon^*(\omega) = \varepsilon'(\omega) - i\varepsilon''(\omega) = \frac{C^*(\omega)}{C_0} \qquad (33)$$

where $C^*(\omega)$ is the complex capacitance and C_0 is the vacuum capacitance respectively. Applying a sinusoidal electric voltage $U^*(\omega) = U_0 \exp(i\omega t)$ to the capacitor, the complex dielectric permittivity can be derived by measuring the complex impedance $Z^*(\omega)$ of the sample as follows:

$$\varepsilon^*(\omega) = \frac{1}{i\omega\varepsilon_0 Z^*(\omega)C_0} \qquad (34)$$

where $Z^*(\omega) = U^*(\omega)/I^*(\omega)$, and $I^*(\omega)$ is the complex current through the capacitor. However, the measuring cells require correction for the residual inductance and capacitance arising from the cell itself and the connecting leads [4,94]. If a fringing field at the edges of parallel plate electrodes causes a serious error, the three-terminal method is effective for its elimination [95].

In general, the "Wave Methods" (b) may be classified in two ways [4,8,91,96]:

(a) They may be traveling-wave or standing wave methods.

(b) They may employ a guided-wave or a free-field propagation medium. Coaxial line, metal and dielectric waveguide, microstrip line, slot line, coplanar waveguide, and optical-fiber transmission lines are examples of guided-wave media while propagation between antennas in air uses a free-field medium.

In guided-wave propagating methods the properties of the sample cell are measured in terms of Scattering parameters or "S-parameters" [4,97], which are the reflection and transmission coefficients of the cell, defined in relation to a specified characteristic impedance, Z_0. In general, Z_0 is the characteristic impedance of the transmission line connected to the cell (50 Ω for most coaxial transmission lines). Note that S-parameters are complex number matrices in the frequency domain, which describe the phase as well as the amplitude of traveling waves. The reflection S-parameters are usually given by the symbols $S_{11}^*(\omega)$ for the multiple reflection coefficients and S_{12}^* for the forward multiple transmission coefficients. In the case of single reflection $S_{11}^*(\omega) = \rho^*(\omega)$ the simplest formula gives the relationship between the reflection coefficient $\rho^*(\omega)$ and impedance of the sample cell terminated by a transmission line with characteristic impedance Z_0:

$$\rho^*(\omega) = \frac{Z^*(\omega) - Z_0}{Z^*(\omega) + Z_0} \qquad (35)$$

In all wave methods the transmission line is ideally matched except the sample holder. If the value of $Z^*(\omega)$ is differ from Z_0, one can observe the reflection from a mismatch of the finite magnitude. A similar type of wave analysis also applies in free space and in any other wave systems, taking into account that in free space $Z_0 \approx 377\,\Omega$ for plane waves.

Several comprehensive reviews on the BDS measurement technique and its application have been published recently [3,4,95,98], and the details of experimental tools, sample holders for solids, powders, thin films, and liquids were described there. Note that in the frequency range 10^{-6}–3×10^{10} Hz the complex dielectric permittivity $\varepsilon^*(\omega)$ can be also evaluated from time-domain measurements of the dielectric relaxation function $\phi(t)$ which is related to $\varepsilon^*(\omega)$ by (14). In the frequency range 10^{-6}–10^5 Hz the experimental approach is simple and less time-consuming than measurement in the frequency domain [3,99–102]. However, the evaluation of complex dielectric permittivity in the frequency domain requires the Fourier transform. The details of this technique and different approaches including electrical modulus $M^*(\omega) = 1/\varepsilon^*(\omega)$ measurements in the low-frequency range were presented recently in a very detailed review [3]. Here we will concentrate more on the time-domain measurements in the high-frequency range 10^5–3×10^{10}, usually called time-domain reflectometry (TDR) methods. These will still be called TDS methods.

B. The Basic Principles of the TDS Methods

TDS is based on transmission line theory in the time domain that allows one to study heterogeneities in coaxial lines as a function of the shape of a test signal [79–86]. As long as the line is homogeneous, the shape of this pulse will not change. However, in the case of a heterogeneity in the line (the inserted dielectric, for example) the signal is partly reflected from the interface and partly passes through it. Dielectric measurements are made along a coaxial transmission line with the sample mounted in a sample cell that terminates the line. The simplified block diagram of the setup common in most TDS methods (except transmission techniques) is presented in Fig. 2. Main differences include the construction of the measuring cell and its position in the coaxial line. These lead to different kinds of expressions for (a) the values that are registered during the measurement and (b) the dielectric characteristics of the objects under study.

A rapidly increasing voltage step $V_0(t)$ is applied to the line and recorded, along with the reflected voltage $R(t)$ returned from the sample and delayed by

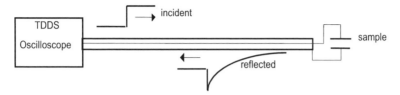

Figure 2. Illustration of the basic principles of the TDS system. (Reproduced with permission from Ref. 113. Copyright 2000, Marcel Dekker, Inc.)

the cable propagation time (Fig. 3). Any cable or instrument artifacts are separated from the sample response due to the propagation delay, thus making them easy to identify and control. The entire frequency spectrum is captured at once, thus eliminating drift and distortion between frequencies.

The complex permittivity is obtained as follows: For nondisperse materials (frequency-independent permittivity), the reflected signal follows the exponential response of the RC line-cell arrangement; for disperse materials, the signal follows a convolution of the line-cell response with the frequency response of the sample. The actual sample response is found by writing the total voltage across the sample as follows:

$$V(t) = V_0(t) + R(t) \tag{36}$$

and the total current through the sample [80,86,103]

$$I(t) = \frac{1}{Z_0}[V_0(t) - R(t)] \tag{37}$$

where the sign change indicates direction and Z_0 is the characteristic line impedance. The total current through a conducting dielectric is composed of the displacement current $I_D(t)$ and the low-frequency current between the capacitor electrodes $I_R(t)$. Since the active resistance at zero frequency of the

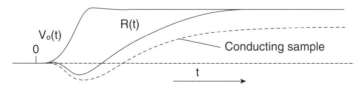

Figure 3. Characteristic shape of the signals recorded during a TDS experiment, $V_0(t)$, incident pulse; $R(t)$, reflected signal. (Reproduced with permission from Ref. 86. Copyright 1996, American Institute of Physics.)

sample-containing cell is [86] (Fig. 3):

$$r = \lim_{t\to\infty} \frac{V(t)}{I(t)} = Z_0 \lim_{t\to\infty} \frac{V_0(t) + R(t)}{V_0(t) - R(t)} \qquad (38)$$

the low-frequency current can be expressed as

$$I_R(t) = \frac{V(t)}{r} = \frac{V_0(t) + R(t)}{Z_0} \lim_{t\to\infty} \frac{V_0(t) - R(t)}{V_0(t) + R(t)} \qquad (39)$$

Thus relation (37) can be written as

$$I_D(t) = \frac{1}{Z_0} \left\{ [V_0(t) - R(t)] - [V_0(t) + R(t)] \lim_{t\to\infty} \frac{V_0(t) - R(t)}{V_0(t) + R(t)} \right\} \qquad (40)$$

Relations (36) and (40) present the basic equations that relate $I(t)$ and $V(t)$ to the signals recorded during the experiment. In addition, (40) shows that TDS permits one to determine the low-frequency conductivity σ of the sample directly in time domain [84–86]:

$$\sigma = \frac{\varepsilon_0}{Z_0 C_0} \lim_{t\to\infty} \frac{V_0(t) - R(t)}{V_0(t) + R(t)} \qquad (41)$$

Using $I(t)$, $V(t)$ or their complex Fourier transforms $i(\omega)$ and $v(\omega)$, one can deduce the relations describing the dielectric characteristics of a sample being tested either in the frequency or the time domain. The final form of these relations depends on the geometric configuration of the sample cell and its equivalent representation [79–86].

The admittance of the sample cell terminated by the coaxial line is then given by

$$Y(\omega) = \frac{i(\omega)}{v(\omega)} \qquad (42)$$

and the sample permittivity can be presented as follows:

$$\varepsilon(\omega) = \frac{Y(\omega)}{i\omega C_0} \qquad (43)$$

To minimize line artifacts and establish a common time reference, (42) is usually rewritten in differential form, to compare reflected signals from the sample and a calibrated reference standard and thus eliminate $V_0(t)$ [79–86].

If one takes into account the definite physical length of the sample and multiple reflections from the air–dielectric or dielectric–air interfaces, relation (43) must be written in the following form [79–84,103]:

$$\varepsilon^*(\omega) = \frac{c}{i\omega(\gamma d)} Y(\omega) X \cot X \qquad (44)$$

where $X = (\omega d/c)\sqrt{\varepsilon^*(\omega)}$, d is the length of the inner conductor, c is the velocity of light, and γ is the ratio between the capacitance per unit length of the cell to that of the matched coaxial cable. Equation (44) in contrast to (43) is a transcendental one, and its exact solution can be obtained only numerically [79–84]. The key advantage of TDS methods in comparison with frequency methods is the ability to obtain the relaxation characteristics of a sample directly in time domain. Solving the integral equation, one can evaluate the results in terms of the dielectric response function $\Phi(t)$ [86,103,104]. It is then possible to associate $\varphi(t) = \Phi(t) + \varepsilon_\infty$ with the macroscopic dipole correlation function $\Psi(t)$ [2,105,106] using linear response theory.

1. Experimental Tools

a. Hardware. The standard time-domain reflectometers used to measure the inhomogeneities of coaxial lines [80, 86,107,108] are the basis of the majority of modern TDS setups. The reflectometer consists of a high-speed voltage step generator and a wide-band registering system with a single- or double-channel sampling head. In order to meet the high requirements of TDS measurements, such commercial equipment must be considerably improved. The main problem is because the registration of incident $V(t)$ and reflected $R(t)$ signals is accomplished by several measurements. In order to enhance the signal–noise ratio, one must accumulate all the registered signals. The high level of drift and instabilities during generation of the signal and its detection in the sampler are usually inherent to the serial reflectometry equipment.

The new generation of digital sampling oscilloscopes [109–111] and specially designed time-domain measuring setups [86] offer comprehensive, high precision, and automatic measuring systems for TDS hardware support. They usually have a small jitter-factor (<1.5 ps), important for rise time; small flatness of incident pulse ($<0.5\%$ for all amplitudes); and in some systems a unique option for parallel time nonuniform sampling of the signal [86].

The typical TDS setup consists of a signal recorder, a two-channel sampler, and a built-in pulse generator. The generator produces 200-mV pulses of 10-µs duration and short rise time ($\sim$30 ps). Two sampler channels are characterized by an 18-GHz bandwidth and 1.5-mV noise (RMS). Both channels are triggered by one common sampling generator that provides their time correspondence

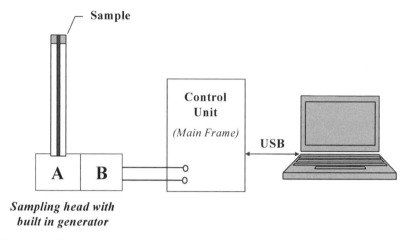

Figure 4. Circuit diagram of a TDS setup. Here A and B are two sampler channels.

during operation. The form of the voltage pulse thus measured is digitized and averaged by the digitizing block of time-domain measurement system (TDMS). The time base is responsible for the major metrology TDMS parameters. The block diagram of the described TDS setup is presented in Fig. 4 [86].

b. Nonuniform Sampling. In highly disordered complex materials, the reflected signal $R(t)$ extends over wide ranges of time and cannot be captured on a single time scale with adequate resolution and sampling time. In an important modification of regular TDR systems, a nonuniform sampling technique (parallel or series) has been developed [86,112].

In the series realization, consecutive segments of the reflected signal on an increasing time scale are registered and linked into a combined time scale. The combined response is then transformed using a running Laplace transform to produce the broad frequency spectra [112].

In the parallel realization a multi-window sampling time scale is created [86]. The implemented time scale is the piecewise approximation of the logarithmic scale. It includes $nw \subseteq 16$ sites with a uniform discretization step determined by the following formula:

$$\delta_{nw} = \delta_1 \times 2^{nw} \qquad (45)$$

where $\delta_1 = 5$ ps is the discretization step at the first site, and the number of points in each step except for the first one is equal to $npw = 32$. At the first site, the number of points $npw_1 = 2^*npw$. The doubling of the number of points at the first site is necessary in order to have the formal zero time position, which is impossible in the case of a strictly logarithmic structure of the scale. In addition, a certain number of points located in front of the zero time position are added. They serve exclusively for the visual estimation of the stability of the time position of a signal and are not used for the data processing.

The described structure of the time scale allows the overlapping of the time range from 5 ps to 10 μs during one measurement, which results in a limited number of registered readings. The overlapped range can be shortened, resulting in a decreasing number of registered points and thus reducing the time required for data recording and processing.

The major advantage of the multi-window time scale is the ability to get more comprehensive information. The signals received by using such a scale contain information within a very wide time range, and the user merely decides which portion of this information to use for further data processing. Also, this scale provides for the filtration of registered signals close to the optimal one already at the stage of recording.

c. Sample Holders. Unfortunately, a universal sample holder that can be used for both liquid and solid samples in both the low- and high-frequency regions of the TDS as well as BDS methods is not yet available. The choice of a sample holder configuration depends on the method, value of dielectric permittivity, dc-conductivity and dielectric losses, state of the substance (liquid, solid, powder, film, etc.), and data treatment procedure. In the lumped capacitor approximation, one can consider three main types of sample holders [86,103,113] (see Fig. 5):

(a) A cylindrical capacitor filled with sample. This cell (a cutoff cell) can also be regarded as a coaxial line segment with the sample having an effective γd length characterized in this case by the corresponding propagation parameters. This makes it possible to use practically identical cells for various TDS and BDS method modification [114]. For the total reflection method the cutoff cell is the most frequent configuration [79–85]. The recent theoretical analysis of the cutoff sample cell (Figure 5a) showed that a lumped-element representation enables the sample cell properties to be accurately determined over a wide frequency range [114].

(b) Another type of sample holder that is frequently used is a parallel plate capacitor terminated to the central electrode at the end of the coaxial line (Figure 5b) [86,103,114–116].

(c) The current most popular sample holder for different applications is an open-ended coaxial line sensor (Figure 5c) [117–125].

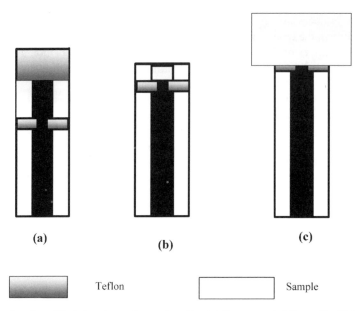

(a)

(b)

(c)

Teflon   Sample

Figure 5. Simplified drawings of sample cells. (a) Open coaxial line cell; (b) lumped capacitance cell; (c) open-ended coaxial cell. (Reproduced with permission from Ref. 113. Copyright 2000, Marcel Dekker, Inc.)

In the case of the lumped capacitance approximation the configurations in Figure 5a,b have high-frequency limitations, and for highly polar systems one must take into account the finite propagation velocity of the incident pulse [82–86,113]. The choice of sample cell shape is determined to a high extent by the aggregate condition of the system under study. While cell (a) is convenient to measure liquids, configuration (b) is more suitable for the study of solid disks, crystals [115,126], and films. Both cell types can be used to measure powder samples. While studying anisotropic systems (liquid crystals, for instance) the user may replace a coaxial line by a strip line or construct the cell with the configuration providing measurements under various directions of the applied electric field [84,85]. The (c) cell type is used only when it is impossible to put the sample into the (a) or (b) cell types [86,121–125,127–129]. The fringing capacity of the coaxial line end is the working capacity for such a cell. This kind of cell is widely used now for investigating the dielectric properties of biological materials and tissues [122–125], petroleum products [119], constructive materials [110], soil [129], and numerous other nondestructive permittivity and permeability measurements. Theory and calibration procedures for such open-coaxial probes are well-developed [129–131], and the results meet the high standards of other modern measuring systems.

2.   Data Processing

Measurement procedures, registration, storage, time referencing, and data analyses are carried out automatically in modern TDS systems. The process of operation is performed in an online mode, and the results can be presented in both frequency and time domain [86]. There are several features of the modern software that control the process of measurement and calibration. One can define the time windows of interest that may be overlapped by one measurement. During the calibration procedure the precise determination of the front edge position is carried out, and the setting of an internal auto center on these positions applies to all subsequent measurements. The precise determination and settings of horizontal and vertical positions of calibration signals are also carried out. All parameters may be saved in a configuration file, allowing for a complete set of measurements using the same parameters without additional calibration.

The data processing software includes the options of signal correction, correction of electrode polarization and dc-conductivity, and different fitting procedures both in time and frequency domain [86].

C.   Data Analysis and Fitting Problems

The principal difference between the BDS and TDS methods is that BDS measurements are accomplished directly in the frequency domain while the TDS operates in time domain. In order to avoid unnecessary data transformation, it is preferable to perform data analysis directly in the domain, where the results were measured. However, nowadays there are no inherent difficulties in transforming data from one domain to another by direct or inverse Fourier transform. We will concentrate below on the details of data analysis only in the frequency domain.

1.   The Continuous Parameter Estimation Problem

Dielectric relaxation of complex materials over wide frequency and temperature ranges in general may be described in terms of several non-Debye relaxation processes. A quantitative analysis of the dielectric spectra begins with the construction of a fitting function in selected frequency and temperature intervals, which corresponds to the relaxation processes in the spectra. This fitting function is a linear superposition of the model functions (such as HN, Jonscher, dc-conductivity terms; see Section II.B.1) that describes the frequency dependence of the isothermal data of the complex dielectric permittivity. The temperature behavior of the fitting parameters reflects the structural and dynamic properties of the material.

However, there are several problems in selecting the proper fitting function, such as the limited frequency and temperature ranges of the experiment, distortion influences of the sample holder, and the overlapping of several physical processes with different amplitudes in the same frequency and temperature ranges. The latter is the most crucial problem, because some of the relaxation processes are "screened" by the others. For such a "screened" process, the confidence in parameter estimations can be very small. Here, the behavior of the parameters may be inconsistent. Despite these discontinuities, there may still be some trends in the parameter behavior of the "screened" processes which may reflect some tendencies of the physical processes in the system. Therefore, it is desirable to obtain a continuous solution of the model parameters via temperature. This solution is hardly achievable if the estimation of the parameters is performed independently for the different temperature points on the selected fitted range. Post-fitted parameter smoothing can spoil the quality of the fit. A new procedure for smooth parameter estimation, named the "global fit," was proposed recently [132]. It obtains a continuous solution for the parameter estimation problem. In this approach, the fitting is performed simultaneously for all the temperature points. The smoothness of the solution is obtained by the addition of some penalty term to the cost function in the parameter minimization problem. Coupled with a constraint condition for the total discrepancy measure between the data and the fit function [132], the desired result is achieved.

The penalized functional approach for obtaining a continuous solution of the minimization problem is a well-known regularization technique in image restoration problems such as image de-noising or image de-blurring [133].

In the field of dielectric spectroscopy, such regularization procedures have been used by Schäfer et al. [14] for extracting the logarithmic distribution function of relaxation times, $G(\tau)$. In contrast to the parametric description of the broadband dielectric spectra considered in our work, the approach of Schäfer et al. is essentially nonparametric. These authors used a regularization technique for the construction of the response function through the Fredholm integral equation solution. The approach proposed in Ref. 132 deals with the problem of (a) finding fitting parameters that describe dielectric data in the frequency domain in a wide frequency band, (b) obtaining a continuous estimation of the fitting parameters via temperature, or (c) finding any other external parameters.

2. dc-Conductivity Problems

The dielectric spectroscopy study of conductive samples is very complicated because of the need to take into account the effect of dc-conductivity. The dc-conductivity σ_0 contributes, in the frequency domain, to the imaginary part of the complex dielectric permittivity in the form of additional function $\sigma_0/(\varepsilon_0\omega)$. The presence of dc-conductivity makes it difficult to analyze relaxation processes especially when the contribution of the conductivity is much greater than the amplitude of the process. The correct calculation of the dc-conductivity is important in terms of the subsequent analysis of the dielectric data. Its evaluation by fitting of the experimental data does not always give correct results, especially when relaxation processes are present in the low-frequency range. In particular, the dc-conductivity function has frequency power-law dependence similar to the Jonscher terms in the imaginary part of the complex dielectric permittivity, and this makes the computation of the dc-conductivity even more difficult.

It is known that in some cases the modulus representation $M^*(\omega)$ of dielectric data is more efficient for dc-conductivity analysis, since it changes the power law behavior of the dc-conductivity into a clearly defined peak [134]. However, there is no significant advantage of the modulus representation when the relaxation process peak overlaps the conductivity peak. Moreover, the shape and position of the relaxation peak will then depend on the conductivity. In such a situation, the real component of the modulus, containing the dc-conductivity as an integral part, does not help to distinguish between different relaxation processes.

Luckily, the real and imaginary parts of the complex dielectric permittivity are not independent of each other and are connected by means of the Kramers–Kronig relations [11]. This is one of the most commonly encountered cases of dispersion relations in linear physical systems. The mathematical technique entering into the Kramers–Kronig relations is the Hilbert transform. Since dc-conductivity enters only the imaginary component of the complex dielectric permittivity the static conductivity can be calculated directly from the data by means of the Hilbert transform.

3. Continuous Parameter Estimation Routine

The complex dielectric permittivity data of a sample, obtained from DS measurements in a frequency and temperature interval can be organized into the matrix data massive $\varepsilon \equiv [\varepsilon_{i,j}]$ of size $M \times N$, where $\varepsilon_{i,j} \equiv \varepsilon(\omega_i, T_j)$, M is the number of measured frequency points, and N is the number of measured temperature points. Let us denote by $f = f(\omega; \mathbf{x})$ the fitting function of n parameters $\mathbf{x} = \{x_1, x_2, \ldots, x_n\}$. This function is assumed to be a linear superposition of the model descriptions (such as the Havriliak–Negami function or the Jonscher function, considered in Section II.B.1). The dependence of f on temperature T can be considered to be via parameters only: $f = f(\omega; \mathbf{x}(T))$. Let us denote by $X \equiv [x_i(T_j)]$ the $n \times N$ matrix of n model parameters x_i, computed at N different temperature points T_j.

The classical approach to the fit parameter estimation problem in dielectric spectroscopy is generally formulated in terms of a minimization problem: finding values of X which minimize some discrepancy measure $S(\varepsilon, \hat{\varepsilon})$ between the measured values, collected in the matrix ε and the fitted values $\hat{\varepsilon} = [f(\omega_i, \mathbf{x}(T_j))]$ of the complex dielectric permittivity. The choice of $S(\varepsilon, \hat{\varepsilon})$ depends on noise statistics [132].

4. Computation of the dc-Conductivity Using the Hilbert Transform

The coupling between the real and imaginary components of the complex dielectric permittivity $\varepsilon^*(\omega)$ is provided by the Kramers–Kronig relations, one of the most general cases of dispersion relations in physical systems. The mathematical technique used by the Kramers–Kronig relations, which allows one component to be defined through another, is the Hilbert transform since $\varepsilon'(\omega)$ and $\varepsilon''(\omega)$ are Hilbert transform pairs. Performing a Hilbert transform of $\varepsilon'(\omega)$ and subtracting the result from $\varepsilon''(\omega)$, the dc-conductivity, $\sigma_0/(\varepsilon_0\omega)$, can be computed directly from the complex dielectric permittivity data. The simulated and experimental examples show very good accuracy for calculating dc conductivity by this method.

The Kramers–Kronig dispersion relations between the real and imaginary parts of the dielectric permittivity can be written as follows [11]:

$$\varepsilon'(\omega) = \varepsilon_\infty + \frac{1}{\pi}\hat{P}\int_{-\infty}^{\infty} \frac{\hat{\varepsilon}''(\omega')}{\omega'-\omega}\, d\omega' \qquad (46)$$

$$\hat{\varepsilon}''(\omega) = \varepsilon''(\omega) - \frac{\sigma_0}{\varepsilon_0\omega} = -\frac{1}{\pi}\hat{P}\int_{-\infty}^{\infty} \frac{\varepsilon'(\omega')}{\omega'-\omega}\, d\omega' \qquad (47)$$

where the symbol $\hat{P}$ denotes the Cauchy principal value of the integral. The Hilbert transform $H[g]$ of a real function $g(t)$ is defined as

$$H[g] = -\frac{1}{\pi}\hat{P}\int_{-\infty}^{\infty} \frac{g(\xi)}{\xi-\omega}\, d\xi \qquad (48)$$

Therefore the conductivity term in the second dispersion relation (47) can be presented as follows:

$$\frac{\sigma_0}{\varepsilon_0\omega} = \varepsilon''(\omega) - \hat{\varepsilon}''(\omega) \qquad (49)$$

The result shows that the dc-conductivity can be computed by using the Hilbert transform applied to the real components of the dielectric permittivity function and subtracting the result from its imaginary components. The main obstacle to the practical application of the Hilbert transform is that the integration in Eq. (48) is performed over infinite limits; however, a DS spectroscopy measurement provides values of $\varepsilon^*(\omega)$ only over some finite frequency range. Truncation of the integration in the computation of the Hilbert transform can yield a serious computational error in calculating $H[\varepsilon'(\omega)]$ in the measured frequency range. This problem cannot be overcome unless the "missing" dielectric data is supplied. However, the computational error can be reduced by extending $\varepsilon'(\omega)$ into a frequency domain outside the measuring frequency range. Although this is a rather crude data treatment, computer simulations show that computational error due to the truncation of the measuring frequency range is greater near the borders of the range. Far from the extrema of the frequency range the relative error is much smaller and is of the order of 10^{-4}.

While our method works well with most situations, it is limited when $\varepsilon'(\omega)$ exhibits a low-frequency tail. Such a situation is characteristic of percolation, electrode polarization, or other low-frequency process, where the reciprocal of the characteristic relaxation time for the process is just below our frequency window. In this case, aliasing effects distort the transform result.

Practically, the Numerical Hilbert transform can be computed by means of the well-known Fast Fourier Transform (FFT) routine. It is based on the following property of the Hilbert transform [135]. If

$$g(t) = \frac{1}{\pi} \int\limits_0^\infty [a(\omega) \cos \omega t + b(\omega) \sin \omega t]\, d\omega \qquad (50)$$

is the Fourier transform of a real function $g(t)$, then the Fourier-transform of the function $H[g]$ is the following:

$$H[g] = \frac{1}{\pi} \int\limits_0^\infty [b(\omega) \cos \omega t - a(\omega) \sin \omega t]\, d\omega \qquad (51)$$

Thus, by performing the Fourier transform of the data with an FFT algorithm, the Hilbert transform is computed by inverse-FFT operation to the phase-shifting version of the Fourier transform of the original data. Such an approach was realized in the work of Castro and Nabet [136], where the real component of the dielectric permittivity was calculated from its imaginary component, using the Hilbert transform. For the Hilbert transform computation, the authors used a procedure included in the MATLAB package. This methodology was also based on the FFT technique, requiring uniform sampling over the frequency interval. If the data are not measured uniformly, it should be interpolated to frequency points, evenly spaced with an incremental frequency equal to or less than the start frequency. However, this routine cannot be used for a wide spectral range. For example, in order to cover an interval of 12 decades with an incremental frequency equal to the start frequency, 10^{12} points are required. This, of course, is not practical. To overcome this limitation, a procedure based on a moving frequency window has been developed, where the scale inside the window is linear, but the window jump is logarithmic. This kind of methodology employing moving windows with FFT has been used in the past [137].

5. Computing Software for Data Analysis and Modeling

Software for dielectric data treatment and modeling in the frequency domain has been developed recently [132]. This program (MATFIT) was built around the software package MATLAB (Math Works Inc.), and it is available through an intuitive visual interface. Key features of the program include:

- Advanced data visualization and preprocessing tools for displaying complex dielectric permittivity data and selecting appropriate frequency and temperature intervals for modeling
- A library of standard relaxation fit functions
- Simultaneous fit of both real and imaginary components of the complex dielectric permittivity data
- Linear and nonlinear fitting methods, from least-squares and logarithmic to fitting procedures based on the entropy norm
- Global fit procedure on all selected temperature ranges for continuous parameter estimation
- Hilbert transform for computing dc-conductivity
- Parameter visualization tool for displaying fitting parameter functions via temperature and subsequent analysis of the graphs

The methodology described above, utilized in the presented program [132], reduces the problem of dielectric data analysis to choosing the appropriate model functions and an estimation of their model parameters. The penalized maximum likelihood approach for obtaining these parameters as a function of temperature has proven to be a consistent method for accurately obtaining the global minimum in this estimation. This methodology is a phenomenological approach to obtaining the underlying temperature dependence of the parameter space, while not presupposing a particular physical model. The risk is present that such a regularization routine may perturb the result if used excessively. For this reason a regularization parameter is used to control the smoothing term. However, this risk is far less than the risk of *a priori* conclusion of the result according to the researcher's personal belief in a preferred analytical model.

IV. DIELECTRIC RESPONSE IN SOME DISORDERED MATERIALS

In this contribution we present some concepts of modern dielectric spectroscopy. We are going to illustrate these ideas by experimental examples related to different complex systems. In many cases we will refer different features of dielectric response observed in the same samples. Parallel with the description of morphology for each system, the phenomenological identification of the relaxation processes and their relation to dynamic of structural units is established. In spite of the structural and dynamic differences, the general features of the relaxation processes will be indicated.

A. Microemulsions

Microemulsions are thermodynamically stable, clear fluids, composed of oil, water, surfactant, and sometimes co-surfactant that have been widely investigated during recent years because of their numerous practical applications. The chemical structure of surfactants may have a low molecular weight as well as being polymeric, with nonionic or ionic components [138–141]. For a water/oil-continuous (W/O) microemulsion, at low concentration of the dispersed phase, the structure consists of spherical water droplets surrounded by a monomolecular layer of surfactant molecules whose hydrophobic tails are oriented toward the continuous oil phase (see Fig. 6). When the volume fractions of oil and water are high and comparable, random bicontinuous structures are expected to form.

The structure of the microemulsion depends on the interaction between droplets. In the case of repulsive interaction, the collisions of the droplets are short and no overlapping occurs between their interfaces. However, if the

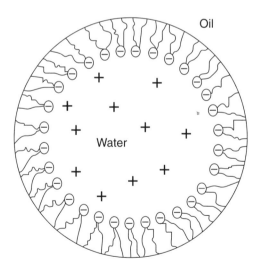

Figure 6. Schematic presentation of spherical water droplet surrounded by a monomolecular layer of ionic surfactant molecules.

interactions are attractive, transient random droplet clusters are formed. The number and sizes of such clusters are increasing with the temperature, the pressure, or the water-to-surfactant ratio, leading to a percolation in the system [113,142–145].

The majority of the different chemical and physical properties, as well as the morphology of microemulsions, is determined mostly by the micro-Brownian motions of its components. Such motions cover a very wide spectrum of relaxation times ranging from tens of seconds to a few picoseconds. Given the complexity of the chemical makeup of microemulsions, there are many various kinetic units in the system. Depending on their nature, the dynamic processes in the microemulsions can be classified into three types:

The first type of relaxation processes reflects characteristics inherent to the dynamics of single droplet components. The collective motions of the surfactant molecule head groups at the interface with the water phase can also contribute to relaxations of this type. This type can also be related to various components of the system containing active dipole groups, such as cosurfactant, bound, and free water. The bound water is located near the interface, while "free" water, located more than a few molecule diameters away from the interface, is hardly influenced by the polar or ion groups. For ionic microemulsions, the relaxation contributions of this type are expected to be related to the various processes associated with the movement of ions and/ or surfactant counterions relative to the droplets and their organized clusters and interfaces [113,146].

For percolating microemulsions, *the second* and the *third types* of relaxation processes characterize the collective dynamics in the system and are of a *cooperative nature*. The dynamics of the *second type* may be associated with the transfer of an excitation caused by the transport of electrical charges within the clusters in the percolation region. The relaxation processes of the *third type* are caused by rearrangements of the clusters and are associated with various types of droplet and cluster motions, such as translations, rotations, collisions, fusion, and fission [113,143].

Dielectric spectroscopy can be successful in providing unique information about the dynamics and structure of microemulsions on various spatial and temporal scales. Being sensitive to percolation, DS is expected to provide unambiguous conclusions concerning the stochastic type, the long time scale cooperative dynamics, and the imposed geometric restrictions of molecular motions before, during, and after the percolation threshold in microemulsions. It also can give valuable information about fractal dimensions and sizes of the percolation clusters.

The percolation behavior is manifested by the rapid increase in the dc electrical conductivity σ and the static dielectric permittivity ε_s as the system approaches the percolation threshold (Fig. 7).

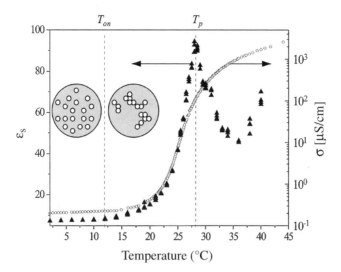

Figure 7. The percolation behavior in AOT–water–decane microemulsion (17.5:21.3:61.2 vol%) is manifested by the temperature dependences of the static dielectric permittivity ε_s (▲ left axis) and conductivity r (◯ right axis). T_{on} is the temperature of the percolation onset; T_p is the temperature of the percolation threshold. Insets are schematic presentations of the microemulsion structure far below percolation and at the percolation onset. (Reproduced with permission from Ref. 149. Copyright 1998, Elsevier Science B.V.)

The dielectric relaxation properties in a sodium bis(2-ethylhexyl) sulfosuccinate (AOT)–water–decane microemulsion near the percolation temperature threshold have been investigated in a broad temperature region [47,143,147]. The dielectric measurements of ionic microemulsions were carried out using the TDS in a time window with a total time interval of ~ 1 μs. It was found that the system exhibits a complex nonexponential relaxation behavior that is strongly temperature-dependent (Figure 8).

An interpretation of the results was accomplished using the dynamic percolation model [148]. According to this model, near the percolation threshold, in addition *to the fast relaxation* related to the dynamics of droplet components ($\tau_1 \cong 1$ ns) [149], there are at least two much longer characteristic time scales. *The longest process* has a characteristic relaxation time greater than a few microseconds and is associated with the rearrangement of the typical percolation cluster. The temporal window of the intermediate process is a function of temperature. This *intermediate process* reflects the *cooperative relaxation phenomenon* associated with the transport of charge carriers along the percolation cluster [148,150,151]. Thus, due to the cooperative nature of relaxation, the DCF decay behavior (see Figure 9) contains information

YURI FELDMAN ET AL.

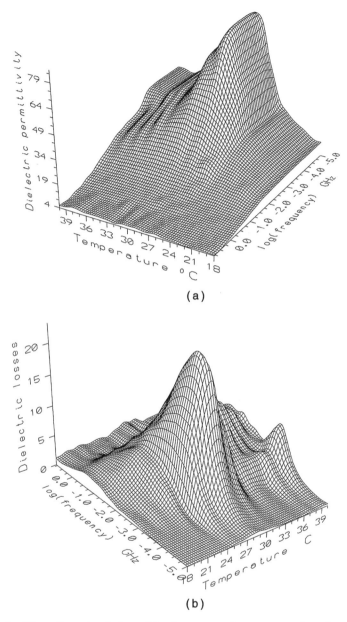

(a)

(b)

Figure 8. Three-dimensional plots of the frequency and temperature dependence of the dielectric permittivity ε' (a) and dielectric losses ε'' (b) for AOT–water–decane microemulsion. (Reproduced with permission from Ref. 143. Copyright 1995, The American Physical Society.)

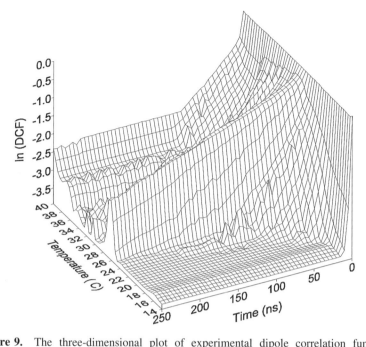

Figure 9. The three-dimensional plot of experimental dipole correlation function versus time and temperature. The percolation threshold temperature $T_p = 26.5°C$. (Reproduced with permission from Ref. 47. Copyright 1996, The American Physical Society.)

regarding the transient cluster morphology on the mesoscale reflecting the dynamical character of percolation.

The type of the relaxation law seen in time domain is strongly dependent on the distance from the percolation threshold. Figure 10 shows in log–log coordinates that at the percolation onset temperature ($\sim14°C$) the relaxation follows a fractional power law: $\Psi(t) \sim (t/\tau_1)^{-\mu}$.

By the same token, in the variables $\log\Psi$ versus $\log(t/\tau_1)$ near the percolation threshold $T_p = 26.5°C$, the relaxation law changes from a power law to stretched exponential behavior, that is, $\Psi(t) \sim \exp[-(t/\tau_m)^{\nu}]$ (see Fig. 11). In the crossover region the relaxation law is considered to be a product of both the power law and stretched exponential terms described by (25). The results of the fitting of the experimental dipole correlation functions to Eq. (25) are shown in Figs. 12 and 13.

One can see (Fig. 12) that the magnitude of parameter μ decreases to almost zero as the temperature approaches the percolation threshold temperature. This effect confirms the statement mentioned above that at the percolation threshold

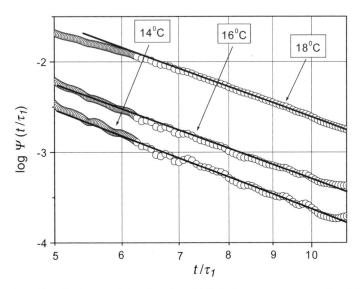

Figure 10. The dipole correlation function $\psi(t/\tau_i)$ demonstrates power-law behavior for the temperature region near the percolation onset ($T_{on} = 12°C$). (Reproduced with permission from Ref. 2. Copyright 2002, Elsevier Science B.V.)

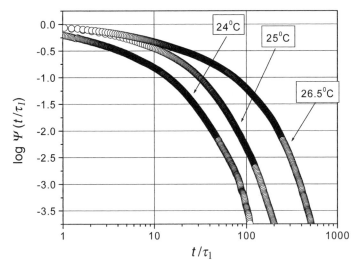

Figure 11. The dipole correlation function $\psi(t/\tau_1)$ demonstrates KWW behavior near the temperature of the percolation threshold ($T_p = 26.5°C$). (Reproduced with permission from Ref. 2. Copyright 2002, Elsevier Science B.V.)

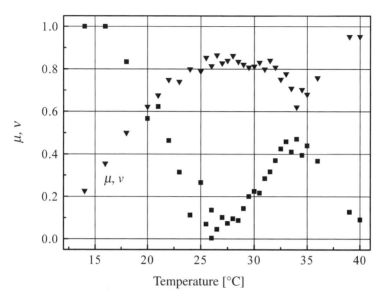

Figure 12. The temperature dependence of the exponents μ (■) and ν (▼) illustrate the transformation of the dipole correlation function $\psi(t/\tau_1)$ from the power-law pattern to the KWW behavior at the percolation threshold. (Reproduced with permission from Ref. 2. Copyright 2002, Elsevier Science B.V.)

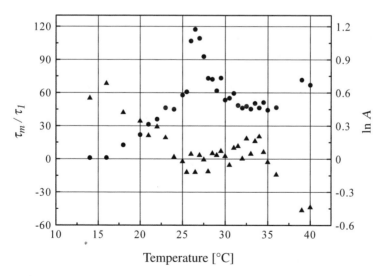

Figure 13. The temperature dependence of the parameter τ_m (• left axis) and A (▲ right axis), obtained by the fitting of the relaxation law (25) to the experimental data. (Reproduced with permission from Ref. 2. Copyright 2002, Elsevier Science B.V.)

temperature the behavior of the dipole correlation function is of the KWW type. The stretched parameter ν changes its value from ~ 0.2 near the percolation onset to ~ 0.8 in the vicinity of the percolation threshold. Notwithstanding that the value of ν is not equal to zero at percolation onset, note that the stretched exponential term with $\nu = 0.2$ changes imperceptibilty in the considerable time interval ($\sim 1\,\mu s$) and the decay of the DCF $\psi\,(t)$ is governed mainly by the power law.

Figure 13 plots the relaxation times ratio τ_m/τ_1 and the amplitude A corresponding to the macroscopic relaxation time of the decay function determined by (25). Near the percolation threshold, τ_m/τ_1 exhibits a maximum and exhibits the well-known *critical slowing down* effect [152]. The description of the mechanism of the cooperative relaxation in the percolation region will be presented in Section V.B.

B. Porous Materials

Non-Debye dielectric relaxation in porous systems is another example of the dynamic behavior of complex systems on the mesoscale. The dielectric properties of various complex multiphase systems (borosilicate porous glasses [153–156], sol–gel glasses [157,158], zeolites [159], and porous silicon [160,161]) were studied and analyzed recently in terms of cooperative dynamics. The dielectric response in porous systems will be considered here in detail using two quite different types of materials, namely, porous glasses and porous silicon.

1. Porous Glasses

Porous silica glasses obtained from sodium borosilicate glasses are defined as bicontinuous random structures of two interpenetrating percolating phases, the solid and the pore networks. The pores in the glasses are connected to each other, and the pore size distribution is narrow. The characteristic pore spacing depends on the method of preparation and can be between 2 and 500 nm [156,162]. A rigid SiO_2 matrix represents the irregular structure of porous glasses. Water can be easily adsorbed on the surface of this matrix. Recently, the dielectric relaxation properties of two types of silica glasses have been studied intensively over broad frequency and temperature ranges [153–156]. The typical spectra of the dielectric permittivity and losses associated with the relaxation of water molecules of the adsorptive layer for the studied porous glasses versus frequency and temperature are displayed in Fig. 14.

Figure 14 shows that the complex dielectric behavior can be described in terms of the four distributed relaxation processes as follows:

The first relaxation process, which is observed in the low-temperature region from $-100°C$ to $+10°C$ is due to the reorientation of the water molecules in ice-like water cluster structures. It was shown that the hindered dynamics of the water molecules located within the pores reflects the interaction of the absorptive layer with the inner surfaces of the porous matrix [153,155].

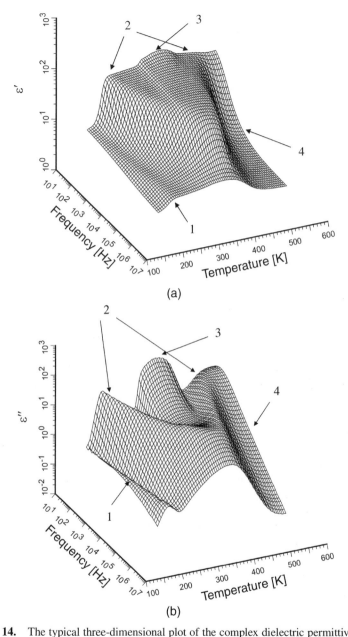

Figure 14. The typical three-dimensional plot of the complex dielectric permittivity real ε' (a) and imaginary part ε'' (b) versus frequency and temperature for porous glass (sample *E*). All the details of different sample preparation and their properties are presented in Ref. 156. (Reproduced with permission from Ref. 2. Copyright 2002, Elsevier Science B.V.)

YURI FELDMAN ET AL.

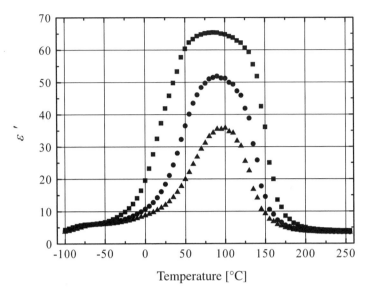

Figure 15. Typical temperature dependence (for sample E [156]) of the complex dielectric permittivity of the real part of different frequencies (■ 8.65 kHz; • 32.4 kHz; ▲ 71.4 kHz). (Reproduced with permission from Ref. 2. Copyright 2002, Elsevier Science B.V.)

The second relaxation process has a specific saddle-like shape and manifests itself in the temperature range of −50°C to +150°C. This relaxation process is thought to be a kinetic transition due to water molecule reorientation in the vicinity of a defect [155].

The third relaxation process is located in the low-frequency region and the temperature interval 50°C to 100°C. The amplitude of this process essentially decreases when the frequency increases, and the maximum of the dielectric permittivity versus temperature has almost no temperature dependence (Fig 15). Finally, the low-frequency ac-conductivity σ demonstrates an *S-shape* dependency with increasing temperature (Fig. 16), which is typical of percolation [2,143,154]. Note in this regard that at the lowest-frequency limit of the covered frequency band the ac-conductivity can be associated with dc-conductivity σ_0 usually measured at a fixed frequency by traditional conductometry. The dielectric relaxation process here is due to percolation of the apparent dipole moment excitation within the developed fractal structure of the connected pores [153,154,156]. This excitation is associated with the self-diffusion of the charge carriers in the porous net. Note that as distinct from dynamic percolation in ionic microemulsions, the percolation in porous glasses appears via the transport of the excitation through the geometrical static fractal structure of the porous medium.

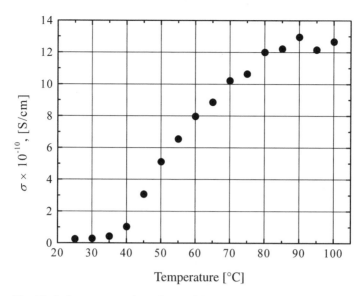

Figure 16.   Typical temperature dependence of the low frequency ac-conductivity σ of the porous glass sample E. (Reproduced with permission from Ref. 2. Copyright 2002, Elsevier Science B.V.)

The fourth relaxation process is observed in the high-temperature region, above 150°C, where the glasses become markedly electrically conducting and show an increase of dielectric constant and dielectric losses in the low-frequency limit. This relaxation process is most probably related to the interface polarization process as a result of the trapping of free charge carriers at the matrix–air interface inside the sample. It causes a buildup of macroscopic charge separation or space charge with a relatively long temperature-independent relaxation time. As shown [153,156], the value of the relaxation time is closely correlated with the pore sizes: The larger the pores, the slower the relaxation process. The description of the mechanism of cooperativity in confinement (Process II) and relaxation in the percolation region (Process III) will be presented below.

2.   *Porous Silicon*

Non-Debye dielectric relaxation was also observed in porous silicon (PS) [25,160,161]. PS has attracted much attention recently, mainly due to its interesting optical and electro-optical properties that can be utilized for device applications [164,165]. So far, most of the activity in this field has focused on the intense visible photoluminescence (PL) from nano-PS and the underlying physical mechanism that is responsible for the generation of light. In addition, transport and dielectric relaxation phenomena in PS have also attracted considerable attention for injection-type PS devices. It was mentioned in the previous section that the correlation between the morphology of porous media and their dielectric properties has already been studied [153,154,166]. In many porous media, the dielectric response is directly related to the fractality and the nano- and meso-structural properties of these disordered systems [153,156–159]. In principle, one would expect to find a similar correlation between the micro-geometry and the dielectric properties of PS media. For example, dc-conductivity measurements demonstrate a dual transport mechanism that has been assigned to thermally activated hopping and excited charged carrier tunneling [167]. As a result, carriers excited to the band tail would give rise to a thermally activated dc-conduction with activation energy of about 0.5 eV [163,168]. This activation energy is less than half the optical bandgap of PS deduced from PL experiments [164]. The ac-conductivity measurements in PS revealed complex transport properties due to a random walk in fractal geometry and thermally activated hopping, as in the case of dc-conductivity [163]. Therefore, it is commonly accepted that both the nano-geometry, the nature of the Si nanocrystallites that form the PS medium and their surfaces as well as the host matrix all contribute to the electrical and optical properties of PS.

The dielectric relaxation properties in nano-PS with different thicknesses have been investigated recently over a wide range of frequency and temperature [160,161]. The dielectric properties of the PS samples were measured in the 20-Hz to 1-MHz frequency range and in the 173 to 493 K temperature interval. For all the dielectric measurements, the amplitude of a sinusoidal ac-voltage source was maintained 1 V so that the average electric field across our sample was of the order of hundreds of volts per centimeter depending on the sample thickness. It was verified that the response was linear with respect to the ac-voltage amplitude such that a linear response analysis could be utilized for our sample.

Three-dimensional plots of both the measured real part ε' and the imaginary part ε'' of the complex dielectric permittivity versus frequency and temperature for 20-μm-thickness PS sample are shown in Fig. 17a,b. From the figure, one can identify three distinct processes, marked by I, II, and III, defined as follows:

Low-temperature process I: This process extends over low temperatures (170–270 K). Despite the fact that both the real and imaginary parts of the dielectric function display this process, it can be appreciated most by examining ε'' at high frequencies and low temperatures where the local maximum, which shifts to higher frequencies with increasing temperature, can be easily detected. Process I represents the existence of additional groups of exited states in PS, which contribute to a thermally activated transport processes [160].

Mid-temperature process II: This process extends over mid-range temperatures (300–400 K) and over low to moderate frequencies (up to 10^5 Hz). The mid-temperature process was associated with the percolation of charge excitation within the developed fractal structure of connected pores at low

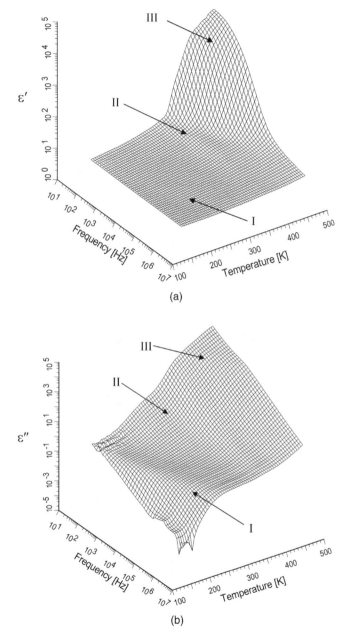

Figure 17. Three-dimensional plots of the frequency and temperature dependence of the real ε' (a) and imaginary part ε'' (b) of the complex dielectric permittivity for the 20-μm PS sample. (Reproduced with permission from Ref. 2. Copyright 2002, Elsevier Science B.V.)

frequencies and with an activated hopping conductivity between neighboring Si crystallites at high-frequency tail [160].

High-temperature process III: This process is very significant at high temperatures (>400 K). Its amplitude increases very rapidly with decreasing frequency for both the real and the imaginary part of the dielectric function. One of the sources of its behavior is the large dc-conductivity of the sample that appears at high temperatures.

Similar processes were also observed for 30-μm-thickness samples. The detailed description of the mechanism of the mid-temperature process in the percolation region and porosity determination in PS will be presented below.

C. Ferroelectric Crystals

Amongst the perovskite family of ferroelectrics KTN crystals ($KTa_{1-x}Nb_xO_3$) have attracted great interest both as model systems and for their optical properties. They were first systematically studied in 1959 [169], and a complete microscopic description of the ferroelectric phase transition was given in 2001 [170]. They display both displacive-like and order–disorder-like properties [171–174]. For niobium (Nb) concentrations of $x > 0.2$ the ferroelectric phase transition is of the first order and follows the linear rule for ferroelectric phase transition temperature $T_c \approx 682x$ [175]. At the ferroelectric phase transition the crystal structure is transformed from cubic to tetragonal. Further cooling incurs two additional structural transitions: tetragonal to orthorhombic, and orthorhombic to rhombohedral [175]. It was recently shown that the interaction of the off-center Nb ion with the soft mode phonon of the crystal in fact governs the phase transition [170]. At the phase transition the distortion of the crystal lattice caused by the off-center position of the Nb ion leads to the creation of virtual dipoles in the crystal lattice which are randomly distributed throughout the crystal. The resultant local fields strongly polarize the lattice leading to long-range cooperativity, frequently of length scales 1000–10,000 Å [176], indicative of the ferroelectric phase. The addition of transition metal ions to this system leads to further novel properties. Of particular interest is the addition of small quantities of copper (Cu) ions. Copper, in small concentrations, approximately 1.5×10^{-4} molar concentration, is known to greatly enhance the photorefractive effect in these crystals [177], leading to important electroholographic applications [178]. Generally, Cu ions occupy the potassium site in the lattice and sit off-center. This virtual dipole is the source of further relaxations in the crystal.

The dielectric behavior of copper-doped and pure KTN crystals were compared over a wide range of temperature and frequency in order to study the effect of such small Cu ion concentrations on the dielectric landscape [179]. The two KTN crystals studied were grown using the top seeded solution growth method [180]. The Ta/Nb ratio in both crystals was estimated by Perry's linear relation [175] linking T_c to the concentration of Nb, $T_c = 682x + 33.2$, and was found to be approximately 62/38 per mole. The first crystal (crystal #1) was doped with copper. The doping level was 2% in the flux yielding approximately 1.5×10^{-4} per mole in the grown crystal. The second crystal (crystal #2) was a pure KTN crystal. Samples of $1 \times 1 \times 2 \, \text{mm}^3$ were cut from the grown bole along the crystallographic [001] axes. The $x-y$ faces of the samples (perpendicular to the growth direction z) were polished and coated with gold electrodes. Cu concentration was established by using inductively coupled plasma optical emission spectrometry (ICP-OES) [181] on similarly grown crystals with the same flux concentrations of Cu. It was found that 2% Cu in the flux produced a nearly constant concentration of 1.5×10^{-4} per mole independent of the ratios of other constituents, with an accuracy of 0.2%.

Dielectric measurements were carried out in the frequency range of $10^{-2}-$ 10^6 Hz and the temperature interval 133–473 K [179]. The crystals were cooled from 297 K to 190 K with a temperature step of 4 K. In the region of the phase transition, 292–297 K, the temperature step was reduced to 0.5 K. Reheating from 190 K to 440 K was done with a step of 5 K. As before, in the region of the ferroelectric phase transition the step was reduced to 0.5 K. The dielectric landscape of crystal #1 is presented in Fig. 18.

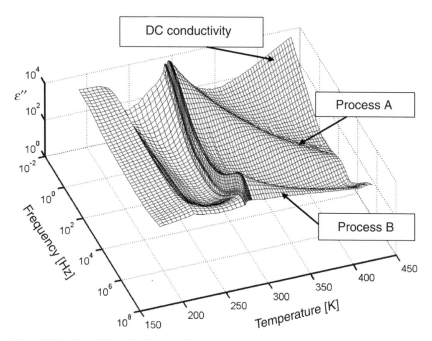

Figure 18. The dielectric losses, ε'', for crystal 1. The three phase transitions are evident at $T = 295.6 \, \text{K}$, 291.1 K, and 230 K, respectively. (Reproduced with permission from Ref. 179. Copyright 2004, The American Physical Society.)

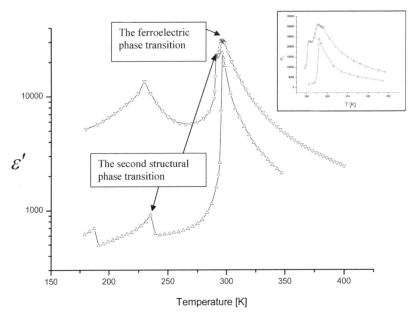

Figure 19. The dielectric permittivity of crystals 1(∇) and 2 (Δ) measured at 12 Hz. The shift in the second- and third-phase transition temperatures, due to Cu ion doping, is clearly evident. The insert shows the detail of the ferroelectric phase transition in both crystals. (Reproduced with permission from Ref. 257. Copyright 2005, Elsevier Science B.V.)

The complex dielectric response of crystal #1 can be described in terms of a number of distributed dynamic processes separated by different frequency and temperature ranges. The ferroelectric phase transition is observed at 295.6 K. It is followed by tetragonal to orthorhombic and orthorhombic to rhombohedral transitions occurring at 291.1 K and 230 K, respectively. A comparison of these transition temperatures with the undoped crystal #2 revealed that the second and third transitions (tetragonal to orthorhombic and orthorhombic to rhombohedral) were shifted by approximately 40 K toward the lower temperatures (see Fig. 19).

In the paraelectric phase, above 295 K, a thermally activated process (process **A**) exists starting in the low frequencies at the phase transition and shifting toward higher frequencies as the temperature increases. The quantitative nature of Process **A** was established by examining the temperature dependence of τ^A, obtained as the inverse value of the characteristic frequency $f_m(T)$ along the crest representing process **A** in the $\varepsilon''(T, f)$ landscape [6]. It was found to be Arrhenius in nature with an activation energy of $E_a^A = 0.94 \pm 0.01$ eV and the high temperature limit of the relaxation time $\tau_o^A = 1.7 \pm 0.4 \times 10^{-15}$ s. Additionally, it was found to be correlated with the well-pronounced

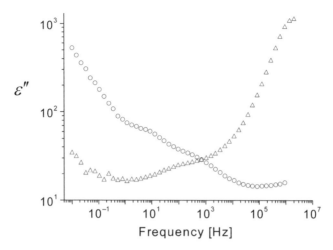

Figure 20. A comparison of the dielectric losses measured at $T = 250$ K: $\triangle$ represents crystal #1 (KTN doped with Cu ions), and $\circ$ represents crystal #2 (pure KTN). (Reproduced with permission from Ref. 179. Copyright 2004, The American Physical Society.)

dc-conductivity. The dc-conductivity, σ, was found to follow Arrhenius behavior, namely, $\sigma = \sigma_0 \exp(-E_\sigma/k_B T)$, with an activation energy of $E_\sigma = 0.9 \pm 0.01$ eV and the high temperature limit of conductivity $\sigma_0 = 42 \pm 7$ s m^{-1}. The similarity in nature and activation energy of Process **A** and the dc-conductivity suggests that they originate from the same physical mechanism, most likely electron mobility. While Process **A** was observed in both crystal #1 and crystal #2, Process **B** was observed only in crystal #1 (see Fig. 20).

Process **B** was observed to pass through all three phase transitions, and it was found to be non-Arrhenius with distinct changes in its relaxation behavior delineated by the phase transitions. In the same manner as for Process **A** the characteristic relaxation time τ^B was extracted from the peak maximum of the dielectric losses, $\varepsilon''(\omega, T)$, (Fig. 21).

In the high-temperature range, above 354 K, Process **B** exhibits Arrhenius behavior with an activation energy $E_a^B = 0.37 \pm 0.01$ eV and $\tau_o^B = 2.8 \pm 0.9 \times 10^{-12}$ s. Below $T_x = 354$ K Process **B** obeys a Vogel–Fulcher–Tammann (VFT) relaxation in which τ^B is given by (28) $\ln(\tau^B) = \ln(\tau_0^B) + \frac{E_{VFT}}{k_B(T-T_{VFT})}$ with $T_{VFT} = 228$ K, and $E_{VFT} = 0.02$ eV. Following the onset of the ferroelectric phase transition at $T_c = 295.6$ K, τ^B decreases until it reaches a minimum at 264 K, exhibiting a small cusp at the second (tetragonal to orthorhombic) phase transition. Upon further cooling, τ^B increases until it reaches a maximum at the third (orthorhombic to rhombohedral) transition at 230 K, exhibiting a "saddle" [78] that will be discussed in detail below.

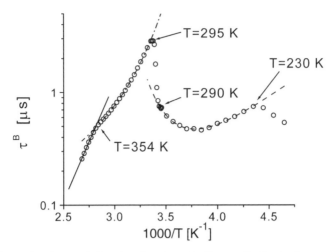

Figure 21. The characteristic relaxation times for Process **B**. On the figure are marked the three phase transitions and the critical crossover temperature, $T = 354$ K, between Ahrrenius and VFT behaviors. The circles are the experimental data and the lines are the fitting functions: Ahrrenius, VFT, and "Saddle-like" [78,179]. (Reproduced with permission from Ref. 257. Copyright 2005, Elsevier Science B.V.)

D. H-Bonding Liquids

The study of slow dynamics in glass-forming liquids is currently a significant challenge in the research field of soft condensed matter science [182–186]. Hydrogen-bonding liquids and their mixtures occupy a special place among complex systems due to existence of directed H-bonds (in contrast to Van der Waals and ionic systems) that can be rearranged relatively easily (in contrast to covalent bonds). Although an enormous amount of literature exists which relates to the investigation of hydrogen-bonding systems (see, for example, Refs. 3, 183, and 185) there is still a lack of clear understanding of their dynamics, structure, and glass transitions. Among them, glycerol ($C_3H_8O_3$) [17,74,187–189] and its mixtures with water [190] is widely used as excellent model to study their cooperative dynamics.

Usually, glycerol exists only in liquid, supercooled, or glassy states. However, after special treatment, pure dehydrated glycerol can be crystallized [191,192]. Uncrystallized glycerol is a common system used for studying glass-forming dynamics [3,184,187–189,193–195] while crystallized glycerol, until now, has not been investigated.

Under normal conditions, glycerol does not undergo crystallization; however, during cooling, it becomes a supercooled liquid which can be vitrified [184,187,196,197] at $T_g = 190$ K. In contrast, anhydrous glycerol, cooled down below the glass transition point T_g and then slowly heated up, can be crystallized. However, crystallization of glycerol is a very unusual and unstable process, which depends on the temperature history and impurities of the sample. The main features of glycerol crystallization were studied recently [186] by comparing the glass-forming dynamics of anhydrous glycerol (sample A) with those of glycerol that was not specially treated to prevent water absorption (sample B). To reach crystallization, sample A was cooled from room temperature to 133 K. Then measurements of the complex dielectric permittivity $\varepsilon^*(\omega)$ were performed by a Novocontrol Alpha Analyzer in the frequency interval 0.01 Hz to 3 MHz and the temperature range 133–325 K (see Fig. 22). Thus, the overall experimental time was 30 h and average heating rate was about 0.1 K min^{-1}. Considerable changes in $\varepsilon^*(\omega)$ behavior in the

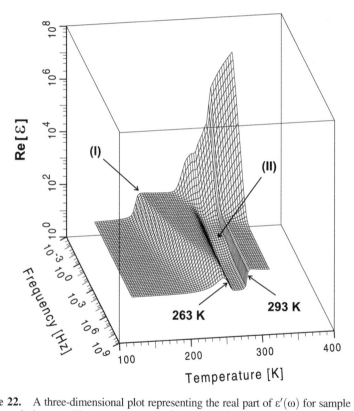

Figure 22. A three-dimensional plot representing the real part of $\varepsilon'(\omega)$ for sample A. The arrows mark the crystallization temperature ($T_x = 263$ K), the melting point ($T_m = 293$ K), and the principal relaxation process, before (I) and after (II) the crystallization. (Reproduced with permission from Ref. 186. Copyright 2003, The American Physical Society.)

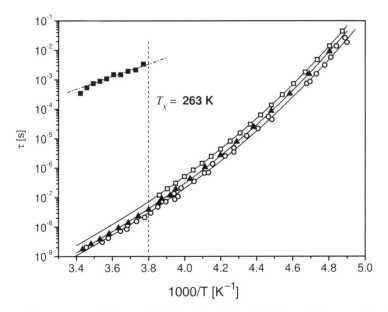

Figure 23. Sample A before (unfilled boxes) and after (filled boxes) the crystallization compared with sample B (triangles), and literature data [195] (circles). In the supercooled phase, all samples obey a VFT law (full lines), while the relaxation process in sample A above 263 K obeys an Arrhenius law (dash–dotted line). (Reproduced with permission from Ref. 186. Copyright 2003, The American Physical Society.)

measured frequency range are observed in the temperature interval from 263 K to 293 K. The transition at 293 K is known as the glycerol melting point [198]. Thus, the transition near 263 K is thought to be attributed to the glycerol crystallization. Note that the relaxation process (I) of the supercooled glycerol in this temperature interval disappears and the relaxation process (II), with a reduced strength, appears in the low-frequency region.

The data presented in Fig. 23 were analyzed as a set of isothermal spectra in the framework of (21):

$$\varepsilon^*(f) = \varepsilon_\infty + \frac{\varepsilon_s - \varepsilon_\infty}{\left(1 + (i\omega\tau)^\alpha\right)^\beta}$$

It is remarkable that in the liquid and supercooled phase glycerol exhibits an assymmetrical relaxation peak broadening ($\alpha \approx 1, \beta \approx 0.6$), whereas for process (II) the broadening is rather symmetric ($0.6 \leq \alpha \leq 0.7, \beta = 1$). The Arrhenius plot in Fig. 23 shows that relaxation dynamics in the supercooled glycerol phase before crystallization (I) obey the VFT relationship while after crystallization (II) follow the Arrhenius law. Note that in the considered frequency and temperature landscape, another low-frequency process related to the crystallization is observed. Unfortunately, it was not resolved properly in the mentioned study [186].

The VFT behavior of supercooled glycerol is well known from studies of liquid and supercooled glycerol [3,186–190], while the Arrhenius dependence of the dielectric relaxation time is more relevant for crystals. For example, the temperature dependence of the dielectric relaxation time of ice I also obey the Arrhenius law with the activation energy about $60\,kJ\,mol^{-1}$ [198,199].

Therefore, the observed process (I) could be related to the cooperative dynamics of glycerol in the supercooled phase, while process (II) is most likely related to the crystalline phase of glycerol and is the result, similar to water, of the mobility of defects in the crystalline lattice [200]. The temperature dependence of the relaxation time for dehydrated glycerol is compared in Fig. 23 with those for the usual behavior of glycerol, which has absorbed some water from the atmosphere.

In Fig. 23, we compared temperature dependence of fitted relaxation time for samples A and B and data recently published by Lunkenheimer and Loidl [195]. The fitting indicates that the process (I) in the supercooled phase for sample A, the relaxation in sample B, and literature data [195] all obey the VFT law $\tau_{max} = \tau_v \exp\{DT_v/(T - T_v)\}$ where T_v is the Vogel–Fulcher temperature, τ_v is the preexponential factor, and D is the fragility parameter. From the data presented in Fig. 23, one can see that values of the VFT temperature T_v and fragility D are very close for all the samples where $D = 22 \pm 2$ and $T_v = 122 \pm 2\,K$, while the preexponential factors τ_v are remarkably different. For the anhydrous sample A, $\tau_v = 3.93 \times 10^{-16}\,s$, for sample B, $\tau_v = 2.3 \times 10^{-16}\,s$, and for the literature data [195] $\tau_v = 1.73 \times 10^{-16}\,s$. Taking into account the fact that sample A was specially protected from water absorption, it is strongly suspected that this significant difference in τ_v is caused by the water absorbed from the atmosphere. This observation signifies that even very small water content can result in significantly different dynamics in the supercooled phase for the anhydrous glycerol and for the glycerol samples usually studied [186].

As mentioned above, the frequency dependence of the complex dielectric permittivity (ε^*) of the main relaxation process of glycerol [17,186] can be described by the Cole–Davidson (CD) empirical function [see (21) with $\alpha = 1$, $0 < \beta_{CD} \leq 1$]. Now τ_{CD} is the relaxation time which has non-Arrhenius type temperature dependence for glycerol (see Fig. 23). Another well-known possibility is to fit the BDS spectra of glycerol in time domain using the KWW relaxation function (23) $\phi(t)$ (see Fig. 24):

The CD function indicates that the dielectric loss (ε'') of glycerol follows the power law $\varepsilon'' \sim f^{\beta_{CD}}$ at high frequencies $(f \gg f_{max})$, where f_{max} is the frequency corresponding to the dielectric loss peak. However, the high-frequency experimental data in Fig. 24 demonstrate a significant deviation from the expected asymptotic behavior both for CD and KWW functions. ε'' values

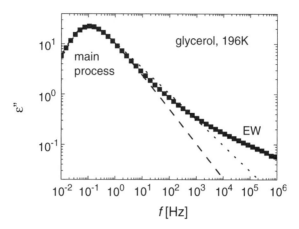

Figure 24. The imaginary parts of the dielectric spectrum for anhydrous glycerol in the supercooled state at 196 K [186]. The dotted and dashed line show descriptions of the main relaxation process by CD [Eq. (21)] with $\tau_{CD} = 2.61$ s, $\Delta\varepsilon = 63.9$, and $\beta_{CD} = 0.51$) and KWW [Eq. (23)] with $\tau_K = 1.23$ s, $\Delta\varepsilon = 62.0$, and $\beta_K = 0.69$) functions, respectively. (The half-width of the loss curve were fixed for both CD and KWW functions.) (Reproduced with permission from Ref. 208. Copyright 2005, American Chemical Society.)

larger than predicted by the spectral function of the main relaxation are known as the excess wing (EW). Although there is no unique interpretation of the EW, it is known that this phenomenon is characteristic of many glass-forming systems [195,201–203]. Note that deviations from a power law slope for ε'' were also observed for pure water [204–206]. These features were treated as an additional Debye-type relaxation process without any relation to glass-forming properties. However, with given glass forming properties of water, they can be discussed in the context of the EW feature as well. There is some speculation [201] that the EW is most probably a Johari–Goldstein (JG) [207] mode. However, there are still open questions because analysis of pressure and temperature effects leads to the conclusion that EW and JG modes "cannot be treated on the same footing" [202]. The non-Arrhenius temperature dependence of the EW found recently [203] is in contradiction to the original idea of Johari and Goldstein, who argued that "the Arrhenius plots in the secondary relaxation region are linear" [207]. Thus the relaxation mechanism of the EW is still unclear, and it would be helpful to study how the water concentration in glycerol influences the relaxation dynamics of the mixture, since the EW could be related to H-bond networks of glycerol affected by the presence of water.

The typical results of recent BDS studies of glycerol–water mixtures of 75 mol% of glycerol at different temperatures are presented in Fig. 25 [208]. Figure 26 shows that the temperature dependencies of the main relaxation

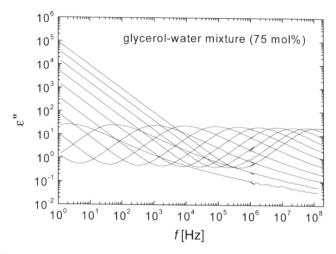

Figure 25. The imaginary parts of the dielectric spectra for a glycerol–water mixture (75 mol%) at various temperatures from 197 to 290 K with an interval of 3 K. (Reproduced with permission from Ref. 208. Copyright 2005, American Chemical Society.)

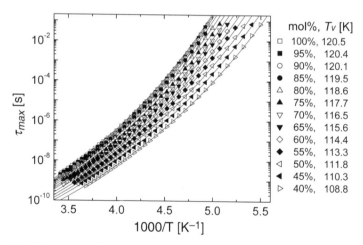

Figure 26. Temperature dependence of $\tau_{max} = 1/(2\pi f_{max})$ for a glycerol sample and its mixtures with water. The solid curves show the description by the VFT model, where values of the Vogel–Fulcher temperature (T_v) are shown with the legend. The preexponential factor (τ_v) is almost independent with concentration: $\ln(\tau_v) = -35.9$ for 100 mol% of glycerol, -36.1 for 95 to 65%, -36.2 for 60%, -36.3 for 60% to 45%, and -36.4 for 40 mol% of glycerol–water mixtures. $D = 22.7$ for all concentrations presented here. (Reproduced with permission from Ref. 208. Copyright 2005, American Chemical Society.)

process in the terms of $\tau_{max} = 1/(2\pi f_{max})$ for concentrations from 100 to 40 mol% of glycerol were well-described by the VFT law:

It is worth noting that value of D was the same for all concentrations presented in Figure 26 and values of τ_v were also almost the same in this concentration range.

Let us start the examination of the dynamics of glycerol–water mixtures with an inspection of the main dielectric loss peak $\varepsilon''_{max} = \varepsilon''(f_{max})$ with f_{max} as the characteristic frequency that was observed in the considered experimental ranges for glycerol and its mixtures with water. Figure 27 presents a so-called "master plot" that presents the normalized plots of $\varepsilon''(f)/\varepsilon''_{max}$ versus the dimensionless variable f/f_{max}. Namely, each data point in the dielectric spectrum was normalized by only one point such as the loss peak. It is remarkable that for all temperatures, the dielectric spectra of glycerol have the same single curve (Fig. 27a). This indicates that both the EW and low-frequency contribution of dc-conductivity follow the same VFT temperature dependence of the main relaxation process (Fig. 25). For glycerol–water mixtures whose concentrations are from 60 up to 95 mol% of glycerol (e.g., Fig. 27b), the master plots are similar to those presented in Fig. 27a for glycerol. These results demonstrate that the EW and the main dielectric relaxation process have the same dependency on the composition of the glycerol–water mixture over a comparatively wide range of concentrations:

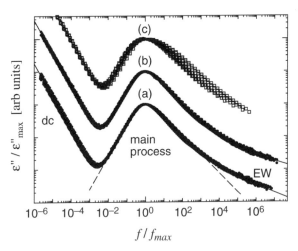

Figure 27. Master plots of the imaginary parts of the dielectric spectra for 100 mol% (a), 75 mol% (b), and 35 mol% (c) of glycerol measured in the temperature intervals 202–292 K, 197–290 K, and 176–221 K, respectively. The dashed line represents CD law with $\beta_{CD} = 0.58$. (Reproduced with permission from Ref. 208. Copyright 2005, American Chemical Society.)

from 60 to 100 mol%. This observation indicates that the EW, the main relaxation process, and dc-conductivity in these mixtures most probably have the same origin.

Further decrease of glycerol content caused the universality in the master plot to disappear. At 55 mol% or lower concentrations of glycerol, a different behavior of the main dielectric relaxation and the EW was observed (e.g., Fig. 27c). This region with small glycerol molar content has been investigated in detail by Sudo et al. [190], who ascribed the main and secondary relaxation processes to the relaxation dynamics of the so-called cooperative domains of glycerol and water, respectively. The consideration of the mesoscopic dynamics and glycerol and water glycerol–water mixtures results will be discussed in detail in later sections.

V.   COOPERATIVE DYNAMIC AND SCALING PHENOMENA IN DISORDERED SYSTEMS

Following the general plan of our chapter, we will discuss here a universal view of scaling which is widely employed in modern dielectric spectroscopy. We will show how this concept can be applied to the description of disordered materials and how these ideas can be useful in the determination of the topological parameters of these systems.

A.   Static Percolation in Porous Materials, Fractal Concept, and Porosity Determination

In this section, we will consider in more detail the non-Debye dielectric response associated with percolation in porous disordered materials: silica glasses and porous silicon where the pores form topologically connected pore channels. The movement of charge carriers along the inner pore surface results in a transfer of the electric excitation within the channels along random paths [154]. Note that in the general case, such transfers through the porous medium can occur even for closed pores, which are topologically not connected one to another. However, the distance between the neighboring pores filled with the dielectric or conductive material should be small enough in order to provide "physical" pore coupling via the electric interaction.

It was shown recently that disordered porous media can been adequately described by the fractal concept, where the self-similar fractal geometry of the porous matrix and the corresponding paths of electric excitation govern the scaling properties of the DCF $\Psi(t)$ (see relationship (22)) [154,209]. In this regard we will use the model of electronic energy transfer dynamics developed by Klafter, Blumen, and Shlesinger [210,211], where a transfer of the excitation of a donor molecule to an acceptor molecule in various condensed media through many parallel channels was considered.

A detailed description of the relaxation mechanism associated with an excitation transfer based on a recursive (regular) fractal model was introduced earlier [47], where it was applied for the cooperative relaxation of ionic microemulsions at percolation.

According to this model, an elementary act of the excitation transfer along the length L_j is described by the microscopic relaxation function $g(z/z_j)$, where L_j is the "effective" length of a channel of the relaxation in the jth stage of self-similarity. In this function, z_j is a dimensional variable characterizing the jth stage of the self-similarity of the fractal system considered, z is the dimensionless time, $z = t/\tau$, where the parameter τ is the minimal relaxation time needed for an excitation to hop from one excitation center to its nearest neighbor.

The following assumption is invoked: $z_j = aL_j$, where a is a coefficient of proportionality. For each stage of the self-similarity j, the time of relaxation $\tau_j = \tau z_j$ is proportional to the length L_j. From fractal geometry [212,213], L_j can be expressed as

$$L_j = lk^j \tag{52}$$

where l is the minimal scale and k is a scaling factor ($k > 1$). We assume that the total number of activation centers located along the segment L_j also obeys the scaling law

$$n_j = n_0 p^j \tag{53}$$

where p is the scaling factor ($p > 1$), and n_0 is the number of the nearest neighbors near the selected center (i.e., $j = 0$).

The macroscopic correlation function can be expressed as a product of the relaxation functions $g(z/z_j)$ at all stages of self-similarity of the fractal system considered [47,154]:

$$\Psi(z) = \prod_{j=0}^{N} [g(z/z_j)]^{n_j} = \prod_{j=0}^{N} [g(Z\xi^j)]^{n_0 p^j} \tag{54}$$

where $Z = t/al\tau$, $\xi = 1/k$, and $N = \frac{1}{\ln k}\ln(L_N/l)$. Here L_N is the finite geometrical size of the fractal cluster and N refers to the last stage of self-similarity.

The estimations of the product (5.4) for various values of $\xi < 1$ and $p > 1$ are given in Ref. 47. The results of the calculations may be written in the form of a modified Kohlrausch–Williams–Wats (KWW) stretched-exponential relaxation law:

$$\Psi(Z)/\Psi(0) = \exp[-\Gamma(v)Z^v + B(v)Z] \tag{55}$$

where the parameters $\Gamma(v)$ and $B(v)$ are given by

$$\Gamma(v) = \frac{n_0}{\ln(1/\xi)} \int_0^\infty y^{-v} \left| \frac{g'(y)}{g(y)} \right| dy \tag{56}$$

$$B(v) = \frac{n_0 a_1}{\ln(1/\xi)(1-v)} \kappa^{1-v} \tag{57}$$

Here

$$v = \ln p / \ln(1/\xi) \tag{58}$$

with $0 < v < 1$ and for $\kappa = \xi^N \ll 1$. We note that the parameter Γ depends on the relaxation function g and affects the macroscopic relaxation time $\tau_M = \tau a l \, \Gamma^{-1/v}$, and the term $B(v)Z$ in the exponent corrects the KWW function at large times.

The temporal bounds of the applicability of (55) are determined by the expression

$$\left| \frac{A_1 n_0}{\bar{g}} \left(\frac{1}{2} - \frac{1}{\ln(1/\xi)(1+v)} \right) \right| \ll \frac{t}{al\tau} \ll \left| \left(\frac{2\ln(1/\xi)(2-v)}{n_0(2a_2 - a_1^2)\kappa^{2-v}} \right)^{1/2} \right| \tag{59}$$

The parameters $\bar{g}$, A_1, a_1, and a_2 in (55)–(59) are related to the asymptotic properties of the elementary relaxation function $g(y)$:

$$g(y) = 1 - a_1 y + a_2 y^2 + \cdots \qquad \text{for} \quad y \ll 1 \tag{60}$$

$$g(y) = \bar{g} + A_1/y + A_2/y^2 + \cdots \qquad \text{for} \quad y \gg 1 \tag{61}$$

The relationship between the exponent v, ($v = \ln p / \ln k$), and the fractal dimension D_p of the excitation transfer paths may be derived from the proportionality and scaling relations by assuming that the fractal is isotropic and has spherical symmetry. The number of pores that are located along a segment of length L_j on the jth step of the self-similarity is $n_j \sim p^j$. The total number of pores in the cluster is $S \sim n_j^d \sim (p^j)^d$, where d is the Euclidean dimension

($d = 3$). The similarity index, η, which determines by how much the linear size of the fractal is enlarged at step j, is $\eta \sim L_j \sim k^j$. Here, we obtain the following simple relationship between ν and the fractal dimension D_p:

$$D_p = \ln S / \ln \eta = 3j \ln p / j \ln k = 3\nu \tag{62}$$

Furthermore, we will focus our attention only on the time-dependent behavior of the dipole correlation function $\Psi(t)$ defined by (55), which is given by

$$\Psi(t) \approx C(t) \exp[-(t/\tau)^\nu] \tag{63}$$

where $C(t)$ is a slowly increasing function of time. By taking into account (62) and ignoring the slow variation with time of $C(t)$, we obtain the asymptotic stretched-exponential term

$$\Psi(t) \sim \exp[-(t/\tau)^{D_p/3}] \tag{64}$$

that can be further fitted to the experimental correlation functions in order to determine the value of the fractal dimension of the paths of excitation transfer within the porous medium. If the fractal dimension of these paths coincides with the fractal dimension of the pore space, then it can be used in the asymptotic equations derived below for the porosity determination.

1. Porous Glasses

The dielectric relaxation at percolation was analyzed in the time domain since the theoretical relaxation model described above is formulated for the dipole correlation function $\Psi(t)$. For this purpose the complex dielectric permittivity data were expressed in terms of the DCF using (14) and (25). Figure 28 shows typical examples of the DCF, obtained from the frequency dependence of the complex permittivity at the percolation temperature, corresponding to several porous glasses studied recently [153–156].

As mentioned earlier, typical three-dimensional plots of ε' and ε'' versus frequency and temperature (see Fig. 14) suggest superimposing two processes (percolation and saddle-like) in the vicinity of the percolation. Therefore, in order to separate the long-time percolation process, the DCF was fitted as a sum of two functions. The KWW function (64) was used for fitting the percolation process and the product (25) of the power law and the stretched exponential function (as a more common representation of relaxation in time domain) was applied for the fitting of the additional short-time process. The values obtained for D_p of different porous glasses are presented in the Table I. The glasses studied differed in their preparation method, which affects the size of the pores, porosity and availability of second silica and ultra-porosity [153–156].

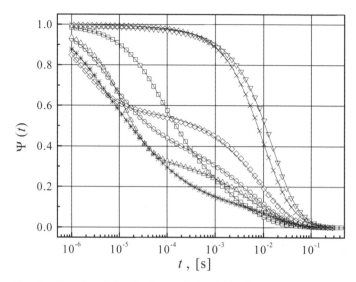

Figure 28. Semilog plot of the dipole correlation $\psi(t)$ of all the samples studied at the temperature corresponding to percolation ($\square$ sample A; $\bigcirc$ sample B; $\triangle$ sample C; $\triangledown$ sample D; $\diamondsuit$ sample E; $\times$ sample F; $*$ sample G). The solid lines are the fitting curves of the sum of the KWW and the product of KWW and the power-law relaxation function. (Reproduced with permission from Ref. 2. Copyright 2002, Elsevier Science B.V.)

Apparently the fractal dimension of the excitation paths in sample A is close to unity. Topologically, this value of D_p corresponds to the propagation of the excitation along a linear path that may correspond to the presence of second silica within the pores of the sample A. Indeed, the silica gel creates a subsidiary tiny scale matrix with an enlarged number of hydration centers within the pores.

TABLE I
The Values of KWW Exponent ν, Fractal Dimension D_p, Porosity Φ_m Obtained from the Relative Mass Decrement (A, B, C, and D Glasses) and BETA (E, F, and G Glasses) Measurements and Average Porosity $\langle \Phi_p \rangle$ Estimated from Dielectric Spectra for Porous Glasses Samples.

Sample	ν	D_p	Φ_m	$\langle \Phi_p \rangle$
A	0.33	0.99	0.38	0.33
B	0.63	1.89	0.48	0.47
C	0.44	1.31	0.38	0.37
D	0.83	2.50	0.50	0.68
E	0.65	1.96	0.27	0.49
F	0.80	2.40	0.43	0.63
G	0.73	2.20	0.26	0.56

Source: Reproduced with permission from Ref. 2. Copyright 2002, Elsevier Science B.V.

Since these centers are distributed in the pore volume, the excitation transmits through the volume and is not related to the hydration centers located on the pore surface of the connective pores. Due to the large number of hydration centers, and the short distance between the neighboring centers, the path can be approximated by a line with a fractal dimension close to unity (see Fig. 29a).

The fractal dimensions of the excitation paths in samples B, C, and E have values between 1 and 2. In contrast with sample A, the silica gel in these samples has been leached out; that is, water molecules are adsorbed on the inner pore surface (see the details of different porous glass preparations in Ref. 156). The values of D_p observed in samples B, C, and E can be explained in one of two ways. On one hand, the surface can be defractalized upon deposition of an adsorbed film of water, resulting in "smoothing" of the surface. On the other hand, the transfer of the excitation in these samples occurs along the inner pore surface from one hydration center to another. The distance between the centers is significantly larger than the small-scale details of the surface texture (see Fig. 29b). Therefore, the fractal dimension observed is that of the chords connecting the hydration centers and should be less than 2, which is in agreement with the data obtained from the energy-transform measurements [214,215].

The fractal dimensions of the excitation paths in samples D, F, and G lie between 2 and 3. Thus, percolation of the charge carriers (protons) is also moving through the SiO_2 matrix because of the availability of an ultra-small porous structure that occurs after special chemical and temperature treatment of the initial glasses [156].

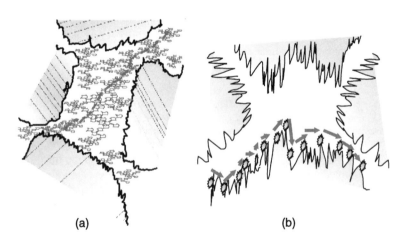

(a)   (b)

Figure 29. The schematic presentation of the percolation pass in porous glasses; (a) The glass whose porous are filled with second silica; (b) The porous glasses where the silica gel is leached out.

Note that the fractal dimensions discussed here are the fractal dimensions of the excitation transfer paths connecting the hydration centers located on the inner surface of the pores. Due to the low humidity, all of the water molecules absorbed by the materials are bound to these centers. The paths of the excitation transfer span along the fractal pore surface and "depict" the backbone of clusters formed by the pores on a scale that is larger than the characteristic distance between the hydration centers on the pore surface. Thus the fractal dimension of the paths D_p approximates the real surface fractal dimension in the considered scale interval. For random porous structures, D_p can be also associated with the fractal dimension D_r of the porous space: $D_p \cong D_r$. Therefore, the fractal dimension D_p can be used for porosity calculations in the framework of the fractal models of the porosity.

The porosity Φ_p of a two-phase solid-pore system can be defined as the ratio of the volume of the empty space, V_p, to the whole volume, V, of a sample [166]:

$$\Phi_p = \frac{V_p}{p} \tag{65}$$

Disordered porous media have been adequately described by the fractal concept [154,216]. It was shown that if the pore space is determined by its fractal structure, the *regular fractal* model could be applied [154]. This implies that for the volume element of linear size Λ, the volume of the pore space is given in units of the characteristic pore size λ by $V_p = G_g(\Lambda/\lambda)^{D_r}$, where D_r is the regular fractal dimension of the porous space, Λ coincides with the upper limit, and λ coincides with the lower limit of the self-similarity. The constant G_g is a geometric factor. Similarly, the volume of the whole sample is scaled as $V = G_g(\Lambda/\lambda)^d$, where d is the Euclidean dimension ($d = 3$). Hence, the formula for the macroscopic porosity in terms of the regular fractal model can be derived from (65) and is given by

$$\Phi_p = \left(\frac{\lambda}{\Lambda}\right)^{d-D_r} \tag{66}$$

In general, in order to include more types of porous media the *random fractal* model can be considered [2,154,216]. Randomness can be introduced in the fractal model of a porous medium by the assumption that the ratio of the scaling parameters $\xi = \lambda/\Lambda$ is random in the interval $[\xi_0, 1]$, but the fractal dimension D in this interval is a determined constant. Hence, after statistical averaging, (66) reads as follows:

$$\langle \Phi_p \rangle = \int_{\xi_0}^{1} \Phi_p(\xi, D_r) w(\xi) d\xi \tag{67}$$

where ξ_0 is the minimal value of the scaling parameter ξ and $w(\xi)d\xi$ measures the probability to find some scaling parameter in the range from $\xi + d\xi$.

For a percolating medium the generalized exponential pore-size distribution function of the scale for porous medium can be written as

$$w(\xi) \sim \xi^{\alpha_w} \exp(-a_w \xi^{b_w}) \tag{68}$$

This function accounts for the mesoscale region and comprises most of the listed distribution functions [154]. It includes three empirical parameters, α_w, β_w, and a_w. Having ascertained the relationships between these parameters and the properties of anomalous self-diffusion, fractal morphology, and polydispersity of the finite pore-size, physical significance can be assigned to these parameters in the framework of the percolation models [152].

On a length scale larger than the pore sizes, the morphology of the glass porous space can be modeled as a random-packed assembly of clusters formed by pores connected to each other [203,217]. In order to find the macroscopic porosity in such systems we must assume that the pore structure has a fractal character in a rather narrow scale range. Hence, in the interval $[\xi_0, 1]$ the uniform distribution function, $w(\xi) = w_0$, can be chosen as an approximation to the function given in (67). The value of w_0 is determined from the normalization condition $\int_{\xi_0}^1 w(\xi)d\xi = 1$ and is $w_0 = 1/(1 - \xi_0)$. In this approximation, by substituting this uniform distribution function into integral (67) and integrating it, we obtain the relationship for the average porosity as

$$\langle \Phi_p \rangle = \frac{1}{1 + d - D_r} \cdot \frac{1 - \xi_0^{1+d-D_r}}{1 - \xi_0} \tag{69}$$

Taking into account that $\xi_0 \ll 1$, and $1 + d - D_\gamma > 0$ (since topologically $D_\gamma < 3$, and $d = 3$), we obtain a simple approximate relationship between the average porosity of a glass and the fractal dimension of the pore space, which reads

$$\langle \Phi_p \rangle \approx \frac{1}{4 - D_r} \tag{70}$$

Note that in our approximation, due to the randomized character of the fractal medium the average porosity of the disordered porous glasses determined by (70) depends only on the fractal dimension D_r and does not exhibit any scaling behavior. In general, the magnitude of the fractal dimension may also depend on the length scale of a measurement extending from λ to over

Λ, where the minimal scale λ and the maximal scale Λ are determined by the measurement technique.

The results of the porosity calculation for various porous glasses using (70) together with the fractal dimension determined from dielectric measurements are shown in the last column of Table I. These values can be compared with the porosity Φ_m determined from the relative mass decrement (A, B, C, and D glasses) and BETA (E, F, and G glasses) measurements shown in the same Table I. Note that the values obtained from dielectric spectroscopy coincide with the porosity data obtained from the relative mass decrement method only for samples A, B, and C. The porosity values for the other samples obtained through the dielectric measurements are significantly larger. This correlates with the availability of ultra-small porous structures with penetrability for the smallest charge carriers (such as protons) [156]. Thus, in the case of a net of super-small open pores, the dielectric response is more sensitive and accurate in the determination of real porosity than any other conventional method.

2. Porous Silicon

The detailed analysis of mid-temperature relaxation processes observed in porous silicon was provided recently by using a superposition of two Jonscher terms of the form $B_1(i\omega)^{u_1-1} + B_2(i\omega)^{u_2-1}$ [160]. The results of our fitting were in good agreement with those of Ben-Chorin et al. [163] that were discussed in terms of the transport of charged carriers at the different scales. The high-frequency Jonscher exponent was associated with the typical size of the Si nano-crystallites, while the low-frequency (LF) exponent was assigned to the transport of charged carriers across a disordered fractal structure of porous silicon [160,161,218,219]. At the same time, the mid-temperature process II demonstrates several specific features that are similar to those observed in other porous systems discussed in the previous section [153–156]. The amplitude of this process essentially decreases when the frequency increases (Fig. 17a). Furthermore, the maximum of the dielectric permittivity versus temperature has almost no temperature dependence. Finally, the low-frequency ac-conductivity increases with the increase in temperature and has an S-shaped dependency (Fig. 30), which is typical for percolation processes [143,154,161]. Thus, we will analyze this process in the same way as we did for percolation in porous glasses (see Section V.A.1) [2,153–159].

The experimental macroscopic DCF for PS samples with porous layers of 20 and 30 μm, obtained by inverse Fourier transforms, are shown in Fig. 31. The correlation functions then were fitted by the KWW expression (23) with

YURI FELDMAN ET AL.

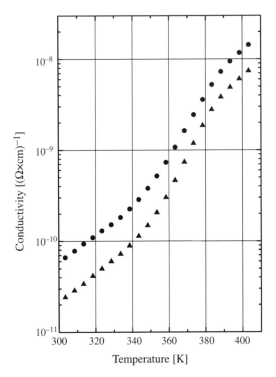

Figure 30. Temperature dependence of the low-frequency conductivity of the 20-μm sample (•), and the 30-μm sample (▲). (Reproduced with permission from Ref. 2. Copyright 2002, Elsevier Science B.V.)

determination of the fractal dimension D_p of the percolation path. Applying the same routine to determine the porosity in other porous systems [154,156–161], the average porosity of the porous silicon was evaluated with the help of relationship (70). The results are presented in Table II. The values of porosity determined from the dipole correlation function analysis are in good agreement with porosity values determined by weight loss measurements during PS preparation (before and after the anodization process).

Thus, the non-Debye dielectric behavior in silica glasses and PS is similar. These systems exhibit an intermediate temperature percolation process associated with the transfer of the electric excitations through the random structures of fractal paths. It was shown that at the mesoscale range the fractal dimension of the complex material morphology (D_r for porous glasses and porous silicon) coincides with the fractal dimension D_p of the path structure. This value can be obtained by fitting the experimental DCF to the stretched-exponential relaxation law (64).

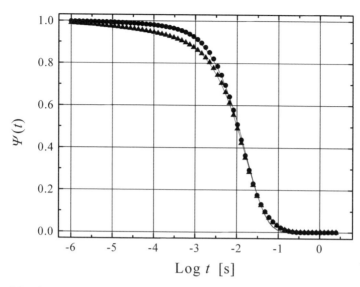

Figure 31. Semilog plot of the macroscopic correlation function of the 20-μm sample (•) and the 30-μm sample (▲) at the temperature corresponding to percolation. The solid lines correspond to the fitting of the experimental data by the KWW relaxation function. (Reproduced with permission from Ref. 2. Copyright 2002, Elsevier Science B.V.)

B. Dynamic Percolation in Ionic Microemulsions

1. Dipole Correlation Function for the Percolation Process

A description of the percolation phenomenon in ionic microemulsions in terms of the macroscopic DCF will be carried out based on the static lattice site percolation (SLSP) model [152]. In this model the statistical ensemble of various

TABLE II

The Values of the KWW Exponent ν, Fractal Dimension D_p, and Porosity Φ_m Obtained from Relative Mass Decrement Measurements and Average Porosity $\langle \Phi_p \rangle$ Estimated from Dielectric Spectra for Porous Silicon Samples of 20- and 30-μm Thickness [2]

Sample Thickness	ν	D_p	Φ_m	$\langle \Phi_p \rangle$
20 (μm)	0.88	2.64	0.78	0.74
30	0.87	2.61	0.75	0.72

Source: Reproduced with permission from Ref. 2. Copyright 2002, Elsevier Science B.V.

size clusters is described by the distribution function [152,213,220]

$$w(s, s_m) = C_w \cdot s^{-\Omega} \exp\left(-\frac{s}{s_m}\right) \tag{71}$$

Here C_w is the normalization constant and Ω is a scaling exponent of the probability density "per lattice site" that the site, chosen randomly, belongs to the s-cluster (a cluster that consist of s lattice sites). The value of s_m is the cutoff cluster size corresponding to the maximum cluster size. Note that clusters having sizes in the interval of $1 < s < s_m$ are referred to as mesoscopic clusters. The scaling properties of the mesoscopic dipole correlation functions related to the s-cluster of the geometrical substrate can be utilized for establishing a link between the static lattice geometrical percolation model and the relaxation functions.

We assumed that the mesoscopic relaxation function has a simple exponential form:

$$g[\bar{z}, \bar{z}_s(s)] = \exp[-\bar{z}/\bar{z}_s(s)] \tag{72}$$

Here the dimensionless time $\bar{z} = t/t_1$ is normalized by the characteristic relaxation time t_1, the time required for a charge carrier to move the distance equal to the size of one droplet, which is associated with the size of the unit cell in the lattice of the static site–percolation model. Similarly, we introduce the dimensionless time $\bar{z}_s = t_s/t_1$ where t_s is the effective correlation time of the s-cluster, and the dimensionless time $\bar{z} = t_m/t_1$. The maximum correlation time t_m is the effective correlation time corresponding to the maximal cluster s_m. In terms of the random walker problem, it is the time required for a charge carrier to visit all the droplets of the maximum cluster s_m. Thus, the macroscopic DCF may be obtained by the averaging procedure

$$\Psi[\bar{z}, \bar{z}_m(s_m)] = \int_1^{\infty} g[\bar{z}, \bar{z}_s(s)] \, w(s, s_m) ds \tag{73}$$

In the context of the SLSP model the relationship between the fractal dimension D_s of the maximal percolating cluster, the value of its size s_m, and the linear lattice size L is determined by the asymptotic scaling law [152,213,220].

$$\frac{L}{l} = c_1 s_m^{1/D} \quad (s_m \to \infty, D > 0) \tag{74}$$

where l is the linear size of the lattice cell and c_1 is a coefficient of proportionality.

Let us assume that on the temporal scale at percolation a scaling relationship between the characteristic relaxation time $\bar{z}_m = t_m/t$ and the size s_m exists (similar to the space scaling relationship (74), that is,

$$\bar{z}_m = c_2 s_m^\alpha \quad (s_m \to \infty, \alpha > 0) \tag{75}$$

where c_2 is a constant. Additionally, we assume that self-similarity in the temporal scales is maintained also for clusters of size $s < s_m$, that is,

$$\bar{z}_s(s) = c_2 s^\alpha \quad (s < s_m) \tag{76}$$

Taking into account relationships (75) and (76), over the long time interval $\bar{z} \gg 1$ the integration of (73) may be performed asymptotically by the saddle-point method [135]. The main term of the asymptotic expansion can be obtained as the product of the power and stretched exponential universal relaxation laws [38]:

$$\Psi(\bar{z}) \cong C\bar{z}^{\frac{1-2\Omega}{2(1+\alpha)}} \exp\left[-Q \cdot \left(\frac{\bar{z}}{c_2}\right)^{\frac{1}{1+\alpha}}\right] = A\bar{z}^{-\mu} \exp\left[-\left(\frac{\bar{z}}{\bar{z}_m}\right)^\nu\right] \tag{77}$$

where

$$C = C_w K, \qquad K = \sqrt{\frac{2\pi}{\alpha(1+\alpha)}}(\alpha s_m)^{\frac{\alpha+2(1-\Omega)}{2(1+\alpha)}}$$

$$\mu = \frac{2\Omega - 1}{2(1+\alpha)}, \qquad \nu = \frac{1}{1+\alpha}, \qquad Q = (\alpha \cdot s_m)^{-\frac{\alpha}{1+\alpha}}(1+\alpha) \tag{78}$$

Thus, theoretically obtained asymptotic behavior (77) concurs with the phenomenological power-stretched exponential law (25). Using (78), it is possible to ascertain the relationship between the structural parameters α, Ω, s_m and the phenomenological fitting parameters ν, μ, $\bar{z}_m$ as follows:

$$\alpha = \frac{1}{\nu} - 1, \qquad \Omega = \frac{\mu}{\nu} + \frac{1}{2}, \qquad s_m = \frac{1}{\alpha}\left(\frac{\bar{z}_m}{c_2}\right)^{\frac{1}{\alpha}} \cdot (1+\alpha)^{\frac{1+\alpha}{\alpha}} \tag{79}$$

The set of structural parameters obtained by fitting and by using relationship (79) allows us to reconstruct the cluster size distribution function $w(s,s_m)$ and to treat the dynamic percolation in ionic microemulsions in terms of the classical static percolation model.

2. Dynamic Hyperscaling Relationship

The values of the exponents Ω, D, and α in the distribution function (71) and scaling laws (74)–(76) depend on the Euclidean dimension d of the system and satisfy hyperscaling relationships (HSR). The HSR may be different for the various models describing the various systems [221–225].

For instance, in the SLSP model, a HSR may be obtained by taking into account both the self-similarity of the percolating cluster and the scale invariance of the cluster size distribution function (71). By utilizing the renormalization procedure [213], in which the size L of the lattice changes to a new size L_c with a scaling coefficient $b = L_c/L$, the relationship between the distribution function of the original lattice and the lattice with the adjusted size can be presented as $w(s, s_m) = b^{d-\Omega D} w(\tilde{s}, \tilde{s}_m)$, where $s = \tilde{s}^{-D}$ and $s_m = \tilde{s}_m^{-D}$.

The scale invariance condition $b^{d-\Omega D} = 1$, leads to the following static HSR:

$$\Omega = \frac{d}{D} \tag{80}$$

which is also valid for a hyperlattice with $d > 3$.

For the dynamic case, the percolation problem can be considered in hyperspace, where a temporal coordinate is introduced complementary to the Euclidean spatial coordinates. Using this approach, we shall obtain the scaling relationships in the case of dynamic percolation and derive the dynamic HSR.

Let us consider two-dimensional square lattice $OABC$ (see Fig. 32), whereby microemulsion droplets occupy a number of sites. A selected separate charge carrier hosted by the droplet starts its motion from a position on the side OA at $t = 0$, and under the combined diffusion-hopping transport mechanism moves within the lattice $OABC$. It is understood that owing to this transport mechanism the trajectory of the individual carrier on the lattice $OABC$ may be very complex and can even include loops. An example of such trajectories on the lattice $OABC$ is shown in Fig. 33.

One way to determine the characteristics of these trajectories is by solving a transport equation with different probabilities of hopping of the charge carrier and the corresponding diffusion parameters of the host droplet. Another way, which we shall use here, is based on a visualization of the equivalent static cluster structure. This approach allows us to interpret the dynamic percolation process in terms of the static percolation.

In the static percolation model the trajectory of the charge carrier passes through both a real and visual percolation cluster. While the real cluster in the dynamic percolation is invisible, the charge carrier trajectory can be drawn. In order to visualize the real cluster in the case of dynamic percolation let us assume that the site of trajectory intersections belongs to the equivalent static

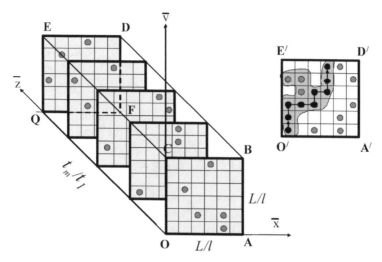

Figure 32.   The visualization of dynamic percolation. A set of the realization of the occupied sites in the two-dimensional lattice over time with fixed time increments.

percolation cluster (ESPC). One must bear in mind that each site in the ESPC may be occupied several times. However, in the static percolation model each site may be occupied only once. In order to exclude the multi-occupation effect, we must increase the dimension of the lattice.

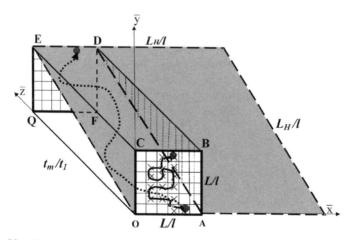

Figure 33.   The charge carrier trajectories in the hyperlattice and the new scale L_H/l of the equivalent lattice formation.

A three-dimensional hyperlattice in the hyperspace formed by the spatial and temporal coordinates corresponds to the two-dimensional square lattice of size L, complemented by the time axis (see Fig. 33). We use nondimensional spatial coordinates ($\bar{x} = x/l$ and $\bar{y} = y/l$), which are normalized by the droplet size l. In turn, the nondimensional temporal coordinate ($\bar{z} = t/t_1$) is normalized by the characteristic time t_1, which is the time needed for a charge carrier to move a distance equal to the size of one droplet. Thus the linear nondimensional size of the hyperlattice along the two spatial axes is L/l, and the nondimensional size of the hyperlattice along the temporal axis is $\bar{z}_m = t_m/t_1$. The characteristic time t_m corresponds to the time needed for a charge carrier to visit all the droplets in the maximal cluster s_m [150].

Note that the classic SLSP model for a square lattice $ABCO$ would provide a percolation trajectory connecting opposite sites OA and CB (see Fig. 32). On the other hand, for visualization of dynamic percolation we shall consider an effective three-dimensional static representation of a percolation trajectory connecting ribs OA and ED spaced distance L_H (see Fig. 33). Such a consideration allows us to return to the lattice $OEDA$, with the initial dimension $d = 2$, which is non-square and characterized by the two dimensionless sizes L_H/l and L/l. The new lattice size of the system can be determined from the rectangular triangle OEQ by

$$L_H/l = \sqrt{(L/l)^2 + (t_m/t_1)^2} \tag{81}$$

where $L_H > L$. A projection of the three-dimensional trajectory on the non-square plane lattice $OADE$ is shown in Fig. 33 by the dotted line. According to the conventional static site percolation model which always deals with square, cubic, and multidimensional one-scale lattices, the maximum scale L_H/l needs to be chosen in our case as the new scale of a square lattice (see Fig. 33). A transformation from the initial lattice with scale size L/l to the lattice with the new scale size L_H/l may be performed by a renormalization procedure. By the substitution of scaling relationships (74) and (75) into (81), the expansion coefficient b_H for dynamic percolation can be presented as

$$b_H = \frac{L_H}{L} = \sqrt{1 + \left(\frac{c_2}{c_1}\right)^2 \cdot s_m^{2\alpha\left(1-\frac{1}{\alpha D}\right)}} \tag{82}$$

For dynamic percolation the renormalization condition $b_H^{d-\Omega D} = 1$ may be satisfied under the conventional condition $d - \Omega D = 0$ and under the additional requirement $b_H = \text{const}$, because $s_m \to \infty$. This is equivalent to the condition $s_m^{2\alpha\left(1-\frac{1}{\alpha D}\right)} = \text{const}$ and leads to the relationship $1 - \alpha D = 0$.

Thus, the dynamic HSR is expressed in the form of a system of two equations:

$$d - \Omega \cdot D = 0 \tag{83}$$

$$1 - \alpha \cdot D = 0 \tag{84}$$

A peculiarity of the dynamic case is the additional equation (83) which must be combined with the static HSR (82). After simple transformation, the solution of the system can be presented in the following way:

$$\Omega = d \cdot \alpha \tag{85}$$

One must bear in mind that the parameter α describes the scaling law (75) for the temporal variable and is equal to the inverse dynamic fractal dimension $\alpha = 1/D_d$. Thus, the dynamic HSR (84) can be rewritten as

$$\Omega = \frac{d}{D_d} \tag{86}$$

and formally coincides with the HSR (80) for the static site percolation model.

Note that the dynamic fractal dimension obtained on the basis of the temporal scaling law should not necessarily have a value equal to that of the static percolation. We shall show here that in order to establish a relationship between the static and dynamic fractal dimensions, we must go beyond relationships (83) and (84) for the scaling exponents.

3. Relationship Between the Static and Dynamic Fractal Dimensions

Let us consider the hyperlattice anisotropy coefficient Θ_H as the ratio of the hyperlattice temporal to the spatial scale

$$\Theta_H = \frac{(t_m/t_1)}{(L/l)} \tag{87}$$

Taking into account (81) and the scaling laws (74), (75), it is easy to show that when $s_m \to \infty$ the validity of the renormalization procedure is similar to the condition that the extension coefficient b_H must be a constant,

$$\Theta_H = \Theta \cdot s_m^{\alpha\left(1-\frac{1}{\alpha D}\right)} = \text{const} \tag{88}$$

where $\Theta = c_2/c_1$. For dynamic percolation the additional equation (84) leads to the relationship $\Theta_H = \Theta = $ const. In general, $\Theta \neq 1$ and the hyperlattice is noncubic as far as the value of the coefficient Θ depends on the type of dynamics in the complex system.

The validity of (88) can be easily proven by assuming that parameter Θ depends on the fractal dimension D. The derivation of both parts of (88) with respect to a variable D leads to the differential equation

$$\frac{d\Theta}{\Theta} = - \ln s_m \frac{dD}{D^2} \qquad (89)$$

In order to establish the relationship between the static and dynamic fractal dimensions, the initial conditions of the classical static percolation model must be considered for the solution of differential equation (89) which can be written as $\Theta = \Theta_s = 1$ for $D = D_s$. Here the notation "s" corresponds to the static percolation model, and the condition $\Theta_s = 1$ is fulfilled for an isotropic cubic "hyperlattice." The solution of (89) with the above-mentioned initial conditions may be written as

$$\ln \Theta = \ln s_m (1/D - 1/D_s) \qquad (90)$$

Taking into account that for dynamic percolation $D = 1/\alpha = D_d$, we can easily obtain the relationship between the dynamic and static fractal dimensions, namely,

$$D_s = \frac{D_d}{1 - D_d \frac{\log \Theta}{\log s_m}} \qquad (91)$$

Note that as is usual for scaling laws, the fractal dimensions $D_d = 1/\alpha$ and D_s do not depend individually on the coefficients c_1 and c_2 entered in the scaling relationships (74), (75), respectively. However, as follows from (91) the relation between D_d and D_s depends on the ratio $\Theta = c_2/c_1$.

For actual experimental setup, $L = 2 \times 10^3$ [86] and $l = 5 \times 10^{-9}$ [143], and the nondimensional lattice size is equal to $L/l = 4 \times 10^5$. Then, taking into account that $\tau_m/\tau_1 = 1.2 \cdot 10^2$, form (87) we obtain $\Theta_H = 3 \cdot 10^{-4}$. Using this value and relationship (88), static fractal dimension, D_s, can be calculated.

Figure 34 shows the temperature dependencies of the static fractal dimensions of the maximal cluster. Note that at percolation temperature the value of the static fractal dimension D_s is extremely close to the classical value 2.53 for a three-dimensional lattice in the static site percolation model [152]. Moreover, the temperature dependence of the stretch parameter ν (see Fig. 34) confirms the validity of our previous result [see (62)] $D_s = 3\nu$ obtained for the regular fractal model of the percolation cluster [47].

Thus, the non-Debye dielectric behavior in silica glasses, PS and AOT microemulsions, has similar properties. These systems exhibit an intermediate temperature percolation process that is associated with the transfer of the electric

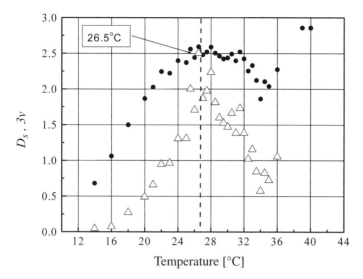

Figure 34.   The temperature dependence of the static D_s ($\triangle$) fractal dimensions and the product 3v ($\bullet$). At the percolation threshold temperature ($T_p = 26.5°C$) $D_s = 3v$. (Reproduced with permission from Ref. 2. Copyright 2002, Elsevier Science B.V.)

excitations through the random structures of the fractal paths. It was shown that in the mesoscale range the fractal dimension of the complex material morphology (D_r for porous glasses and D_s for microemulsions) coincides with the fractal dimension D_p of the path structure. This value can be evaluated by the fitting of the experimental DCF to the stretched-exponential relaxation law (23).

C. Percolation as Part of "Strange Kinetic" Phenomena

As shown above, the dielectric percolation phenomena in different systems such as ionic microemulsions, porous silicon, and porous glasses can actually be analyzed using one universal approach based on the idea of electric excitation transport through the fractal network clusters due to the motion of the charge carriers. This model describes the growth of fractal pre-percolation clusters. Nevertheless, the model is applicable only to one particular phenomenon and does not address any basic theories of "strange kinetic" behavior [2]. Obviously, "strange kinetics"comprises a very wide class of phenomena, which cannot be covered by the limited number of existing models, and has several distinctive features. Most famous among these characteristics is the power-law (or stretched exponent) asymptote with a fractional exponent for the dipole correlation function in time domain [see (23)–(25)]. Another property of the percolation as part of "strange kinetics" is that this class of phenomena is inherent in many-particle cooperative systems. Such complex matter cannot be considered as a simple sum of elementary units but rather should be regarded as a whole system due to the interactions between the elementary components. In a wide sense the word "interaction" not only represents the interaction with some kind of physical far-ranging field (say an electromagnetic field) but also can mean a geometrical (or even a quantum) constraint, or other type of coupling.

Since percolation is a property of macroscopic many-particle systems, it can be analyzed in terms of statistical mechanics. The basic idea of statistical mechanics is the relaxation of the perturbed system to the equilibrium state. In general the distribution function $\rho(p, q; t)$ of a statistical ensemble depends on the generalized coordinates q, momentum p, and time t. However, in the equilibrium state it does not depend explicitly on time [226–230] and obeys the equation

$$\frac{\partial}{\partial t} \rho(p, q; t) = 0 \tag{92}$$

The evolution of the distribution function to the equilibrium state is governed by the Liouville equation (evolution equation)

$$\frac{\partial \rho(p, q, t)}{\partial t} = -i\mathbf{L}\rho(p, q, t) \tag{93}$$

where $\mathbf{L}$ is the Liouville operator. Thus, by virtue of (93) the evolution operator $\mathbf{L}$ determines the dynamical properties of the statistical system. The specific form of this operator is dependent on the Hamiltonian function H [226–230] as

$$\mathbf{L}g = -i\{H, g\} \tag{94}$$

where $\{H, g\}$ are the Poisson brackets. In the classical statistical mechanics

$$\{H, g\} = \sum_k \left(\frac{\partial H}{\partial q_k} \frac{\partial g}{\partial p_k} - \frac{\partial H}{\partial p_k} \frac{\partial g}{\partial q_k} \right) \tag{95}$$

In quantum mechanics the functions H and g become operators $\hat{H}$ and $\hat{g}$, and $\{\hat{H}, \hat{g}\}$ is the commutator form $\{\hat{H}, \hat{g}\} = \frac{2\pi}{ih}(\hat{H}\hat{g} - \hat{g}\hat{H})$, where h is Planck's constant.

Therefore, the consistent study of many-particle system dynamics should start by establishing the Hamiltonian H and then solving the evolution equation (93). Unfortunately, examples of such calculations are very rare and are only valid for limited classes of model systems (such as the Ising model) since these are quite extended calculations. In particular, to the best of our knowledge, the relaxation patterns (23)–(25) have not as yet been derived in this manner (see, however, Chapter 8 in Part B).

In this section, we consider the problem from a different aspect. We will assume that (23)–(25) are given and try to guess what statistical properties leads to the "strange kinetic" behavior. Let us first examine the equilibrium state (92), where

$$\rho(\boldsymbol{p}, \boldsymbol{q}; t) = \text{const} \tag{96}$$

Recently the new concept of fractional time evolution was introduced [45]. In addition to the usual equilibrium state (96), this concept leads to the possibility of the existence of an equilibrium state with power-law long-time behavior. Here the infinitesimal generator of time evolution is proportional to the Riemann–Liouville fractional differential operator $_0D_t^\nu$. By definition of the Riemann–Liouville fractional differentiation operator [231,232] we have

$$_0D_t^\nu[h(t)] = \frac{d}{dt}\,_0D_t^{\nu-1}[h(t)], \qquad 0 < \nu \le 1 \tag{97}$$

where

$$_0D_t^{-\nu}[g(t)] = \frac{1}{\Gamma(\nu)}\int_0^t (t - t')^{\nu-1} g(t')dt', \qquad 0 < \nu \le 1 \tag{98}$$

is the Riemann–Liouville fractional integration operator and $\Gamma(\nu) = \int_0^\infty y^{\nu-1}\exp(-y)\,dy$ is the Gamma function. Obviously, the derivation of order υ should depend on the properties of the cooperative system, although at present there is no clear understanding of how υ depends on these properties.

Nevertheless, let us call the "fractional equilibrium state" the state of the statistical system that under the condition of time independence obeys the following:

$$_0D_t^\upsilon\rho(\boldsymbol{p}, \boldsymbol{q}; t) = 0 \tag{99}$$

We will discuss this state in relation to the recent approaches of the anomalous diffusion theory [31]. It is well known [226–230] that by virtue of the divergent form of Poisson brackets (95) the evolution of the distribution function $\rho(\boldsymbol{p}, \boldsymbol{q}; t)$ can be regarded as the flow of a fluid in phase space. Thus the Liouville equation (93) is analogous to the continuity equation for a fluid

$$\frac{\partial\rho}{\partial t} + \text{div}(\rho\mathbf{v}) = 0 \tag{100}$$

where the distribution function ρ is interpreted as the density of a fluid and v is its local velocity. Let us extend this analogy. The continuity equation accompanied by the relationship between the gradient of the fluid density and its flux (Darcy's law for the liquid flow or Fourier's law for heat flow for instance) [233,234]

$$\text{grad}(\rho) = -\vartheta\rho\mathbf{v} \tag{101}$$

gives

$$\frac{\partial\rho}{\partial t} = \vartheta\Delta\rho \tag{102}$$

where ϑ is the appropriate constant that characterizes permeability of the space and Δ is the Laplace differential operator. From a mathematical point of view, (102) is a diffusion equation where ρ is regarded as the density of diffusing particles and ϑ is proportional to the appropriate diffusion coefficient.

It is well known that the diffusion equation can be obtained in two ways. The first is based on the equation of continuity and the relationship between the fluid density gradient and its flux (100–102). The second way is the probabilistic approach developed from the theory of Brownian motion [31,226,227,235]. This approach does not appeal to local differential equations like (100) and (101), but considers the probability of jumps between the sites of some lattice. There is an extension of this approach for the case when the lengths of the jumps as well as the waiting times between jumps are random. This is the so-called continuous time random walk (CTRW) scheme [31,236–238]. By applying different probability distribution functions for waiting time and jump length, one can obtain different types of diffusion patterns [31]. In particular, if the characteristic waiting time diverges because of a long-tailed waiting time probability distribution function (proportional to $t^{-(1+v)}$), but the jump length variance is still kept finite, then the diffusion (102) has a fractional derivative [31] instead of the first derivative in time on its left-hand side.

$$_0D_t^v\rho = \vartheta_v\Delta\rho \tag{103}$$

Here the parameter ϑ_v has a physical meaning similar to ϑ in (102), but with a different physical dimension. In the general case a term proportional to $\rho(t = 0)$ should be added to the right-hand side of (103), but by choosing appropriate initial conditions it can be subtracted. Thus, we will not discuss this term.

Let us reiterate that diffusion defined by (102) can be derived using the Brownian motion approach and a continuity equation. Moreover, one can imply that (103), a generalization of (102), can be derived not only in the framework of the CTRW scheme but also by using some analogy of the continuity equation as well. The difference between (102) and (103) lies only in the time derivatives. Thus, the analogous of continuity equation (100) corresponding to anomalous diffusion (103) is

$$_0D_t^\nu \rho + \text{div}(\rho \mathbf{v}_\nu) = 0 \tag{104}$$

Let us call this equation the "anomalous continuity equation." There are two main features that distinguish this equation from (100). The first is that (104) becomes nonlocal in time by virtue of the convolution form of the fractional derivation operator $_0D_t^\nu$. Second, in spite of the different physical dimension $[\text{m} \cdot \text{s}^{-\nu}]$, the quantity $\mathbf{v}_\nu$ has a physical meaning similar to the local velocity $\mathbf{v}$.

Let us return back to the Liouville equation regarded as a continuity equation. One may also establish an evolution equation not only of the usual form (93) but also based on the anomalous continuity (104)

$$_0D_t^\nu \rho(\mathbf{p}, \mathbf{q}; t) = -iL\rho(\mathbf{p}, \mathbf{q}; t) \tag{105}$$

This equation implies an equilibrium state of the form (99).

Thus, the fractional equilibrium state (99) can be considered as a consequence of anomalous transport of phase points in the phase space resulting in the anomalous continuity equation (104). Note that the usual form of the evolution (93) is a direct consequence of the canonical Hamiltonian form of the microscopic equations of motion. Thus, the evolution of (105) implies that the microscopic equations of motion are not canonical. The actual form of these equations has not yet been investigated. However, there are strong indications that dissipative effects on the microscopic level become important.

If we assume factorization of the time dependence in the distribution function then the formal solution of (99) is

$$\rho_f(\mathbf{p}, \mathbf{q}; t) = \rho_f(\mathbf{p}, \mathbf{q}; \tau_f) \left(\frac{t}{\tau_f}\right)^{\nu-1} \tag{106}$$

where $t \geq \tau_f$, $\rho_f(\mathbf{p}, \mathbf{q}; \tau_f)$ and τ_f depend on the initial conditions. Obviously, the assumption about factorization of the time dependence for the distribution function is not universal. However, this type of factorization is justified when the equilibrium and nonequilibrium (in the ordinary sense) parts of distribution function $\rho(\mathbf{p}, \mathbf{q}; t)$ are orthogonal to each other in the phase space [230].

The interpretation of any distribution function as a probability density function in phase space leads to the requirement

$$\iint \rho(\boldsymbol{p}, \boldsymbol{q}; t) \frac{d\boldsymbol{p}d\boldsymbol{q}}{N!h^{3N}} = 1 \tag{107}$$

where $3N$ is the number of degrees of freedom. This normalization is based on the uncertainty relation establishing the minimum phase cell as $dp_k dq_k \geq h$. Substitution of the solution (106) in (107) gives

$$\iint \rho_f(\boldsymbol{p}, \boldsymbol{q}; t) d\boldsymbol{p}d\boldsymbol{q} = \frac{\Xi}{N!h^{3N}} \left(\frac{t}{\tau_f}\right)^{\nu-1} = 1 \tag{108}$$

where $\Xi = \iint \rho_f(\boldsymbol{p}, \boldsymbol{q}; \tau_f) d\boldsymbol{p}d\boldsymbol{q}$ is a constant. Thus from (108) we have

$$\frac{N!h^{3N}}{\Xi} = \left(\frac{t}{\tau_f}\right)^{\nu-1} \tag{109}$$

which for $\nu < 1$ indicates a decrease in the number of degrees of freedom.

From one point of view, (109) can be interpreted as a manifestation of the noncanonical nature of the microscopic equation of motion and supports the idea of dissipative effects on the microscopic level (for time scale $t < \tau_f$). From another point of view (109) can be related to the "coarse graining" of the phase volume minimum cells. The concept of fractional evolution is due to the action of the averaging operator [45]. Each application of the averaging operator is equivalent to a loss of information regarding the short time mobility and is closely associated with the renormalization approach ideas [239].

A simple example of this "coarse graining" is two masses in a viscous medium connected by a spring. The spring here represents an interaction between the microscopic particles, while viscosity reflects the dissipative effects. Now let us discuss the situation when one mass is subject to some mechanical perturbation with a wide spectrum (say a δ-function force). Initially the motions of the masses are almost independent of each other. The viscosity effect then leads to a decay of the high-frequency modes of motion so that the motions become more and more correlated. Initially, one should be able to observe the motion of two centers of gravity separately, while for longer time interval it is sufficient to know the position of the joint center of gravity. Thus, the "coarse graining" effect leads to a reduction in the number of degrees of freedom [240].

The reduction of degrees of freedom can also be regarded as the transition from a noncorrelated state to a state with long-range space correlations and may be accompanied by a phase transition. This is the reason the renormalization approach was initially used to describe phase transition phenomena [226,228,239]. The theory of phase transitions investigates the dependence of macroscopic physical quantities (like sample magnetization or polarization vectors) on an external parameter (like external fields or temperature) values. The changes of the degrees of freedom are the result of competition between external perturbations (say temperature) and internal interactions.

In contrast to phase transitions in the fractional equilibrium state (99) the statistical system loses degrees of freedom during evolution in time. The degrees of freedom, independent at the short time limit, become dependent later due to the interactions (in the wide sense coupling, constraint, etc.) and dissipative effects. The distribution function for the fractional equilibrium state (106) can be utilized to calculate the macroscopic dipole correlation function (22). The statistical averaging designated $\langle \cdots \rangle$ in (22) is performed over the equilibrium ensemble (in the usual sense) with a distribution function that does not depend explicitly on time. If we regard the evolution of the DCF as a fractional equilibrium state (97), then by statistical averaging and the Liouville operator [229,230] we can transfer the time dependence of the dynamic variable $M(t)$ to the time dependence of the distribution function $\rho_f(p, q; t)$. Thus, instead of $\langle M(0)M(t) \rangle$ in (22) we use $\langle M(0)M(0) \rangle_f$, where the subscript f means that statistical averaging was performed using the distribution function (106). Here we have

$$
\begin{aligned}
\phi(t) &\cong \frac{\langle M(0)M(t) \rangle}{\langle M(0)M(0) \rangle} = \frac{\langle M(0)M \rangle}{\langle M(0)M(0) \rangle} \\
&= \frac{\int M(p, q; 0)M(p, q; 0)\rho_f(p, q; t)dp\,dq}{M(p, q; 0)M(p, q; 0)\rho(p, q; 0)dp\,dq} \sim \left(\frac{t}{\tau_f} \right)^{\nu - 1}, \qquad t \geq \tau_f
\end{aligned}
\tag{110}
$$

Thus, one can regard the power law dependence of relaxation function (24) as a result of the fractional equilibrium state (99).

In order to understand the stretched exponential behavior of DCF (23), let us discuss Gibb's phase exponent $\eta_G = -\ln \rho(p, q; t)$. This quantity plays a special role in statistical mechanics and relates to the entropy of the system. If Gibb's exponent obeys the fractional evolution equation

$$
{}_0D_t^\mu \eta_G = 0
\tag{111}
$$

then the distribution function is

$$
\rho_\eta(p, q; t) = \rho_\eta(p, q; 0) \exp\left(-\left(\frac{t}{\tau_\eta} \right)^{\mu 1} \right)
\tag{112}
$$

where $\rho_\eta(\boldsymbol{p},\boldsymbol{q};0)$ and τ_η depend on the initial conditions. A calculation analogous to (110) shows that this distribution function leads to the KWW dependency of the DCF. Thus, in the framework of the fractional time evolution concept the power-law time dependence of the relaxation function behavior (24) and the stretched exponential relaxation (23) can be regarded as two different realizations: the fractal equilibrium state of the distribution function and the fractal evolution of Gibb's phase exponent.

A similar idea can also be the basis of the relaxation pattern (25) when these two different types of fractal evolution coexist for two subspaces $(\boldsymbol{p}_f,\boldsymbol{q}_f)$ and $(\boldsymbol{p}_\eta,\boldsymbol{q}_\eta)$ of the total statistical system phase space $(\boldsymbol{p},\boldsymbol{q})$. Here the total distribution function $\rho_{f\eta}(\boldsymbol{p},\boldsymbol{q};t)$ is the product of two statistically independent distribution functions $\rho_f(\boldsymbol{p}_f,\boldsymbol{q}_f;t)$ and $\rho_\eta(\boldsymbol{p}_\eta,\boldsymbol{q}_\eta;t)$:

$$\rho_{f\eta}(\boldsymbol{p},\boldsymbol{q};t) = \rho_\eta(\boldsymbol{p}_\eta,\boldsymbol{q}_\eta;0)\rho_f(\boldsymbol{p}_f,\boldsymbol{q}_f;\tau_f)\left(\frac{t}{\tau_f}\right)^{\nu-1}\exp\left(-\left(\frac{t}{\tau_\eta}\right)^{\mu 1}\right) \qquad (113)$$

which may be related to the relaxation pattern (25).

The relaxation law (113) has been observed for dynamic percolation in ionic microemulsions (see Section V.B). Below the percolation threshold, the relaxation process is provided by two types of mobility: (1) mobility of the pre-percolation clusters and (2) mobility of the charge carriers inside these clusters. The first type of mobility is governed by the power-law (24), while the second type exhibits stretched exponential behavior (23) [152]. Thus, there are two subspaces of degrees of freedom: The first is related to the mobility of pre-percolation clusters as a whole, and the second reflects the mobility inside the clusters. Approaching the percolation threshold, pre-percolation clusters grow and become an infinite (or very large for a real finite-size system) percolation cluster at the threshold. At this point, mobility of the cluster is impossible, because the first subspaces of degrees of freedom disappear and the relaxation function attains the stretched exponential pattern (23). Far from the percolation, one finds the opposite situation. The pre-percolation clusters are small, second subspace degrees of freedom are not yet developed, and the relaxation function obeys the power-law (24).

In this section we have made an attempt to interpret our main results in static and dynamic percolation in terms of statistical mechanics and to find the link to several other approaches and models that describe different aspects of "strange kinetic" phenomena. Obviously, the arguments represented in this section are heuristic. Nevertheless, our attempts to find a unified ideology can result in a deeper understanding of the statistical background of "strange kinetics" and allows us to draw the following general picture.

As we have discussed above, the noncanonical nature of the motion on the microscopic level can be considered as the initial basis of "strange kinetics." In general, this noncanonical motion of one elementary microscopic unit reflects the cooperative effect of the various kinds of interactions in the statistical ensembles. Such interactions, which are specific for each complex system, lead to energy dissipation and can be reduced to the integral term associated with the memory effect in the different kinds of evolution equations. For processes that display scaling properties (such as percolation, or phase transitions), the noncanonical motion can be described in terms of the fractional Liouville equation (105). According to the integral representation of the fractional derivation by (97) and (98) the scaling properties are contained in the kernel of the convolution integral. As was shown, the fractional Liouville equation can be considered as the basic equation for different approaches and models (percolation, anomalous diffusion, damping oscillations, renormalization approach, etc.).

D. Universal Scaling Behavior in H-Bonding Networks

The properties of H-bonded liquids were already reviewed in Section IV.D. Here we would like to continue the consideration of these samples this time with regard to their universal scaling behavior. We have separated this discussion into two parts, which consider two different kinds of glycerol–water mixtures.

1. Glycerol-Rich Mixtures

The analyses of the master plots presented in Fig. 27 in the glycerol-rich region showed that the use of known phenomenological relations and their super-position for the simultaneous fitting of the EW, the main process, and dc-conductivity provide satisfactory results only with a significant number of fitting parameters. Therefore, we propose a new phenomenological function with a fewer fitting parameters as follows:

$$\varepsilon^*(\omega) - \varepsilon_\infty = \frac{\Delta\varepsilon \cdot [1 + A(\omega\tau)^q]}{(1 + i\omega\tau)^\beta} + \frac{B\Delta\varepsilon}{i\varepsilon_0\omega\tau} \tag{114}$$

where $\sigma = B\Delta\varepsilon/\tau$ is the dc-conductivity, B is a constant, β is a parameter describing asymmetrical relaxation peak broadening (similar to β_{CD}), and A and q are parameters to describe the EW, respectively [208]. For glycerol–water mixtures in the present temperature and concentration range, we have

$$q = 1 - \beta + \gamma \tag{115}$$

where γ is equal to 0.08 for the entire range of the observed concentrations. The nature of this parameter is not yet clear. Figure 35 shows a good fitting of both real and imaginary parts of the raw dielectric spectra using relationship (114).

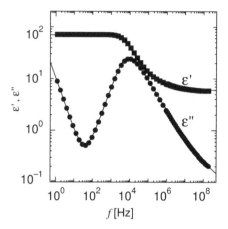

Figure 35. A typical fitting result by the function (114) for the dielectric spectrum (75 mol% at 224 K, both real and imaginary parts are shown) where parameters β and A were fixed to be the same values obtained in the master plot of 75 mol% of glycerol (see Fig. 27b) with $\tau = 25\,\mu s$ and $\Delta\varepsilon = 64.1$. (Reproduced with permission from Ref. 208. Copyright 2005, American Chemical Society.)

Moreover, this fitting shows that in the considered temperature–frequency landscape, only the temperature dependence of $\Delta\varepsilon = \Delta\varepsilon(T)$ and $\tau = \tau(T)$ is observed, while the other parameters B, β, A, and q are not temperature-dependent. Thus, the proposed phenomenological model (114) as a modification of the CD function (21) with a conductivity term could be successfully applied for simultaneous fitting of dc-conductivity, the main process, and the EW presented in the master plots (see Fig. 27).

At the same time, parameters β, A, and q have a concentration dependence where A and β decrease with a decrease of glycerol content (Fig. 36), while parameter q increases according to (115). Note, that the parameters A and q can be associated with some characteristic mesoscopic relaxation time τ_0 as follows:

$$A(\omega t)^q = (\omega\tau_0)^q, \qquad \tau/\tau_0 = A^{-1/q} \tag{116}$$

Since the parameters A and q of the master plot are not dependent on temperature, the ratio τ/τ_0 is also independent of temperature that is, the scale τ_0 follows the same VFT law as the main relaxation time τ (Fig. 37).

Temperature dependencies of τ_0 that were obtained by (114) and (115) and representation by the VFT model (full lines) where D and T_v have the same values as those for τ_{max} presented in Fig. 26 as well as in the legend (100 to 60 mol%). The preexponential factor such as $\ln(\tau_{v0})$ is shown in the legend.

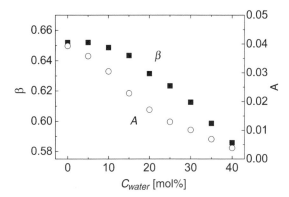

Figure 36. Water concentration dependence of the shape parameters (β, filled box; A, unfilled circle) in (114) obtained by curve fittings of the master plots. (Reproduced with permission from Ref. 208. Copyright 2005, American Chemical Society.)

In this context, the value τ_0 should be associated with the smallest cooperative relaxation time reflecting the EW dynamics obeying the same VFT temperature law as the main relaxation times τ and τ_{max} (see Fig. 26). At the same time, the ratio τ/τ_0 is strongly dependent on concentration (Fig. 38). The increase of the ratio τ/τ_0 with increase of water content indicates that the EW of glycerol is eroded by water molecules much faster than the main relaxation process. In order to clarify such a mechanism, let us assume that the EW is the result of some fast short-range cooperative dynamics that can be associated with

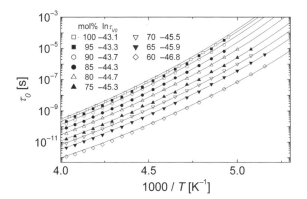

Figure 37. Temperature dependence of τ_0 obtained by (114) and (115) and representation by the VFT model (full lines) where D and T_v are the same values as those for τ_{max} presented in Fig. 26 as well as in the legend (100 to 60 mol%). The pre-exponential factor such as $\ln(\tau_{v0})$ is shown with the legend.

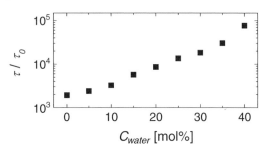

Figure 38. Water concentration dependence of the ratio of characteristic times (τ/τ_0) obtained by (16). (Reproduced with permission from Ref. 208. Copyright 2005, American Chemical Society.)

the so-called "cage level" [183]. Thus for anhydrous glycerol at the mesoscopic scale we are considering only the glycerol–glycerol interaction. For glycerol–water mixtures in the range of the master plot universality, the EW is caused mostly by two kinds of interactions: glycerol–glycerol and glycerol–water interactions. It is manifested by the decreasing of τ_0 due to fast mobility of water molecules contributing to the cooperative dynamics at the "cage level." However, the glycerol–glycerol molecular interactions remain the dominant relaxation mechanism at the scale of the whole H-bond network. A simplified picture of interaction for the 60 mol% of glycerol in water where the master plot universality is still observed could be presented as follows. It is well known [241] that in the liquid phase three OH groups of a glycerol molecule may be involved in approximately $n_g \cong 6$ hydrogen bonds with neighboring molecules, while a water molecule may establish $n_w = 4$ hydrogen bonds. The critical mole fraction where glycerol H-bonds are balanced with water H-bonds may be roughly estimated as $x = 100\% \cdot n_w(n_w + n_g) = 40$ mol%. However, water–water interactions may still appear at higher concentrations of glycerol because of density fluctuations, as is suggested by the breakdown of the universality only at 55 mol%. Most probably the mechanism of the observed universality is related to the H-bond dynamic structures of the glycerol–water mixture. Here the EW is a projection of the short-range fast dynamics of H-bond structures, the main relaxation process results from a larger scale of H-bond clusters and dc-conductivity relates to the transfer of charge carriers through the percolated H-bond network.

Figure 39 shows the water content dependence of normalized conductivity on the master plots evaluated by using the relation

$$\sigma_m = \frac{2\pi\tau_{max}}{\varepsilon''_{max}} \sigma \qquad (117)$$

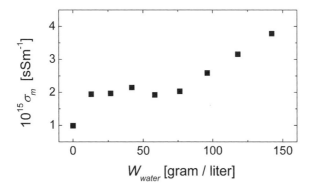

Figure 39. Water content dependence of the normalized conductivity on the master plots (117), where $\sim 1.6 \times 10^{-16}$ s Sm^{-1}. Note that our data of density measurements for glycerol–water mixtures at 288 K were used to obtain water content, W_{water}, [gram/liter] from the mole fractions. (Reproduced with permission from Ref. 208. Copyright 2005, American Chemical Society.)

The nature of the charge carriers is not yet clear and there could be at least two possibilities: ionic impurities or H-bond network defects (so-called orientation and ionic defects) similar to those considered in the conduction and relaxation of ice [242,243]. Conductivity behavior due to ionic impurity is expected to increase linearly with the impurity concentration per unit volume, and the impurities would be proportional to water content. However, the normalized conductivity did not show such a linear behavior (Fig. 39), although some tendency of increase with water content can be observed. This increase of the normalized conductivity does not immediately contradict the concept of defect-conductivity, because the number of defects can also be increased by increasing water content. At the very least, diffusion of ions should also be accompanied by breaking or changing of H-bond networks around the ion molecule. Such rearrangements of H-bond networks around ions can affect the EW and may influence the main relaxation process on a cluster level. Thus, the existence of universality is most likely due to the presence of an H-bond network and its defect structure.

Moreover, the experimental data through dc-conductivity, the main process and the EW could be described by the new phenomenological function (114). If the concentration remains the same, only two variable parameters, $\Delta\varepsilon$ and τ, with a set of constants, can describe the whole temperature dependence of spectra. The universality in the master plots of dielectric spectra for glycerol and glycerol–water mixtures has the same origin as elementary molecular processes in so far as it is most likely due to a "defect" formation and its percolation in an H-bond network.

2. Water-Rich Mixtures

Such universal dielectric behavior disappeared at higher water concentrations (lower than 60 mol% of glycerol) due to the appearance of water–water interaction coexisting with the glycerol–glycerol and glycerol–water interactions in the glycerol domains. These water–water interactions were also observed recently by Sudo et al. [190]. Therefore, we can assume that H-bond networks in water-rich mixtures are inhomogeneous on a mesoscopic level, although the detailed features of such heterogeneous structures are not fully known. With increasing of amounts of water, the water pools grow in size, leading to their freezing at low temperatures. For H-bonding liquid mixtures, however, the dynamical and structural properties in frozen states have not yet been extensively studied. BDS and differential scanning calorimetry (DSC) measurements of glycerol–water mixtures in the high water concentration range and temperature intervals including the water-frozen state were reported recently [244].

Let us consider the dynamics of glycerol–water mixtures in terms of the characteristic time $\tau_{max} = 1/(2\pi f_{max})$ and the temperature dependence of the main relaxation process. In the glycerol-rich region (100–40 mol%), as mentioned in Section IV.D and as shown in Fig. 40, τ_{max} is well-described by the VFT law with the same value of fragility ($D = 22.7$) and almost the same values of preexponent factor ($\ln \tau_v = -35.9$ to -36.4). The plots in Fig. 40a represent the experimental data, for glycerol–water mixtures, where the glycerol concentration is systematically changed in steps of 5 mol%, and the solid lines show the fitting results according to the VFT law. The results indicate that the relaxation mechanisms and H-bond networks for glycerol-rich mixtures

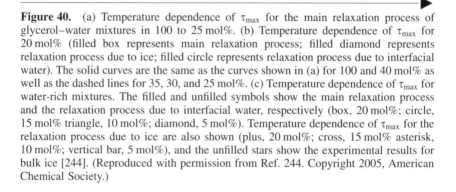

Figure 40. (a) Temperature dependence of τ_{max} for the main relaxation process of glycerol–water mixtures in 100 to 25 mol%. (b) Temperature dependence of τ_{max} for 20 mol% (filled box represents main relaxation process; filled diamond represents relaxation process due to ice; filled circle represents relaxation process due to interfacial water). The solid curves are the same as the curves shown in (a) for 100 and 40 mol% as well as the dashed lines for 35, 30, and 25 mol%. (c) Temperature dependence of τ_{max} for water-rich mixtures. The filled and unfilled symbols show the main relaxation process and the relaxation process due to interfacial water, respectively (box, 20 mol%; circle, 15 mol% triangle, 10 mol%; diamond, 5 mol%). Temperature dependence of τ_{max} for the relaxation process due to ice are also shown (plus, 20 mol%; cross, 15 mol% asterisk, 10 mol%; vertical bar, 5 mol%), and the unfilled stars show the experimental results for bulk ice [244]. (Reproduced with permission from Ref. 244. Copyright 2005, American Chemical Society.)

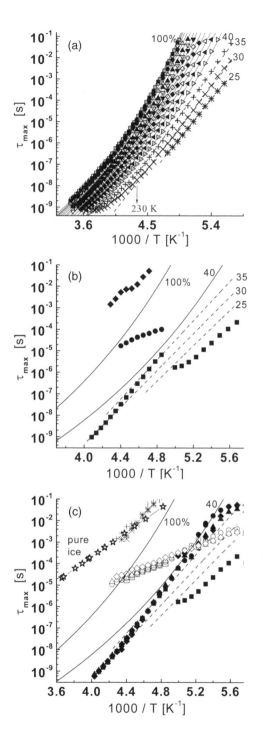

are similar to those of pure glycerol. Below 40 mol% of glycerol and at lower temperatures (below $\sim$230 K), the temperature dependence of the τ_{max} was continuously changed from VFT to Arrhenius shape, as shown in Fig. 40a (dashed lines). Our analyses and comparison with the literature [190] shows that τ_{max} temperature behavior in the water-rich region can depend on the temperature history, although no differences were found in glycerol-rich regions above T_g (here T_g is the glass transition temperature). These results may indicate that mesoscopic structures of H-bond networks in glycerol-rich regions are quite homogeneous. In contrast, in water-rich regions, such mixtures have a more complicated dynamic behavior. Glycerol molecules cannot provide all water molecules with H-bond networks, and correspondingly the system forms so-called glycerol and water cooperative domains, respectively. The Dynamic structures of such domains can depend on the temperature hysteresis. The critical molar fraction (x_g) would then relate to the number of H-bonds of glycerol and would be approximately 40 mol% as it was shown above (see Section V.D.1).

The existence of this critical concentration of 40 mol% of glycerol is well supported by the results obtained from further study of the water-rich region. Figure 40b shows the temperature dependence of τ_{max} for 20 mol% of glycerol. At lower temperatures, τ_{max} demonstrates a behavior similar to that of higher glycerol concentrations. However, when the temperature was increased and approached $T = 206$ K crystallization of the extra water occurred and τ_{max} increased to nearly the same value as that of 35 mol% of glycerol. After crystallization, two additional Arrhenius-type relaxation processes were observed (Fig. 41 shows the typical dielectric spectra before and after crystallization). One of these relaxation processes is due to the presence of ice cores, since the relaxation time and its activation energy ($\sim$77 kJ mol^{-1}) is similar to the well-known values for bulk ice which were reported by Auty and Cole [245] (Fig. 40c). Another relaxation process may relate to the presence of some interfacial water between the ice core and the mesoscopic glycerol–water domain. The glycerol concentration of this domain was enriched up to the critical concentration $\sim$40 mol% by the freezing of the extra water that was free from the glycerol H-bond networks.

For of 15, 10, and 5 mol% of glycerol–water mixtures, the ice cores have already been formed during the quenching down to the starting temperature of the BDS measurements, and τ_{max} of the main relaxation process for all these mixtures shows values that are similar to those of 35 or 40 mol% glycerol–water mixtures (see Figure 40c). Furthermore, the two additional relaxation processes (due to ice and interfacial water) also follow the same curves. It is worth noting that the activation energy of $\sim$33 kJ mol^{-1} of the relaxation process resulting from interfacial water is similar to the reported values $\sim$28 kJ mol^{-1} for bound water on protein surfaces in aqueous solutions [246] and $\sim$30–40 kJ mol^{-1} for

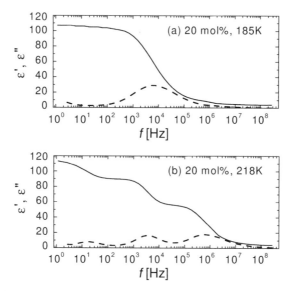

Figure 41. Typical dielectric spectra of 20 mol% of glycerol—water mixtures at (a) 185 K (supercooled state) and (b) 218 K (frozen state), where solid and dashed curves show the real and imaginary parts of complex dielectric permittivity. Each relaxation process in the frozen state was fitted by (114) and by Cole–Cole and Debye relaxation functions, respectively, in order to separate the main process, the process due to interfacial water, and the process due to ice. (Reproduced with permission from Ref. 244. Copyright 2005, American Chemical Society.)

surface water on porous glasses, depending on the porous glass preparation [256]. The relaxation time for this interfacial water is ~100 times larger than that for the bound water reported in protein solutions [246]. For bound water in protein solutions, this interfacial water is surrounded by fast-mobile water molecules from the solvent. For water-rich glycerol–water mixtures the interfacial water molecules would be confined between two regions such as the ice and glycerol–water domains; therefore, mobility of the interfacial water in water–glycerol mixtures would be highly restricted.

The glass transition temperature (T_g), which is defined as the temperature where the relaxation time is 100 s [247,248], was evaluated for glycerol–water mixtures using the temperature dependences of τ_{max} for the main dielectric relaxation process (see Figure 40a). As shown in Fig. 42a, the glass transition temperatures (T_g) obtained by DSC measurements were in good agreement with BDS data. It is worth pointing out that T_g for 2.5, 10, and 15 mol% of glycerol–water mixtures in the frozen state all have the same values as T_g for 40 mol% of the glycerol–water mixture. This result indicates that the glass transition

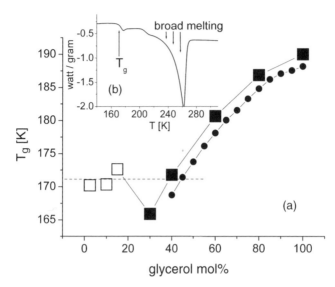

Figure 42. (a) Glycerol concentration dependence of the glass-transition temperature obtained by DSC measurements (filled box represnts supercooled state, unfilled box represents frozen state) and by BDS data (filled circle) (b) Typical raw data of DSC measurements (10 mol% of glycerol), where the glass transition and a broad melting behavior were observed. (Reproduced with permission from Ref. 244. Copyright 2005, American Chemical Society.)

observed in the frozen state is due to the mesoscopic glycerol–water domains of 40 mol% glycerol. Moreover, even simple dc-conductivity data obtained by BDS at particular temperatures show concentration independent behaviors in the frozen state, and the values are close to that of 40 or 35 mol% of glycerol (Fig. 43). Note that these results support the hypothesis that the mechanism of dc-conductivity in a supercooled state is most likely that of defect translocation as discussed recently [208], possible because the increase of ionic impurities did not affect the frozen state as shown in Fig. 43.

In order to estimate the ratio between the amounts of water in the mesoscopic glycerol–water domains, interfacial water, and ice in glycerol–water mixtures, one can use the melting-enthalpy ΔH obtained from DSC data. Note that in this specific case, the total melting-enthalpy (ΔH) was obtained by integration of the transition heat capacity (ΔC_p) over the broad melting temperature interval:

$$\Delta H = \int_{T_1}^{T_2} \Delta C_p \, dT = \int_{T_1}^{T_2} (C_P - C_{P(baseline)})_p dT \qquad (118)$$

where T_1 is the starting temperature of the broad melting (223 K for 10 mol% glycerol; see Fig. 42b), T_2 is the temperature of the phase transition, and

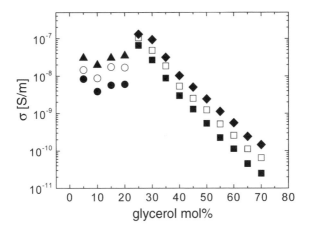

Figure 43. Glycerol concentration dependence of dc-conductivity (filled box represents supercooled state at 209 K, filled circle represents frozen state at 209 K, open box represents supercooled state at 212 K, open circle represents frozen state at 212 K, filled diamond represents supercooled state at 215 K, and filled triangle represents frozen state at 215 K). (Reproduced with permission from Ref. 244. Copyright 2005, American Chemical Society.)

$C_{p(\text{base line})}$ is the extrapolated heat capacitance base line in the same temperature interval [226]. Thus, the amount of ice below the starting temperature of the broad melting was estimated by the ratio $(\Delta H / \Delta H_0)$, where ΔH_0 is the melting-enthalpy of bulk ice [249]. Using the known total amount of water and the critical 40 mol% concentration, we can estimate that in the 10 mol% of glycerol–water mixture, for example, we have approximately 50 mol% of ice, 15 mol% of water in the mesoscopic domains, and 25 mol% of interfacial water. This estimation using DSC data also indicates the presence of a significant amount of interfacial water in the frozen state of glycerol–water mixtures.

These facts support the existence of three water states: water in glycerol H-bond networks ($H_2O_{(GW)}$), water in ice structure ($H_2O_{(ice)}$) and interfacial water ($H_2O_{(interface)}$). Let us discuss a possible kinetic mechanism of the broad melting behavior. The relations between the three states of water can be described by

$$H_2O_{(GW)} \underset{k_2}{\overset{k_1}{\rightleftharpoons}} H_2O_{(interface)} \underset{k_4}{\overset{k_3}{\rightleftharpoons}} H_2O_{(ice)} \qquad (119)$$

where k_1, k_2, k_3, and k_4 are the reaction rates (e.g., the reaction velocity from $H_2O_{(GW)}$ to $H_2O_{(interface)}$ is described by $v_1 = k_1[H_2O_{(GW)}]$, where $[H_2O_{(GW)}]$ is the mole concentration of $H_2O_{(GW)}$). If the system is in an equilibrium state, apparently $[H_2O_{(interface)}]$ does not change. Therefore it can be described as

$$k_1 \lfloor H_2O_{(GW)} \rfloor + k_4 \lfloor H_2O_{(ice)} \rfloor = (k_2 + k_3) \lfloor H_2O_{(interface)} \rfloor, \qquad (120)$$

and it can be rewritten:

$$[H_2O_{(interface)}] = \frac{k_1}{k_2} [H_2O_{(GW)}] = \frac{k_4}{k_3} [H_2O_{(ice)}]. \qquad (121)$$

Monotonous increase of ratios k_1/k_2 and k_4/k_3, causes continuous increase of $[H_2O_{(interface)}]$ and decrease of $[H_2O_{(ice)}]$. This is a possible mechanism of the broad melting behavior.

The extra-interfacial water, which is produced by melting of ice, can penetrate into the mesoscopic glycerol-water domains. Nevertheless, these water molecules should be classified as $H_2O_{(interface)}$ because they are out of the glycerol H-bonding network within a mesoscopic glycerol domain. This extra-interfacial water can affect the dynamic properties of the mesoscopic glycerol-water domain, and it is a possible reason why τ_{max} does not follow the 40 mol% curve at higher temperatures (Figure 40c). In this temperature range, a faster replacement between $H_2O_{(GW)}$ and $H_2O_{(interface)}$ allows the mixture to retain more water in the supercooled phase than the expected amount based on the critical concentration of 40 mol%.

Hence, we can assume the following physical picture of dynamic and structural behavior of water–glycerol mixtures in the considered frequency–temperature landscapes for different concentrations of components (see Fig. 44). As one can see, for high concentrations of glycerol including pure dehydrated glycerol, the universality evident in the master plots of temperature and concentration has the same origin as the elementary molecular process. It is most likely the result of defect formation and the resultant percolation though the H-bond network.

Below 40 mol% of glycerol–water, domains appear. This critical concentration is related to the numbers of H-bonds of glycerol and of water respectively. DSC studies also confirm the same critical concentration. In water-rich mixtures, some water is frozen; as a result, three relaxation processes where observed. These were related to ice-like structures, interfacial bound-water, and glycerol–water mixtures in the mesoscopic domains, where the concentration remains at 40 mol%.

Thus, concluding this section, we can summarize that H-bonded liquids can be regarded as model systems that reproduce many basic features of liquid-

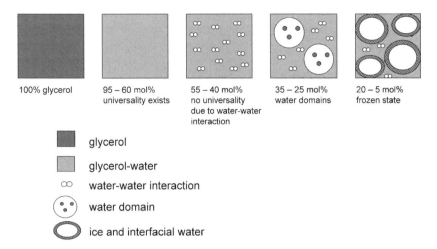

| 100% glycerol | 95 – 60 mol%
universality exists | 55 – 40 mol%
no universality
due to water-water
interaction | 35 – 25 mol%
water domains | 20 – 5 mol%
frozen state |

- ▪ glycerol
- ▫ glycerol-water
- ∞ water-water interaction
- (∵) water domain
- ⬭ ice and interfacial water

Figure 44. A schematic model of glycerol–water H-bonding liquids in different concentration of components.

glass-forming materials. The reasoning behind this is that the interaction between their constituents provides cooperative properties of the system as a whole. In particular, glycerol–water mixtures, depending on their composition, may exhibit a variety of relaxation properties: from properties of homogeneous solutions to those of binary mixtures.

E. Liquid-like Behavior in Doped Ferroelectric Crystals

In the previous section we considered in detail the behavior of liquid-glass-forming substances. However, such liquid-like behavior can be also inherent in the dielectric properties of crystalline samples.

Now let us discuss the behavior recently observed in doped ferroelectric crystals in the paraelectric phase (see Section IV.C). Because process B was observed only in crystal #1 (see Fig. 18), it must be attributed to the presence of the Cu impurities embedded at random in the KTN crystal. The Arrhenius nature of the process at elevated temperatures above 354 K indicates normal relaxation of the independent Cu^+ ions [179]. These ions are significantly smaller than the K^+ sites in which they reside (the radii of the Cu^+ and the K^+ ions are 0.77 Å and 1.52 Å, respectively [250]), leading to off-center displacements. The Cu^+ ions can therefore hop between the eight symmetrical minima of their potential wells. Indeed, the energy of activation of $E_a^B = 0.37\,eV$ corresponds to the activation energies for the hopping of transition metal ion impurities in $KTaO_3$ [251].

VFT behavior has been noted in crystal systems with a dipole glass phase
[176]. However, the VFT behavior in crystal #1 is related to the relaxation of the
Cu^+ ions in the paraelectric phase and is more akin to relaxations in glass-
forming liquids [179]. The formalism of Adam and Gibbs [63] (A G) has been
successfully applied to such systems. According to the A G theory, this behavior
originates from the cooperative rearrangement of some clusters. It is well
established that in KTN crystals in the paraelectric phase the Nb^{+5} ions are
displaced from the center of inversion of the unit cells [172]. These
displacements form dipolar clusters with a correlation length that increases
as the system is cooled towards the ferroelectric phase transition. We propose
that the cooperative relaxation observed at $T < 354$ K is produced by the
interaction between such dipolar clusters that form around the relaxing Cu^+ ions
and that act as the rearranging clusters in the A G model. Adopting the formalism
of A G [63], the minimum cluster size z is related to the rate of relaxation $P(T, z)$
by

$$P(T, z) \propto \exp\left(-z\frac{\Delta\mu}{k_B T}\right) \qquad (122)$$

VFT behavior is obtained by equating $z = T/(T - T_{VFT})$ and noting that
increment of chemical potential $\Delta\mu = E_{VFT}$ [65]. Fitting the VFT model to the
experimental results of τ^B in the paraelectric phase gives $T_{VFT} = 228$ and
$\Delta\mu = 0.02$ eV. Identifying the temperature at which τ^B deviates from the
Arrhenius model with the onset of cooperativity yields a minimum cluster size
given by

$$z = \frac{T_x}{T_x - T_k} = 2.8 \qquad (123)$$

This compares well with an estimation of the minimum size of the dipolar cluster
formed around nearest neighbors Nb^{5+} ion given by $[Nb^{5+}]^{-1} = [0.38]^{-1} \approx 2.63$.
At elevated temperatures $(T > 354$ K$)$, the Nb^{5+} ions hop at random between
equivalent minima of their potential wells within their site in the unit cell. As
the phase transition is approached, these ions form a dipolar cluster around the
Cu^+ ions that are randomly distributed far apart from each other. These clusters
are at first of a minimum size containing only the Nb^{5+} ions that are closest to
the Cu^+ ion. As the phase transition is approached, the cluster size around the
Cu^+ ion grows accordingly. These represent the rearranging clusters of the A G
theory.

An independent assessment of the validity of the A G interpretation to the Cu^+-induced relaxation (process B) was provided by direct estimation of the configurational entropy at the ferroelectric phase transition. The phase transition is dominated by a shift in the position of the Nb^{5+} ion in relation to the lattice. The coupling between the Cu^+ ions and the Nb^{5+} polar clusters, evident in the paraelectric phase, will further contribute to the configuational entropy of the phase transition.

It is well known that the configurational entropy can be evaluated from the heat capacitance of the crystal, derived from DSC measurements, by the integral

$$S_C(\Delta T) = \int_{\ln T_1}^{\ln T_2} [C_p(\ln T) - C_{p(\text{baseline})}]d(\ln T) \qquad (124)$$

where T_1 is the onset and T_2 is the completion of the phase transition, and $C_{p(\text{baseline})}$ is the extrapolated baseline heat capacitance in the temperature interval [226]. The results for the pure KTN crystal (#2) and the Cu-doped KTN crystal (#1) are presented in Fig. 45. The resulting difference was found to be $\Delta S = 0.79 \times 10^{-3} \, J \, g^{-1}$. Normalized to the Cu^+ ion content we have per Cu^+ ion $\Delta S_{Cu} = 2.068 \times 10^{-21} \, J \approx 0.013 \, eV$. This result is in a good agreement with $\Delta \mu = 0.02 \, eV$ derived from fitting the VFT model to $\tau^B(T)$, given the spatial variation of the Cu concentration in the crystal.

At the onset of the ferroelectric phase transition the Cu^+ ions are frozen and no longer constitute the seed of the relaxation process. In this respect the ferroelectric phase transition "quenches" the glass-forming liquid. Below the phase transition temperature, the dipolar clusters surrounding the Nb^{5+} ions merge to yield the spontaneous polarization of the (now ferroelectric) crystal, and due to the strong crystal field, the off-center potential minima of the Cu^+ ions are no longer symmetrical.

Thus, one may summarize the physical picture of the relaxation dynamics in KTN crystal-doped with Cu^+ ions in the following way: In the paraelectric phase, as the ferroelectric phase transition is approached, the Nb^{5+} ions form dipolar clusters around the randomly distributed Cu^+ impurity ions. The interaction between these clusters gives rise to a cooperative behavior according to the AG theory of glass-forming liquids. At the ferroelectric phase transition the cooperative relaxation of the Cu^+ ions is effectively "frozen."

F.   Relaxation Kinetics of Confined Systems

It was already mentioned in Section II.C that the kinetics associated with relaxation parameters measured by dielectric spectroscopy can provide

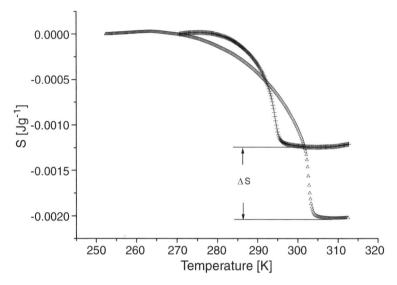

Figure 45. The configurational entropies of the ferroelectric phase transition for crystals # 1 ($\triangle$) and # 2 (+). The difference in entropy between the two crystals, ΔS, is $0.79 \times 10^{-3}\,\mathrm{J\,g^{-1}}$; this is assigned to the Cu^+ dopants. The DSC measurements were made with the cooling and heating rates 5 K/min in an interval ranging from 373 K to 220 K using a DSC 2920 calorimeter (TA Instruments) [179]. (Reproduced with permission from Ref. 179. Copyright 2004, The American Physical Society.)

significant information about the substance under study. Even without the quantitative data, having only qualitative information, one can, for example, identify a glassy material by nonlinear VFT relaxation time dependence. In this section we will discuss another type of relaxation kinetics which is associated with confined systems, and we will illustrate this type using some experimental examples.

1. Model of Relaxation Kinetics for Confined Systems

The model we utilize here was first introduced to describe the relaxation properties of water adsorbed on the inner surfaces of porous glasses [155] and was then reviewed in Ref. 78. The main idea of this model is that the relaxation kinetics is now provided by a process that must satisfy two statistically independent conditions. Thus, if one assigns the probability p_1 to satisfy the first condition and p_2 to satisfy the second condition, then the probability p for a relaxation event to occur in such a system is

$$p = p_1 p_2 \tag{125}$$

Let us discuss a system that consists of a number of particles where their relaxation is provided by the reorientations (a jump or another type of transition) of particles between two local equilibrium states. In the spirit of the Arrhenius model (26), the first requirement for the relaxation is that the particles have enough energy to overcome the potential barrier E_a between the states of local equilibrium for the elementary constituents of the system under consideration. Thus,

$$p_1 = \exp\left(-\frac{E_a}{k_B T}\right) \tag{126}$$

The essential idea of the new model is that p_2 is the probability that there will be enough free volume, v_f, in the vicinity of a relaxing particle with its own volume, v_0, to allow reorientation [69,70,78,155]. In other words, we imply that in order to participate in a relaxation a particle must have a "defect" in its vicinity. Then,

$$p_2 = \exp\left(-\frac{v_0}{v_f}\right) \tag{127}$$

In itself this probability represents a kind of constraint for the entire relaxation process and slows down the relaxation. Combining (124)–(127) and taking into account the relationship $\tau \sim 1/p$, the following expression can be obtained:

$$\ln\left(\frac{\tau}{\tau_0}\right) = \frac{E_a}{k_B T} + \frac{v_0}{v_f} \tag{128}$$

As mentioned in Section II.C, it is usually assumed that the free volume grows with increasing temperature. This idea reflects thermal expansion; that is, if the number of relaxing particles in the system is kept constant, then the thermal expansion leads to an increase of the free volume with increase in temperature growth. However, this assumption may be wrong for a confined system where the total volume is kept constant, but the number of relaxing particles varies. In our case we implied earlier that the microscopic act of relaxation is conditional on the presence of a defect in the vicinity of the relaxing particle. As a first approximation, one may assume that the number of defects, n, obeys the Boltzmann law $n = n_0 \exp(-E_b/k_B T)$. Then, the free volume per relaxing particle is $v_f = V/n$, where V is the total volume of the system. Thus, instead of

(128), one immediately obtains [78,155]

$$\frac{\tau}{\tau_0} = \exp\left[\frac{E_a}{k_B T} + C\exp\left(-\frac{E_b}{k_B T}\right)\right] \tag{129}$$

where E_b is the energy that would enable an "inert" particle to participate in relaxation (or alternatively the energy required to form a so-called defect), $C = v_0 n_0 / V$ is a confinement factor, and n_0 is the maximal possible number of defects in the considered system. In contrast to all other kinetic models, (129) exhibits a nonmonotonic temperature dependence since it is related to two processes of a different nature: the Arrhenius term reflecting the activated character of the relaxation process, and the exponential term reflecting a decrease of free volume per relaxing particle with temperature growth. This second term is a consequence of the constant volume constraint and the implication that the number of relaxing particles obeys the Boltzmann law. If the total volume of the system V is sufficiently large and the maximum possible concentration of relaxing particles is sufficiently small $(n_0/V \ll 1/v_0)$, then the free volume arguments become irrelevant and the relaxation kinetics retain the Arrhenius form. However, in the case of a constraint, when the volume of a system is small and $n_0/V \approx 1/v_0$, an increase of temperature leads to a significant decrease of free volume and slows down the relaxation. As we will show, this situation usually occurs for "small" systems where relaxing particles are able to participate in the relaxation due to the formation of defects in the ordered structure of the system.

2. Dielectric Relaxation of Confined Water

It is known [52] that the dielectric relaxation of water is due to the reorientation of water molecules having a permanent dipole moment (6×10^{-30} C or 1.8 D). It is also known that in bulk water and ice water molecules are embedded in the network structure of hydrogen bonds. Thus, the reorientation of a water molecule leading to dielectric relaxation may occur only in the vicinity of a defect in the hydrogen bonds network structure. This mechanism fits perfectly the relaxation model described in the section above. Thus, one may expect that a confined water sample could exhibit the nonmonotonous relaxation kinetics predicted by (129). E_a could now be regarded as the activation energy of reorientation of a water molecule, and E_b could be regarded as the defect formation energy. Here we discuss the dielectric relaxation of water confined in porous glasses [153,155]. The main features of this relaxation have been described already in Section IV.B.1. In Figure 14 one can clearly see the nonmonotonic saddle-like process (Process II), which will be discussed here.

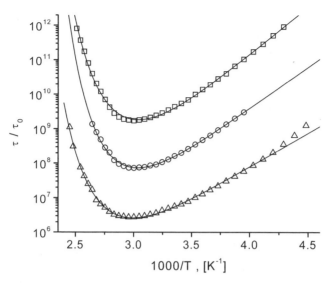

Figure 46. Temperature dependence of the dielectric relaxation time of water confined in porous glasses [153,155]. Symbols represent experimental data. Full lines correspond to the best fit according to Eq. (129). Sample A (cycles) $\ln \tau_0 = -27 \pm 0.5$, $E_a = 46 \pm 1\,\text{kJ mol}^{-1}$, $E_b = 33 \pm 1\,\text{kJ mol}^{-1}$, $C = 27 \times 10^4 \pm 9 \times 10^4$. Sample B (boxes) $\ln \tau_0 = -33 \pm 0.5$, $E_a = 53 \pm 1\,\text{kJ mol}^{-1}$, $E_b = 29 \pm 1\,\text{kJ mol}^{-1}$, $C = 7 \times 10^4 \pm 2 \times 10^4$. Sample C (triangles) $\ln \tau_0 = -26 \pm 0.3$, $E_a = 38 \pm 1\,\text{kJ mol}^{-1}$, $E_b = 32 \pm 1\,\text{kJ mol}^{-1}$, $C = 12 \times 10^4 \pm 3 \times 10^4$. (Reproduced with permission from Ref. 78. Copyright 2004, The American Physical Society.)

Similar nonmonotonic relaxation behavior was also observed for water confined in zeolites [252,253].

Figure 46 shows three curves corresponding to three different porous glasses samples A, B, and C that differ in their structure (average pore diameter) and chemical treatment [153]. The fitting curves presented in these figures show that the model (129) is in good agreement with the experimental data. The fitted values of E_a and E_b for all the samples are in fair agreement with the energies attributed to the water molecule reorientation and defect formation for the bulk *ice I*, which are evaluated as $55.5\,\text{kJ mol}^{-1}$ and $32.9\,\text{kJ mol}^{-1}$, respectively [78]. This fact leads to the conclusion that most probably the water confined in small pores is quite immobile and represents a kind of ice-like structure. For further discussion, more detailed information about the samples should be recalled. For example, let us discuss the relationship between the pore network structure and fitted activation energies E_a and E_b. The pore diameters of samples A and B are nearly the same ($\sim$50 nm), while the chemical treatment of these samples is different. Glass B was obtained from Glass A via an additional immersion in

KOH solution. Sample C has pores with an average diameter of 300 nm and, just as in sample A, was not especially purified with KOH [153,155].

It follows from the fit presented in Fig. 46 that E_b energies for all porous glass samples are about the same value of 33 kJ mol^{-1}. However, for sample B the value of E_b is about 10% less than those for samples A and C. This fact can most likely be explained by the additional chemical treatment of sample B with KOH, which removes the silica gel from the inner surfaces of the pore networks. It is reasonable to assume that the defects generally form at the water interfaces, and only then penetrate into the water layer. Thus, it seems that the KOH treatment decreases the interaction between the water and inner pore surfaces and, consequently, decreases the defect formation energy E_b.

3.   Dielectric Relaxation in Doped Ferroelectric Crystal

As discussed in Section V.E, the appearance of the "saddle-like" relaxation process in KTN:Cu crystal is linked to two facts: (i) It appears only in the Cu-doped crystals, and (ii) its appearance coincides with the onset of the ferroelectric phase transition.

We propose that following the onset of the phase transition the small Cu$^+$ ions residing in the large K$^+$ sites are shifted to an off-center position and produce a distortion in the neighboring unit cells. This imparts to each of these unit cells the ability to behave as a relaxing dipole. In a pure (copper-free) crystal these dipoles are closely interlaced with the complementary part of their respective unit cells and, hence, are unable to reorient. Thus the relaxation process that is linked with these dipoles is not observed in the pure KTN crystal.

In order to derive the temperature dependence of the relaxation time associated with this process, we shall consider a one-dimensional model in which the Cu$^+$ ions reside in an asymmetric double potential well as illustrated qualitatively in Fig. 47.

At a given temperature, the Cu$^+$ ions are distributed between two states: "high" and "low." If [Cu$^+$] is the molar concentration of the Cu$^+$ ions in the crystal, then according to Boltzman statistics the concentrations of the Cu$^+$ ions in their respective states are given by

$$[\text{Cu}^+]_H = [\text{Cu}^+] \cdot \exp\left(\frac{-E_b}{k_B T}\right) \tag{130}$$

$$[\text{Cu}^+]_L = [\text{Cu}^+] \cdot \left[1 - \exp\left(\frac{-E_b}{k_B T}\right)\right] \tag{131}$$

where [Cu$^+$]$_H$ and [Cu$^+$]$_L$ are the molar concentration of the Cu$^+$ ions in the "high" and "low" states respectively.

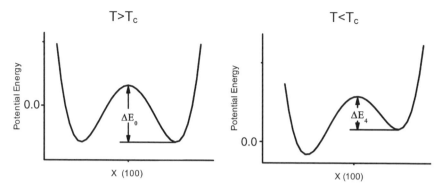

Figure 47. A schematic representation of the double-well potential of the Cu^+ ions. In the ferroelectric phase, local fields distort it and it becomes asymmetrical. (Reproduced with permission from Ref. 257. Copyright 2005, Elsevier Science B.V.)

As the crystal is cooled below the phase transition temperature, the relaxation time decreases until it reaches a minimum, thenceforth it follows an Arrhenius law. It is therefore reasonable to assume that the minimum point coincides with the temperature at which the great majority of the Cu^+ ions are at the "low" state. From this point onwards, as the crystal continues to be cooled, all the dipoles that are available to participate in the process are relaxing, yielding an Arrhenius relaxation time. Based on this argumentation, we claim that (v_0/v_f) in Eq. (127) is inversely proportional to the number of Cu^+ ions that are in the low state such that

$$p_2 = \exp(-Ce^{-E_b/k_BT}) \tag{132}$$

This leads to the saddle-like behavior of the relaxation time as described in Eq. (129). The fitting of (129) to the experimental data of the ferroelectric phase is presented in Figure 21. As can be seen, in the temperature range 230–290 K the model fits the experimental data very well, yielding $\tau_0 = 1.5 \pm 0.7 \times 10^{-9}$ s, $E_a = 0.12 \pm 0.01$ eV, and $E_b = 0.34 \pm 0.12$ eV. A comparison of E_b to the energetic barrier separating equivalent sites for Cu^+ ions in the paraelectric phase, $\Delta E = 0.37 \pm 0.01$ eV [179], strongly indicates that the Cu^+ ions assume the role of the defect.

In summary, we maintain that while the free volume concept is an intuitively natural concept for liquids confined in a host matrix, it is not an obvious model for ionic/covalent crystals. Saddle-like relaxation is not a universal feature of all crystals. Hence, v_f, the free volume necessary for the relaxation to occur, cannot be attributed to the presence of any defect that causes an irregularity in the crystal structure. In the special case of

KTN:Cu crystal, the Cu^+ ion has an effective ionic radius that is smaller than that of the original occupant of its site (K^+), while having the same ionization state. Below the ferroelectric phase transition temperature the Cu^+ ions are shifted to an off-center position in their respective sites. This allows some of the constituents of the original KTN unit cells in the vicinity of the Cu^+ ions to assume the behavior of relaxing dipoles. Because the number of these dipoles is constrained by the number of Cu^+ ions that are frozen in the off-center position, this yields a saddle-like relaxation process. The question remaining is, What is the relaxing dipole? A number of authors have noted dynamic relaxation of the octahedra in various perovskites and related structures [254–256]. In general the relaxation is Arrhenius in nature, with energies of activation ranging between 0.05 eV and 0.24 eV. The soft mode phonon is dominated by the oxygen octahedra and its interaction with the Nb ion. As noted above, there is a strong coupling of the Cu^+ ions to the soft mode phonon. Given that the activation energy of the Arrhenius tail of the saddle is in the range noted, it is not unreasonable to designate dynamic relaxations of the octahedra as the relaxing dipole [257].

4. Possible Modifications of the Model

All the examples described above show that confinement in different cases may be responsible for nonmonotonic relaxation kinetics and can lead to a saddle-like dependence of relaxation time versus temperature. However, this is not the only possible reason for nonmonotonic kinetics. For instance, work [258] devoted to the dielectric study of an antiferromagnetic crystal discusses a model based on the idea of screening particles. Starting from the Arrhenius equation and implying that the Arrhenius activation energy has a linear dependence on the concentration of screening charge carriers, the authors of Ref. 258 also obtained an expression that can lead to nonmonotonic relaxation kinetics under certain conditions. However, the experimental data discussed in that work does not show clear saddle-like behavior of relaxation time temperature dependence. The authors of Ref. 258 do not even discuss such a possibility.

At the same time, model (129) is also open to modifications. This model is based on the assumptions (126) and (127) regarding temperature dependences for the probabilities p_1 and p_2. Assuming, instead of the Arrhenius law (26), a cooperative term of the VFT (28) type for p_1, the temperature dependence of the relaxation time is obtained in the form

$$\ln\left(\frac{\tau}{\tau_0}\right) = \frac{DT_k}{T - T_k} + C \exp\left(-\frac{E_b}{k_B T}\right) \tag{133}$$

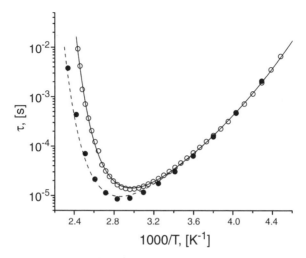

Figure 48. Temperature dependence of dielectric relaxation time for water confined in sample C. The data were measured under different conditions and contain a different amount of water: Unfilled circles correspond to the data presented early in Fig. 46; filled circles represent the experiment with reduced water content [78]. Full line is the best fit according to (133): $\ln \tau_0 = -17.8 \pm 0.5$, $E_b = 39 \pm 1\,\text{kJ}\,\text{mol}^{-1}$, $T_k = 124 \pm 7\,\text{K}$, $D = 10 \pm 2$, $C = 9 \times 10^5 \pm 3 \times 10^5$. The dashed line was simulated from (133) for the same $\ln \tau_0$, E_b, T_k, and D, but with C divided by a factor 1.8 (explanation in the text). (Reproduced with permission from Ref. 78. Copyright 2004, The American Physical Society.)

where the first term of the VFT type on the right-hand side of Eq. (129) could express the idea of cooperative behavior in accordance with the Adam–Gibbs model [63].

a. Confined Glassy Water. The experimental data for water confined in the porous glass sample C that was discussed previously (see Fig. 46) is well-fitted with (133) as presented in Fig. 48. Compared to the other samples, this porous glass has the largest pore diameter and humidity [153,156]. Therefore, it is reasonable to assume that the cooperative relaxation properties, described by the VFT term, should be more pronounced for this sample. It is worth noting that the fitted value of the Kauzmann temperature is $T_k = 124 \pm 7\,\text{K}$ (see caption for Fig. 48). From T_k, using the empirical rule $T_g \sim 1.1 \div 1.2 T_k$ [14], the estimation $T_g \sim 145\,\text{K}$ of the water glass transition temperature can be obtained. This value is in fair agreement with usual estimations of T_g for water that are expressed by interval $T_g \approx 130 \pm 6\,\text{K}$ [259,260]. The fitted value of fragility $D = 10 \pm 2$ is close to the estimations of this parameter $D \approx 8$ that have been derived from the diffusivity data for the amorphous solid water in work [260]. These findings may also support the idea that the porous glass samples treated in the preceding discussion dealt with a kind of noncrystalline state of water. The most probable reason for this is that the pore walls provide the necessary confinement. Note that in porous glasses, the glassy properties of water can be observed at comparatively high temperatures such as room temperature, whereas quite low temperatures and special treatment [259,260] are usually required in order to obtain glassy water. However, further investigations are required to support these experimental findings.

5. Relationships Between the Static Properties and Dynamics

Let us now consider the relationship between the data obtained from the kinetics and the amplitude of this process. The data for water confined in porous glass C can be discussed in terms of equilibrium properties of the relevant relaxation processes obtained in the two different experiments [78]. The two experimental runs presented in Fig. 49 are quite similar at low temperatures. However, in the high-temperature range, they exhibit a remarkable difference from each other.

In Fig. 49, the temperature dependences of the so-called dielectric strength $\Delta\varepsilon$ for these experiments are presented. The dielectric strength is the difference between the high- and low-frequency limits of the real part of the complex

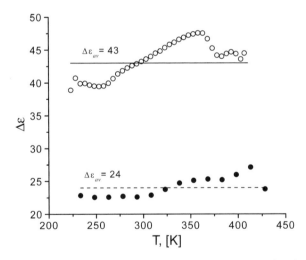

Figure 49. Temperature dependence of the dielectric strength $\Delta\varepsilon$ for sample C. Symbols represent experimental data corresponding to the relaxation times presented in Fig. 48: unfilled circles correspond to the data presented early in Fig. 46; filled circles represent the experiment with reduced water content [78]. The lines mark values of the averaged dielectric strength $\Delta\varepsilon_{av}$ for these experiments. (Reproduced with permission from Ref. 78. Copyright 2004, The American Physical Society.)

dielectric permittivity of the process under consideration. This quantity reflects the concentration of dipole moments n_d in a sample and, in its simplest approximation, is in linear proportion to the concentration $\Delta\varepsilon \sim n_d$ [6,7]; that is, the dielectric strength is proportional to the water content. Thus, the two experimental runs presented in Fig. 49, as well as the data in Figure 48, correspond to the two different amounts of water in sample C.

Recall that the preexponential factor $C = v_0 n_0 / V$, where n_0 is the maximum possible number of defects. This number is proportional to the water content and $C \sim n_0 \sim n_d \sim \Delta\varepsilon$. The dielectric strength $\Delta\varepsilon$ for these runs is almost constant (the variations of $\Delta\varepsilon$ are about 5% of the averaged value for both runs). By comparing the averaged values of $\Delta\varepsilon$ for these runs, the difference in water content between these experiments has been estimated to be 1.8 times (see Fig. 49). The preexponential factors C for these two runs should also differ by the same factor. The comparison in Fig. 48 shows that this is indeed so.

Finally we can conclude that confinement could be responsible for nonmonotonic relaxation kinetics and could provide a specific saddle-like temperature dependence of the relaxation time. The experimental examples discussed show that this type of kinetics may be inherent in systems of completely different natures: confined liquids, ferroelectric crystals, and it was even demonstrated recently macromolecular folding kinetics [78]. In each case, the specific interpretation of the parameters of model (129) depends on the discussed experimental situation. We are far from the opinion that confinement is the only reason for nonmonotonic relaxation kinetics. However, for all the examples discussed in this paper, the nonmonotonic dependence of the relaxation time on temperature has the same origin, that is, confinement either in real or configurational space.

G.   Dielectric Spectrum Broadening in Disordered Materials

In the previous sections, we have presented several examples of the nonexponential dielectric response in time domain. We have also discussed the dielectric response in H-bonding networks and liquid-like behavior in doped ferroelectric crystals. The models we have presented in the time domain enable us to determine some of the topological properties of the investigated complex systems. We have also observed that the frequency representation has its own advantages. In particular, the nondissipative part of the system response to an external perturbation and the dissipative part are clearly separated in the frequency domain as the real and imaginary part of the complex permittivity, while in the time domain these effects are "mixed" together in the relaxation function. The separation of the response itself and the dissipative effects in the time domain are not obvious, although it is possible to do in principle by using the appropriate integral transform. In spite of the existence of a single mapping between the time and frequency domain by the Laplace transform, the separation in frequency representation may at times be more convenient for analysis of the dynamic processes in complex systems.

In Section II.B we have classified several types of non-Debye relaxation and have mentioned a few particular approaches that have been developed in order to explain the origins of these relaxation patterns. We shall now discuss a model that considers one particular case of nonexponential relaxation.

1. Symmetric Relaxation Peak Broadening in Complex Systems

As mentioned in Section II.B, the dielectric response in the frequency domain for most complex systems cannot be described by a simple Debye expression (17) with a single dielectric relaxation time. In a most general way this dielectric behavior can be described by the phenomenological Havriliak–Negami (HN) formula (21).

Usually, the exponents α and β are referred to as measures of symmetrical and unsymmetrical relaxation peak broadening. This terminology is a consequence of the fact that the imaginary part of the complex susceptibility for the HN dielectric permittivity exhibits power-law asymptotic forms $\text{Im}\{\varepsilon^*(\omega)\} \sim \omega^\alpha$ and $\text{Im}\{\varepsilon^*(\omega)\} \sim \omega^{-\alpha\beta}$ in the low- and high-frequency limits, respectively.

The experimental data show that α and β are strictly dependent on temperature, structure, composition, pressure, and other controlled physical parameters [87,261–265].

In this section we will consider the specific case of the HN formula $\beta = 1$, $0 < \alpha < 1$, corresponding to symmetric relaxation peak broadening or to the so-called Cole–Cole (CC) law [16]. The complex dielectric permittivity $\varepsilon^*(\omega)$ for the CC process is represented in the frequency domain as

$$\varepsilon^*(i\omega) = \varepsilon_\infty + \frac{\varepsilon_s - \varepsilon_\infty}{1 + (i\omega\tau)^\alpha} \qquad (134)$$

In order to explain the non-Debye response (134) it is possible to use the memory function approach [22,23,31,266–268]. Thus, the normalized dipole correlation function $\Psi(t)$ (22) corresponding to a nonexponential dielectric relaxation process obeys the equation

$$\frac{d\Psi(t)}{dt} = -\int_0^t m(t - t')\Psi(t')dt' \qquad (135)$$

where $m(t)$ is the memory function, and t is the time variable. The specific form of the memory function is dependent on the features of relaxation.

On Laplace transformation of (15), using the convolution theorem, Eq. (135) reads as

$$pF(p) - 1 = -M(p)F(p) \tag{136}$$

where $F(p)$ and $M(p)$ are the Laplace transforms of $\Psi(t)$ and $m(t)$. Combining (136) with (134) and taking into account the relationship between the complex susceptibility and the correlation function (14), one can obtain the Laplace transform of the memory function for the CC process in the form

$$M(p) = p^{1-\alpha}\tau^{-\alpha} \tag{137}$$

Since $0 < \alpha < 1$ the exponent in Eq. (137) is $1 - \alpha > 0$. The mathematical implication is that $M(p)$ (137) is a multivalued function of the complex variable p. In order to represent this function in the time domain, one should select the schlicht domain using supplementary physical reasons [135]. These computational constraints can be avoided by using the Riemann–Liouville fractional differential operator $_0D_t^{1-\alpha}$ [see definitions (97) and (98)]. Thus, one can easily see that the Laplace transform of $_0D_t^{1-\alpha}[\Psi(t)]$ is

$$\hat{L}[_0D_t^{1-\alpha}[\Psi(t)]] = p^{1-\alpha}F(p) - C_\alpha, \qquad \text{where} \qquad C_\alpha = D_0^{-\alpha}[\Psi(t)]|_{t=+0} \tag{138}$$

Taking (138) into account, we can rewrite Eq. (135) with the memory function (137) as follows:

$$\frac{d\Psi(t)}{dt} = -\tau_0^{-\alpha}D_t^{1-\alpha}[\Psi(t)] \tag{139}$$

Note that the relationship between the complex susceptibility and correlation function (14), together with Eq. (134) leads directly to the requirement that $C_\alpha = 0$.

Equation (139) was already discussed elsewhere [22,23,31] as a phenomenological representation of the dynamic equation for the CC law. Thus, Eq. (139) shows that since the fractional differentiation and integration operators have a convolution form, it can be regarded as a consequence of the memory effect. A comprehensive discussion of the memory function (137) properties is presented in Refs. 22 and 23. Accordingly, Eq. (139) holds for some cooperative domain and describes the relaxation of an ensemble of microscopic units. Each unit has its own microscopic memory function $m_\delta(t)$, which describes the interaction between this unit and the surroundings (interaction with the statistical reservoir). The main idea of such an interaction was introduced in Refs. 22 and 23 and suggests that $m_\delta(t) \sim \sum_i \delta(t_i - t)$ (see Fig. 50).

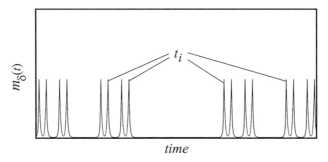

Figure 50. Schematic picture of $m_\delta(t)$ dependence. t_i are the time moments of the interaction that construct in time a fractal Cantor set with dimension $d_f = \ln 2 / \ln 3 \cong 0.63$. (Reproduced with permission from Ref. 2. Copyright 2002, Elsevier Science B.V.)

This term reflects the idea of interrupted interaction between the relaxing unit and its neighbors. The time instants t_i (the time position of the delta functions) are the instants of the interaction. The sequence of t_i, constructs a fractal set (the Cantor set, for example) with a fractal dimension $0 < d_f \leq 1$. This statement is related to the idea that cooperative behavior provides both a degree of ordering and long-lasting scaling. Following these assumptions the memory function $m(t)$ for a cooperative domain can be obtained as a result of averaging over the ensemble of $m_\delta(t)$ (see Fig. 51, where for more convenient representation $I(t) = \int_0^t m(t')\, dt'$ is plotted instead of $m(t)$). The requirements of

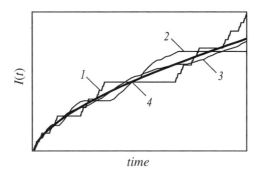

Figure 51. Schematic presentation which illustrates averaging of $m(t)$ over an ensemble of microscopic units. Here $I(t) = \int_0^t m(t')\, dt'$. Curve 1 corresponds to the cooperative ensemble of a single microscopic unit with t_i distributed by the Cantor set. Curve 2 represents the ensemble of 3 units of the same type, Curve 3 represents 10 units. Curve 4 corresponds to 1000 units in the ensemble. The latter exhibits the power-law behavior $I(t) \sim t^{\ln 2/\ln 3}$. (Reproduced with permission from Ref. 2. Copyright 2002, Elsevier Science B.V.)

measure conservation in the interval $[0, 1/\zeta]$ and conservation of the fractal dimension d_f for all $m_\delta(t)$ give this averaging as

$$m(t) = \int_{-1/2}^{1/2} m_\delta(\zeta^{-u}t)\zeta^{-u(1-d_f)}du \quad \text{and} \quad M(p) \sim p^{1-d_f} \qquad (140)$$

Thus, the memory function (137) is a cooperative one and the CC behavior appears on the macroscopic level after averaging over the ensemble of microscopic dipole active units. Comparing (137) and (140), one can establish that $\alpha = d_f$. This result once again highlights the fact that in this model the fractal properties on a microscopic level induce the power-law behavior of memory functions (137), (140), and CC permittivity (134) on a macroscopic level.

By definition [213,220], the fractional dimension is given by

$$d_f = \alpha = \frac{\ln(N)}{\ln(\zeta)} \qquad (141)$$

Here the scaling parameter ζ is the dimensionless time interval size and N is the number of delta functions (relaxation acts) in that interval. However, a characteristic time constant of the CC process is the relaxation time τ. Thus, the scaling parameter ζ and the relaxation time should be proportional to each other:

$$\zeta = \frac{\tau}{\tau_0} \qquad (142)$$

The minimum τ_0 which is a constant, is the cutoff time of the scaling in time.

In the general case, different physical conditions can determine the fractal properties of the microscopic memory function $m_\delta(t)$ and, consequently, the power-law time dependence of the macroscopic memory function (140). However, there is a computer simulation proof [269] that an anomalous relaxation on a fractal structure exhibits CC behavior. Therefore, one may assume that the memory function (140) has its origin in the geometrical self-similarity of the investigated system. Thus, the scaling parameter N is actually the number of points where the relaxing units are interacting with the statistical reservoir (i.e., by the ergodic assumption—the number of relaxation acts on a microscopic level for a cooperative domain). The assumption of geometrical self-similarity of the considered system suggests that this number is

$$N = G\left(\frac{R}{R_0}\right)^{d_G} \qquad (143)$$

Here, d_G is a spatial fractal dimension of the point set where relaxing units are interacting with the surroundings. R is the size of a sample volume section where movement of one relaxing unit occurs. R_0 is the cutoff size of the scaling in the space or the size of the cooperative domain. G is a geometrical coefficient of order unity, which depends on the shape of the system heterogeneity. For example, the well-known two-dimensional recurrent fractal Sierpinski carpet has $d_G = \ln(8)/\ln(3) \approx 1.89$, $G = \sqrt{3}/4 \approx 0.43$ [213].

The relaxation process may be accompanied by diffusion. Consequently, the mean relaxation time for such kinds of disordered systems is the time during which the relaxing microscopic structural unit would move a distance R. The Einstein–Smoluchowski theory [226,235] gives the relationship between τ and R as

$$R^2 = 2d_E D \tau \tag{144}$$

where D is the diffusion coefficient and d_E is the Euclidean dimension. Thus, combining the relationships (141)–(144), one can obtain the relationship between the Cole–Cole parameter α and the mean relaxation time τ in the form

$$\alpha = \frac{d_G}{2} \frac{\ln(\tau \omega_s)}{\ln(\tau/\tau_0)} \tag{145}$$

where $\omega_s = 2d_E G^{2/d_G}(D/R_0^2)$ is the characteristic frequency of the diffusion process. This equation establishes the relationship between the CC exponent α, the relaxation time τ, the geometrical properties (fractal dimension d_G), and the diffusion coefficient (through ω_s).

2. Polymer–Water Mixtures

The first mention of the $\alpha(\tau)$ dependence was in experimental work [265]. The dielectric relaxation data of water in mixtures of seven water-soluble polymers was presented there. It was found that in all these solutions, relaxation of water obeys the CC law, while the bulk water exhibits the well-known Debye-like pattern [270,271]. Another observation was that α is dependent not only on the concentration of solute but also on the hydrophilic (or hydrophobic) properties of the polymer. The seven polymers were: poly(vinylpyrrolidone) (PVP; weight average molecular weight (MW) = 10,000), poly(ethylene glycol) (PEG; MW = 8000), poly(ethylene imine) (PEI; MW = 500,000), poly(acrylic acid) (PAA; MW = 5000), poly(vinyl methyl ether) (PVME; MW = 90,000), poly(allylamine) (PAlA; MW = 10,000), and poly(vinyl alcohol) (PVA; MW = 77,000). These polymers were mixed with different ratios (up to 50% of polymer in solution) to water and measured at a constant room temperature (25°C) [265].

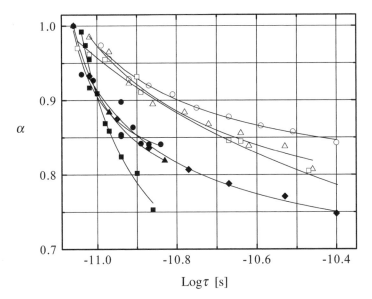

Figure 52. Cole–Cole exponent α versus relaxation time τ for PVA ($\bullet$), PAlA ($\blacktriangle$), PAA ($\blacksquare$), PEI ($\blacklozenge$), PEG ($\bigcirc$), PVME ($\triangle$), and PVP ($\square$) samples. The curves correspond to the model described in this section. The filled symbols correspond to the hydrophilic polymers and the open symbols correspond to the hydrophobic samples. (Reproduced with permission from Ref. 2. Copyright 2002, Elsevier Science B.V.)

We would like to mention a recent application [46] of model (145) to these systems. In Fig. 52 the experimental dependences of CC exponent α versus τ together with the fitting curve are presented. The values of the fitting parameters are listed in Table III.

TABLE III
The Space Fractional Dimension d_G, the Cutoff Time of the Scaling in the Time Domain τ_0, the Characteristic Frequency ω_s, and Estimated Self-Diffusion Coefficient for the Polymer–Water Mixtures[a]

Sample	d_G	τ_0[ps]	$\omega_s \times 10^{-11}$ [Hz]	$D \times 10^9$ [m^2 s^{-1}]
PVA	1.56 ± 0.09	7.18 ± 0.74	1.47 ± 0.21	3.31
PAlA[a]	1.43	6.46	1.74	3.92
PAA	1.12 ± 0.17	6.34 ± 0.83	2.08 ± 0.68	4.68
PEI	1.33 ± 0.02	4.89 ± 0.45	2.67 ± 0.40	6.01
PEG	1.54 ± 0.04	4.45 ± 0.74	2.78 ± 0.63	6.26
PVME	1.38 ± 0.10	3.58 ± 1.23	4.24 ± 2.47	9.54
PVP	1.00 ± 0.01	0.79 ± 0.11	127 ± 34	286

[a]For the sample PAlA there are only three experimental points. Thus it is impossible to determine the mean square deviation value and consequently the confidence intervals for the fitting parameters.

Source: (Reproduced with permission from Ref. 46. Copyright 2002, American Institute of Physics.

It is well known [54,270] that the macroscopic dielectric relaxation time of bulk water (8.27 ps at 25°C) is about 10 times greater than the microscopic relaxation time of a single water molecule, which is about one hydrogen bond lifetime [206,272–274] (about 0.7 ps). This fact follows from the associative structure of bulk water where the macroscopic relaxation time reflects the cooperative relaxation process in a cluster of water molecules.

In the context of the model presented above, the microscopic relaxation time of a water molecule is equal to the cutoff time of the scaling in time domain τ_0. For the most hydrophilic polymer, PVA, the strong interaction between the polymer and the water molecule results in the greatest value of microscopic relaxation time τ_0, only 10% less than the macroscopic relaxation time of the bulk water. The most hydrophobic polymer, PVP, has the smallest value of a single water molecule microscopic relaxation time, which is almost equal to the microscopic relaxation time of bulk water (see Table III). Therefore, weakening the hydrophilic properties (or intensifying the hydrophobic properties) results in a decreasing of interaction between the water and the polymer and consequently in the decrease of τ_0.

The interaction between the water and the polymer occurs in the vicinity of the polymer chains, and only the water molecules situated in this interface are affected by the interaction. The space fractal dimension d_G is now the dimension of the macromolecule chain. If a polymer chain is stretched as a line, then its dimension is 1. In any other conformation, the wrinkled polymer chain has a larger space fractal dimension, which falls into the interval $1 < d_G < 2$. Thus, it is possible to argue that the value of the fractal dimension is a measure of polymer chain meandering. Straighter (probably more rigid) polymer chains have d_G values close to 1. More wrinkled polymer (probably more flexible) chains have d_G values close to 2 (see Table III).

The presence of a polymer in the water affects not only the relaxation, but also the diffusion, of the solvent. For an estimation of the diffusion coefficient, we can use the following expression

$$D \cong \frac{\omega_s R_0^2}{2 d_E} \tag{146}$$

which is directly derived from the definition of the characteristic frequency ω_s. It was assumed in this last expression that the geometrical factor $G = 1$. In our case the scaling cutoff size in space is equal to the diameter of a water molecule $R_0 \approx 3 \,\text{Å}$ [52]. The Euclidean dimension of the space where diffusion occurs is the nearest integer number greater than the fractal dimension. Thus, $d_E = 2$. The results of this estimation are in Table III. The diffusion coefficient for bulk water [52] at 25°C is $2.57 \times 10^{-9} \,\text{m}^2\,\text{s}^{-1}$. The presence of a polymer in the water prevents clusterization of the water and facilitates the diffusion. However, the strong interaction between polymer and water for hydrophilic samples slows down the diffusion. The competition between these two effects leads to the clear tendency of the diffusion coefficient to increase with an increase of hydrophility (see Table III).

Note that the polymer affects only water molecules situated in the vicinity of the polymer chains. Thus, the estimated diffusion coefficient corresponds only to these water molecules and is not dependent on the polymer concentration. The averaged self-diffusion coefficient estimated for the entire polymer–water mixture should be different depending on the polymer concentration.

3.   Microcomposite Material

Another example of an application of Eq. (145) is on microcomposite polymer materials. We have performed dielectric measurements of the glass transition relaxation process in a nylon-6,6 sample quenched in amorphous (QN), a crystalline nylon-6,6 sample (CN), and a microcomposite sample (MCN), which is the same crystalline nylon-6,6 but with incorporated kevlar fibers [275,276].

The quantitative analysis of the dielectric spectra of the glass-transition process was carried out by fitting the isothermal dielectric loss data according to the HN law (21). It was found from the fitting that $\beta = 1$ for the glass-transition process in all of the samples measured. The glass-transition relaxation process in these systems is due to the motion of a polymer chain that is accompanied by diffusion. In general, the diffusion of a polymer chain is more complex than the Brownian model for diffusion [277,278]. However, in all the models the dependence of $\langle R^2 \rangle$ on time t is linear in the time scales associated with a monomeric link and in the time scale associated with the mobility of the entire chain. For this particular example, Eq. (144) describes the mobility of the polymer groups at the microscopic levels—that is, at the scale of a monomeric link.

The experimental α versus τ dependence for these samples, together with the fitting curves, are shown in Fig. 53. Note that in contrast to the previous example, these data are obtained at a constant sample composition. Now, Variations of the parameters α and τ are induced by temperature variation. As mentioned above, the exponents α as well as the relaxation time τ are functions of different experimentally controlled parameters. The same parameters can affect the structure or the diffusion simultaneously. In particular, both α and τ are functions of temperature. Thus, the temperature dependence of the diffusion coefficient in (144) should be considered. Let us consider the temperature dependence of the diffusion coefficient D:

$$D = D_0 K(T) \tag{147}$$

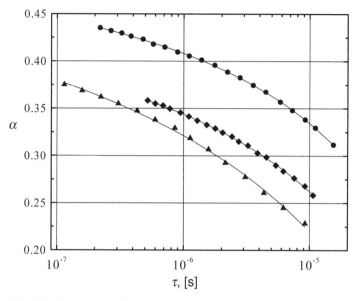

Figure 53. The dependence of α versus relaxation time τ for QN (•), CN (▲), and MCN (◆) samples. The curves correspond to the model described in this section. (Reproduced with permission from Ref. 2. Copyright 2002, Elsevier Science B.V.)

where $K(T)$ is a dimensionless function that represents the temperature dependence of the diffusion coefficient. D_0 is the appropriate constant with dimensions $[\mathrm{m}^2\,\mathrm{s}^{-1}]$. An increase of the diffusion coefficient with increasing temperature also signifies an increase of the characteristic spatial scale R_0 (cutoff size of the scaling in the space). Let us assume that R_0^2 is proportional to the diffusion coefficient D and obeys the Einstein–Smoluchowski theory:

$$R_0^2 = 2d_E D \tau_{\max} \tag{148}$$

where $\tau_{\max}$ is the long-time limit of the scaling. Thus, combining together (147) and (148) with (142)–(144) we can obtain the relationship between the CC exponent α and relaxation time τ in a form similar to (145):

$$\alpha = \frac{d_G}{2} \frac{\ln(\tau \omega_0)}{\ln(\tau/\tau_0)} \tag{149}$$

with substitution of ω_s by $\omega_0 = G^{2/d_G}/\tau_{\max}$. The latter relationship shows that with the assumption (148) the temperature dependence of the diffusion coefficient does not alter the form of the α versus τ relationship.

TABLE IV

The Space Fractional Dimension d_G, the Cutoff Time of the Scaling in the Time Domain τ_0, and the Characteristic Frequency ω_0 for Quenched Polymer (QN), Crystalline (CN) and Microcomposite Samples (MCN).

Sample	d_G	τ_0 [ms]	ω_0 [kHz]
QN	1.12 ± 0.01	1.1 ± 0.1	5.9 ± 0.3
CN	1.20 ± 0.05	5.8 ± 4.4	9.7 ± 1.9
MCN	1.04 ± 0.02	1.5 ± 0.4	8.1 ± 0.7

Source: Reproduced with permission from Ref. 2. Copyright 2002, Sage Publications.

The average length of a nylon-6,6 polymer chain is about $50-100\,\mu m$ (each polymer chain contains about 10^5 groups while the length of a polymer group r_g is about $10\,\overset{\circ}{A}$). This length is comparable to the thickness of a sample $120-140\,\mu m$ [275]. Thus, the movement of the chains is most likely occurring in the plane of the sample. This fact correlates with the values of the space fractional dimension d_G. For all of the samples $d_G \in (1,2)$ (see Table IV). Thereby, the Euclidean dimension of the space in which chain movement occurs is $d_E = 2$.

Although unambiguous data for the mesoscale structure of the samples under investigation does not presently exist, it is possible to estimate the order of magnitude of some physically significant quantities from the cutoff time τ_0 and characteristic frequency ω_0 values. Despite the fact that ω_0 and ω_s have different physical significance, for the estimation one can neglect the temperature dependence of the diffusion coefficient and assume that $R_0^2 \approx 10^{-16}\,m^2$ (R_0 is the cube root of the volume occupied by one polymer chain), $G \approx 1$. Then, the self-diffusion coefficient evaluated by expression (66) falls in the interval 10^{-14} to $10^{-13}\,m^2\,s^{-1}$, which is typical of such polymer materials [279–281].

The cutoff time τ_0 is related to the size of the cooperative domain l_c by $l_c^2 = 4D\tau_0$. Thus, in the two-dimensional case one can estimate the number of polymer groups in the cooperative region as $n_g = l_c^2/r_g^2 \approx 10^2$, which is in fair agreement with the results obtained in the Ref. 282.

One can also see from Table IV that the presence of either the crystalline phase or kevlar fibers in a sample leads to an increase of the cutoff time τ_0, indicating a slowdown of the relaxation process. Their presence also leads to an increase of the ω_0 value as well. This is a manifestation of a decreasing mobile polymer chain length.

Concluding this section, we can summarize as follows: The model of symmetric relaxation peak broadening presented here is not universal. However, it was illustrated through two examples that this model can describe some common features of relaxations of this type. It provides the relationship between microscopic parameters of experimental samples and properties of experimentally measured macroscopic correlation functions. Thus, we hope that it may be useful in other cases where dielectric relaxation of the Cole–Cole type exists.

VI. SUMMARY

In conclusion, we would like to recapitulate the main goals of this chapter. We have demonstrated the effectiveness of dielectric spectroscopy as a tool for investigating complex materials. The unique capabilities of DS allowed us to investigate materials in a broad time scale range, to identify general relaxation phenomena, and to discover new ones for the complex systems studied, independent of the particularity of the material. DS makes it possible to cover extremely wide continuous ranges of experimentally controlled parameters, a feature that no other existing spectroscopic method today can offer. Moreover, this method, in comparison to conventional radio spectroscopic techniques such as electron paramagnetic or nuclear magnetic resonance, is much simpler in terms of hardware realization. We have also presented current state-of-the-art applications of DS including the resolution of some of the critical issues related to hardware and modern data treatment procedures, which make possible the recent achievements of this method. Last, but not least, in this chapter we have discussed some contemporary ideas that are currently employed by the dielectric community. In this regard the DS studies of different complex materials presented in this chapter demonstrate the capability to generate new knowledge about the cooperative dynamics in quite different complex structures. It has been shown that despite the morphological differences of the systems studied, the most general feature of the relaxation processes observed by DS is the non-Debye dielectric response, which is a consequence of the cooperative dynamics and confinement that also lead to non-Arrhenius kinetics.

Acknowledgments

We would like to thank Dr. Anna Gutina and Dr. Yoshihito Hayashi for numerous discussions and stimulating advices. We are also grateful to current graduate students Mrs. Ekaterina Axelrod, Mr. Noel Axelrod, Mr. Paul Ben Ishai, and Mr. Shimon Lerner for their valuable help and technical assistance, as well as to Ms. Elianna Zour for technical editing of the manuscript.

References

1. M. F. Shlesinger, G. M. Zaslavsky, and J. Klafter, *Nature* **363**, 31 (1993).

2. Y. Feldman, A. Puzenko, and Ya. Ryabov, *Chem. Phys.* **284**, 139 (2002).

3. F. Kremer and A. Schönhals, eds., *Broadband Dielectric Spectroscopy*, Springer-Verlag, Berlin, 2003.

4. R. N. Clarke, ed., *A Guide to the Characterization of Dielectric Materials of RF and Microwave Frequencies*, NPL, London, 2004.

5. Y. Feldman, I. Ermolina, and Y. Hayashi, *IEEE Trans. Dielectr. Electr. Insul.* 10, 728 (2003).

6. H. Fröhlich, *Theory of Dielectrics. Dielectric Constant and Dielectric Loss*, 2nd ed., Clarendon Press, Oxford, 1958.

7. C. J. F. Böttcher, *Theory of Electric Polarization*, Vol. 1, 2nd ed., Elsevier Science B.V., Amsterdam, 1993.

8. N. E. Hill, W. E. Vaughan, A. H. Price, and M. Davis, *Dielectric Properties and Molecular Behavior*, Van Nostrand, London, 1969.

9. N. Kozlovich, A. Pusenko, Y. Alexanadrov, and Y. Feldman, *Colloids and Surfaces A* **140**, 229 (1998).

10. A. Volmari and H. Weingärtner, *J. Mol. Liq.* **98–99**, 293 (2002).

11. C. F. Böttcher and P. Bordewijk, *Theory of Electric Polarisation*, Vol. 2, 2nd ed., Elsevier Science B.V., Amsterdam, 1992.

12. B. K. P. Scaife, *Principles of Dielectrics*, Oxford University Press, Oxford, 1989.

13. A. N. Tikhonov and V. Y. Arsenin, *Solutions of Ill-Posed Problems*, John Wiley & Sons, New York, 1977.

14. H. Schäfer, E. Sternin, R. Stannarius, M. Arndt, and F. Kremer, *Phys. Rev. Lett.* **76**, 2177 (1996).

15. S. Havriliak and S. Negami, *J. Pol. Sci.: Part C* **14**, 99 (1966).

16. K. S. Cole and R. H. Cole, *J. Chem. Phys.* **9**, 341 (1941).

17. D. W. Davidson and R. H. Cole, *J. Chem. Phys.* **19**, 1484 (1951).

18. A. K. Jonscher, *Dielectric Relaxation in Solids*, Chelsea Dielectric Press, London, 1983.

19. A. K. Jonscher, *Universal Relaxation Law*, Chelsea Dielectric Press, London, 1996.

20. J. C. Dyre and T. B. Schrøder, *Rev. Mod. Phys.* **72**, 873 (2000).

21. N. F. Mott and E. A. Davis, *Electronic Processes in Non-Crystalline Materials*, Clarendon Press, Oxford, 1978.

22. R. R. Nigmatullin and Y. E.. Ryabov, *Phys. Solid State* **39**, 87 (1997).

23. L. Nivanen, R. Nigmatullin, and A. LeMehaute, *Le Temps Irreversible a Geometry Fractale*, Hermez, Paris, 1998.

24. S. Alexander, J. Bernasconi, W. R. Schneider, and R. Orbach, *Rev. Mod. Phys.* **53**, 175 (1981).

25. M. Ben-Chorin, F. Möller, F. Koch, W. Schirmacher, and M. Eberhard, *Phys. Rev. B* **51**, 2199 (1995).

26. W. T. Coffey, *J. Mol. Liq.* **114**, 5 (2004).

27. G. Williams, *Chem. Rev.* **72**, 55 (1972).

28. R. H. Cole, in *Mol. Liq. NATO ASI Ser. C* **135**, 59 (1984).

29. R. Kohlrausch, *Ann. Phys.* **12**, 393 (1847).

30. G. Williams and D. C. Watts, *Trans. Farad. Soc.* **66**, 80 (1970).

31. R. Metzler and J. Klafter, *Phys. Rep.* **339**, 1 (2000).

32. A. Blumen, J. Klafter, and G. Zumofen, in: *Optical Spectroscopy of Glasses*, ed., I. Zschokke, Reidel, Dordrecht, Netherlands, 1986, p. 199.

33. G. H. Weiss, *Aspects and Applications of the Random Walk*, North-Holland, Amsterdam, 1994.

34. E. Barkai, R. Metzler, and J. Klafter, *Phys. Rev. E* **61**, 132 (2000).

35. T. Zavada, N. Südland, R. Kimmich, and T. F. Nonnenmacher, *Phys. Rev. E* **60**, 1292 (1999).

36. R. Kimmich, *Chem. Phys.* **284**, 253 (2002).

37. T. Wadayama, S. Yamamoto, and A. Hatta, *Appl. Phys. Lett.* 65, 1653 (1994).

38. V. Degiorgio, R. Piazza, F. Mantegazza, and T. Bellini, *J. Phys.: Condens. Matter* 2, SA69 (1990).

39. J. E. Martin, D. Adolf, and J. P. Wilcoxon, *Phys. Rev. A* **39**, 1325 (1989).

40. A. Bello, E. Laredo, and M. Grimau, *Phys. Rev. B* **60,** 12764 (1999).

41. F. Alvarez, A. Alegría, and J. Colmenero, *Phys. Rev. B* 44, 7306 (1991).

42. G. Katana, E. W. Fischer, T. Hack, V. Abetz, and F. Kremer, *Macromolecules* **28**, 2714 (1995).

43. R. V. Chamberlin, *Phys. Rev. Lett.* **82**, 2520 (1998).

44. K. Weron and A. Klauzer, *Ferroelectrics* **236**, 59 (2000).

45. *Application of Fractional Calculus in Physics*, R. Hilfer, ed., World Scientific, Singapore, 2000.

46. Y. E. Ryabov, Y. Feldman, N. Shinyashiki, and S. Yagihara *J. Chem. Phys.* **116**, 8610 (2002).

47. Y. Feldman, N. Kozlovich, Y. Alexandrov, R. Nigmatullin, and Y. Ryabov, *Phys. Rev. E.* **54** 5420 (1996).

48. S. Arrhenius, *Z. Phys. Chem.* **4**, 226 (1889).

49. H. Eyring, *Chem. Rev.* **17**, 65 (1935).

50. H. Eyring, *J. Chem. Phys.* **3**, 107 (1935).

51. H. Eyring, *Trans. Faraday. Soc.* **34**, 41 (1938).

52. D. Eisenberg and W. Kauzmann, *The Structure and Properties of Water*, Clarendon, Oxford, 1969.

53. J. Barthel, R. Buchner, and B. Wurm, *J. Mol. Liq.* **98–99**, 51 (2002).

54. R. Buchner, J. Barthel, and J. Stauber, *Chem. Phys. Lett.* **306**, 57 (1999).

55. C. A. Angell, J. Non-Cryst. Solids 73, 1 (1985).

56. C. A. Angell, J. Non-Cryst. Solids **131–133**, 13 (1991).

57. R. Böhmer, K. L. Ngai, C. A. Angell, and D. J. Plazek, J. Chem. Phys. **99**, 4201 (1993).

58. R. Richert and C. A. Angell, J. Chem. Phys. **108**, 9016 (1998).

59. R. Brand, P. Lunkenheimer, and A. Loidl, J. Chem. Phys. **116**, 10386 (2002).

60. H. Vogel, *Phys. Z.* **22**, 645 (1921).

61. G. S. Fulcher, *J. Am. Ceram. Soc.* **8**, 339 (1925).

62. G. Tammann and W. Hesse, *Z. Anorg. Allg. Chem.* **156**, 245 (1926).

63. G. Adam and J. H. Gibbs, *J. Chem. Phys.* **43**, 139 (1965).

64. W. Kauzmann, *Chem. Rev.* **43**, 219 (1948).

65. C. A. Angell, *Chem. Rev.* **102**, 2627 (2002).

66. T. G. Fox and P. J. Flory, *J. App. Phys.* **21**, 581 (1950).

67. T. G. Fox and P. J. Flory, *J. Phys. Chem.* **55**, 221 (1951).

68. T. G. Fox and P. J. Flory, *J. Pol. Sci.* **14**, 315 (1954).

69. A. K. Doolittle, *J. Appl. Phys.* **22**, 1471 (1951).

70. D. Turnbull and M. H. Cohen, *J. Chem. Phys.* **34**, 120 (1961).

71. J. T. Bendler and M. F. Shlesinger, *J. Mol. Liq.* **36**, 37 (1987).

72. J. T. Bendler and M. F. Shlesinger, *J. Stat. Phys.* **53**, 531 (1988).

73. J. T. Bendler, J. J. Fontanella, and M. F. Shlesinger, *Phys. Rev. Lett.* **87**, 195503 (2001).

74. W. Götze and L. Sjögren, *Rep. Prog. Phys.* **55**, 241 (1992).

75. U. Bengtzelius, W. Götze, and A. Sjölander, *J. Phys. C: Solid State Phys.* **17**, 5915 (1984).

76. P. Taborek, R. N. Kleiman, and D. J. Bishop, *Phys. Rev. B* **34**, 1835 (1986).

77. R. Richert and H. Bässler, *J. Phys.: Condens. Matter* **2**, 2273 (1990).

78. Y. E. Ryabov, A. Puzenko, and Y. Feldman, *Phys. Rev B.* **69**, 0142041 (2004).

79. R. H. Cole, J. G. Berberian, S. Mashimo, G. Chryssikos, A. Burns, and E. Tombari, *J. Appl. Phys.* **66**, 793 (1989).

80. J. G. Berberian *J. Mol. Liq.* **56**, 1, (1993).

81. J. G. Berberian and E. King, *J. Non.-Cryst. Solids*, **305**, 10 (2002).

82. S. Mashimo, T. Umehara, T. Ota, S. Kuwabara, N. Shinyashiki, and S. Yagihara, *J. Mol. Liq.* **36**, 135 (1987).

83. D. Bertolini, M. Cassettari, G. Salvetti, E. Tombari, and S. Veronesi, *Rev. Sci. Instrum.* **62**, 450 (1991).

84. R. Nozaki and T. K. Bose, *IEEE Trans. Instrum. Meas.* **39**, 945 (1990).

85. B. Gestblom, E. Noreland, and J. Sjöblom, *J. Phys Chem.* **91**, 6329 (1987).

86. Y. Feldman, A. Andrianov, E. Polygalov, I. Ermolina, G. Romanychev, Y. Zuev, and B. Milgotin, *Rev. Sci. Instrum.* **67**, 3208 (1996).

87. A. Schönhals, F. Kremer, and E. Schlosser, *Phys. Rev. Lett.* **67**, 999 (1991).

88. U. Kaatze, V. Lonneckegabel, and R. Pottel, *Z. Phys. Chem.-Int. J. Res. Phys.* **175**, 165 (1992).

89. K. Folgerø, T. Friisø, J. Hilland, and T. Tjomsland, *Meas. Sci. Technol.* **6**, 995 (1995).

90. F. Kremer, D. Boese, G. Meier, E. W. Fischer, *Prog. Colloid Polym. Sci.* **80**,129 (1989).

91. U. Kaatze and K. Giese, *J. Phys. E: Sci. Instrum.* **13**, 133 (1980).

92. R. Böhmer, M. Maglione, P. Lunkenheimer, and A. Loidl, *J. Appl. Phys.* **65**, 901 (1989).

93. J. R. Birch and T. J. Parker, *Dispersive Fourier Transform Spectroscopy* in K. J. Batton, ed., *Infrared and Millimeter Waves*, Vol 2: Instrumentation, Chapter 3 Academic Press, New York, pp. 137–271, 1979.

94. H. P. Schwan, *Determination of biological impedances*, in: *Physical Techniques in Biological Research*, Vol. VI, Part B, W. L. Nastuk, edr., Academic Press, New York, pp. 323–407, 1963.

95. K. Asami, *Prog. Polym. Sci.* **27**, 1617 (2002).

96. E. H. Grant, R. J. Sheppard, and G. P. South, *Dielectric Behavior of Biological Molecules in Solution*, Oxford University Press, Oxford, 1978, Chapter 3.

97. R. E. Collin, *Foundations for Microwave Engineering*, 2nd ed., McGraw-Hill, New York, 1966.

98. F. Kremer, *J. Non.-Cryst. Solids* **305**, 1 (2002).

99. F. I. Mopsic, *Rev. Sci. Instrum.* **55**, 79 (1984).

100. R. Richert, *Rev. Sci. Instrum.* **67**, 3217 (1996).

101. R. Richert, *Physica A* **287**, 26 (2000).

102. R. Richert, *J. Non.-Cryst. Solids*, **305**, 29 (2002).

103. Y. D. Feldman, Y. F. Zuev, E. A. Polygalov, and V. D. Fedotov, *Colloid Polym. Sci.* **270**, 768 (1992).

104. B. Gestblom and P. Gestblom, *Macromolecules* **24**, 5823 (1991).

105. R. H. Cole, Dielectric polarization and relaxation, in NATO ASI Series, Series C, Vol. 135, *Molecular Liquids*, 1984, pp. 59–110.

106. Y. D. Feldman and V. V. Levin, *Chem. Phys. Lett.* **85**, 528 (1982).

107. A. M. Bottreau and A. Merzouki, *IEEE Trans. Instrum. Meas.* **42**, 899 (1993).

108. *Time Domain Measurements in Electromagnatics*, E. K. Miller, ed., Van Nostrand Reinhold, New York, 1986.

109. R. Nozaki and T. K. Bose, *IEEE Trans. Instrum. Meas.* **39**, 945 (1990).

110. N. Miura, N. Shinyashiki, S. Yagihara, and M. Shiotsubo, *J. Am. Ceramic Soc.* **81**, 213 (1998).

111. R. Buchner and J. Barthel, *Ber. Bun. Phys. Chem. Chem. Phys.* **101**, 1509 (1997).

112. N. E. Hager III, *Rev. Sci. Instrum.* **65**, 887 (1994).

113. Y. Feldman, T. Skodvin, and J. Sjöblom, Dielectric spectroscopy on emulsion and related colloidal systems—A review, in *Encyclopedic Handbook of Emulsion Technology*, Marcel Dekker, New York, 2001, pp. 109–168.

114. O. Göttmann, U. Kaatze, and P. Petong, *Meas. Sci. Tech.* **7**, 525 (1996).

115. A. Oka, R. Nozaki, and Y. Shiozaki, *Ferroelectrics* **230**, 459 (1999).

116. J. G. Berberian, and R. H. Cole, *Rev. Sci. Instr.* **63**, 99 (1992).

117. J. P. Grant, R. N. Clarke, G. T. Symm, and N. M. Spyrou, *Phys. Med. Biol.* **33**, 607 (1988).

118. J. P. Grant, R. N. Clarke, G. T. Symm, and N. M. Spyrou, *J. Phys. E: Sci.* Istrum. **22**, 757 (1989).

119. B. Gestblom, H. Førdedal, and J. Sjöblom, *J. Disp. Sci. Tech.* **15**, 449 (1994).

120. T. Skodvin, T. Jakobsen, and J. Sjöblom, *J. Disp. Sci. Tech.* **15**, 423 (1994).

121. G. Q. Jiang, W. H. Wong, E. Y. Raskovich, W. G. Clark, W. A. Hines, and J. Sanny, *Rev. Sci. Instrum.* **64**, 1614 (1993).

122. T. Jakobsen and K. Folgerø, *Meas. Sci. Tech.* **8**, 1006 (1997).

123. Y. J. Lu, H. M. Cui, J. Yu, and S. Mashimo, *Bioelectromagnetics* **17**, 425 (1996).

124. S. Naito, M. Hoshi, and S. Mashimo, *Rev. Sci. Instr.* **67**, 3633 (1996).

125. J. Z. Bao, S. T. Lu, and W. D. Hurt, *IEEE Trans. Microw. Theory Tech.* **45**, 1730 (1997).

126. G. Bitton, Y. Feldman, and A. J. Agranat, *J. Non.-Cryst. Solids* **305**, 362 (2002).

127. T. Skodvin and J. Sjöblom, *Colloid Polym. Sci.*, **274**, 754 (1996).

128. M. S. Seyfried and M. D. Murdock, *Soil Science* **161**, 87 (1996).

129. J. M. Anderson, C. L. Sibbald, and S. S. Stuchly, *IEEE Trans. Microw. Theory Tech.* **42**, 199 (1994).

130. D. V. Blackham and R. D. Pollard, *IEEE Trans. Instrum. Meas.* **46**, 1093 (1997).

131. S. Jenkins, A. G. P. Warham, and R. N. Clarke, *IEE Proc.-H Microw. Antennas Propag.* **139**, 179 (1992).

132. N. Axelrod, E. Axelrod, A. Gutina, A. Puzenko, P. Ben Ishai, and Y. Feldman, *Meas. Sci. Tech.* **15**, 755 (2004).

133. P. J. Verveer, M. J. Gemkow, and T. M. Jovin, *J. Microsc.* **193**, 50 (1999).

134. M. Wübbenhorst, and J. van Turnhout, *J. Non-Crys. Solids*, **305**, 40 (2002).

135. G. A. Korn and T. M. Korn *Mathematical Handbook*, McGraw-Hill, New York, 1968.

136. F. Castro and B. Nabet, *J. Franklin Inst. B* **336**, 53 (1999).

137. B. G. Sherlock, *Signal Process.* **74**, 169(1999).

138. S. E. Friberg, *Microemulsions: Structure and Dynamics*, CRC Press, Boca Raton, FL, 1987.

139. *Surfactants, Adsorption, Surface Spectroscopy and Disperse Systems*, B. Lindman, ed., Steinkopff, Darmstadt, 1985.

140. D. Langevin, *Annu. Rev. Phys. Chem.*, **43**, 341 (1992).

141. J. Sjöblom, R. Lindberg, and S. E. Friberg, *Adv. Colloid Interface Sci.* **65**, 125 (1996).

142. M. A. van Dijk, G. Casteleijn, J. G. H. Joosten, and Y. K. Levine, *J. Chem. Phys.* **85**, 626 (1986).

143. Y. Feldman, N. Kozlovich, I. Nir, and N. Garti, *Phys. Rev. E.* **51**, 478 (1995).

144. C. Cametti, P. Codastefano, P. Tartaglia, S.-H. Chen, and J. Rouch, *Phys. Rev. A*, **45**, R5358 (1992).

145. F. Bordi, C. Cametti, J. Rouch, F. Sciortino, and P. Tartaglia, *J. Phys. Condens. Matter* **8**, A19 (1996).

146. Y. Feldman, N. Kozlovich, I. Nir, and N. Garti, *Colloids Surfaces A* **128**, 47 (1997).

147. Y. Feldman, N. Kozlovich, I. Nir, N. Garti, V. Archipov, Z. Idiyatullin, Y. Zuev, and V. Fedotov, *J. Phys. Chem.* **100**, 3745 (1996).

148. A. L. R. Bug, S. A. Safran, G. S. Grest, and I. Webman, *Phys. Rev. Lett.* **55** 1896 (1985).

149. N. Kozlovich, A. Puzenko, Y. Alexandrov, and Y. Feldman, *Colloids and Surfaces A* **140**, 299 (1998).

150. G. S. Grest, I. Webman, S. A. Safran, and A. L. R. Bug, *Phys. Rev. A* **33**, 2842 (1986).

151. C. Cametti, P. Codastefano, A. Di Biasio, P. Tartaglia, and S.-H. Chen, *Phys. Rev. A* **40**, 1962 (1989).

152. D. Stauffer and A. Aharony, *Introduction to Percolation Theory*, revised 2nd ed., Taylor & Francis, London, 1994.

153. A. Gutina, E. Axelrod, A. Puzenko, E. Rysiakiewicz-Pasek, N. Kozlovich, and Y. Feldman, *J. Non-Cryst. Solids* **235–237**, 302 (1998).

154. A. Puzenko, N. Kozlovich, A. Gutina, and Y. Feldman, *Phys. Rev. B* **60**, 14348 (1999).

155. Y. Ryabov, A. Gutina, V. Arkhipov, and Y. Feldman, *J. Phys. Chem. B* **105**, 1845 (2001).

156. A. Gutina, T. Antropova, E. Rysiakiewicz-Pasek, K. Virnik, and Y. Feldman, *Microporous Mesoporous Mater.* **58**, 237 (2003).

157. A. Gutina, Y. Haruvy, I. Gilath, E. Axelrod, N. Kozlovich, and Y. Feldman, *J. Phys. Chem. B*, **103**, 5454 (1999).

158. T. Saraidarov, E. Axelrod, Y. Feldman, and R. Reisfeld, *Chem. Phys. Lett.* **324**, 7 (2000).

159. G. Øye, E. Axelrod, Y. Feldman, J. Sjöblom, and M. Stöcker, *Colloid Polym. Sci.* **278**, 517 (2000).

160. E. Axelrod, A. Givant, J. Shappir, Y. Feldman, and A. Sa'ar, *Phys. Rev. B* **65**, 165429 (2002).

161. E. Axelrod, A. Givant, J. Shappir, Y. Feldman, and A. Sa'ar, *J. Non-Cryst. Solids*, **305**, 235 (2002).

162. E. Rysiakiewicz-Pasek and K. Marczuk, *J. Porous Mater.* **3**, 17 (1996).

163. M. Ben-Chorin, F. Moller, and F. Koch, *Phys. Rev. B* **49**, 2981 (1994).

164. L. T. Canham, *Appl. Phys. Lett.* **57**, 1046 (1990).

165. A. G. Cullis, L. T. Canham, and P. D. J. Calcott, *J. Appl. Phys.* **82**, 909 (1997).

166. R. Hilfer, in: *Advances in Chemical Physics*, Vol. XCII, I. Prigogine, and A. Rice, eds., John Wiley & Sons, New York, 1996, p. 299.

167. Y. Lubianiker and I. Balberg, *Phys. Rev. Lett.* **78**, 2433 (1997).

168. J. Kočka, J. Oswald, A. Fejfar, R. Sedlačik, V. Železný, Ho The-Ha, K. Luterová, and I. Pelant, *Thin Solid Films* **276**, 187 (1996).

169. S. Triebwasser, *Phys. Rev.* **114**, 63 (1959).

170. Y. Girshberg and Y. Yacoby, *J. Phys. Condens. Matter* **13**, 8817 (2001).

171. R. Comes, R. Currat, F. Denoyer, M. Lambert, and A. M. Quittet, *Ferroelectrics* **12**, 3 (1976).

172. Y. Yacoby, *Z. Phys. B* **31**, 275 (1978).

173. M. D. Fontana, G. Metrat, J. L. Servoin, and F. Gervais, *J. Phys. C: Solid State Phys.* **17**, 483 (1984).

174. M. D. Fontana, H. Idrissi, and K. Wojcik, *Europhys. Lett.* **11**, 419 (1990).

175. C. H. Perry, R. R. Hayes, and N. E. Tornberg, in *Proceeding of the International Conference on Light Scattering in Solids*, M. Balkansky, ed., John Wiley & Sons, New York, 1975, p. 812.

176. B. E. Vugmeister and M. D. Glinchuk, *Rev. Mod. Phys.* **62**, 993 (1990).

177. A. J. Agranat, V. Leyva, and A. Yariv, *Opt. Lett.* **14**, 1017 (1989).

178. A. J. Agranat, in *Infrared Holography for Optical Communications*, Topics in Applied Physics, Vol. 86, P. Boffi, P. Piccinin, and M. C. Ubaldi, eds., Springer-Verlag, Berlin, 2003, p. 129.

179. P. Ben Ishai, C. E. M. de Oliveira, Y. Ryabov, Y. Feldman, and A. J. Agranat, *Phys. Rev. B* **70**, 132104 (2004).

180. R. Hofineister, S. Yagi, A. Yariv, and A. J. Agranat, *J. Cryst. Growth* **131**, 486 (1993).

181. I. Segal, E. Dorfman, and O. Yoffe, *Atomic Spect.* **21**, 46 (2000).

182. O. Mishima and H. E. Stanley, *Nature* **396**, 329 (1998).

183. E. Donth, *The Glass Transition: Relaxation Dynamics in Liquids and Disordered Materials*, Springer-Verlag, Berlin, 200.

184. K. L. Ngai, R. W. Rendell, and D. J. Plazek, *J. Chem. Phys.* **94**, 3018 (1991).

185. C. A. Angell,; K. L. Ngai, G. B. McKenna, P. F. McMillan, and S. W. Martin, *J. Appl. Phys.* **88**, 3113 (2000).

186. Y. E. Ryabov, Y. Hayashi, A. Gutina, and Y. Feldman, *Phys. Rev. B*, **67**, 132202 (2003).

187. K. L. Ngai and R. W. Rendell, *Phys. Rev. B* **41**, 754 (1990).

188. P. Lunkenheimer, A. Pimenov, M. Dressel, Y. G. Goncharov, R. Böhmer, and A. Loidl, *Phys. Rev. Lett.* **77**, 318 (1996).

189. G. P. Johari and E. Whalley, *Faraday Symp. Chem. Soc.* **6**, 23 (1972).

190. S. Sudo, M. Shimomura, N. Shinyashiki, and S. Yagihara, *J. Non-Cryst. Solids* **307–310**, 356 (2002).

191. A. Van Hook, *Crystallization Theory and Practice*, Reinhold, New York, 1961.

192. Van Koningsveld, *Recl. Trav. Chim. Pays-Bas.* **87**, 243 (1968).

193. A. Schönhals, F. Kremer, A. Hofmann, E. W. Fischer, and E. Schlosser, *Phys. Rev. Lett.* **70**, 3459 (1993).

194. A. P. Sokolov, W. Steffen, and E. Rössler, *Phys. Rev. E* **52**, 5105 (1995).

195. P. Lunkenheimer and A. Loidl, *Chem. Phys.* **284**,205 (2002).

196. H. Sillescu, *J. Non-Cryst. Solids* **243**, 81 (1999).

197. R. Richert, *J. Phys. Condens. Mater.* **14**, R703 (2002).

198. *Handbook of Chemistry and Physics*, R. C. Weast, ed., CRC Press, Cleveland, 1974.

199. F. Petrenko and R. W. Whitworth, *Physics of Ice*, Oxford University Press, Oxford, 1999.

200. P. V. Hobbs, *Ice Physics*, Clarendon Press, Oxford, 1974.

201. U. Schneider, R. Brand, P. Lunkenheimer, and A. Loidl, *Phys. Rev. Lett.* **84**, 5560 (2000).

202. S. Hensel-Bielowka and M. Paluch, *Phys. Rev. Lett.* **89**, 025704 (2002).

203. S. Adichtchev, T. Blochowicz, C. Tschirwitz, V. N. Novikov, and E. A. Rössler, *Phys. Rev. E*, **68**, 011504 (2003).

204. U. Kaatze, R. Behrends, and R. Pottel, J. *Non-Cryst. Solids* **305**, 19 (2002).

205. C. Rønne, L. Thrane, P. O. Astrand, A. Wallqvist, K. V. Mikkelsen, and S. R. Keiding, *J. Chem. Phys.* **107**, 5319 (1997).

206. V. Arkhipov and N. Agmon, *Israel J. Chem. Phys.* **43**, 363 (2003)

207. G. P. Johari and M. Goldstein, *J. Chem. Phys.* **53**, 2372 (1970).

208. A. Puzenko, Y. Hayashi, Y. E.. Ryabov, I. Balin, Y. Feldman, U. Kaatze, and R. Behrends, *J. Phys. Chem. B.* **109**, 6031 (2005).

209. R. R. Nigmatullin, *Phys. Stat. Sol.* **153**, 49 (1989).

210. J. Klafter and A. Blumen, *Chem. Phys. Lett.* **119**, 377 (1985).

211. J. Klafter and M. F. Shlesinger, *Proc. Natl. Acad. Sci.* USA **83**, 848 (1986).

212. B. B. Mandelbrot, *The Fractal Geometry of Nature*, Freeman, New York, 1982.

213. E. Feder, *Fractals*, Plenum Press, New York, 1988.

214. U. Even, K. Rademann, J. Jortner, N. Manor, and R. Reisfeld, *Phys. Rev. Lett.* **52**, 2164 (1984).

215. U. Even, K. Rademann, J. Jortner, N. Manor, and R. Reisfeld, *Phys. Rev. Lett.* **58**, 285 (1987).

216. R. R. Nigmatullin, L. A. Dissado, and N. N. Soutougin, *J. Phys. D* **25**, 32 (1992).

217. W. D. Dozier, J. M. Drake, and J. Klafter, *Phys. Rev. Lett.* **56**, 197 (1986).

218. I. Webman, *Phys. Rev. Lett.* **47**, 1496 (1981).

219. T. Odagaki and M. Lax, *Phys. Rev. B*, **24**, 5284 (1981).

220. S. Havlin and A. Bunde, *Percolation II*, in *Fractals in Disordered Systems*, edited by A. Bunde and S. Havlin, Springer-Verlag Berlin, Heidelberg, 1996.

221. F. David, B. Duplantier, and E. Guitter, *Phys. Rev. Lett.* **72**, 311 (1994)

222. G. A. Baker and N. Kawashima, *Phys. Rev. Lett.* **75**, 994 (1995).

223. F. F. Assaad and M. Imada, *Phys. Rev. Lett.* **76**, 3176 (1996).

224. K. B. Lauritsen, P. Fröjdh, and M. Howard, *Phys. Rev. Lett.* **81**, 2104 (1998).

225. Y. Alexandrov, N. Kozlovich, Y. Feldman, and J. Texter, *J. Chem. Phys.* **111**, 7023 (1999).

226. P. K. Pathria, *Statistical Mechanics*, Pergamon Press, Oxford, 1972.

227. L. D. Landau and E. M. Lifshitz, *Statistical Mechanics*, Pergamon Press, Oxford, 1980.

228. S. K. Ma, *Statistical Mechanics*, World Scientific, Singapore, 1985.

229. D. N. Zubarev, *Nonequilibrium Statistical Thermodynamics*, Consultant Bureau, New York, 1974.

230. R. W. Zwanzig, in W. E. Brittin, B. W. Downs, and J. Downs, eds., *Lectures in Theoretical Physics*, Vol. III, Interscience Publishers, New York, 1996.

231. K. Oldham and J. Spanier, *The Fractional Calculus*, Academic Press, New York, 1974.

232. K. Miller and B. Ross, *An Introduction to the Fractional Calculus and Fractional Differential Equations*, John Wiley & Sons, New York, 1993.

233. H. Darcy, *Fontanies Publiques de la Ville de Dijon*, Libraire des Corps Imperiaux des Ponts et Chausses et des Mines, Paris, 1856.

234. J. B. J. Fourier, *Théorie Analytique de la Chaleur, Firmin Didot*, Père et Fils, Paris, 1822.

235. A. Einstein, *Ann. Phys.* **322**, 549 (1905).

236. E. W. Montroll and G. H. Weiss, *J. Math. Phys.* **6**, 167 (1965).

237. H. Scher and M. Lax, *Phys. Rev. B* **7**, 4491 (1973).

238. H. Scher and M. Lax, *Phys. Rev. B* **7**, 4502 (1973).

239. K. G. Wilson and J. Kogut, *Phys. Rep.* 2, 75(1974).

240. Y. E. Ryabov and A. Puzenko, *Phys. Rev. B*, **66**, 184201 (2002).

241. J. A. Padró, L. Saiz, and E. Guàrdia, *J. Mol. Struct.* **416**, 243 (1997).

242. R. Podeszwa and V. Buch, *Phys. Rev. Lett.* **83**, 4570 (1999).

243. R. Mittal and I. A. Howard, *Phys. Rev. B* **53**, 14171 (1996).

244. Y. Hayashi, A. Puzenko, I. Balin, Y. E. Ryabov, and Y. Feldman, *J. Phys. Chem. B* **109**, 9174 (2005).

245. R. P. Auty and R. H. Cole, *J. Chem. Phys.* **20**, 1309 (1952).

246. N. Miura, Y. Hayashi, N. Shinyashiki, and S. Mashimo, *Biopolymers*, **36**, 9 (1995).

247. C. T. Moynihan, P. B. Macedo, C. J. Montrose, P. K. Gupta, M. A. DeBolt, J. F. Dill, B. E. Dom, P. W. Drake, A. J. Easteal, P. B. Elterman, R. P. Moeller, H. Sasabe, and J. A. Wilder, *Ann. N. Y. Acad. Sci.* **279**, 15 (1976).

248. Y. Hayashi, N. Miura, N. Shinyashiki, S. Yagihara, and S. Mashimo, *Biopolymers*, **54**, 388 (2000).

249. S. Naoi, T. Hatakeyama, and H. Hatakeyama, *J. Therm. Anal. Calorim.* **70**, 841 (2002).

250. R. D. Shannon, *Acta Cryst. A* **32**, 751 (1976).

251. K. Leung, *Phys. Rev. B* **65**, 012102 (2001).

252. L. Frunza, H. Kosslick, S. Frunza, R. Fricke, and A. Schönhals, *J. Non-Cryst. Solids* **307–310**, 503 (2002).

253. L. Frunza, H. Kosslick, S. Frunza, R. Fricke, and A. Schönhals, *J. Phys. Chem. B* **106**, 9191 (2002).

254. A. Hushur, J. H. Ko, S. Kojima, S. S. Lee, and M. S. Jang, *J. Appl. Phys.* **92**, 1536 (2002).

255. J. H. Ko and S. Kojima, *J. Korean Phys. Soc.* **41**, 241 (2002).

256. F. Cordero, A. Campana, M. Corti, A. Rigamonti, R. Cantelli, and M. Ferretti, *Intl. J. Mod. Phys. B* **13**, 1079 (1999).

257. P. Ben Ishai, C.E.M. de Oliveira, Y. Ryabov, A. J. Agranat, and Y. Feldman, *J. Non-Cryst. Solids* (2005).

258. A. A. Bokov, M. Mahesh Kumar, Z. Xu, and Z.-G. Ye, *Phys. Rev. B* **64**, 224101 (2001).

259. G. P. Johari, A. Hallbruker, and E. Mayer, *Nature* **330**, 552 (1987).

260. R. S. Smith and B. D. Kay, *Nature* **398**, 788 (1999).

261. J. Kakalios, R. A. Street, and W. B. Jackson, *Phys. Rev. Lett.* **59**, 1037 (1987).

262. E. W. Fischer, E. Donth, and W. Steffen, *Phys. Rev. Lett.* **68**, 2344 (1992).

263. J. Colmenero, A. Alegía, A. Arbe, and B. Frick, *Phys. Rev. Lett.* **69**, 478 (1992).

264. M. Salomon, M. Z. Xu, E. M. Eyring, and S. Petrucci, *J. Phys. Chem.* **98**, 8234 (1994).

265. N. Shinyashiki, S. Yagihara, I. Arita, and S. Mashimo, *J. Phys. Chem. B* **102**, 3249 (1998).

266. H. Mori, *Prog. Theor. Phys.* **33**, 423 (1965).

267. H. Mori, *Prog. Theor. Phys.* **34**, 399 (1965).

268. I. M. Sokolov, *Phys. Rev. E* **63**, 011104 (2001).

269. S. Fujiwara and F. Yonezawara, *Phys. Rev. E* **51**, 2277 (1995).

270. J. B. Hasted, in F. Franks, ed., *Water, A Comprehensive Treaties*, Plenum Press, New York, 1972, p. 255.

271. U. Kaatze, *J. Chem. Eng. Data* **34**, 371 (1989).

272. J. Barthel, K. Bachhuber, R. Buchner, and H. Hetzenauer, *Chem. Phys. Lett.* **165**, 369 (1990).

273. U. Kaatze, *J. Mol. Liq.* **56**, 95 (1993).

274. C. Rønne, P.-O Åstrad, and S. R. Keiding, *Phys. Rev. Lett.* **82**, 2888 (1999).

275. H. Nuriel, N. Kozlovich, Y. Feldman, and G. Marom, *Composites: Part A* **31**, 69 (2000).

276. Y. E. Ryabov, G. Marom, and Y. Feldman, *J. Thermoplast. Compos. Mater.* **17**, 463 (2004).

277. P. G. DeGennes, *J. Chem. Phys.* **55**, 572 (1971).

278. W. W. Graessley, *J. Polym. Sci.: Polym. Phys.* **18**, 27 (1980).

279. F. Joabsson, M. Nyden, and K. Thuresson, *Macromolecules* **33**, 6772 (2000).

280. J. E. M. Snaar, P. Robyr, and R. Bowtell, *Magn. Reson. Imag.* **16**, 587 (1998).

281. P. Gribbon, B. C. Heng, and T. E. Hardingham, *Biophys. J.* **77**, 2210 (1999).

282. E. Donth, *J. Non-Cryst. Solids* **131–133**, 204 (1991).

CHAPTER 2

EVOLUTION OF THE DYNAMIC SUSCEPTIBILITY IN SUPERCOOLED LIQUIDS AND GLASSES

THOMAS BLOCHOWICZ

Institut für Festkörperphysik, Technische Universität Darmstadt, D 64289 Darmstadt, Germany

ALEXANDER BRODIN, and ERNST A. RÖSSLER

Experimentalphysik II, Universität Bayreuth, D 95444 Bayreuth, Germany

CONTENTS

Fractals, Diffusion, and Relaxation in Disordered Complex Systems: A Special Volume of Advances in Chemical Physics, Volume 133, Part A, edited by William T. Coffey and Yuri P. Kalmykov. Series editor Stuart A Rice.

I. INTRODUCTION

Amorphous or glassy materials show distinct physical properties in comparison to crystalline matter. Many of these materials are in widespread use in modern life. There are several ways of producing amorphous matter, of which the simplest and most commonly used is to supercool a liquid. Though not always easily possible, supercooling a liquid below the melting point results in a strong continuous increase of the viscosity. For many liquids, the viscosity increases by more than 10 orders of magnitude within some tens of degrees Kelvin. Eventually, the viscous flow becomes so slow that the system can be regarded as a solid body. This viscous freezing is referred to as the glass transition, and liquids that have passed the glass transition are called glasses. Conventionally, the glass transition temperature T_g is defined as the temperature where the viscosity equals 10^{12} Pa·s. Due to the absence of a first-order phase transition, no essential structural change occurs upon cooling, except for a gradual increase in density; glasses and liquids exhibit virtually the same structure. Since around T_g the relaxation time quickly exceeds that of a typical laboratory experiment upon cooling, quite an abrupt change is observed in many physical properties; for instance, a nearly step-like behavior of the thermal expansion coefficient and the heat capacity is observed. Well below T_g, structural relaxation no longer occurs, and the system is trapped in a nonequilibrium, nonergodic state. Despite its being a ubiquitously observed phenomenon, the nature of the glass transition is still a matter of vigorous scientific debate; however, in spite of the lack of a consensus on the driving forces of the vitreous solidification, important progress has been made in the last few decades.

Presumably all liquids can be vitrified if they are cooled at a sufficiently high rate. Thus, glasses can be prepared from polymeric, metallic, and molecular melts. Molecular liquids are however of particular interest, since their glass transition can be studied by a large variety of experimental techniques. Moreover, they are models of simple van der Waals liquids that are often examined in molecular dynamics simulations and other theoretical approaches. In fact, only a relatively small group of molecular liquids can be easily supercooled. These liquids have been investigated thoroughly in recent years and are the focus of the present chapter. Other glass formers, such as polymers, ionic, or mixed systems, are usually omitted from the present discussion. Also, we restrict ourselves to the results obtained at normal ambient pressure, although the pressure effects are important in order to clarify the origin of certain relaxation processes.

One of the distinguishing properties of glasses, not found in crystals, is the occurrence of molecular motion on virtually all time scales. In the supercooled liquid state, the primary α-relaxation of the structure towards thermal equilibrium slows down from picosecond times at $T \gg T_g$ to about 100 s near T_g. Contrary to the exponential time dependence expected in classical liquid theories, one normally observes markedly nonexponential relaxation processes. This is usually referred to as *relaxation stretching* and concerns the long-time ultimate decay of structural fluctuations. At short times, relaxation sets in as oscillation de-phasing with typical attempt frequencies in the terahertz range (picosecond times) at all temperatures. Therefore, upon cooling a liquid, a large gap opens up between the time/frequency scales of the α-relaxation and microscopic dynamics, in which secondary relaxation processes emerge. These relaxation features have attracted particular interest in recent years; and beyond determination of the time constants, the aim has been to monitor the evolution of the susceptibility spectra in detail. Near T_g, the relaxation spectrum spreads over more than 14 decades in time or frequency. Indeed, it is a tremendous experimental challenge to cover this range.

Overall, the susceptibility spectrum of supercooled liquids is qualitatively different in different temperature intervals. For instance, secondary processes usually appear only close to T_g. One can phenomenologically distinguish four regimes of spectrum evolution, which are sketched in Fig. 1: (a) a low-density fluid, where relaxation and microscopic dynamics have essentially merged, (b) a liquid of moderate viscosity with its characteristic two-step relaxation and a stretched long-time part, (c) a highly viscous liquid close to T_g where additional, slow secondary relaxations have emerged, and (d) the glassy state below T_g. In the glass, structural relaxation no longer occurs, and the susceptibility spectra are solely determined by the vibrational and secondary relaxational dynamics.

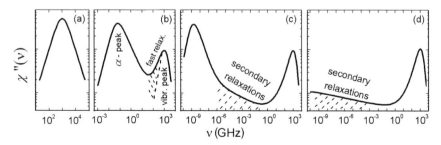

Figure 1. A sketch of the temperature evolution of the susceptibility spectra of simple molecular liquids upon passing from low-density fluid (a) to the glass (d). We anticipate two distinguishable temperature regimes for the evolution of "glassy dynamics," namely, a high-temperature regime (b) and a low-temperature regime (c), the latter characterized by the emergence of slow secondary relaxation processes.

The double-peaked spectrum with a stretched α-relaxation part in Fig. 1b is often taken as characteristic of the glassy dynamics. It turns out, however, that in almost all cases such a dynamical pattern sets in significantly above the melting point, thus being typical of the "normal" liquid state in thermodynamic equilibrium. In this context, using the term "glassy dynamics" appears somewhat misleading. It is in fact hardly possible, as will be demonstrated later, to reach such high temperatures that this kind of dynamics has completely disappeared.

In simple glass formers, the molecular slowing down has been studied using a variety of techniques. Several review articles cover the topic from various points of view, experimental [1–15] as well as theoretical [3,16–21]. In the present contribution we will mainly focus on experimental results concerning molecular reorientation dynamics, specifically the results obtained by dielectric spectroscopy (DS), depolarized light scattering (LS), and nuclear magnetic resonance (NMR) spectroscopy. In most cases, molecular reorientations are strongly coupled to the structural dynamics and thus may be taken as a probe of the structural relaxation and its slowing down.

Many of the experiments performed in the last two decades were inspired by the development of the mode coupling theory (MCT), starting in 1984. In contrast to other theoretical efforts, MCT describes, in particular, the dynamics of a simple liquid in the gigahertz range adjacent to the high-frequency region, the main features of which are presumably well understood from conventional liquid theories. Moreover and in contrast to most other approaches, MCT makes detailed predictions about the evolution of the dynamic susceptibility, the latter actually being a solution of the MCT equations. Consequently, several experimental techniques have been extended and/or refined in order to gain better access to the relevant gigahertz frequency range. Neutron scattering (NS) studies were among the first to address the results of the theory. In light scattering experiments, an important breakthrough was achieved by using tandem Fabry-Perot interferometer spectroscopy in combination with traditional grating spectroscopy, allowing one to study the evolution of the dynamics in the range 0.3 GHz to 10 THz. Similarly, optical Kerr effect studies were used to probe the dynamics in the time domain in a comparable dynamic window. Finally, the most prominent technique probing the dynamics of glasses in the kilohertz-to-megahertz range, the dielectric spectroscopy, was extended to fully cover the gigahertz and reach the far-IR range. Combining DS and LS spectroscopy, the susceptibility spectra of molecular glass formers can nowadays be obtained over more than 18 decades in frequency. In this review, we intend to describe common features regarding the temperature evolution of these spectra, in particular, in terms of scaling properties, from a phenomenological point of view as well as from a viewpoint given by the predictions of MCT. In addition, we discuss NMR results and demonstrate that this technique provides unique information on the details of the molecular reorientation processes, information that is only indirectly accessible by LS and DS methods. Thus, only by combining the results from different methods may a coherent picture of the dynamics of molecular glasses emerge.

This chapter is organized as follows. We start with a brief description of the three methods—dielectric, light scattering, and NMR spectroscopies—that we have chosen in order to explore molecular reorientation dynamics (Section II). We then briefly review theoretical approaches to the glass transition and discuss in some detail the mode coupling theory (Section III). Section IV reports the major experimental results compiled in recent years. This is the central section of the chapter, containing six parts. Due to the controversial debate about the applicability of MCT to molecular glass formers, a strictly phenomenological approach is chosen to describe the evolution of the susceptibility spectra over the temperature range from well above the melting point down to T_g (Sections II.A–II.D). We try to provide firm arguments that the "glassy dynamics" in molecular glass formers is qualitatively different in two different temperature intervals, which we call the high- and the low-temperature regimes of the dynamics and which are separated by a discernible crossover temperature T_x. In the same section we also include results from multidimensional NMR that shed light on the mechanism of molecular reorientations close to T_g. A separate section is dedicated to reviewing experimental tests of MCT (Section II.E). In Section II.F we attempt to reflect on the results of molecular liquids at $T > T_g$ in a broader context. We would like to stress that we focus on experimental results of the evolution of the relaxation spectra in molecular liquids upon cooling and on the search for a dynamic crossover at $T > T_g$. Therefore, we shall not dwell on such topics as molecular dynamics simulation, dynamic heterogeneities, aging effects, or thermodynamic aspects of the glass transition. Finally, Section V briefly discusses selected results obtained at temperatures below T_g. Section VI provides general conclusions. We fully realize that such a review cannot be complete and that the particular choice of the material covered reflects the taste of the authors. Moreover, the field is still subject to intense development. Nevertheless, we hope that the review will help the prospective reader to develop a coherent picture of the subject.

II. SELECTED METHODS TO STUDY MOLECULAR REORIENTATION DYNAMICS

A. Molecular Reorientation: Correlation Function, Spectrum, and Susceptibility

A simple (atomic) liquid is adequately described by its time-dependent microscopic density $\rho(\mathbf{r}, t) = \sum_{j=1}^{N} \delta(\mathbf{r} - \mathbf{r}_j(t))$ or in the reciprocal space

$\rho(\mathbf{q}, t) = \Sigma_{j=1}^{N} \exp(i \mathbf{q} \mathbf{r}_j(t))$. The corresponding macroscopic observables are the autocorrelation function

$$C_\rho(\mathbf{q}, t) = N^{-1} \langle \rho^*(\mathbf{q}, t) \rho(\mathbf{q}, 0) \rangle \tag{1}$$

also known as the intermediate scattering function, and its Fourier conjugate $S(\mathbf{q}, \omega)$, the dynamic structure factor [22]. Experimental determination of these properties is the domain of inelastic neutron scattering that provides information on the translational dynamics at microscopic scales. There are, however, other variables pertaining to the structure of laboratory (molecular) liquids, notably the molecular orientations. In fact, it is the dynamics of the orientational degrees of freedom that is most readily accessible experimentally—for example, by light scattering or dielectric spectroscopy. One can formally generalize the microscopic density to include orientations of the molecules in addition to their positions in space [23]. The time correlation function of this generalized particle density involves both orientational and translational coordinates whose individual contributions cannot, in general, be experimentally disentangled. However, in the practically important case $\mathbf{q} \approx 0$ appropriate to dielectric spectroscopy and depolarized (anisotropic) light scattering in molecular liquids, translational motion of the molecules is relatively unimportant, which leaves only orientational correlations.

The central quantity, characterizing fluctuations in a liquid, is thus the correlation function of the relevant dynamical variable. Correlation functions of different properties can be experimentally accessed either directly, such as in a scattering experiment, or through their relation to the corresponding response functions, such as in a relaxation experiment. In most cases, a macroscopic property is involved—for example, the macroscopic polarization in a dielectric relaxation experiment. The corresponding correlation function is therefore in general a *collective* property [24], which is not simply related to the microscopic correlation function characterizing a tagged molecule in the liquid. Evidently, every experimental technique should correctly identify the properties that are being assessed (cf. Sections II.B. and II.C). On the other hand, there is a long tradition of describing reorientation of a tagged molecule in condensed matter in terms of relatively simple stochastic models [25–28]. Such models provide an analytical expression for the conditional probability $P(\Omega, t|\Omega_0)$ of finding the orientation Ω at time t provided it was Ω_0 at $t = 0$. Then, the correlation of two functions of the orientation of a tagged molecule, say $X(\Omega)$ and $Y(\Omega)$, can be calculated according to

$$\langle X(\Omega(t)) Y(\Omega(0)) \rangle = \int d\Omega \int d\Omega_0 X(\Omega) Y(\Omega_0) P(\Omega, t|\Omega_0) p^{\text{eq}}(\Omega_0) \tag{2}$$

where $p^{eq}(\Omega_0)$ is the equilibrium probability of finding Ω_0. For example, if the assumption of a first-order Markov process applies to the dynamics, such calculations can be performed by setting up the exchange matrix for a given motional model and then solving the master equation. Alternatively, a dynamical process can be simulated in a computer by producing a large number of, for example, random walk trajectories $\Omega(t)$. Such random walk simulations are commonplace in multidimensional NMR data analyses, where it is straightforward to calculate the time-dependent NMR frequencies from the corresponding trajectories and thus to compute the results of the NMR experiments (cf. Section IV.C.3 and V.B).

The functions $X(\Omega)$ and $Y(\Omega)$ are specified by the choice of the particular experiment. Prominent orientational correlation functions result when setting $X(\Omega) = Y(\Omega) = P_l(\cos\theta)$, where P_l is the Legendre polynomial of rank l and the angle θ specifies the orientation of the molecule with respect to some fixed axis. For example, consider a molecule that possesses a vector property, say the molecular electric dipole $\mu_j = \mu \mathbf{u}_j$ ($\mathbf{u}$ is a unit vector). Then, one defines the *dipole autocorrelation function* $g_1(t) = \langle \mathbf{u}_i(t)\mathbf{u}_i(0)\rangle$. Similarly, one defines a correlation function $g_2(t)$ for second rank tensorial molecular properties. In general the normalized $(g_l(0) = 1)$ orientational correlation function of rank l is given by

$$g_l(t) = \langle P_l(\cos\theta(t))P_l(\cos\theta(0))\rangle / \langle P_l(\cos\theta(0))^2\rangle \qquad (3)$$

Integrals of these functions give the corresponding single-molecule correlation times τ_l. The functions $g_l(t)$ represent the so-called self-part of more general collective correlation functions $C_l(t)$, which contain also cross correlation terms involving pairs of molecules. For example, $C_1(t)$ is given by $C_1(t) = \Sigma_{i,j}\langle \mathbf{u}_i(t)\mathbf{u}_j(0)\rangle / \Sigma_{i,j}\langle \mathbf{u}_i(0)\mathbf{u}_j(0)\rangle$. It is exactly the function $C_1(t)$ that is related to the properties assessed in dielectric relaxation experiments of dipolar molecular liquids, while $C_2(t)$ is relevant in depolarized light scattering experiment. Other methods allow one to access single-molecule correlations, such as $g_2(t)$ in ^{2}H NMR experiments. We also note that some (in particular, NMR), experiments provide information on more complicated angular correlation functions, such as those involving trigonometric functions of the Legendre polynomials (Section II.D.3).

There is no general relation between the collective correlation functions and their single-molecule (self-) counterparts, except for approximate relations for the relaxation times [24–26,29]. However, it is usually argued that, since orientational correlations in molecular liquids are presumably rather short-ranged, the functions $C_l(t)$ and $g_l(t)$ are similar. Experimentally, a direct comparison is not an easy task because, in most cases, $C_l(t)$ and $g_l(t)$ cannot be studied in the same temperature range. Discernible differences are, however, observed in computer simulations [27,30]. As a note of further caution on relating dielectric and depolarized light scattering data to the orientational correlation functions, one should in addition consider the so-called interaction-induced contributions to these data, whereby the relevant property of a molecule (e.g., its dipole moment) is influenced by the surrounding molecules through dipolar interactions, collisions, or other intermolecular interactions. In dielectric spectroscopy, such effects are usually ignored (see, however, Ref. 24), while their importance in light scattering was a matter of some controversy (cf. Section II.C).

Given a correlation function $C(t)$ of some physical property, its Fourier conjugate $S(\omega)$ is the fluctuation power spectrum of this property (Wiener–Khinchin Theorem, cf. Refs. 22 and 26). $S(\omega)$ is directly measured in, for example, a scattering experiment. For instance, the depolarized light scattering spectrum is given by the Fourier transform of the scattered electric field correlation function, which, in turn, is proportional in its major part to $C_2(t)$. Fourier conjugates of the orientational correlation functions $g_{1,2}(t)$ will be denoted $S_{1,2}(\omega)$ and called spectral densities. Finally, the relation between the spectral density and the corresponding linear spectral response of the related physical property to an external perturbation, or the susceptibility, is given by the fluctuation–dissipation theorem, which states that in the linear response regime the reaction of the system to an external perturbation is governed by the same microscopic mechanism that determines thermal fluctuations about the equilibrium in the system [22,31–33]. Explicitly,

$$S(\omega) = \frac{k_B T}{\omega} \chi''(\omega) \tag{4}$$

where $\chi''(\omega)$ is the imaginary (loss) part of the susceptibility, and the relation is valid in the classical limit $\hbar\omega \ll k_B T$ [cf. Eq. (13) for a quantum mechanical result]. Thus, for instance, the dielectric permittivity $\varepsilon''(\omega)$ of a dipolar liquid is related to the Fourier transform of $\mu^2 C_1(t)$, introduced above.

B.   Dielectric Spectroscopy

Since the late nineteenth century, dielectric spectroscopy has been used to monitor dynamical properties of solid and liquid materials. At that time, dielectric measurements were performed either at a single frequency or in a very limited frequency range; now, however, measurement technique and instrumentation have developed to such an extent that dielectric spectroscopy is today a well-established method to probe molecular dynamics over a broad range in frequency or time (cf. reviews by Johari [1], Böttcher and Bordewijk [34], Williams [35,36], and Kremer and Schönhals [37]), even with commercially available equipment. Including the latest developments, one can even say that nowadays dielectric spectroscopy is the only method that is fully able to realize the idea of 0- to 1-THz spectroscopy. In data sets that cover the range of up to 10^{-6}–10^{13} Hz—that is, from ultra-low frequencies up to the far infrared—the full range of reorientational dynamics in molecular glass formers has become accessible and was systematically studied by Loidl, Lunkenheimer, and co-workers [9,10].

In dielectric spectroscopy the polarization response $P(t)$ of a dipolar material is monitored, which is subject to a time-dependent electric field (Maxwell field), $E(t)$. For a linear and isotropic dielectric one can write (e.g., Ref. 34):

$$P(t) = \varepsilon_\infty E(t) - \varepsilon_0 \Delta\varepsilon \int_{-\infty}^{t} \dot{\Phi}(t - t')E(t')dt' \qquad (5)$$

with $\Phi(t)$ being the step response function, ε_0 the vacuum permittivity, and $\Delta\varepsilon = \varepsilon_s - \varepsilon_\infty$ the relaxation strength—that is, the difference between the static dielectric constant ε_s and ε_∞, the dielectric constant of induced polarization. The frequency-dependent complex dielectric permittivity $\hat{\varepsilon}(\omega)$, which can be determined in a frequency-domain experiment according to

$$P(\omega) = \varepsilon_0(\hat{\varepsilon}(\omega) - 1)E(\omega) = \varepsilon_0\hat{\chi}(\omega)E(\omega) \qquad (6)$$

and which is closely related to the complex dielectric susceptibility $\hat{\chi}(\omega)$, is then given by the Fourier transform

$$\frac{\hat{\varepsilon}(\omega) - \varepsilon_\infty}{\Delta\varepsilon} = 1 - i\omega \int_0^{\infty} \Phi(t)e^{-i\omega t}dt \qquad (7)$$

As one wishes to deduce from such experiments the microscopic equilibrium properties of the material under study, it is essential to first establish a link to the equilibrium fluctuations $\Delta\mathbf{M}(t)$ of the macroscopic dipole moment $\mathbf{M}(t) = \Sigma_{i=1}^{N}\boldsymbol{\mu}_i(t)$—that is, the sum of the permanent molecular moments $\boldsymbol{\mu}_i(t)$, which may be expressed in terms of a correlation function: $G_M(t) = \langle\Delta\mathbf{M}(t)\Delta\mathbf{M}(0)\rangle/\langle\Delta\mathbf{M}^2\rangle$. Note that in some cases such fluctuations may be extracted directly from polarization noise data [38,39]. $G_M(t)$ is most directly related to the step response of the system to an external electric field [i.e. a so-called vacuum field, which is entirely due to an external charge distribution as opposed to the Maxwell field in Eq. (5), which may also depend on the state of the dielectric itself] via the fluctuation dissipation theorem of statistical mechanics [31,32]. However, the relation between $G_M(t)$ and $\hat{\varepsilon}(\omega)$ (or $\Phi(t)$) in general depends on the particular shape of the dielectric sample, as $G_M(t)$ is shape-dependent due to the long-range dipole–dipole interactions, whereas $\hat{\varepsilon}(\omega)$ is a thermodynamically intensive quantity and as such is independent of the sample shape. However, once such a relation is established for any particular shape of a dielectric sample, a shape-independent relation can be obtained by reducing $G_M(t)$ to a microscopic correlation function. For the latter, one may, for example, consider n dipole moments in a spherical volume around a given reference dipole $\mathbf{\mu}_i$. The size of the volume is chosen small compared to the overall sample size so that the average moment of the sphere for fixed $\mathbf{\mu}_i$ depends on short-range interactions, and the remaining $N - n$ molecules are treated as a dielectric continuum. The correlation function reads as follows:

$$C_1(t) = \frac{\left\langle \mathbf{\mu}_i(0) \cdot \sum_{j}^{n} \mathbf{\mu}_j(t) \right\rangle}{\left\langle \mathbf{\mu}_i(0) \cdot \sum_{j}^{n} \mathbf{\mu}_j(0) \right\rangle} \tag{8}$$

where the sum extends over the moments contained in the sphere around $\mathbf{\mu}_i$. Note that the sum over all reference dipoles $\mathbf{\mu}_i$ was omitted, because it simply leads to a factor of N (i.e., number of molecules) in numerator and denominator alike. More complications arise when in addition the effects of induced dipole moments are considered; however, finally a relation between the correlation function $C_1(t)$ and the permittivity may be obtained as

$$\frac{\hat{\varepsilon}(\omega) - \varepsilon_\infty}{\Delta\varepsilon} g(\omega) = 1 - i\omega \int_0^\infty C_1(t) e^{-i\omega t} dt \tag{9}$$

with $g(\omega)$ being an internal-field factor, which is calculated in a slightly different manner depending on the approach used [41–43] and others (for a review, cf. Williams [40] or Böttcher and Bordewijk [34]).

In most supercooled liquids, however, $g(\omega) \approx 1$ holds to a reasonable degree of approximation; moreover, the overall evolution of the spectral shape seems to be basically unaffected by the deviations of $g(\omega)$ from unity, which may in particular appear at low values of the dielectric loss $\varepsilon''(\omega)$.

The contributions to the microscopic correlation function $C_1(t)$ for an arbitrary reference molecule (i) may be split into an autocorrelation and a cross-correlation part as follows:

$$C_1(t) = \frac{\langle \mathbf{\mu}_i(0) \cdot \mathbf{\mu}_i(t) \rangle + \langle \mathbf{\mu}_i(0) \cdot \sum_{i \neq j} \mathbf{\mu}_j(t) \rangle}{\langle \mathbf{\mu}_i(0) \cdot \mathbf{\mu}_i(0) \rangle + \langle \mathbf{\mu}_i(0) \cdot \sum_{i \neq j} \mathbf{\mu}_j(0) \rangle} \tag{10}$$

With the static cross-correlations contained in the Kirkwood–Fröhlich correlation factor (e.g., Ref. 44) $g_k = 1 + \langle \mathbf{\mu}_i(0) \cdot \Sigma_{i \neq j}\mathbf{\mu}_j(0) \rangle/\mu^2$, which is finally absorbed in the relaxation strength, $C_1(t)$ reduces to

$$C_1(t) = [\langle \mathbf{\mu}_i(0)\mathbf{\mu}_i(t) \rangle + \langle \mathbf{\mu}_i(0) \sum_{i \neq j} \mathbf{\mu}_j(t) \rangle]/(g_k\mu^2) \tag{11}$$

In molecular glass-forming systems it is usually argued that the dynamic cross-correlations are negligible or that at least their time dependence is similar to that of the dipolar autocorrelation function. Evidence for that view is given, for example, by comparing the spectral densities attained by different methods, like dielectric and NMR spin–lattice relaxation, where the latter technique (in the case of deuteron nuclei) is strictly related to the autocorrelation function of the second-order Legendre polynomial (cf. Section IV.C.3). In such a case, Eq. (9) indeed establishes a relation between the complex dielectric permittivity and the microscopic dipolar autocorrelation function $g_1(t) = \langle \mathbf{\mu}_i(0) \cdot \mathbf{\mu}_i(t) \rangle / \mu^2 \approx C_1(t)$. Note that $g_1(t)$ is equivalent to the definition given in Eq. (3), because on average the unit vector of the dipole moment may be reduced to its projection in the direction of the electric field. Under such conditions, Eq. (9) becomes

$$\frac{\hat{\varepsilon}(\omega) - \varepsilon_\infty}{\Delta\varepsilon} \approx 1 - i\omega \int_0^\infty g_1(t) e^{-i\omega t} dt \qquad (12)$$

In order to actually cover 19 decades in frequency, dielectric spectroscopy makes use of different measurement techniques each working at its optimum in a particular frequency range. The techniques most commonly applied include time-domain spectroscopy, frequency response analysis, coaxial reflection and transmission methods, and at the highest frequencies quasi-optical and Fourier transform infrared spectroscopy (cf. Fig. 2). A detailed review of these techniques can be found in Kremer and Schönhals [37] and in Lunkenheimer [45], so that in the present context only a few aspects will be summarized.

1. *Time-Domain Spectroscopy.* This technique is mostly used at the lowest frequencies from microhertz to some kilohertz. Here, the time-dependent charging or discharging of a sample capacitor is monitored after a voltage step. After a numerical Fourier transform the data can be combined with frequency domain measurements. A frequently applied experimental configuration is the so-called modified Sawyer–Tower setup suggested by Mopsik [46] and variations thereof. Also among the time domain techniques is one suggested by Richert and Wagner [47], where the electric field relaxation is monitored after a step in the dielectric displacement; that is, at $t = 0$ a charge q is applied to the sample capacitor and the time evolution of the voltage across the capacitor plates is measured at constant charge. The Fourier transform of the relaxation function in this case yields the complex electric modulus function, which is just the inverse of the permittivity: $\hat{M}(\omega) = 1/\hat{\varepsilon}(\omega)$. Although time-domain techniques can in principle be extended up to frequencies in the gigahertz regime [48], their advantages are most prominent at lowest frequencies, where frequency-domain measurements can become extremely time-consuming, as for each data

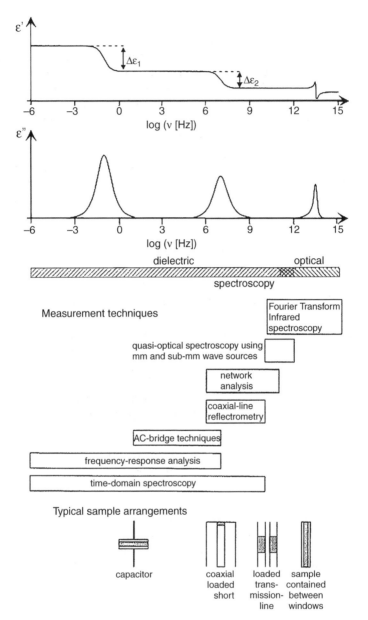

Figure 2.   Different techniques used in dielectric spectroscopy in order to cover the full dynamic range of relaxational features at the glass transition; for details see text. (From Ref. 37.)

point several cycles $\sin(\omega t)$ have to be acquired, whereas a time-domain measurement contains the complete spectrum in one single sweep. However, the resolution of impedance analysis in the frequency domain is far superior (due to the use of phase-sensitive detection and lock-in techniques) and so, apart from specialized experimental techniques, time-domain measurements are usually restricted to the ultra-low-frequency regime.

2. *Frequency Response Analysis.* This technique is used in the region between 10^{-4} and 10^7 Hz. The voltage and current in the sample capacitor are measured directly with a frequency response analyzer, which works on the same principles (Fourier correlation analysis) as lock-in amplifiers. In many cases a two-channel vector voltage analyzer is used in conjunction with a dielectric converter, which converts the sample current into a voltage in the appropriate range so that both quantities may be measured at the same time by the analyzer. To improve accuracy, for each frequency the sample impedance is compared to a similar internal reference impedance of the converter so that linear amplitude or phase errors cancel out in the measurement. With modern equipment a resolution of $\tan \varepsilon''/\varepsilon' > 10^{-4}$ is reached. Alternatively, even higher resolution may be obtained by high-precision autobalance capacitance bridges which, however, usually only operate at a single fixed frequency. But the latest development in instrumentation provides devices, where a number of such bridges is combined in a single unit so that in a frequency range of 50 Hz to 20 kHz a resolution of $\tan \delta \approx 10^{-7}$ may be reached with data being obtained at 10 points per decade. These novel devices offer a unique oportunity to provide reliable data even in the very low loss regimes far below T_g [49,50].

3. *Coaxial Reflection and Transmission Techniques.* In the 1-MHz to 10-GHz range, coaxial line reflectometry may be used. Here, the sample terminates a precision coaxial line, thereby bridging the inner and outer conductor, and the complex reflection coefficient is measured by an impedance or a network analyzer: Reflected and incoming waves are separated by means of two-directional couplers, and a phase-sensitive measurement is performed. In general the impedance introduced by cabling, lines, and connectors plays an important role in measurements at frequencies above 1 MHz. Thus, usually sophisticated calibration procedures have to be applied to take account of the nonideal conditions given even by precision-machined coaxial lines, connectors, and sample cells. In a different setup, not only the reflected but also the transmitted wave can be measured in terms of phase and amplitude by a network analyzer. In that technique the sample fills the space between inner and outer conductor in a coaxial line, and measurements may be extended typically up to some tens of gigahertz. In the present frequency range, typical resolutions of $\tan \delta \approx 10^{-2}$ are achieved.

4. *Quasi-optical and FTIR Spectroscopy*. As investigations of glass-forming liquids in the regime of highest frequencies have become increasingly important in particular due to detailed predictions of dynamical properties made by the mode coupling theory, attempts were made to fully cover the spectral range up to highest frequencies. The frequency scale may be covered with coaxial techniques up to some tens of gigahertz and again from about 300 GHz on with commercial Fourier transform IR spectrometers. In the intermediate range, one could in principle use resonant cavity systems, which, however, operate at a single frequency only. Alternatively, quasi-optical methods are used in this range, where an unguided monochromatic electromagnetic wave propagates in a setup of a quasi-optical Mach–Zehnder interferometer [51,52]. The monochromatic wave is produced in a so-called backward-wave oscillator, which may be tuned over a limited frequency range. Phase-sensitive detection of the transmitted wave is achieved by inserting the sample into one arm of the interferometer. With that technique a typical resolution limit is again of the order of $\tan \delta \approx 10^{-2}$.

Although at the time of writing the full range of dielectric methods has been applied only for two molecular glass-forming liquids, namely glycerol and propylene carbonate, presently quite a large database exists for the dielectric behavior of molecular glass formers in the frequency range from the millihertz up to the gigahertz regime so that dielectric data provide a good starting point to look for common features in the evolution of the dynamic susceptibility in supercooled liquids.

C. Depolarized Light Scattering

While using the scattering of light to study material media dates back at least to the mid-nineteenth century (Tyndall scattering), light scattering spectroscopy in its full glory emerged only after the advent of the laser in 1960. Soon thereafter, it proved a potent tool for structural relaxation studies, supplementing and at times superseding the traditional ultrasonic and dielectric techniques [25]. In contrast to dielectric spectroscopy, which determines the dielectric response of a system to an external perturbing field and thus assesses a *susceptibility* function $\chi(\omega)$, a spontaneous light scattering experiment determines the *spectral density* of spontaneous thermal fluctuations $S(\omega)$. It is common to present LS data in the form of light scattering susceptibility $\chi''_{LS}(\omega)$ (cf. Section II.A), which is related to the Stokes (energy loss) part of the spectrum as follows:

$$\chi''(\omega) = \frac{1}{\hbar} \frac{S_{LS}(\omega)}{n(\omega, T) + 1} \approx \frac{\omega}{k_B T} S_{LS}(\omega) \tag{13}$$

where $n(\omega, T) = [\exp(\hbar\omega/k_B T) - 1]^{-1}$ is the Bose distribution function, and the latter equality holds at low frequencies and high temperatures, $\hbar\omega/k_B T \ll 1$.

Generally, scattering of electromagnetic radiation at optical frequencies (light scattering) occurs through the secondary emission of electromagnetic radiation from a fluctuating electronic polarization induced by the excitation field. Therefore, it reflects fluctuations of the electronic polarizability tensor that occur due to fluctuations of the molecular positions and orientations. Compared to dielectric spectroscopy, it involves a second rank tensor property (electronic polarizability) rather than the electric polarization vector. The two main sources of the fluctuating polarizability are (a) rotations of anisotropic molecules that modulate the off-diagonal elements of the tensor and (b) fluctuations of the particle density that contribute to its diagonal (isotropic) part. In addition, one has to consider the so-called "interaction induced" mechanisms of second order, generally weak scattering, whereby the polarizability of a molecule is influenced by nearby molecules due to, for example, molecular collisions or dipole–dipole interactions. The anisotropic scattering, by its mechanism, is thus comparable with the dielectric absorption in nonionic molecular liquids, which predominantly occurs due to rotating molecular dipoles. Isotropic scattering, on the contrary, has no direct counterpart in the dielectric spectroscopy, except for relatively weak interaction-induced effects. Isotropic and anisotropic parts of the scattered field can be separated experimentally owing to their different polarization properties. Isotropic scattering always occurs with the same polarization as the (linearly polarized) excitation field, while the anisotropically scattered field has components in all polarizations. Choosing therefore the orthogonal, so-called depolarized scattering component, one excludes isotropic scattering of the density fluctuations. Selecting furthermore the back-scattered radiation, one also excludes scattering from shear waves, which is then forbidden by symmetry. Otherwise, transverse waves give rise to a relatively weak anisotropic scattering through orientation-to-sheer coupling. The resulting scattering geometry is commonly referred to as "depolarized back scattering." We will assume this scattering geometry in the following and use the abbreviation LS (light scattering) to mean *depolarized* scattering, unless otherwise noted.

There has been certain controversy in the past as to the physical nature of the fluctuating property whose spectrum is detected in LS experiments, with discussions sometimes couched in terms of the "light scattering mechanism." These discussions were stimulated, among other things, by the advent of the mode coupling theory and the urge to test its predictions for the spectral shapes, which, however, were formulated for *density* fluctuation spectra. Thus, it was often implicitly or explicitly assumed that experimental spectra, even taken in depolarized configurations, reflected density fluctuations. A systematic discussion of the importance of interaction-induced mechanisms, in particular of the dipole-induced–dipole (DID) mechanism, was given by Madden [24]. Sometimes depolarized scattering was assumed to mainly originate from the DID and/or other interaction-induced mechanisms (cf. Refs. 53 and 54 and references therein).

For instance, Cummins and co-workers initially interpreted LS spectra of liquid salol as being mainly due to DID [55] while later showing that the DID contribution in this liquid was actually negligible [53]. Others argue that the relative importance of the DID mechanism may be different in different spectral regions [54]. Currently, we believe it is commonly accepted that depolarized scattering from optically anisotropic molecules mainly occurs due to their rotating polarizability tensors [53]. Thus, the depolarized scattering spectra predominantly reflect the correlation

$$C_2(t) = \Sigma_{i,j}\langle P_2(\mathbf{u}_i(t)\mathbf{u}_j(0))\rangle / \Sigma_{i,j}\langle P_2(\mathbf{u}_i(0)\mathbf{u}_j(0))\rangle.$$

The momentum $\mathbf{q}$ of the excitations that are probed in LS experiments is of the order $2\pi/\lambda$. Since the optical wavelength λ is much larger than the intermolecular distances and since there are no conceivable intermolecular interactions in liquids that could lead to orientational correlations between different molecules on such long length scales,[1] depolarized light scattering is essentially $\mathbf{q}$-independent, which is commonly referred to as the $\mathbf{q} = 0$ limit. In this respect, too, it is comparable with the dielectric spectroscopy, which also operates in the $\mathbf{q} = 0$ limit. Of course, light scattered in any other geometry than the depolarized back-scattering will include coherent, momentum-allowed contributions from long-wavelength shear and/or longitudinal waves, which naturally are $\mathbf{q}$-dependent.

The dynamics of the glass transition occurs on a frequency scale from zero to ~1 THz. A spectroscopic setup that is used to detect such dynamics should therefore offer sufficient spectral contrast and resolution in order to be able to spectrally resolve light whose frequency is shifted by, say, some gigahertz relative to the excitation frequency of ~500 THz. In addition, since spontaneous light scattering is a weak process, a sensitive detection system is required. The latter requirement becomes critical for investigations at low temperatures of the deeply supercooled or vitrified samples, for the following additional reasons. First, the amplitude of spontaneous polarizability fluctuations that give rise to the nonresonant scattering is directly proportional to the temperature [cf. Eq. (4)]. Second, the LS susceptibility in the range of secondary relaxations— that is, between the main α-relaxation peak and the microscopic vibration (boson) peak—often becomes exceedingly small with decreasing the temperature (see Fig. 12b), which also makes this spectral range especially prone to artifacts due to imperfect technique. In practice, therefore, a useful spectroscopic setup to study deeply supercooled liquids is to be equipped with a low-noise photon counting detector, such as a high-efficiency photon multiplier tube

---

[1]The long-range nature of the Coulomb forces in dipolar systems implies, however, the existence of collective excitations, so-called *dipolarons* [56], that are analogous to the plasma oscillations (plasmons). Such excitations were indeed found in computer simulations [57]. However, the permittivities of the laboratory liquids are usually too small to render this effect important [29].

or an avalanche diode. Overall, no single spectroscopic method meets such demanding requirements, which is why one commonly combines two or more techniques to cover a wider frequency range with an adequate S/N. Individual spectroscopic techniques that find common use are briefly described below.

1. Conventional scanning grating monochromators, such as Raman (scattered light) spectrometers, offer frequency resolution as high as ~5 GHz with the useful frequency range starting at ~50–100 GHz. The classical design combines two identical monochromators that are simultaneously scanned in the additive dispersion mode, thus doubling the resolution. The halves are normally separated by an optical spatial filter that blocks stray light and increases the spectral contrast by several orders. A typical example is the double-grating monochromator U1000 from Jobin-Yvon. Such monochromators can be equipped with a multichannel CCD detector for higher detection efficiency; however, such designs are usually inferior in terms of the spectral contrast and thus are of lesser use for the purposes discussed here. It is also possible to significantly increase the resolution by increasing the optical base of the monochromator to 2 m or more and using specially designed gratings working in high diffraction orders, such as in SOPRA DMDP2000 monochromator [58]. The increased resolution comes, however, at the cost of a significantly decreased optical throughput. Such instruments are therefore only useful for strongly scattering media. Fourier transform spectrometers equipped with a scanning Michelson interferometer are also of limited utility in the present context, mainly because of their rather poor contrast.

2. A much higher spectral resolution that is required below ~50 GHz can be achieved by employing interferometric techniques. The basic element is a Fabry–Perot interferometer (FPI) composed of two flat parallel mirror plates separated by a distance L. FPI transmits collimated light along its axis with the wavelength $\lambda = 2nL/m$ (n is the refractive index of the medium between the plates, m an integer interference order), according to the Airy function:

$$T\left(1 + (4F^2/\pi^2)\sin^2\left(\frac{2\pi v}{c}nL\right)\right)^{-1} \qquad (14)$$

where T is the transmission, F is the interferometer *finesse* that is determined by mirror quality and reflectivity, and v is the light frequency. Spectral tuning can thus be achieved by varying the optical length nL, either by changing n (e.g., through changing the ambient gas pressure) or by scanning L within about a micrometer with piezoelectric actuators. The frequency response of FPI, (i.e., its instrumental function) given by the Airy function of Eq. (14), is shown in Fig. 3a. It consists of sharp equally spaced peaks of width δv. The frequency difference between adjacent transmission peaks is known as the free spectral range FSR $= c/(2L)$ (c is the speed of light). The frequency resolution is then $\delta v =$ FSR$/F$. Currently attainable mirror quality limits F to ~100, which translates,

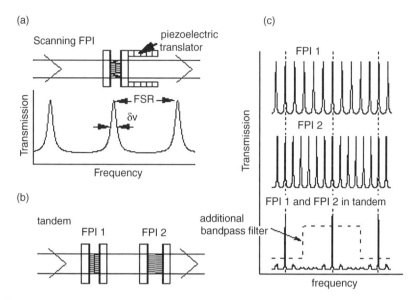

Figure 3. Fabry–Perot interferometer (FPI). (a) Single FPI and its transmission. (b) Two FPIs in tandem. (c) Transmission of the tandem.

for a given plate separation L, into a spectral range of approximately $3/L$ to $150/L$ GHz (L in mm). If the plate separation is varied within the technically feasible range of approximately 0.1–50 mm, FPI is capable to cover the spectral range of approximately 60 MHz to 1500 GHz. The practical utility of a scanning FPI for continuous spectral sources is, however, severely limited by its multipeaked transmission and the narrow free spectral range between the different interference orders (see Fig. 3a), which leads to so-called spectral folding, whereby spurious additional intensity appears in the working interference order because of the transmission in other orders. The total bandwidth of the incoming radiation should therefore be limited to one FSR, which is not feasible in practice. The shortcomings of a single FPI can be greatly circumvented by combining two FPIs with slightly different spacing in series and scanning them synchronously (Fig. 3b). The tandem now transmits only at frequencies where transmission peaks of individual FPIs coincide (Fig. 3c). If the difference in L between the interferometers is small, then the frequency separation of the resulting transmission peaks (the "effective" FSR) is large, much larger than the individual FSRs. Small features that remain between the main transmission peaks are of little concern, because they are orders of magnitude weaker than the main peaks (in Fig. 3c they are greatly exaggerated). To avoid "spectral folding," the total bandpass still has to be limited to the "effective" FSR by an additional filter.[2] Now, however, this can be readily achieved with dielectric multilayer filters that are available with FWHM as narrow as 100 GHz and a peak transmission of 50–70%.

A practical interferometer design has to address a number of issues, of which the most important are (a) maintaining the mirror parallelness and stability of the scan, (b) possibility to change the mirror separation, and (c) circumventing the "spectral folding" effect. The most useful commercial device that has been in wide use in the last two decades is the scanning multipass tandem interferometer.[3] It features two simultaneously scanned FPIs with slightly different L in series with a prism filter, a variable 0.1 to 30 mm plate separation, a three-pass operation to improve the resolution, and a unique vibrationally isolated scanning stage. The practically attainable spectral range spans from approximately 200 MHz to 1000 GHz, which, however, can only be achieved by using several instrument settings. With the optical throughput of $\sim$30% and a low-noise photon counting detector, it is perhaps the most efficient spectroscopic system for the mentioned spectral range to date. Importantly, its spectral range partially overlaps with that of the grating spectrometers, so that the two techniques can be conveniently combined. Scanning FPIs with confocal mirrors and larger plate separations, up to tens of centimeters, provide much better resolution up to about 10 MHz. Their use in light scattering spectroscopy is, however, fairly limited due to low sensitivity.

Most of the frequency-domain light scattering data of supercooled liquids are obtained by the two spectroscopic techniques discussed above. As an example, Fig. 4 shows partial spectra of glycerol obtained with a grating monochromator and a tandem FPI, which are combined into a single composite spectrum that extends from $\sim$400 MHz some five decades up in frequency. We will therefore be mostly concerned with these techniques in the following.

3. Further substantial increase of the spectral resolution can be achieved using optical heterodyne, or light beating techniques. These operate in much the same way as heterodyne radio receivers, whereby the scattered light is mixed on a nonlinear detector (typically a photodiode) with a reference light derived from the same laser excitation source. The photodiode output current at the light beating difference frequency is then examined with a spectrum analyzer [62]. Using a single-frequency laser, the method is capable of covering the frequency range up to gigahertz, thereby reaching the

---

[2]Spectral artifacts, arising in the absence of such a filter, were discussed by several authors [59–61].
[3]Available from JRS Scientific Instruments (Dr. John R. Sandercock, Zwillikon, Switzerland).

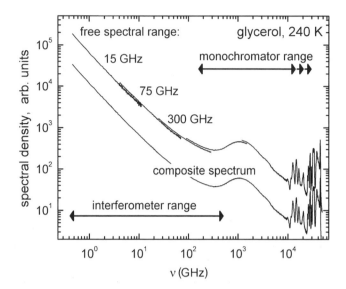

Figure 4. Splicing partial interferometer and monochromator spectra of glycerol. Interferometer spectra, obtained with three different instrument settings (FSR = 15, 75, 300 GHz), together cover the range 0.4 GHz through 300 GHz. Monochromator spectra, obtained with two different instrument settings, cover 100 GHz through 50 THz. Merging the partial spectra produces a composite spectrum covering some five decades in frequency. (Adapted from Ref. 64.)

interferometer range, with a frequency resolution as good as ~1 kHz [63]. Unfortunately the low sensitivity of the method severely limits its utility.

4. Slow fluctuations of the scattered field occurring on time scales of up to thousands of seconds can be detected directly in the time domain using the so-called photon correlation spectroscopy (PCS). In a PCS setup, the fluctuating *total* scattered intensity is detected with a fast photon counter without any frequency selection. The detector output is then digitally autocorrelated, using a dedicated digital processor. The correlator output is the normalized light scattering intensity autocorrelation function, which under certain conditions can be transformed into the normalized electric field correlation function, whose Fourier transform is the usual power spectrum. A typical state-of-the-art example, the ALV-5000 correlator,[4] provides up to 288 approximately log-spaced correlation channels with lag times from 0 to 3.2×10^3 s and an initial time resolution of ~10 ns. In the frequency domain, this corresponds to the millihertz through 100-MHz range. In practice, however, the range is more limited.

---

[4]Available from ALV-GmbH, Germany. Alternative design available from Brookhaven Instruments Corporation, New York.

Significantly, the short time limit is usually reduced to microseconds by the attainable signal intensity. This technique has certain specific limitations, for example, it has rather limited utility for investigations of nonergodic samples. Applied to supercooled liquids, PCS has been mostly used to detect the main α-relaxation at temperatures close to T_g, where it occurs on the time scale from milliseconds through hundreds of seconds. Using PSC to access weaker intermediate relaxation features is possible [65], yet seldom utilized, since the required high data quality is not easily achieved.

5. Finally, we briefly describe time-domain optical spectroscopic techniques that are not based on spontaneous scattering but rather measure response to external perturbing electric field, utilizing the Kerr effect. In classical Kerr-effect relaxometry a strong electric field, applied in stepwise manner, produces transient birefringence that is related to molecular reorientations and whose decay is detected with a probe light beam [34]. This method is well-suited for long relaxation times typical of supercooled liquids, and it was used in the past to compile a substantial body of relaxation time data— for example, by Williams and co-workers [35,66]. Currently it is not widely used, perhaps partly because the static electric field interacts with molecules through both their anisotropic polarizability *and* the dipole moment, the two effects being impossible to disentangle. As a modern extension of the method, one uses the electric field of light to orient the molecules in an optical pump-probe experiment, thus utilizing the nonlinear *optical* Kerr effect (OKE) [67]. In an OKE experiment, one detects transient anisotropic change of the refractive index (birefringence) induced by a short intense light pulse that partly orients the molecules along its electric field through coupling with the anisotropic part of the molecular polarizability tensor. The magnitude of the decaying birefringence after the end of the excitation pulse is determined as a function of time, using a delayed probe pulse and a polarization-selective detector. If the excitation pulse is sufficiently short, say it is approximately a δ-function, then one essentially determines the pulse-response susceptibility function $\chi(t)$ that is the Fourier conjugate of the frequency-domain susceptibility of Eq. (13) (for the theory, see, e.g., Ref. 68). Using the excitation and probe pulses as short as $\sim$50 fs, the useful time range of the method starts at $\sim$300 fs. The longest measurable decay times are limited to microseconds by the vanishing signal, so that the method can, under favorable conditions, cover some six decades in time [69,70]. While conceptually straightforward, OKE spectroscopy is far from being simple in practice. For that reason, this powerful method, particularly as applied to supercooled liquids, is only practiced in a few laboratories worldwide.[5]

---

[5]In Europe, for instance, in the European Laboratory for Nonlinear Spectroscopy (LENS) in Florence.

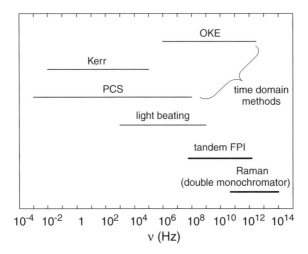

Figure 5. Frequency scales of the optical spectroscopic techniques discussed in the text. The most widely used Raman and interferometric methods are shown in bold.

Figure 5 presents the frequency ranges that, by optimistic estimates, can be covered by the optical techniques discussed above. We have intentionally left out of the discussion several other optical spectroscopic methods, such as coherent anti-Stokes resonant scattering (CARS), or induced transient grating (TG) spectroscopy, mainly because they are rarely used in the present context and/or are much less straightforward to interpret.

D. Nuclear Magnetic Resonance Spectroscopy

A number of nuclear magnetic resonance (NMR) textbooks and review articles that focus on the dynamics in supercooled liquids comprised of organic molecules [2,11,12,15], on polymer specific dynamics [71–75], and on ionic or inorganic glasses [76–79] exist. In these contributions, the theoretical background of NMR techniques and models of molecular motion have been comprehensively discussed. Therefore, we curtail the theoretical part and concentrate on selected NMR techniques applied most frequently to the investigation of molecular glass formers.

1. NMR Time Windows and Relevant Spin–Lattice Interactions

Investigating molecular dynamics using NMR, in contrast to DS and LS, involves the application of several conceptually quite different techniques. For example, in spin–lattice relaxation studies one is concerned with familiar time correlation functions that are probed as spectral density point by point (Section II.D.2). In the case of line-shape analysis, usually a two-pulse echo sequence is applied, and the

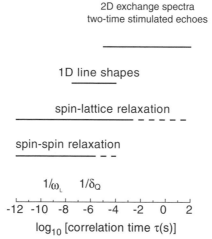

Figure 6. Time windows of different NMR techniques; Larmor frequency ω_L and coupling constant δ typical for ^{2}H NMR are indicated. (Adapted from Ref. 11.)

resulting signal reflects the evolution of the dynamics during the pulse delays (Section II.D.2). Stimulated echo techniques allow one to directly probe different kinds of reorientational correlation functions, and two-dimensional spectra taken at successive times provide a direct visualization of the evolution of reorientational dynamics of an ensemble of molecules (Section II.D.3). Even higher-dimensional NMR techniques were developed in order to gain a deeper understanding of the non-Markovian nature of molecular reorientations near the glass transition.

The dynamic window of a given NMR technique is in many cases rather narrow, but combining several techniques allows one to almost completely cover the glass transition time scale. Figure 6 shows time windows of the major NMR techniques, as applied to the study of molecular reorientation dynamics, in the most often utilized case of the ^{2}H nucleus. Two important reference frequencies exist: The Larmor frequency ω_L determines the sensitivity of spin–lattice relaxation experiments, while the coupling constant δ_Q determines the time window of line-shape experiments. ^{2}H NMR, as well as ^{31}P and ^{13}C NMR, in most cases determines single-particle reorientational dynamics. This is an important difference from DS and LS, which access collective molecular properties.

The important nuclear spin interaction in the context of probing molecular reorientation is provided by the quadrupolar interaction (Q) or the chemical shift anisotropy (CSA), and we assume that the spin system is prepared by, for example, isotopic labeling in such a way that only a single interaction is relevant. The ubiquitous presence of dipolar broadening is assumed to be small. For most NMR experiments it suffices to consider the secular part of the

Hamiltonian determined by terms that commute with the Zeeman Hamiltonian $H_z = -\omega_L I_z$, where ω_L denotes the Larmor frequency. This secular Hamiltonian can be written in the form $H = T_{20}^\lambda \omega_\lambda$ [72,80], with the "NMR frequency" being

$$\omega_\lambda = \frac{1}{2}\delta_\lambda(3\cos^2\theta_\lambda - 1) \tag{15}$$

T_{20}^λ contains spin operators that are not specified here, and δ_λ is the anisotropy parameter of the interaction tensor. The latter is assumed to be axially symmetric, which is in most cases a good approximation for the quadrupolar interaction tensor of ^2H in organic compounds. The angle θ_λ specifies the orientation of the magnetic field with respect to the unique axis of the tensor. For ^2H NMR in molecular systems, this is the direction of the chemical bond. Thus, the NMR frequencies reflect orientations of the molecules, provided that the latter are rigid. In powder samples with immobile molecules, this leads to an inhomogeneous distribution of NMR frequencies in the form of characteristic powder spectra, in the case of ^2H NMR the famous Pake spectrum [81]. Usually, the NMR spectrum is obtained by Fourier transforming the time signal obtained after a single pulse excitation, the so-called free induction decay (FID). However, due to a broad spectrum in the solid state, an echo-pulse sequence has to be applied in most cases (cf. Section II.D.2).

2. Spin–Lattice Relaxation and Line-Shape Analysis

One of the classical NMR methods used to determine molecular correlation times is provided by spin–lattice relaxation experiments. The spin–lattice relaxation rate $1/T_1$ is determined by transitions among the Zeeman levels. For a liquid, the expression for the spin-lattice relaxation rate [81] is

$$\frac{1}{T_1} = K_\lambda[S_2(\omega_L) + 4S_2(2\omega_L)] \tag{16}$$

Here, K_λ is a constant related to the NMR coupling constant, S_2 is the spectral density associated with the second-rank orientational correlation function $g_2(t)$. If $g_2(t)$ is an exponential function, such as in rotational diffusion, Eq. (16) reduces to the famous Bloembergen–Purcell–Pound (BPP) expression for the spin–lattice relaxation rate [81,82].

In supercooled liquids the BPP expression usually fails to reproduce the observed spin–lattice relaxation times. Instead of a single correlation time τ_2, it is found that a distribution of correlation times $G(\ln\tau_2)$ exists. Indication for a distribution $G(\ln\tau_2)$ resulting from a superposition of subensembles with different τ_2 were reported for supercooled liquids close to T_g [11,349,350]. This implies the existence of a distribution of spin-lattice relaxation rates. Correspondingly, the relaxation function $\phi_M(t)$ of the z-magnetization is expected to decay nonexponentially. In deuterated supercooled liquids, such a nonexponential decay is observed near and below T_g, since there the relaxation rate is governed by secondary processes faster than the α-process, and the latter is not sufficiently fast to average the relaxation [83,84]. Then, the mean powder-averaged relaxation rate $\langle 1/T_1 \rangle$ is given by the initial slope of $\phi_M(t)$. At higher temperatures in the supercooled liquid, one usually has a situation in which an effective exchange mechanism leads to an exponential decay of $\phi_M(t)$. This exchange mechanism is provided by the structural relaxation itself. In other words, during T_1 every spin has "seen" all correlation times τ_2 and therefore all spins relax with a mean relaxation rate given by substituting in Eq. (16) a spectral density $S_2(\omega)$ described by an appropriate distribution $G(\ln \tau_2)$.

Usually, spin–lattice relaxation experiments are performed at one or a few magnetic fields. The spectral density can thus be determined at only a few Larmor frequencies, so that a detailed analysis of its temperature variation is not possible. Here, an analysis of the spin–spin relaxation times, T_2, can provide further information about the spectral density, since $T_2^{-1} \sim S_2(\omega = 0)$. Often, the Cole–Davidson distribution $G_{CD}(\ln \tau_2)$ [34] is chosen to interpolate the relaxation around the maximum. However, one has to keep in mind that the spectral density close to T_g contains additional contributions from secondary relaxations, such as the excess wing and/or the β-process discussed in the following sections. In Section IV.C we give an example of a quantitative description of $T_1(T)$ at $T \geq T_g$ obtained by approximating the spectral density $S_2(\omega)$ using dielectric data.

Upon cooling, when the correlation time reaches the limit $\tau_\alpha \delta_\lambda \cong 1$, the central Lorentzian line typical of a liquid ($\delta_\lambda \tau_\alpha \ll 1$) broadens and eventually transforms into a powder spectrum. These solid-state NMR spectra are usually broad ($\cong 100$ kHz). The corresponding measured FID time signals are therefore short (microsecond range) and thus partly lost due to the dead time of the receiver system. This can be circumvented by using an echo technique—for example, applying two pulses separated by a delay t_p (cf. Fig. 7). After the first pulse, the spin system evolves under the relevant interaction (Q or CSA) until a suitable second pulse at time $t = t_p$ reverses the time evolution between the first and the second pulse, leading to the formation of an echo at $t = 2t_p$, and an undistorted spectrum is obtained from Fourier transformation of the time signal at $t \geq 2t_p$. ^{2}H NMR spectra can be measured with the solid-echo pulse sequence, $(\pi/2)_x - t_p - (\pi/2)_y$. In the case of CSA interaction a Hahn-echo sequence $(\pi/2)_x - t_p - (\pi)_y$ can be utilized. Neglecting relaxation effects, the resulting ^{2}H NMR time signal at $t \geq 2t_p$ can be written as [85]

$$I_{tp}(t) = \langle \cos[\Phi(0, t_p) - \Phi(t_p, 2t_p) - \Phi(2t_p, t)] \rangle \qquad (17)$$

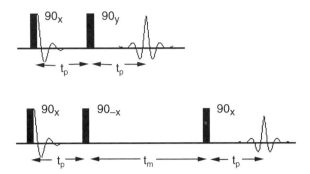

Figure 7. Typical pulse sequences applied for obtaining an ^{2}H NMR solid echo (top) and a stimulated echo (bottom).

where the phases Φ are calculated according to

$$\Phi(t_1, t_2) = \int\limits_{t_1}^{t_2} \omega_Q(t)dt \qquad (18)$$

The angular brackets stand for an ensemble and powder average. If molecular dynamics are absent, ω_Q is time-independent, and the signal can be rewritten as $I_{tp}(t) \propto \langle \cos[(t - 2t_p)\omega_Q] \rangle$. Then, the signal is refocused at $t = 2t_p$, and the time evolution at $t \geq 2t_p$ reflects the FID not depending on the interpulse delay. On the other hand, molecular reorientation during the pulse sequence—that is, dynamics with correlation times in the microsecond range—leads to an attenuation of the echo height and to deviations from the rigid-lattice spectrum (spectrum of immobile molecules) so that analysis of the line-shape may reveal details about the type of the motion. As will be demonstrated for the β-process (cf. Section V.B), the interpulse delay t_p is of particular importance in studies of highly restricted motions since variation of t_p allows one to adjust the spatial resolution of the experiment.

3. Stimulated Echo Experiments and Two-Dimensional NMR

Experiments described in this section are suited to investigate ultraslow motion with correlation times in the millisecond-to-second range. Here, the NMR spectra are given by their rigid-lattice limit and one correlates the probability to find given NMR frequencies at two different times separated by the so-called mixing time t_m [11,72]. A two-dimensional (2D) spectrum results, being a function of two NMR frequencies at $t = 0$ and $t = t_m$, respectively. Since the NMR frequency reflects the orientation of the molecule, 2D spectra provide a visual representation of the reorientational process. Time- and frequency-domain varieties of the technique exist. We start with discussing the time domain or, equivalently, stimulated-echo experiments, and we consider the three-pulse sequence $(\pi/2)_x - t_p - (\pi/2)_{-x} - t_m - (\pi/2)_x$, where the various delays are chosen according to $t_p \ll t_m \approx \tau \ll T_1$ (cf. Fig. 7). Here, a stimulated echo is created at a time t_p after the third pulse. Evaluating the height of the stimulated echo for different mixing times t_m, but for a constant evolution time t_p, the following correlation function is probed.

$$C_{\cos}(t_m; t_p)$$
$$= \langle \cos[\omega_\lambda(t = 0)t_p] \cdot \cos[\omega_\lambda(t = t_m)t_p] \rangle / \langle \cos[\omega_\lambda(t = 0)t_p] \cdot \cos[\omega_\lambda(t = 0)t_p] \rangle \tag{19}$$

The NMR frequencies at two times separated by the time t_m, and thus the corresponding orientations [cf. Eq. (15)] are correlated via cosine functions. The correlation function $C_{\sin}(t_m; t_p)$, where the cosine functions are replaced by sine functions, may also be accessible modifying pulse lengths and pulse phases in an appropriate way. This is possible for both CSA and Q interactions. Rotational jumps of the molecules during the mixing time t_m lead to $\omega_\lambda(0) \neq \omega_\lambda(t_m)$, and hence to a decay of $C_{\cos,\sin}(t_m; t_p)$. Therefore, these correlation functions provide access to the details of the molecular reorientation dynamics.

Two limiting cases of $C_{\cos,\sin}(t_m; t_p)$ can be distinguished. In the limit $t_p \to 0$ it reduces to the second-rank orientational correlation function:

$$C_{\sin}(t_m; t_p \to 0) \propto \langle \omega_\lambda(0) \cdot \omega_\lambda(t_m) \rangle \propto g_2(t) \tag{20}$$

thus allowing to determine the time constant and the stretching parameter of the α-relaxation part of the correlation function $g_2(t)$ close to T_g, as will be shown in Section IV.C. The second limit $t_p \to \infty$ provides the so-called angular jump correlation function $C_{AJ}(t)$. This function is supposed to decay to zero with any single elementary jump, and thus it reflects the fraction of the molecules that have not jumped during the time t [11]. It is the possibility to measure the crossover of $C_{\cos,\sin}(t_m; t_p)$ from the limit $t_p \to 0$ to the limit $t_p \to \infty$ by varying t_p that renders the analysis of the stimulated echo correlation functions especially powerful in accessing details of the molecular reorientation dynamics [11,86,87]. The resulting time constants exhibit characteristic t_p dependences, which are fingerprints of the geometry of the motion. In addition, analyzing the limit $C_{\sin}(t_m \to \infty; t_p)$ provides information on the overall geometry of the reorientation and bears a certain analogy with the determination of the q-dependence of the incoherent structure factor [88,89].

A simple interpretation of the stimulated echo experiments performed in highly viscous liquids is possible by using models developed by Ivanov [90,91] and Anderson [92], in which the molecular reorientation is considered as a stochastic process proceeding via finite angular jumps with magnitude α_J and negligible duration [11]. Since inertial, as well as quantum mechanical, effects can be neglected in viscous liquids, the motion can be modeled as a continuous random walk in computer simulations [86,93]. A large number of trajectories, describing the molecular orientation as a function of time, is generated in such simulation. Based on these trajectories, one calculates the time dependences of the NMR frequencies, which enables to compute the results of NMR experiments as ensemble averages.

In the limit of random jump reorientation, there is no t_p dependence of the correlation function $C_{\cos,\sin}(t_m; t_p)$. The orientation of a molecule changes then by sudden jumps of random angular magnitude, so that the orientational correlation functions $g_l(t)$ coincide with the angular jump correlation function $C_{AJ}(t)$; that is, they are independent of rank l, in particular $\tau_1 = \tau_2$ [72,94]. In contrast, a reorientation proceeding via small jumps leads to pronounced t_p dependence. In this case, many elementary α_J-steps are needed for a decay of $g_2(t)$, and one finds $\tau_1/\tau_2 = 3\cos^2(\alpha_J/2)$. The rotational diffusion model [81,95] represents the limit $\alpha_J \to 0$ yielding the well-known result $\tau_1/\tau_2 = 3$. Assuming a single jump angle α_J the ratio of the time constants observed at $t_p \to 0$ and $t_p \to \infty$, namely τ_2 and τ_{AJ}, respectively, provides the jump angle [92],

$$\alpha_J = \arcsin[(2\tau_{AJ}/3\tau_2)^{1/2}] \qquad (21)$$

Thus, the stimulated echo technique allows one to estimate the elementary angular jump angle α_J. For selected α_J, the crossover in the stimulated echo correlation function, observed between the $t_p \to 0$ to $t_p \to \infty$ limits in a simulation, is displayed in Fig. 8.

Regarding the α-process in supercooled liquids close to T_g the situation is more involved since in addition to a distribution of correlation times also a distribution of elementary angles is expected (see Section IV.C.3).

Reorientational dynamics that involve only small angular excursions restricted in amplitude to $\ll 2\pi$, such as secondary relaxation processes in glasses, give rise to only a relatively small decay of the orientational correlation functions $g_l(t)$ and are therefore difficult to detect. The relative magnitude of the correlation decay due to small-angle processes increases, however, with the rank l of the correlation function. In other words, the sensitivity of the correlation function to small-angle processes increases with l. It turns out that similar increases in sensitivity can be achieved in a stimulated echo experiment by increasing t_p. Indeed, one can show that by expanding the correlation function of the stimulated echo experiment [Eq. (19)] in powers of ω_λ, higher rank functions appear in the expansion if t_p is long [12]. It was by exploiting this possibility that important information on β-processes in glasses was obtained in recent NMR experiments [93,96,97] (see Section V.B).

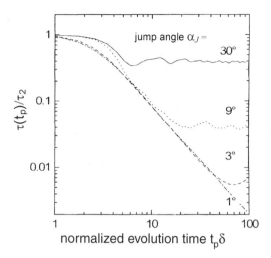

Figure 8. Double logarithmic plot of the ratio of the normalized correlation time $\tau(t_p)/\tau_2$ as a function of the normalized evolution time $t_p\delta$, determined from random walk simulations for different jump angle α_J. (From Ref. 12; cf. also Ref. 189.)

While the merits of the stimulated-echo technique lie in the resolution of small jump angles, measurement of two-dimensional (2D) NMR spectra is suited to the study of large angular displacements. For example, jumps by discrete well-defined angles lead to characteristic elliptical off-diagonal intensity, which yields straightforward access to the jump angle. The method found numerous applications to study rotor-phases, polymers, or side-group motion [72]. The pulse sequence for a 2D exchange spectrum is very similar to the one used for the measurement of stimulated echoes [72]. An important feature of the spectra $S(\omega_1, \omega_2; t_m)$ is their direct relation to the conditional probability $P(\omega_2, t_m|\omega_1)$ to find a frequency ω_2 at time t_m if it was ω_1 at time $t_m = 0$, $S(\omega_1, \omega_2; t_m) \propto P(\omega_2, t_m|\omega_1)$. Because the NMR frequencies are related to the molecular orientation, 2D spectra provide a direct visualization of the reorientations involved in a dynamical process in terms of the corresponding $P(\Omega_2, t_m|\Omega_1)$ [cf. Eq. (2), Section II.A].

III. SOME COMMENTS ON THEORETICAL APPROACHES TO THE GLASS TRANSITION PHENOMENON

A. General Remarks

The glass transition phenomenon is an unsolved problem of condensed matter physics. Most approaches agree that the glass transition temperature T_g—though accompanied by more or less discontinuous changes of, for example, expansion coefficient or heat capacity—is not of major interest due to its cooling rate dependence. Instead, it was argued that a dynamic crossover above T_g or a hidden phase transition below T_g are relevant, the latter being obscured by the divergence of the structural relaxation time when approaching this point. The idea of a hidden transition first appeared from extrapolation of the liquid entropy to low temperatures. Then, the excess entropy ΔS_{ex}—the entropy of the liquid in excess to that of the crystal—may vanish at a temperature $T_K > 0$ [98]. Since an entropy of a disordered state smaller than that of a crystal appears unphysical, a static phase transition at the so-called Kauzmann temperature T_K was proposed. However, the conclusion is not fully compelling [20,99].

The Kauzmann temperature plays an important role in the most widely applied phenomenological theories, namely the configurational entropy [100] and the free-volume theories [101,102]. In the entropy theory, the excess entropy ΔS_{ex} obtained from thermodynamic studies is related to the temperature dependence of the structural relaxation time τ_α. A similar relation is derived in the free-volume theory, connecting τ_α with the excess free volume ΔV_{ex}. In both cases, the excess quantity becomes zero at a distinguished temperature where, as a consequence, $\tau_\alpha(T)$ diverges. Although consistent data analyses are sometimes possible, the predictive power of these phenomenological theories is limited. In particular, no predictions about the evolution of relaxation spectra are made. Essentially, they are theories for the temperature dependence of $\tau_\alpha(T)$ and $\eta(T)$.

In contrast to the approaches described above, the mode coupling theory (MCT) follows a microscopic approach starting from a classical N-body problem. The approach was developed by Kawasaki to describe the critical slowing-down when approaching the critical point of the liquid [103]. In 1984, Bengtzelius, Götze, and Sjölander [104] and Leutheusser [105] extended the theory to dense fluids—that is, to supercooled liquids. Thus, the theory starts from the liquid side to understand the crossover to "glassy dynamics." Many details were elaborated by Götze and co-workers [16,17]. Later the approach was applied to molecular liquids with rotational degrees of freedom [23,106–108]. Reviews on the topic are found in works by Schilling [18,20], Götze [19], Cummins [109], and Das [21]. MCT provides predictions for the temperature evolution of the dynamic structure factor $S(q,t)$. The relevant time window is the gigahertz range, where the "glassy dynamics" separates from the microscopic tetrahertz dynamics. This frequency range can be easily accessed by neutron and light scattering, and also by dielectric spectroscopy, as will be demonstrated in Section IV. The most direct support comes from molecular dynamics simulations [110–117] and from experiments on colloidal systems [19,118,119].

Before we go into some details of MCT, we briefly mention that there exists another microscopic theory of the glass transition phenomenon, the replica theory [120,121], which is inspired by spin glass theory [122] and which lends some justification to the configurational entropy theory. However, the details of the evolution of the dynamic susceptibility are not worked out. An explanation of vitrification processes was also offered in the context of the theory of frustration limited domains [123]. Due to the dominance of "locally preferred structures" that are incompatible with long-range order, a frustration phenomenon occurs. Finally, we remark that light is shed on the glass transition phenomenon by the potential-energy landscape (PEL) approaches [124–128]. In this formalism, the high-dimensional vector of all particle coordinates is considered as a point moving on the potential energy surface. Such a description is useful at sufficiently low temperatures, where the time evolution of the system can be decomposed into vibrations about local minima of the PEL and infrequent transitions between these minima. The shape of the PEL was reported to depend on the fragility of the liquid— that is, on the degree of deviation from Arrhenius behavior. For fragile liquids, the local minima of the PEL are grouped into so-called metabasins. In view of this organization, relations between the features of the PEL and the relaxation processes in supercooled liquids were discussed. Molecular dynamics simulations gave evidence that transitions between metabasins are involved in the α-process [129].

B.   Mode Coupling Theory

In MCT, the nonergodicity parameter $f_q(T)$ is a crucial quantity. It describes the long-time limit of the density correlation function or the so-called intermediate scattering function, in the theory denoted $\phi(q, t)$. In its idealized version, MCT predicts a discontinuous change at a critical temperature T_c.

$$
\begin{aligned}
f_q(T) &= 0, \qquad T > T_c \\
f_q(T) &> 0, \qquad T < T_c
\end{aligned}
\tag{22}
$$

Whereas the correlation function $\phi(q, t)$ decays to zero in the liquid state, this is no longer the case below T_c, where the system becomes nonergodic. Such a scenario results from a generalized oscillator equation with nonlinear damping. The slowing-down of the molecular dynamics creates an enhanced damping, which in turn slows down the correlation function and so on. The density is taken as a control parameter. In Fig. 9a, f (the q-dependence is omitted in the most simple approach) is displayed as a function of the control parameter v [20]. A control parameter $v < v_c$ yields $f_0(v) = 0$, whereas a finite value $f_1(v)$ is observed at $v \geq v_c$. In other words, the nonergodicity parameter f changes discontinuously, although structural changes remain continuous and smooth. This constitutes the paradigm of the glass transition as a *dynamic* phase transition, without diverging static susceptibilities. It turns out that fluctuations with q-values around the first maximum of the structure factor are most important for driving the dynamic transition. Experimentally, $f_q(T)$ can be obtained from the temperature dependence of the Debye–Waller factor (cf. Fig. 42).

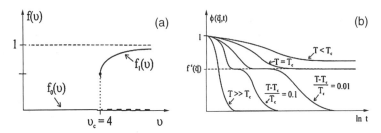

Figure 9. (a) Long-time plateau $f(v)$ of the density correlation function $\phi(\mathbf{q}, t)$ above $(f_0(v))$ and below $(f_1(v))$ the critical value of the control parameter, v_c. (b) Correlation functions at different temperatures with respect to the critical temperature T_c. From a one-step function, a two-step function emerges while cooling; the plateau height $f^c(q)$ is the relaxation strength of the α-process. (Adapted from Ref. 20.)

The corresponding correlation functions are sketched for characteristic temperatures in Fig. 9b. The critical temperature T_c marks the crossover from an ergodic state (liquid) to a nonergodic state (glass). In the liquid state, a two-step process describes the decay of the correlation function, where f is the fraction relaxed by the slow process (α-process) and $1 - f$ is the part decaying due to the fast dynamics. The cage effect provides a physical interpretation of the two-step correlation function [26]. Molecules in a liquid are transiently trapped in cages formed by their neighbours. Inside the cages, the molecules are subject to fast "rattling" motion until the cage decays due to the α-process at much longer times, leading to complete relaxation. The lower the temperature, the longer the molecules are trapped; and eventually at T_c, "self-trapping" leads to a nonergodic state, the glass.

For molecular liquids, it appears that T_c lies well above T_g, and thus these systems remain ergodic even below T_c. In the so-called extended MCT, it is postulated that a phonon-assisted "hopping transport" ensures ergodicity below T_c that is observed experimentally. In other words, at $T > T_g$, the transport mechanism is proposed to change from a liquid-like $(T > T_c)$ to a solid-like behavior $(T < T_c)$. In contrast, for colloidal glasses, it appears that the so-called idealized MCT (without hopping) is sufficient to describe the glass transition [19]. Still, one expects that features of the idealized theory can be identified in the experiment. Because at long times the correlation functions actually always decay to zero at $T > T_g$, an "effective Debye–Waller factor" is defined which represents the amplitude of the α-process; this is what is meant by the nonergodicity parameter f in the following.

Beyond the schematic picture sketched above, the two-step density correlation functions emerging from the idealized theory exhibit characteristic power-law decays toward and from the plateau f. Their experimental identification and analysis allows one to determine the crossover temperature

T_c. The two power-law exponents are not independent but depend on a single parameter, the so-called critical exponent λ, which is specific for a given interaction potential (e.g., hard spheres). Actually, the interaction potential enters the MCT equations only indirectly via the structure factor $S(q)$, which fixes the nonlinear coupling in the generalized oscillator equation. It is important to note that the MCT exponents are not universal in contrast to those of second-order phase transitions. In the case of hard spheres, for example, $S(q)$ can be calculated via the Percus–Yevick approximation [26], and the full time and q-dependence of $\phi(q, t)$ were obtained. As an example, Fig. 10 shows the susceptibility spectra of the hard-sphere system at a particular q. Note that temperature cannot be defined in the hard-sphere system; instead, the packing fraction φ is used as a parameter. Above the critical packing fraction $\varphi_c (\varepsilon > 0)$, which corresponds to $T < T_c$ in systems where T exists, the α-process is absent (frozen) and only the fast dynamics is present. At $\varphi < \varphi_c$ the α-peak and the concomitant susceptibility minimum shift to lower frequencies with increasing φ, so that the closer φ is to the critical value φ_c, the better the critical decay of the fast dynamics can be identified (curve c in Fig. 10).

Close to the crossover temperature T_c, the solution of the idealized MCT equations can be expanded around the nonergodicity parameter f_c at T_c and some generic laws can be derived. These leading-order results establish the so-called asymptotic laws of MCT, which can be regarded as generic features of the MCT dynamics at $T > T_c$. Because almost all experimental tests analyzed the susceptibility spectra in this frame, we are going to briefly summarize the laws [16,17,19]. Examples of tests against experiment are found in Section IV.E.

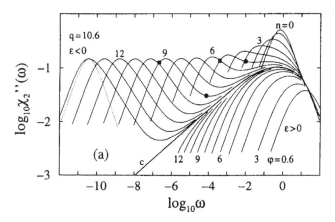

Figure 10.   Solutions of idealized MCT for the hard-sphere system. $\varepsilon = (\varphi - \varphi_c)/\varphi_c$ measures the departure of the packing fraction φ from the critical value φ_c. The solid line marked c represents the critical spectrum at φ_c. (From Ref. 109.)

The asymptotic scaling laws of MCT describe the crossover from the fast relaxation to the onset of the slow relaxation (α-process)—that is, a power-law decay of $\phi(t)$ toward the plateau with an exponent a, and another power-law decay away from the plateau with an exponent b. For the purpose of the present review, we again ignore the q-dependence.

$$\phi(t) = f_c + |\sigma|^{1/2} h(t/t_\sigma)^{-a_{MCT}}, \qquad t_0 \le t \le t_\sigma$$

$$T > T_c \quad (23)$$

$$\phi(t) = f_c - |\sigma|^{1/2} h B(t/t_\sigma)^{b_{MCT}}, \qquad t_\sigma < t \le \tau_\alpha, \quad B > 0 \quad .$$

where f_c is the temperature-independent nonergodicity parameter, h is a temperature-independent amplitude, and $t_\sigma = t_0/|\sigma|^{1/(2a_{MCT})}$, a rescaling time determined by the some microscopic time t_0. The temperature enters via $\sigma = (T_c - T)/T_c$. The exponents a and b are related through the already mentioned critical exponent parameter λ:

$$\lambda = \Gamma^2(1 - a_{MCT})/\Gamma(1 - 2a_{MCT}) = \Gamma^2(1 + b_{MCT})/\Gamma(1 + 2b_{MCT}) \qquad (24)$$

where Γ is the gamma function. Equations (23) constitute the β-scaling regime. At even longer times, $t \gg t_\sigma$, MCT predicts the so-called α-scaling; that is, the correlation function follows a master curve

$$\phi(t) = F(t/\tau_\alpha) \qquad (25)$$

This is none other than the time–temperature superposition principle. However, the exact shape of the function $F(t)$ is not a generic feature of the theory. A good approximation is provided by the Kohlrausch stretched exponential function. In the frequency domain, Eqs. (23) define the shape of the susceptibility minimum usually observed in the gigahertz range, and an interpolation formula follows:

$$\chi''(\nu) = \chi''_{min} \lfloor b_{MCT}(\nu/\nu_{min})^{a_{MCT}} + a_{MCT}(\nu/\nu_{min})^{-b_{MCT}} \rfloor/(a_{MCT} + b_{MCT}) \qquad (26)$$

where ν_{min} and χ''_{min} are the frequency and the amplitude of the minimum. Thus, the minimum has a universal shape that can be described by a master curve. In the present review we call this scaling property the *minimum scaling* (cf. Fig. 11). The exponents a_{MCT} and b_{MCT} determine the low-frequency behavior of the fast dynamics and the high-frequency part of the α-process, respectively. The temperature dependence of ν_{min}, χ''_{min} and the time scale τ_α are expected to be given by

$$\chi''_{min} \propto (T - T_c)^{1/2}$$

$$\nu_{min} \cong 1/t_\sigma \propto (T - T_c)^{1/2a_{MCT}}, \qquad T > T_c \qquad (27)$$

$$\tau_\alpha \propto (T - T_c)^{-\gamma_{MCT}}$$

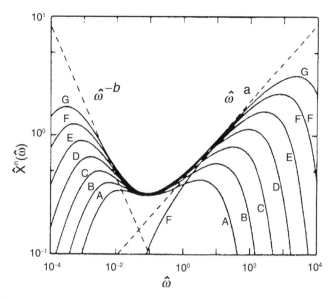

Figure 11. Example of the minimum scaling; plotted is the rescaled quantity $\hat{\chi}'' = \chi''/\chi''_{min}$ as a function of the rescaled frequency $\hat{\omega} = v/v_{min}$; letters denote different temperatures; the respective envelopes are described by two power laws with the exponents b and a. (Adapted from Ref. 130.)

The exponent γ_{MCT} is related to the exponents a_{MCT} and b_{MCT} via $\gamma_{MCT} = 1/(2a_{MCT}) + 1/(2b_{MCT})$. The crossover temperature T_c can thus be determined graphically, by using linearized plots of the experimental χ''_{min}, v_{min}, and τ_α temperature dependences. T_c marks the end of the high-temperature regime of MCT.

Below T_c the fast relaxation spectrum changes its shape. A crossover from a power law with an exponent a at high frequencies to a white noise spectrum $(a = 1)$ at low frequencies is expected. As a consequence of the appearance of this "knee" at T_c and its shifting up in frequency with decreasing the temperature, the strength of the fast relaxation (i.e., its integrated spectral density) decreases upon cooling. This is further reflected in the characteristic temperature dependence, or an anomaly, of the nonergodicity parameter f, since f is the strength of α-relaxation, and therefore $1 - f$ equals the strength of the fast relaxation. Specifically,

$$
\begin{aligned}
f(T) &= f_c, & T > T_c \\
f(T) &= f_c + h(T_c - T)^{1/2}, & T < T_c
\end{aligned}
\tag{28}
$$

This so-called square-root cusp singularity in $f(T)$ is one of the most important forecasts of the theory. For testing the validity of Eq. (28) against experiments, the nonergodicity parameter f (i.e. the strength of α-relaxation) has to be determined above as well as below T_c.

The asymptotic laws are expected to hold when the liquid approaches the critical temperature T_c from above. However, close to T_c, which is still well above T_g, a further relaxation channel (phonon-assisted hopping) is expected to become efficient, which leads to a breakdown of the scaling laws. Thus, regarding the temperature range, no clear-cut prediction for validity of the scaling laws can be given. Assuming that the reorientational dynamics are strongly coupled to the density fluctuation, the asymptotic laws were tested by several groups analyzing DS as well as LS susceptibility spectra (cf. Section IV.E).

An alternative to testing the asymptotic laws is to analyze spectra using simple so-called schematic models. Here, the full correlation function including the α-peak and microscopic dynamics is obtained as the solution of a simplified set of MCT equations that only retain the essential nonlinear coupling [17]. Because the ideal glass transition is not observed in molecular systems, the theory was extended to also include transverse current modes, or equivalently the "hopping" effects mentioned above [131,132], which restore ergodicity below T_c. Then, T_c represents a crossover temperature in the moderately supercooled liquid. Here, a change of the dynamics dominated by the cage effect to the solid-like temperature-activated dynamics is expected. However, below T_c the theory is far from complete.

IV. EXPERIMENTAL RESULTS OF MOLECULAR GLASS FORMERS AT $T > T_g$

In this section, we review the relaxation phenomena observed when a molecular liquid is cooled from above its melting point down to the glass transition temperature T_g. We start in Section IV.A with a general overview of the relaxation features as revealed in dielectric (DS) and depolarized light scattering (LS) susceptibility spectra. In Sections IV.B and IV.C, we discuss separately the evolution of the relaxation spectra in the high-temperature and in the low-temperature regime, respectively. We shall outline arguments, which put this distinction on a firm ground. In Section IV.C, we include results from NMR related to the question of the type of molecular motions involved in the main α-relaxation process close to T_g. Section IV.D discusses the temperature dependence of the nonergodicity parameter in the entire glass transition range. Although the main attention to this quantity was inspired by the mode coupling theory, its experimental determination does not require explicit reference to any theory. Then, we shall focus on tests of mode coupling theory (MCT) addressing the question whether the evolution of the susceptibility can be reproduced by this most advanced theoretical approach (Section IV.E). A conclusion is attempted in Section IV.F.

A. General Overview

1. Evolution of the Dynamic Susceptibility

Exploiting technical progress in instrumentation, it has become possible in recent years to measure the dielectric response of supercooled liquids $(T > T_g)$ in the complete dynamic range associated with the glass transition phenomenon. At least for a few paradigmatic glass formers this was realized by Lunkenheimer and co-workers [9,10] who compiled spectra covering a frequency range of $10^{-6}\,\text{Hz} < \nu < 10^{13}\,\text{Hz}$. This is demonstrated for the glass former glycerol in Fig. 12a where we added also low-temperature data [6,50]. Starting with the pioneering work of Cummins and co-workers (for reviews see Cummins et al. [5,134]), depolarized LS in the gigahertz–terahertz range, made possible by the use of tandem-Fabry–Perot interferometer [135] (cf. Section II.C), was applied to study supercooled liquids. In Fig. 12b we display LS results again for glycerol [64]. Though the spectral range covered by LS is much smaller than that of DS, qualitatively similar features are observed. Cooling the liquid from the highest temperatures, at which essentially a single peak with some high-frequency shoulder is observed, a minimum evolves between the main relaxation peak (α-process) in the gigahertz range and the peak of the vibrational band in the terahertz range. Note that the α-peak is significantly broader than that corresponding to a Debye process for which the full width at half-maximum is 1.14 decades.

With further cooling of the liquid, at a certain temperature there appears a wing with a different slope on the high-frequency side of the α-peak in the DS spectra. This spectral feature is usually called the excess wing because it contains excess spectral intensity with respect to the extrapolated high-frequency flank of the α-peak. In the LS spectra of Fig. 12b, this feature is recognized as a progressive flattening of the minimum. Reaching T_g, the α-peak frequency is close to $10^{-2}\,\text{Hz}$, which corresponds to a viscosity of about $10^{12}\,\text{Pa s}$. Thus, within a small temperature interval, the width of which depends on the fragility of the glass former (cf. Section IV.A.2), the molecular reorientation slows down from the picosecond to the second-range. From the DS spectra it is obvious that the excess wing can be described by a power law extending over several decades with an exponent γ decreasing with T. Below T_g the excess wing degenerates into what is called a nearly constant loss (NCL); that is, one observes approximately a power-law spectrum with a rather small temperature-independent exponent $\gamma = 0.1$–0.2 (cf. Section V.A). Thus, the excess wing persists in the glassy state as a kind of slow secondary process. At even lower temperatures (i.e. below, say, 30 K in Fig. 12a) the exponent γ changes its sign, which is indicative of additional relaxation processes linked to the low-temperature anomalies of glasses [50] (cf. Section V.A). The excess wing phenomenon in DS spectra was already reported

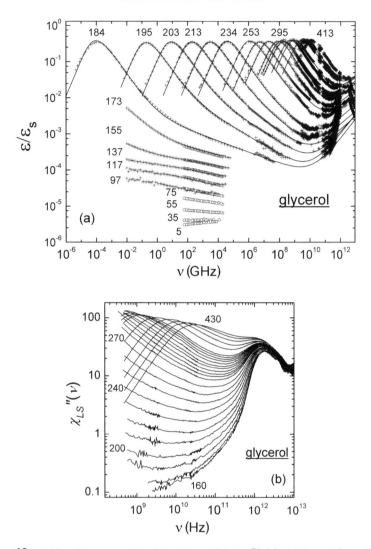

Figure 12. (a) Imaginary part of the dielectric permittivity $\varepsilon''(\omega)/\varepsilon_s$ of glycerol ($T_g = 189\,\mathrm{K}$) normalized by the static susceptibility ε_s measured down to 5 K (data compiled from Refs. 6, 9 and 136). Low-temperature data (circles) are measured by high-precision bridge [137]; numbers indicate temperature in K. Solid lines represent interpolation by using the GGE distribution of correlation times for the slow dynamics [cf. Eq. (36)] and a power-law contribution for the fast dynamics. (b) Dynamic susceptibility $\chi''_{LS}(\nu)$ of glycerol as obtained from light scattering ([64], cf. also Ref. 138.)

by Davidson and Cole in 1951 [139] and systematically characterized above T_g by several groups [6,9,10,140–144].

We note that although similar shapes of the relaxation spectra are observed in LS and DS data of Fig. 12, the peak around 1 THz is found to be significantly stronger in the LS spectra as compared to the DS spectra. The corresponding peak in the spectral density has traditionally been called the boson peak, to emphasize its vibrational character. Without going into details about various approaches which have been used to explain the ubiquitous appearance of the boson peak in LS and neutron scattering (NS) spectra, we mention that it is associated with an excess density of vibrational states, and therefore is most pronounced well below T_g where the relaxational contributions are weak. The *excess* density of vibrational states is also reflected in the anomalous temperature dependence of the heat capacity of glasses at low temperatures [7,145]. A tendency of the boson peak to be relatively stronger in nonfragile [cf. Eq. (29)] glass formers was reported [146].

The spectra of Fig. 12 exhibit the four regimes of temperature evolution sketched in the introduction. At highest temperatures, relaxation and vibrational contributions have almost merged into a single peak (fluid regime), which is, however, still wide; that is, it still exhibits some stretching. At lower temperatures, a pronounced minimum appears in the susceptibility spectra (moderately viscous liquid). Here, the corresponding correlation function has become a two-step function with a slow α-relaxation that is significantly stretched and a fast relaxation part that is essentially temperature-independent and that determines the high-frequency part of the susceptibility minimum. At lower temperatures close to T_g (highly viscous liquid), further secondary relaxation features, such as the excess wing, come into play. These features may be considered as secondary relaxation processes. Finally, in the glass below T_g "isostructural" relaxation processes are observed. Focusing on those dynamics which are characteristic of glass formers at $T > T_g$, it is convenient to further distinguish the high-temperature regime (moderately viscous regime not yet showing slow secondary relaxation processes) and the low-temperature regime (highly viscous regime accompanied by the appearance of slow secondary relaxation processes, e.g. the excess wing) of a glass former. As we shall see, this distinction can be put on a firm basis.

The emergence of the characteristic two-step correlation function with a stretched long-time tail from a simpler one-step function is usually taken to mark the onset of "glassy dynamics." It is observed in molecular dynamics (MD) studies of Lennard-Jones fluids and molecular liquids [30,112–117]; an example is shown in Fig. 13a [30]. As discussed in Section III.B, qualitatively the two-step function with a stretched long-time decay can be explained by the cage effect: Particles are transiently trapped by their neighbors and exercise some rattling motions within the cage. Such dynamics correspond to the first

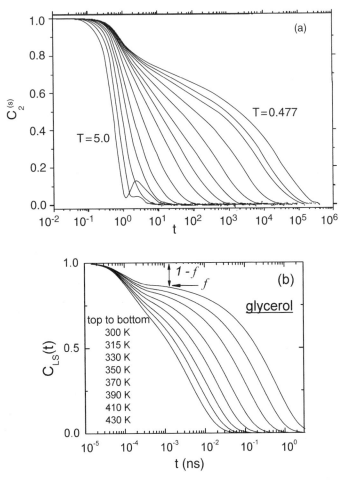

Figure 13. (a) Orientational correlation functions of second rank (self part), obtained from molecular dynamics simulations of a diatomic molecular liquid at different temperatures (From Ref. 30). (b) Fourier-transformed LS data or glycerol taken from Fig. 12b; f specifies the relaxation strength of the α-process (Adapted from Ref. 64); note that the simulation reaches much higher temperatures as compared to the experiment.

step of the correlation function. Only at much longer times a full relaxation occurs due to the decay of the cage. For comparison, Fig. 13b presents correlation functions of glycerol obtained by Fourier transforming the LS data of Fig. 12b. Indeed, quite similar decays are observed as in MD studies, although the crossover to an exponential single-step correlation function seen in the MD simulations at highest temperatures was not reached in the experimental data; instead, a virtually temperature-independent stretching is observed up to highest temperatures, and this constitutes what is called the time–temperature or equivalently the frequency–temperature superposition (FTS) principle; that is, the long-time relaxation functions can be collapsed onto a master curve, a feature also well known from NS studies [147,148]. In the case of glycerol, glassy dynamics is observed up to at least 430 K, which is some 140 K above the melting point. It turns out that it is quite difficult to reach "true liquid dynamics" experimentally [149,150]. Glassy dynamics thus appear to be present in most molecular liquids.

The inflection point f of the two-step correlation functions observed in Fig. 13 may be taken as a measure of what is called the nonergodicity parameter within MCT. It quantifies the relaxation strength of the slow degrees of freedom, that is, the α-process. Correspondingly, the quantity $1 - f$ specifies the strength of the fast dynamics including both fast relaxation processes and vibrational contributions. We emphasize that it is a merit of MCT that it focuses attention on this quantity that exhibits a nontrivial temperature dependence in the low-temperature regime as will be shown (cf. Section IV.D).

The relaxation patterns displayed in Fig. 12 are observed for several molecular glass formers. The susceptibility spectrum of such glass formers exhibits an additional power law, or the excess wing, on the high frequency side of the α-relaxation. Previously, such glass formers were called "type A" to discriminate them from glass formers that show a well-resolved secondary relaxation peak, which were denoted as "type B" systems [6]. Usually the dynamics giving rise to a secondary relaxation peak is called a β-process, and spectra of this kind are presented in Fig. 14. On cooling the liquid, the β-relaxation emerges in the supercooled state and persists below T_g, showing a more or less pronounced peak in the glass, with the temperature dependence of its time constant obeying the Arrhenius law (cf. Fig. 15). At lowest temperatures, say below 50 K (cf. Fig. 14), further relaxation features similar to those in type A glasses (cf. Fig. 12a) are observed and again attributed to the low-temperature anomalies of glasses [137]. Johari and Goldstein first systematically studied this secondary process in molecular glasses [151], and therefore it is often referred to as Johari–Goldstein (JG) β-process. It was also observed in mechanical, thermal relaxation experiments and in neutron scattering (cf. Section IV.C.2) as well as NMR studies (cf. Section IV.C.3). Since it is also found in glasses composed of rigid molecules, the β-process, at least in such cases, must be of intermolecular origin. It is believed to be an intrinsic property of the glassy state as much as the excess wing. Actually, excess wing and β-process have several common properties. The excess wing also emerges at sufficiently deep supercooling and persists in the glass as a nearly constant loss. Because both relaxation phenomena occur in liquids composed of simple rigid molecules, it seems appropriate to call both of them secondary relaxations of JG type. Note however, that sometimes both processes

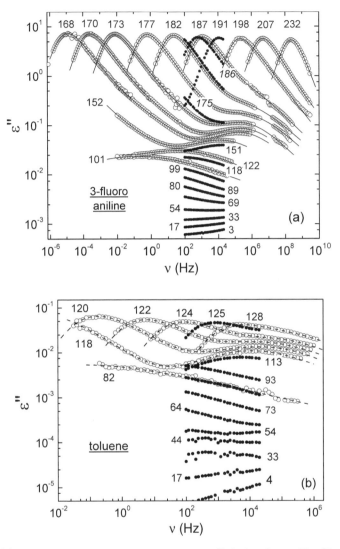

Figure 14. Imaginary part of the dielectric permittivity $\varepsilon''(\omega)$ of (a) fluoroaniline ($T_g = 173$ K) and (b) toluene ($T_g = 117$ K), both type B glass formers showing in addition to the main relaxation (α-process) a secondary relaxation peak (β-process); numbers indicate temperature in K. Unfilled symbols represent data obtained from a broad-band spectrometer [6,153]. Filled symbols represent data from a high-precision bridge [137]; interpolations for fluoroaniline (solid lines) were done by applying the GGE distribution (α-process) and a Gaussian distribution (β-process) of relaxation times [142], and these for toluene (dashed lines) were done by the gamma distribution (α-process) and a Gaussian distribution (β-process) [6] (cf. Section IV.C.2).

are observed simultaneously (cf. Fig. 14a and Fig. 35), although it is unclear what kind of molecular mechanism leads to such a behavior.

We emphasize that the discrimination of "type A" and "type B" glasses is first of all a functional definition, which allows for a simple classification of glasses with phenomenologically distinguishable relaxation behavior. It does not mean that a clear-cut assignment to each of these classes is always possible. For example, recently a debate started on whether the excess wing has to be interpreted as a submerged β-relaxation peak [152], and indeed, a gradual transition between both types of secondary processes can be found in certain systems [154–157]. Moreover, ageing experiments demonstrated that the excess wing may change its spectral shape toward a more or less resolved peak [158,159]. On the other hand, as will be demonstrated below, the excess wing, although clearly being a secondary relaxation process, shows some intriguingly universal properties, which reveal its strong coupling to the main relaxation. In addition, the excess wing is usually discernible in LS experiments, whereas the β-process only seems to be probed weakly if at all by the latter technique [65,160,161]. We shall return to these controversial issues in due course (cf. Section IV.C.2). For the sake of convenience, we will keep the distinction of types A and B and discuss the respective secondary relaxations separately. We emphasize that this distinction only concerns the relaxation behavior in the regime of the highly viscous liquid close to T_g. In the high-temperature regime, α-process and secondary relaxations have merged in most cases (cf. Fig. 15) and, as we shall show, no qualitative difference is found between the two types of glass formers (Section IV.B).

2. Time Constants and Decoupling Phenomenon

One of the major unsolved problems of the glass transition phenomenon is the description of the temperature dependence of the time constant τ_α of the primary relaxation, obtained, for example, by fitting the α-relaxation peak by an appropriate function. Without doubt, in all the cases a super-Arrhenius dependence is observed, and examples are given in Fig. 15. Within a small temperature range, τ_α increases from picoseconds in the fluid liquid to 100 s close to T_g. Although widely applied, the Vogel–Fulcher–Tammann (VFT) equation, $\ln(\tau_\alpha/\tau_0) = B/(T - T_{VFT})$, is only applicable in a restricted temperature interval [162–169]. In addition to the VFT description at low temperatures, often another VFT equation with different parameters applies in the high-temperature regime. The crossover can be identified, for example, by linearizing the temperature coefficient $dlg\tau_\alpha/dT$ in the VFT equation ("Stickel analysis") [170–172]. Furthermore, the VFT equation, derivable in the context of the free-volume and configurational entropy theory (cf. Section III.A), may not constitute a unique interpolation of the data, in particular, in the low-temperature regime as this temperature interval is too small in many glass formers to allow for an unambiguous data analysis. For example, the weak temperature dependence of the apparent activation energy E of the viscous flow close to T_g can be described using "shoving model" relating E to the shear modulus $G_\infty(T)$ [173–175]. Many other expressions for $\tau_\alpha(T)$ were proposed (cf. Ref. 7), but only rarely a crossover temperature is introduced to mark different regimes of dynamics as indicated by the Stickel analysis, for example [169,170–172,176]. We note that the power-law description derived from MCT is such an example (cf. Section III.B). It is one purpose of the present review to provide strong arguments that indeed at least two temperature regimes have to be taken into account for describing $\tau_\alpha(T)$ as well as the evolution of the susceptibility shape.

Another point concerns the question of whether the time constants τ_α obtained by different techniques are similar to each other. In Fig. 15 this is checked for correlation times obtained by DS, LS, NMR, PCS, Kerr effect, and viscosity studies. Indeed, very similar time constants are observed, all over the entire glass transition temperature range. Although collective and self-correlation

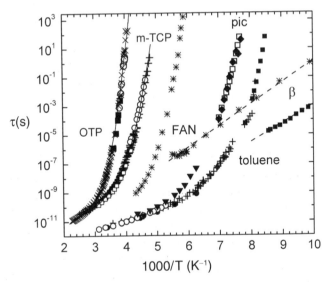

Figure 15. Time constants of the α- and β-processes of several glass formers, as determined by dielectric spectroscopy (DS), light scattering (LS), photon correlation spectroscopy (PCS), NMR, Kerr effect (KE), neutron scattering (NS), and viscosity: *o*-Terphenyl (OTP, type A): NMR (crosses, [177–179]), DS (filled squares [151]), KE (unfilled circles [66]), viscosity (solid line [164]). *m*-Tricresyl phosphate (*m*-TCP, type A): NMR (crosses [15]), LS (unfilled squares [181]), DS (circles [180]) and viscosity (line [182]). *m*-Fluoroaniline (FAN, type B): DS (stars [153]). 2-Picoline (PIC, type A): LS (unfilled circles [183]), NS (filled triangles [184]), PCS (unfilled squares [65], DS (filled diamonds, [181]). Toluene (type B): NMR (+ [11]), DS (filled squares [153]) and LS (filled circles [185]).

times are compared, along with correlation functions of different rank l, no significant differences are observed. This similarity allows one to conclude that in these molecular systems the generalized Kirkwood factors (which are a measure of the orientational cross-correlations that influence the relation between the collective and single particle correlation functions [24,25]) are of the order of unity. Moreover, since the various probes monitor the reorientation of different molecular axes, it suggests that the α-process is associated with a virtually isotropic reorientation of the molecules. Indeed, ^{2}H NMR studies [186] on differently labeled toluene showed that regarding the α-process anisotropic reorientation appears to be suppressed in the viscous regime though being present in the fluid [187]. In the case of glycerol, being a nonrigid molecule, again NMR investigations showed that in the glass transition regime the molecules behave as if they were rigid [188]. As discussed in Section II.D.3, the ratio of the single-particle relaxation times τ_1/τ_2 depends on the mechanism of molecular reorientation and varies between 1 and 3 in the limits of random jump and rotational diffusion, respectively (note, however, that no such simple relation exists for *collective* time constants). Assuming $\tau_1 = \tau_2$ as suggested by comparing DS and Kerr effect data, Williams [35] (cf. also Ref. 66) concluded that molecular reorientation proceeds via large angle jumps. However, multi-dimensional NMR experiments showed that the reorientational process close to T_g essentially involves predominantly small angle jumps (cf. Section IV.C). Thus, a paradox seems to appear in the sense that small angle motion can not explain $\tau_1 \cong \tau_2$. Assuming transient heterogeneous dynamics, Diezemann et al. [189] proposed a free-energy landscape model that allows one to explain $\tau_1 \cong \tau_2$ together with $\gamma_1 > \gamma_2$, where the exponent γ_l characterizes the stretching of the orientational correlation $g_l(t)$ (cf. below).

In contrast, the diffusion coefficient D (including tracer diffusion) may exhibit somewhat weaker temperature dependence than τ_α close to T_g and may also be affected by the probe size [190–197]. Several explanations were offered for this "decoupling phenomenon" leading to a breakdown of the Stokes–Einstein relation (for a discussion, see Refs. 198–200). For example, the difference between $D(T)$ and $\tau_{rot}(T)$ (presumably proportional to the viscosity) is attributed to different averages in the presence of a distribution of correlation times in a dynamically heterogeneous liquid. It turns out that the decoupling sets in only below a certain temperature which is close to the dynamic crossover temperature identified in the phenomenological analysis of the relaxation spectra (cf. Section IV.C). An example is presented in Fig. 16 for the glass former o-terphenyl, where the translational diffusion coefficient is compared with the viscosity [201]. Whereas above say 290 K the time scales are coupled, a rather strong decoupling is observed below. Similar results were observed for salol [198].

From Fig. 15 it is obvious that the temperature dependence of τ_α is quite different among the glass formers. In order to compare the different curves the

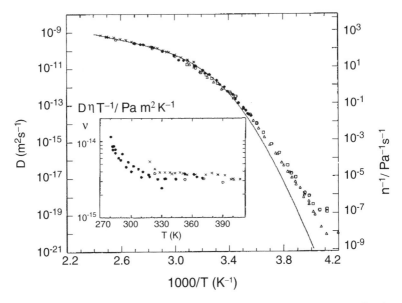

Figure 16. Logarithmic plot of the diffusion coefficient D of o-terphenyl as a function of reciprocal temperature; self-diffusion by gradient NMR (unfilled circles, filled circles, crosses), tracer diffusion data (unfilled triangles and squares) from forced Rayleigh scattering using photochromatic dye tracers, and inverse viscosity $1/\eta$ (line); the insert shows the product $D\eta/T$; note that the decoupling of diffusion and viscosity sets in around 290 K. (From Ref. 201, including data from Refs. 202 and 203.)

most popular way is to plot τ_α as a function of the reduced temperature T/T_g, or reduced reciprocal temperature T_g/T [7,125,164,204,205]. In this plot, fragile and strong glass formers were distinguished, the former exhibiting a strong dependence on T/T_g and the latter a weak dependence (close to T_g). Most molecular glass formers belong to the group of fragile glass formers; an exception is glycerol, which has an intermediate fragility (cf. Fig. 32 for a similar approach). As a quantitative measure of fragility, the fragility index is often used [206,207]:

$$m = \frac{d\log\tau_\alpha}{d(T_g/T)}\bigg|_{T=T_g} \tag{29}$$

Several attempts were made to correlate the degree of fragility with other properties of glass formers. For example, the height of the step in the specific heat temperature dependence at T_g is often found to correlate with m [125]. A remarkable relationship between the fragility m and the bulk and shear moduli of elasticity K_∞ and G_∞ in the glassy state was recently reported: systems with higher m appear to also have higher K_∞/G_∞ ratio [208]. A relation was reported between the degree of fragility m and the parameter α that reflects the slope of the temperature dependence of the nonergodicity parameter $f_q(q \rightarrow 0) = f_0$ in the glass $(T < T_g)$ through $1/f_0 = (1 + \alpha T/T_g)$ [209]. This relationship is reflected in the trend that low m glasses exhibit a small value of the mean-square displacement $\langle u^2 \rangle$ at T_g [8]. Corroborating these findings, it was reported that a higher anharmonicity in the interaction potential corresponds to a higher fragility, which in turn is reflected in a smaller nonergodicity parameter at T_g [210].

Concerning the relaxation spectra, it was found that the stronger the glass former, the higher the ratio of the vibrational (boson peak) to the fast relaxation contribution in LS spectra [146], though this finding was later challenged [211] (however, cf. Refs. 212 and 213). In a survey of many systems, fragility and the degree of nonexponentiality of the primary relaxation at $T = T_g$ were reported to correlate [207,214]. Specifically, more fragile glass formers appear to exhibit more stretching of the relaxation. In Fig. 17(top) we present high-temperature LS spectra of six glass formers with different degrees of fragility [216]. Clearly, the width of the α-peak strongly varies between these systems, as also does the magnitude of the fast dynamics (the second peak) that includes the boson peak and fast relaxations. Possibly, the stretching of the primary relaxation is related to the magnitude of the fast dynamics. In Eq. (30), we shall introduce the apparent von Schweidler exponent γ describing the high-frequency part of the α-peak in the high-temperature regime, which is thus a measure of the α-relaxation stretching. In Fig. 17(middle) the relaxation strength $1 - f$, taken as the integral over the fast dynamics spectrum, is plotted as a function of the stretching parameter γ. Indeed, a correlation is found: Stronger fast dynamics (larger $1 - f$) appears to correspond to smaller values of γ (more stretching). The fragility parameter m in Fig. 17(bottom) also appears to be correlated with γ, with stronger glass formers possessing a more stretched main relaxation (smaller γ). This is the opposite to the above-mentioned correlation reported in Refs. 207 and 214. This apparent contradiction can be partly related to the fact that the latter analysis concerns the α-peak shape at T_g and not at high T, as in Fig. 17. Close to T_g, the width of the α-peak may, however, be influenced by the appearance of the excess wing. Actually, it turns out that the shape of the slow dynamics including α-peak and excess wing appears to be independent of fragility (cf. Section IV.C). Additionally, stretching parameters compiled in Refs. 207 and 214 were obtained by a variety of experimental techniques. The stretching may, however, be different for different relaxing properties that are probed by these techniques. An example of such a difference is given later in Fig. 29.

B.   The High-Temperature Regime $(T \gg T_g)$

In Fig. 18 we inspect in more detail the high-temperature DS spectra of glycerol and LS spectra of 2-picoline, both glass formers of type A. Qualitatively, the

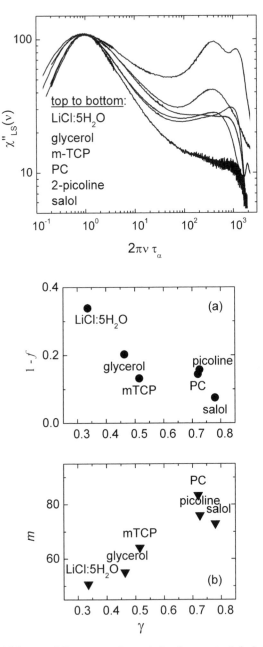

Figure 17. (top) LS susceptibility spectra of several glass formers at relatively high temperatures; note the different width (stretching) of the main relaxation (α-process) and the different amplitude of the fast dynamics (at frequencies above the susceptibility minimum). (middle, a) Strength $1-f$ of the fast dy-namics as a function of the stretching parameter γ; $1-f$ is the integrated intensity above the suscepti-bility minimum of the spectra from the upper panel; (bottom, b) Fragility index m versus stretching γ [216].

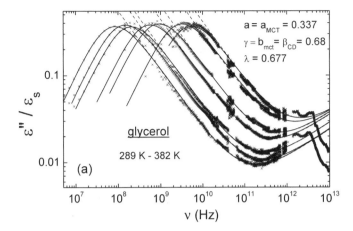

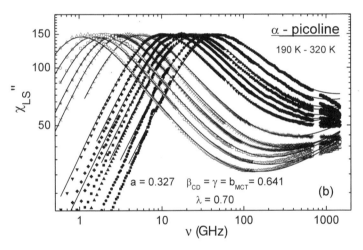

Figure 18.   (a) Normalized imaginary part of the dielectric permittivity $\varepsilon''(\omega)/\varepsilon_s$ of glycerol at high temperatures measured by Lunkenheimer et al. [9] (Adapted from Ref. 136). (b) Light scattering data of 2-picoline in the high-temperature regime (adapted from Ref. 183); interpolations (solid lines) applying a Cole–Davidson susceptibility (α-process) and a power-law contribution (fast dynamics) [cf. Eq. (31)]; exponents β_{CD} and a fixed by the exponent parameter λ of the mode coupling theory (MCT); dashed lines in (a) represent asymptotic laws of MCT.

α-peak shifts with varying temperature without, however, changing its shape. The high temperature LS spectra of toluene, a type B glass former, are plotted in Fig. 19a versus the reduced frequency $\nu\tau_\alpha$. The peaks superpose well, suggesting that their width is independent of temperature. Such behavior, which is one of the

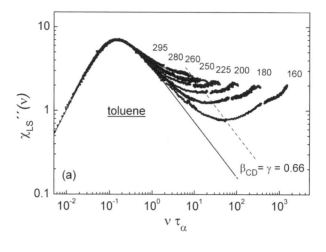

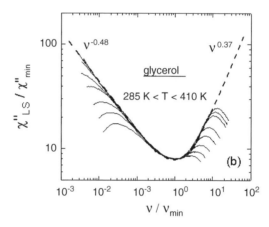

Figure 19. (a) α-Peak scaling for the light scattering (LS) spectra of toluene, interpolation by a Cole–Davidson susceptibility (solid line); dashed line represents power law with exponent $\gamma = \beta_{CD}$ for locus of minimum [cf. Eq. (35)] (adapted from Ref. 185). (b) Minimum scaling for LS data of glycerol, interpolation with a sum of two power laws, exponents as indicated (adapted from Ref. 64). These two scaling properties determine the high-temperature regime of a glass former.

characteristic features of the high-temperature glassy dynamics, shall be called "α-peak scaling" in the following. It implies nothing other than the validity of the frequency–temperature superposition (FTS) principle. In the case of toluene (Fig. 19a) the scaling is not perfect, suggesting that FTS only holds approximately, especially at the highest temperatures, where the α-peak appears to be somewhat broadened when approaching the vibrational band (cf. Fig. 30).

Still, FTS appears to be a good approximation, also recognized in the time-domain data (cf. Fig. 13b).

Another property of the high-temperature spectra is revealed when the spectra of different temperatures are rescaled such that their minima overlap—that is, plotting χ''/χ''_{min} versus ν/ν_{min}. As an example, LS spectra of glycerol are shown in Fig. 19b. Again, the minima superpose well, suggesting that their shape is temperature-independent and thus can be described by a master curve. This behavior shall be called "minimum scaling." Both features, α-peak scaling and minimum scaling, can be easily understood if one assumes that both the spectral shape and the amplitude of both the α- and the fast relaxations are temperature independent, the only change with temperature being the shift of the α-peak [167,168,218]. As will become clear in Section IV.C.3, both scalings are no longer valid in the highly viscous liquid close to T_g. This behavior that is incompatible with the α- and minimum scaling shall be called the low-temperature regime of a molecular glass former.

It is interesting to ask whether the α-peak shape depends on the method applied to probe molecular reorientation. Such a comparison has only been possible for a few systems—for example, glycerol and propylene carbonate (PC), where in addition to the LS spectra also the DS spectra were measured in the gigahertz range [9]. Clearly, the width of the α-peak is much larger in LS as compared to DS (cf. Fig. 20). Similar results were reported for an epoxy glass [219]. Thus, the apparent von Schweidler exponent γ describing the high-frequency part of the α-peak depends on the relaxing property in question, and thus on the experimental technique. Another observation in Fig. 20 is that the terahertz dynamics, comprised of fast relaxations and vibrations, is much stronger in the LS spectra. Concerning the amplitude $1 - f$ of the fast dynamics in the two-step correlation function $g_l(t)$ [cf. Eq. (32)], one expects that $1 - f_{LS} \cong 3(1 - f_{DS})$ provided that only small angular displacements are involved in the fast dynamics [64,220,221]. To a certain approximation, this is found to be correct [64].

The susceptibility minimum observed in the DS and LS relaxation spectra is a consequence of the interplay of the high-frequency tail of the α-relaxation peak and the contribution from certain fast dynamics dominating at frequencies close to but above the minimum. As was demonstrated by several experimental studies as well as by molecular dynamics simulations (cf. Fig. 13a), in addition to the vibrational contribution a fast relaxation process has to be taken into account for describing correctly the susceptibility minimum [5,9,19,55,64, 133,134,136,147]. This spectral contribution may usually be described by a power law with a positive exponent less than unity. In fact, the search for this fast relaxation process was mostly inspired by MCT, where it naturally appears in the solutions and is interpreted as "rattling in the cage" type of dynamics. Some authors discussed in addition a constant loss contribution [10,138,222,

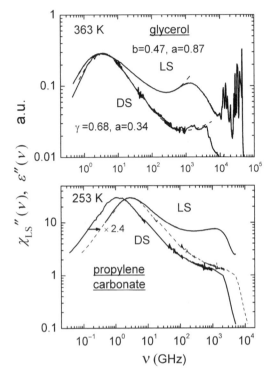

Figure 20. Comparison of light scattering (LS) and dielectric (DS) spectra of glycerol (top) and of propylene carbonate (bottom). Significant differences are observed among the spectra obtained by the different techniques. The spectra are interpolated using the Cole–Davidson susceptibility and a power-law contribution with exponents as indicated [Eq. (31)]; in the case of the DS spectra; the data are compatible with the asymptotic laws of mode coupling theory. (DS data compiled from Ref. [9], LS data compiled from Ref. 64.)

223], in analogy to what is found in ionic systems. A vibrational origin of the fast process was also suggested [224,225].

Our starting point in the spectral shape analysis is describing the susceptibility minimum by a sum of two power laws. Thus, we write in first approximation

$$\chi'' = \chi''_{min}[a(v/v_{min})^{-\gamma} + \gamma(v/v_{min})^{a}]/(a + \gamma) \qquad (30)$$

where we have introduced two exponents, γ describing the high-frequency part of the α-peak and a being the exponent of the fast relaxation contribution. This interpolation formula was first introduced in the framework of idealized MCT, where the exponents have a specific meaning and are not independent of each other. In the phenomenological analysis presented here, they are taken as free fit parameters. Note that in the high-temperature regime $v^{-\gamma}$ is identical with the so-called von Schweidler power law, because it describes the high-frequency part of the α-peak [17]. In the low-temperature regime, γ concerns the shape of the emerging secondary relaxation features, such as the excess wing, rather than of the main peak. The emergence of secondary relaxation features is then revealed as a qualitative change in the temperature dependence of γ (cf. Fig. 29). Thus, only at high temperatures, where the slow processes have merged into an unseparable single relaxation peak, can γ be identified with the von Schweidler exponent.

We may include the α-peak in the interpolation formula by taking the Cole–Davidson (CD) susceptibility $\chi_{CD}(v) = (1 - i2\pi v\tau_\alpha)^{-\beta_{CD}}$ [34,226] instead of the first power law in Eq. (30). Explicitly,

$$\chi''(v) = A\chi''_{CD}(v) + Bv^a \tag{31}$$

where A and B quantify the strength of the α-process and the fast relaxation contribution, respectively. This approach is sometimes called the hybrid model or a "superposition fit" [5,136]. Usually, experimental DS and LS spectra are well-fitted using this ansatz [5,134,64,136,183,185,227–231]. At high frequencies, the CD function asymptotically is a power law with the exponent β_{CD}, explicitly $\chi''_{CD}(\omega) \propto (\omega\tau_\alpha)^{-\beta_{CD}} (\omega\tau_\alpha \gg 1)$. Thus, around the susceptibility minimum, Eq. (31) is asymptotically equivalent to Eq. (30) with $\beta_{CD} = \gamma$. Instead of using the CD function, one can use the generalized gamma (GG) distribution as a model function in the relaxation time domain. Such a distribution offers an extra parameter in addition to β that allows one in some cases to better describe the low-frequency side of the α-peak [142,232] (cf. Section IV.C.1). We emphasize that, since the high-temperature spectra are compatible with using $\beta_{CD} = \gamma$, there is no sign of an extra-high-frequency wing discernible in these spectra. In passing, it also suggests that the Nagel scaling will not be appropriate in that temperature region (cf. Section IV.C.1). In conclusion, in the high-temperature regime the α-relaxation appears to be determined by a single shape parameter, namely the von Schweidler exponent γ.

In the time domain, Eq. (31) corresponds to a two-step correlation function:

$$C(t) = C_{\text{fast}}(t)C_\alpha(t) = [(1-f)\phi_{\text{fast}}(t) + f]\phi_\alpha(t) \cong (1-f)\phi_{\text{fast}}(t) + f(t)\phi_\alpha(t) \tag{32}$$

$(1-f)$ is the correlation losses brought about by the fast processes $\phi_{\text{fast}}(t \to \infty) = 0$, and f is the relaxation strength of the α-process, or the nonergodicity parameter [17] (cf. Section IV.A), and we have assumed that the fast and slow dynamics are statistically independent. One has to keep in mind that the power-law contribution in Eq. (31) is supposed to reflect only the relaxational part determining the susceptibility minimum but not the vibrational contributions at higher frequencies and thus does not allow for a complete description of a spectrum. In the more generally valid expression of Eq. (32), $\phi_{fast}(t)$ contains both contributions, and it is not a simple task to disentangle both parts in a quantitative way [64,147,233–235]. For example, the exponent a extracted from the minimum scaling as well as the determination of the nonergodicity parameter (cf. Section IV.D.2) may be disturbed by the presence of the vibrational contributions; more precisely, applying the simple interpolation formula, Eq. (31), it may not reflect the true relaxational contribution. As already mentioned, at highest temperatures, even the parameter β_{CD} cannot be extracted unambiguously. It should be recalled here that the nonergodicity parameter f is of major interest within MCT, and, in particular, a crossover in its temperature dependence is expected when entering the low-temperature regime (for more details see Section IV.C.2).

Examples of using Eq. (30) and (31) to interpolate the α-peak and the relaxational part of the fast dynamics are shown in Figs. 18, 19b, and 20. The constants A and B, and thus the nonergodicity parameter f, are temperature-independent in good approximation, as also are the two parameters a and $\beta_{CD} = \gamma$ that define the shape of the minimum. Thus, indeed the only effect seen in the high-temperature spectra when temperature is decreased is the shift of the α-peak to lower frequencies, while all shape parameters essentially stay temperature-independent. Moreover, a simple additive superposition fit appears to be sufficient to describe the relaxational part of the high-temperature spectra. Several phenomenological approaches have also been suggested to extend the quantitative description of the LS and DS spectra to include the vibrational peak also [9,64,236,237]. Stochastic models of reorientational motion in condensed matter have also been utilized. For example, the extended rotational diffusion model, or approaches including inertial effects, such as the itinerant oscillator model, have been widely used to describe the high-frequency spectral features [27,238–240]. The latter approaches are usually applied to the liquid at rather low viscosity.

Equation (31) has an unphysical ω^a asymptotic behavior at low frequencies. A physically more correct form is given by the "hybrid function" introduced by Brodin et al. [227].

$$\chi''(\omega) = A \text{Im}[(1 - i\omega\tau_\alpha)^{-\gamma} + i\omega B(\tau_\alpha^{-1} - i\omega)^{a-1}] \tag{33}$$

If a and γ are temperature-independent, Eq. (30) and Eq. (32) imply the α-peak and minimum scaling. The α-peak scaling follows from the fact that τ_α enters the CD (α-peak) part of the susceptibility only in combination $\omega\tau_\alpha$. As for the minimum scaling, in the vicinity of the minimum, Eqs. (30) and (31) yield

$$\chi''(\omega\tau_\alpha) \approx C(\omega\tau_\alpha)^{-\gamma} + D\tau_\alpha^{-a}(\omega\tau_\alpha)^a \tag{34}$$

where C and D are again temperature-independent coefficients. It then follows that the locus of the susceptibility minimum scales as

$$\chi''_{min} = C(1 + \gamma/a))(\omega_{min}\tau_\alpha)^{-\gamma} \tag{35}$$

An example of testing Eq. (35) is given in Figs. 19a and 21, where at high temperatures the minimum position is interpolated by a power-law exhibiting the same exponent as the CD fit.

Phenomenologically, the validity of Eqs. (33)–(35) defines the high-temperature regime of a molecular glass former. As we have stressed, it turns out that these equations, describing a two-step correlation function with a stretched long-time decay, are actually valid up to temperatures above the melting point. In the next section, we shall demonstrate that the equations fail below a certain temperature T_x, and this will define the crossover into the low-temperature regime of a glass former. As further discussed in Section IV.E, the high-temperature regime found particular interest during the last years since MCT drew the attention to the gigahertz regime as the relevant frequency

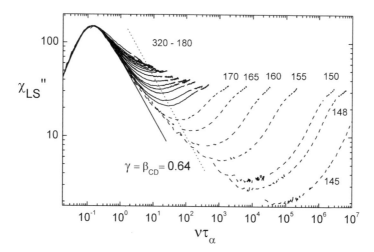

Figure 21. Light scattering (LS) spectra of 2-picoline for different temperatures $T > T_g$ (indicated as numbers) plotted as a function of rescaled frequency $\nu\tau_\alpha$; solid line: interpolation by a Cole-Davidson (CD) susceptibility, dotted line: power-law indicating locus of the minimum ($T > T_g$), cf. Eq. 35 (adapted from [183]); below, say 175 K, the evolution of the spectra changes, i.e. the α-scaling fails.

regime, where the onset of the glassy dynamics is expected to take place [17]. We note that most of the results in the gigahertz-to-terahertz regime were obtained from NS [147,148,241–243] and LS studies [5,55,64,133,134,145,146, 183,185,229,244–246] because dielectric spectra are difficult to obtain in this frequency range. So far, only two molecular glass formers, namely glycerol and propylene carbonate, have been dielectrically measured in the entire frequency range 10^{-6}–10^{13} Hz [9,10] (cf. Fig. 12a). As mentioned, the LS experiments have strongly benefited from the work by Cummins and co-workers introducing the tandem Fabry–Perot interferometry in combination with a double grating monochromator. Recently, optical Kerr effect [69,70,247] as well as impulsive stimulated scattering techniques [248–250] have been applied in order to unravel the evolution of the fast glassy dynamics (cf. Section IV.E).

C. The Low-Temperature Regime

As discussed in Section IV.A.1, the low-temperature dynamics of molecular glass formers is characterized by the appearance of slow secondary relaxation processes, namely the excess wing and β-process; for the sake of convenience we shall discuss these relaxation features separately. As will be demonstrated, the central properties of the relaxation in the high-temperature regime, namely the α-peak and minimum scaling, no longer hold in the low-temperature regime, implying that at temperatures below some crossover temperature T_x the spectral shapes of the slow and fast dynamics change with temperature.

1. The Evolution of the Excess Wing

In the present section, we restrict the discussion to glass forming liquids for which no secondary relaxation peak but rather a simple power-law, the so-called excess wing, is observed on the high-frequency side of the α-relaxation peak. There are several ways to demonstrate that the evolution of the susceptibility spectra of such glass formers qualitatively changes from the above described high-temperature scenario to a behavior determined by the emergence of the excess wing at temperatures close to T_g. In Fig. 21, for the case of 2-picoline, we again plot the LS susceptibility as a function of the reduced frequency $\nu\tau_\alpha$ including now all the data measured at $T > T_g$ [183]. At high temperatures α-peak scaling and minimum scaling are observed but at low temperatures, say below 180 K, the susceptibility minimum flattens and shifts to higher rescaled frequencies. This observation may be taken as reminiscent of the excess wing clearly recognized in the DS spectra but also in the LS spectra of glycerol (cf. Fig. 12(a) and 12(b)). Another way to show that a qualitative change of the spectral shape occurs can be realized when the scaling of the minimum is applied for all temperatures at $T > T_g$ (cf. Fig. 22(a)). Obviously, in the case of the DS spectra of glycerol, the minimum scaling breaks down below ca. 290 K, which

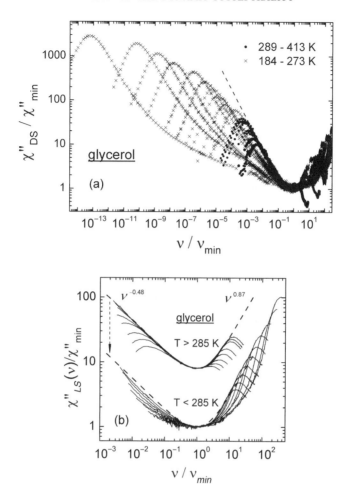

Figure 22. (a) Scaling of the susceptibility minimum for the dielectric (DS) spectra of glycerol measured by Lunkenheimer et al. [9] (adapted from [136]); dashed line: interpolation of the minimum at high temperatures applying Eq. (30) with exponents compatible with mode coupling theory, cf. also Fig. 18(a); (b) Rescaled light scattering (LS) spectra (adapted from [64]). In both data sets, the minimum scaling works above 285 K whereas it fails below.

again is associated with the appearance of the excess wing. The same is observed for the corresponding LS data (cf. Fig. 22(b)) and also for the DS spectra of propylene carbonate [9]. A similar trend can be anticipated from LS data of salol [231]. Thus, both α-peak scaling and minimum scaling break down below a certain temperature T_x. In particular, the exponent γ defining the low frequency side of the susceptibility minimum becomes temperature dependent, specifically, decreases strongly with temperature. Qualitatively, these new spectral features define the low-temperature regime setting in below some crossover temperature T_x.

Usually, the excess wing is probed in the frequency domain by measuring DS spectra. However, the excess wing can also be identified in the time domain as was done in a PCS study [65] and in a DS investigation [142]. PCS is a LS technique able to cover a rather broad time window [25], and it is usually applied to monitor the α-process close to T_g (cf. Sect. II.C., and [161,252,253,254,255,256]). The corresponding DS results are displayed in Fig. 23. In the time domain, often the Kohlrausch stretched exponential decay, $\exp(-(t/\tau_K)^\beta)$, is applied to interpolate the data, however, by doing so subtle deviations are observed at short times. They can be directly identified, when the derivative of the relaxation function is inspected (cf. inset of Fig. 23(a)). Indeed, a well defined power-law shows up at short times. Performing the Fourier transformation of the data an excess wing is well recognized in the spectrum (Fig. 23(b)). Similar results are reported for the PCS data of 2-picoline in Fig. 24 where the time resolution of the technique was pushed to its limits [65]. Concluding, given data covering a sufficient time interval, the Kohlrausch decay is not an appropriate function to fit time domain data close to T_g where the excess wing is most pronounced.

Several approaches were published attempting to quantitatively describe the slow response in type A glass formers including the α-peak and excess wing as obtained by DS. For example, Nagel and coworkers [140,143] introduced a scaling procedure including both the α-peak and excess wing, which works quite well though being not correct in a strict sense [259,260,261]. For further tests of the "Nagel scaling", compare refs. [9,10,262]. The scaling procedure also attracted attention from a theoretical side [263,264,265]. Building on this approach Menon and Nagel claimed that the static susceptibility diverges,

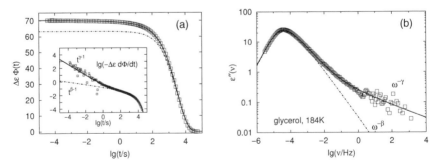

Figure 23. (a) Dielectric time domain data of glycerol (squares) interpolated by an extension of the generalized gamma distribution, cf. Eq. 36 (solid line); for comparison, interpolation without the excess wing (dash-dotted line); inset: derivative of data and fit functions, power-law regimes are indicated; (b) Fourier transform of both data and interpolations (taken from [142].)

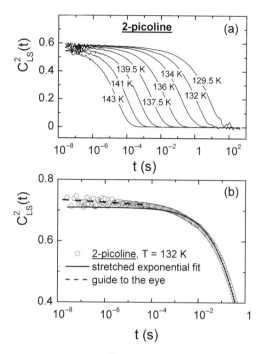

Figure 24. (a) Photon correlation data $C_{LS}^2(t)$ of 2-picoline at selected temperatures; (b) initial part of $C_{LS}^2(t)$ together with a fit to a stretched exponential decay; deviations are recognized at short times (adapted from [65].)

indicating a "hidden phase transition" at a temperature below T_g associated with T_{VFT} [266]. In another attempt a decomposition into two separate relaxation processes was proposed [10,141,257,258]. Some of the latter authors concluded that the excess wing is just a special secondary relaxation process similar to the β-process in type B glass formers but submerged below the α-peak. A microscopic explanation of the excess wing in terms of percolation theory including an interpolation of the DS spectra was provided by Chamberlin [267,268], and another explanation was offered in the frame of a theory of frustration limited domains [123,269]. The excess wing was also interpreted as a consequence of a more or less strong rotational translational coupling within a schematic MCT model, however, this was assumed to be valid in the high-temperature regime of MCT whereas the excess wing is observed only close to T_g [270]. Usually, it was implicitly supposed that two power-law regimes can be identified in the susceptibility spectra on the high-frequency side of the α-peak with one exponent related to the α-peak and the other one to the excess wing, respectively. However, this is a crude approximation, as one can see when inspecting the derivative $D = d \ln \varepsilon'' / d \ln \nu$ of the glycerol data in Fig. 25. A

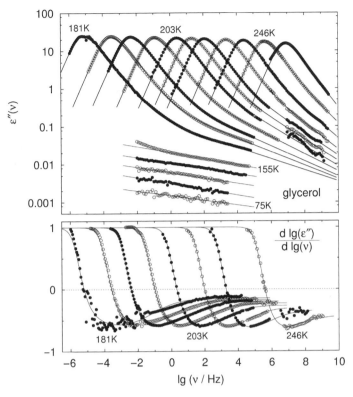

Figure 25. (top) Dielectric loss spectra of glycerol, including time domain data, interpolated by applying the GGE distribution, cf. Eq. 36 (solid lines); (bottom) corresponding derivatives of both the data (points) and fit (solid line) (from [142].)

power-law dependence of the susceptibility corresponds to a plateau of the derivative, but at frequencies just above the susceptibility maximum at $D = 0$ no such plateau is observed. Instead one recognizes a minimum in D, which however does not indicate an actual power-law but may just define some kind of apparent exponent. It is remarkable that in Fig. 25 the minimal D value appears to be almost temperature independent (cf. also Ref. [278]). Only at the highest v and low T does the derivative level off to a plateau value, implying a simple power-law exponent, which can be attributed to the excess wing. In between there is a broad crossover region from the minimum in D to the wing exponent $D = -\gamma$, which extends at 181 K over ca. 7 decades in frequency. Finally, we note that close to T_g it is not always clear, whether a simple power-law is sufficient to describe all the features present in the excess wing region. In some systems it may be possible that, in addition to the excess wing, contributions of a weak β-process may play a role, which partially mask the simple power-law behavior.

Here, for a quantitative analysis of the spectra of the slow dynamics, we use a phenomenological approach introduced by Kudlik et al. [6] and Blochowicz et al. [142]. As we want to demonstrate, the approach allows identifying the crossover temperature T_x dividing the high- and low-temperature regimes as it provides a fit of the slow dynamics spectra in both regimes. The authors described both the α-peak and excess wing by a distribution of correlation times, namely an extension of the generalized gamma distribution (GGE),

$$G_{GGE}(\ln \tau) = N_{GGE}(\alpha, \beta, \gamma) e^{-\frac{\beta}{\alpha}\left(\frac{\tau}{\tau_0}\right)^{\alpha}} \left(\frac{\tau}{\tau_0}\right)^{\beta} \left[1 + \left(\frac{\tau\sigma}{\tau_0}\right)^{\gamma-\beta} \right] \qquad (36)$$

Here, N_{GGE} is a normalizing factor, and τ_0 is proportional to the correlation time τ_α. In addition to the parameters α and β, which describe the α-relaxation peak, two further parameters σ and γ appear which define the onset of the excess wing and its exponent, respectively. Thus, the exponent γ determines the low-frequency part of the susceptibility minimum in the gigahertz region also in the low-temperature regime (cf. Fig. 12(a)). The width parameters α and β can assume values $0 < \alpha, \beta < \infty$. For a given glass former α fixes the distribution at long correlation times and can be chosen constant. The parameter β is the exponent of the α-peak contribution at short correlation times. We note, as β is an exponent in the distribution function, it may become larger than 1 and is in that case related to a behavior ω^{-1} in the susceptibility. However, the onset of the excess wing generally masks this power-law, such that it is never apparent in the spectra. We note that the GGE distribution can be reduced to the generalized gamma distribution (GG) by setting $\sigma = 0$. The GG distribution allows to interpolate quite different relaxation peaks without excess wings [142,232]. From Eq. (1) one calculates the permittivity via:

$$\hat{\varepsilon}(\omega) - \varepsilon_\infty = \Delta\varepsilon \int\limits_{-\infty}^{\infty} G(\ln \tau) \frac{1}{1 + i\omega\tau} d \ln \tau. \qquad (37)$$

As a note of caution, we emphasize that applying a distribution of correlation times does not necessarily imply that excess wing and α-peak are parts of the same relaxation phenomenon. As said, the excess wing has to be interpreted as a secondary relaxation process appearing close to T_g. In that case, the quite universal relation of the parameters of the excess wing (σ and γ) to those of the α-peak disclosed below demonstrates a unique coupling of both processes. We stress that by applying the GGE function one has to optimize three parameters to fit the shape of the slow dynamics spectra, which is one parameter less than in most of the approaches that decompose the spectra into two susceptibility functions. Moreover, as the spectra of excess wing and α-peak strongly overlap a simple additive superposition of susceptibilities does not appear to be justified. We note that a description of the spectra in terms of a GGE distribution may be mapped onto a combination of a function for the α-peak and a function for the secondary relaxation in the framework of the so-called Williams-Watts approach (cf. Eq. 46), by fixing a certain relation between the time constant of the two processes [271]. In this interpretation the parameter $\sigma^{\gamma-\beta}$ is related to the relaxation strength of the excess wing.

As can be checked in Figs. 12(a), 25, 27(a) and 50 an almost perfect interpolation is provided by the GGE distribution for the spectra of the slow dynamics. One may include an additional power-law contribution (with exponent a) for describing the fast dynamics (in analogy to Eq. 31). Then, a complete interpolation of the spectra is possible covering up to 17 decades in frequency; this was actually done for the fits in Fig. 12(a) [136]. No further spectral contribution such as a constant loss contribution discussed by some authors [10,224,225] is needed. For further successful fits applying the GGE distribution we refer to refs. [142,230].

Fig. 26 presents examples of the GGE distribution and the corresponding susceptibility spectra. According to this approach, both the distribution $G(\ln\tau)$ as well as the susceptibility can be decomposed into two contributions, one belonging to the α-peak (described by the parameters α and β), the other to the excess wing (determined by γ and σ). The higher the temperature, the larger become the parameters γ and β fixing the shape of the GGE distribution at $\tau \ll \tau_\alpha$. Simultaneously, the onset parameter σ decreases indicating that the onset of the excess wing approaches τ_α. This marks the crossover to a simple peak (SP) susceptibility (without separate high-frequency wing) which is a characteristic of the high-temperature regime of glass formers (cf. Sect. IV.B). Indeed, due to the properties of the Laplace transform such a non-trivial limit of the GGE distribution exists with $\sigma > 0$ and $\gamma < \beta$ for which a simple peak results in the susceptibility although the GGE distribution still consists of two power-law contributions with exponents β and γ, respectively. Actually, the exponent β never shows up explicitly in the susceptibility except at lowest temperatures (cf. Fig. 26). It turns out that in the SP limit the GGE distribution well interpolates a CD distribution of correlation times with $\gamma = \beta_{CD}$ (cf. Fig. 25; [142]). Thus, the GGE distribution is able to fit both the "excess wing limit" at low temperatures as well as the SP limit at high temperatures, and it offers the possibility to identify the crossover temperature T_x.

Given the good interpolation of the susceptibility spectra by the GGE distribution we can focus on the temperature dependence of the parameters. We compiled the results from analyzing DS spectra of several type A glass formers [6,136,137,142,230,273,275]. Note that the glass formers trimethyl phosphate (TMP, [230,275]), 3-fluoro aniline (FAN, [142]) and methyl tetrahydrofuran (MTHF, [331]) show both the excess wing and a β-process (cf. Fig. 35). In this

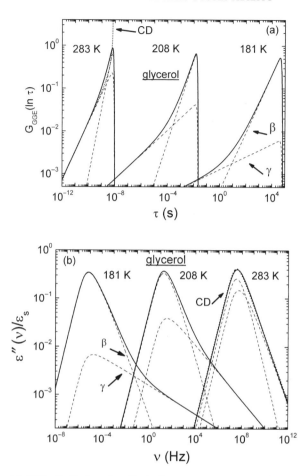

Figure 26. (a) GGE distribution $G_{GGE}(\ln \tau)$, Eq. (36), and (b) corresponding susceptibility spectra as obtained from the fits of the data in Fig. 25, and its decomposition into the respective peak and excess wing contributions as defined by the GGE distribution; the Cole-Davidson function (dots) is almost indistinguishable from the GGE function in its simple peak limit (cf. curves at T = 283 K in (b)).

case, the secondary relaxation peak is well separated from the α-peak, so that a pronounced excess wing is resolved in the spectra. As is obvious from inspecting Fig. 27, the parameters of the GGE distribution are found to be quite similar when plotted as functions of the time constant τ_α [136,142,230,275]. For glycerol and propylene carbonate (PC), we added the results from the high-temperature regime where the exponent γ is virtually not changing with τ_α respectively with temperature as the FTS applies in good approximation. In the low-temperature regime, both γ as well as β change in a unique way, as does the onset parameter $\lg\sigma$ which appears to increase linearly with $\lg\tau_\alpha$. The fact that the excess wing disappears is indicated by $\lg\sigma$ approaching 0 (actually $\sigma \cong 2$ is found in the SP limit [142]). Thus, the crossover from the high-temperature regime with an SP susceptibility to the low-temperature regime characterized by the presence of the excess wing occurs at $\tau_\alpha \cong 10^{-8}$ s associated with the crossover temperature T_x. In the case of glycerol and PC, as high-frequency data are available, the crossover is also indicated when $\gamma(T)$ reaches its high-temperature limit, and again $\tau_\alpha \cong 10^{-8}$ s is observed (cf. Fig. 27(a)). This is close to the "magic correlation time" discussed by Novikov and Sokolov [272], who found in various data indications for a dynamic crossover above T_g (cf. also [274]). Inspecting the small deviations observed for the parameters of the GGE distribution for certain systems with respect to those of the majority (e.g. m-TCP or EG, in Fig. 27) it turns out that by scaling $\gamma(\lg\tau_\alpha)$ with the value of γ at T_g, γ_g, a master curve is obtained for all systems under consideration, as it is demonstrated in Fig. 28. Furthermore it is shown in Refs. [142,275] that a similar scaling leads to a master curve for $\lg\sigma$ as a function of $\lg\tau_\alpha$, again for all systems under study. Finally, we note that the interpretation of the parameter $\sigma^{\gamma-\beta}$ as a measure of the amplitude of the excess wing contribution is supported by the finding that $\sigma^{\gamma-\beta}$ increases linearly with temperature [137], similar to the temperature dependence of the relaxation strength of the JG β-process above T_g [6,277].

Although the results compiled in Fig. 27 were obtained applying a particular distribution $G(\ln\tau)$, some of them can be checked independently. This is directly seen when the spectra of different glass formers are scaled by a vertical shift in order to provide the best agreement at high frequencies (cf. Fig. 29). As expected from the GGE fits, very similar high-frequency contributions are observed provided the time constant τ_α is the same. However, the α-relaxation peak itself is different, which is reflected in different but constant values of the parameter α such as $\alpha = 20$ for PC, $\alpha = 10$ for glycerol and $\alpha = 5$ for 2-picoline. Thus, the dynamics at intermediate time scales is rather universal in the sense that the excess wing manifests itself in a very similar way whereas the slowest dynamics differs along with different values of α. This observation agrees with the finding that around T_g, the width of the α-peak is non-universal even though the apparent high-frequency exponent (given by the minimal $D = d \ln \varepsilon'' / d \ln\nu$) was reported to approach a value close to $D = -0.5$ [278]. However, notice that in Fig. 25 one rather finds $D = -0.65$. So it appears that the minimal value of D varies among type A glass formers [137]. We also emphasize that the findings in Fig. 29 clearly disprove the Nagel scaling [140,143,279], which claims that the excess wing and the shape of the α-peak are uniquely related. However, below, we shall show that some features of the scaling properties can be rediscovered.

The two temperature regimes of molecular glass formers are more clearly recognized when the distribution parameters β and γ are analyzed as functions

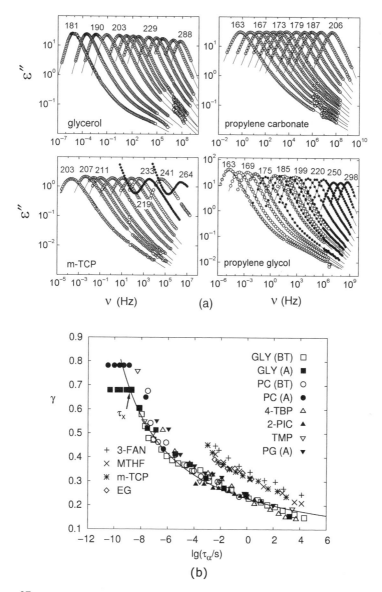

Figure 27. (a) Dielectric spectra and corresponding interpolations (solid lines) with the GGE distribution Eq. (36) of the type A glass formers glycerol (GLY), propylene carbonate (PC), m-tricresyl phosphate (m-TCP, 202–280 K) compiled from [6,142,230,275] and propylene glycol (PG, open symbols [275], full symbols courtesy P. Lunkenheimer [258]). (b) and (c) excess wing parameters γ and σ as functions of the time constant τ_α for ten different glass formers including also 2-picoline (2-PIC), 4-tertbutyl pyridine (4-TBP), trimethyl phosphate (TMP), m-fluoro aniline (3-FAN), methyl tetrahydrofuran (MTHF), and ethylene glycol (EG, courtesy F. Kremer); figures taken from Ref. [275]. Arrows in (b) and (c) indicate the crossover from a susceptibility with excess wing (low-temperature regime) to one without (high-temperature regime). Solid line in (b) represents a fit by Eq. (39), solid line in (c) is a linear interpolation. In cases of two data sets for one substance (A) refers to spectra obtained by the Augsburg group [9,258] and (BT) refers to data measured by the Bayreuth group.

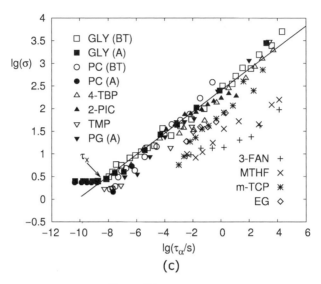

Figure 27. *Continued*

of temperature (cf. Fig. 30(a)). Again we added the high-temperature results for glycerol and PC. Whereas γ does not change at high temperatures a linear decrease is recognized below the crossover temperature T_x. It appears that for each system β and γ extrapolate to zero at a similar temperature. Thus, one may conclude that β and γ are proportional to each other. Indeed, this is demonstrated in Fig. 30(b) where we plotted γ and the scaled parameter β/c, where c is some constant. One finds

$$\beta/\gamma = c = 3.0 \pm 0.3$$
$$\gamma = A(T - T_\gamma). \tag{38}$$

For the exponent γ, such a temperature dependence was reported in the literature [9,10,143,144,258,260]. Below we shall show how the temperature dependence of the exponent γ can explicitly be linked to the temperature dependence of the time constant τ_α to yield the behavior shown in Fig. 27.

The constant ratio β/γ found by analyzing the data with the GGE distribution is to be compared with the finding from the "Nagel scaling". This scaling procedure can be interpreted as operating with the spectral density rather than the susceptibility spectrum [259]. It implies the existence of two power-law regions on the high frequency side of the α-peak, for which the corresponding exponents are connected by $(1 + \beta)/(1 + \gamma) = 1.3$ [143,279]. Inspecting Fig. 30, around T_g the parameter γ is close to 0.2 leading to $\beta/\gamma \cong 2.8$ that is somewhat similar to the factor $c = 3.0$ found by the GGE analysis. However, the higher

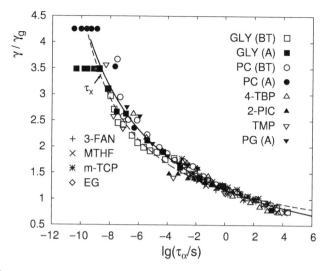

Figure 28.   The excess wing exponent γ for ten different glass formers (cf. Fig. 27) rescaled by $\gamma_g = \gamma(T_g)$ so that all the datasets coincide on a master curve. The solid line represents Eq. (40). For more details see Ref. [275].

the temperature and thus the higher γ, the smaller becomes the ratio expected from the Nagel scaling. As discussed, the drawback of the Nagel analysis is that the assumed "first power-law regime", from which β could be obtained, does not actually exist. Still, the apparent exponent extracted from the master curve becomes the closer to β from the GGE distribution, the lower the temperature. In other words, the lower the temperature the more the α-peak exponent β of the distribution $G_{GGE}(\ln \tau)$ is revealed in the susceptibility. Thus, the approximate validity of the Nagel scaling close to T_g reduces to the fact that the corresponding parameters, β and γ, of the underlying distribution $G(\ln \tau)$ are fixed by a constant ratio.

Clearly, in the low-temperature regime the FTS principle does not hold for the full spectrum of the slow dynamics (α-peak and excess wing) whereas it essentially works in the high-temperature regime. We emphasize that claims such as that the stretching of the primary relaxation decreases monotonically with increasing temperature, which are often made in the literature, should be taken with great care. Instead, indications of a crossover from a weak temperature dependent width of the α-relaxation at high temperatures to a stronger dependence at low temperatures can be anticipated from many studies [9,140,231,258,280,281]. The effect is however not completely clear, probably because poorly defined parameters such as the overall width were used to characterize the α-peak. We also note that fits of the α-peak at low temperature to the CD or the Kohlrausch functions can be misleading because of the

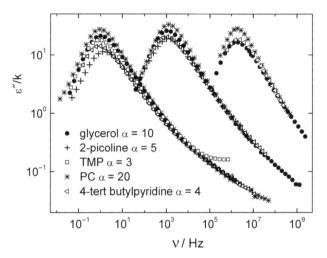

Figure 29. Dielectric spectra of several type A glass formers shifted by a temperature independent factor k to provide coincidence at highest frequencies, i.e., in the region of the excess wing (compiled from [137,142,230]); note that although the spectra exhibit virtually the same excess wing the α-peak itself is different along different α values of the GGE distribution, cf. Eq. 36.

presence of the excess wing that is not described by these functions. Both functions are rather inadequate for describing low-temperature spectra.

The spectral shape analysis reviewed above relies on dielectric data. In order to test whether the above picture is also found when a different probe is chosen we refer to Fig. 20 where the DS and LS spectra of glycerol and PC are compared. Clearly, the apparent exponent γ depends on the method as discussed above. Analyzing the susceptibility minimum with assuming two power-law contributions (cf. Eq. 30), a comparison of γ_{LS} and γ_{DS} is shown in Fig. 31, including data from the high and low-temperature regime. Although different absolute values are found the temperature dependences are similar, actually proportionality could be claimed [64]. Most important, the crossover from a virtually temperature independent γ to a strongly temperature dependent one occurs essentially at the same temperature $T_x \cong 295$ K, indicating that, independent of the method, T_x is the same for a given glass former.

Further comparing the relaxation stretching in DS versus time-domain light scattering PCS data, in some systems strong differences were reported; usually, the stretching is more pronounced in PCS than in DS; in other cases, similar width parameters were found [65,161,251,252,253,254,255,256]. As for the secondary relaxation features, signal-to-noise ratio at short times ($t < 10^{-6}$ s) of most PCS experiments is not sufficient to allow for a quantitative analysis of the short time behavior of the decay, i.e. for determining γ_{LS} (cf. however Fig. 24).

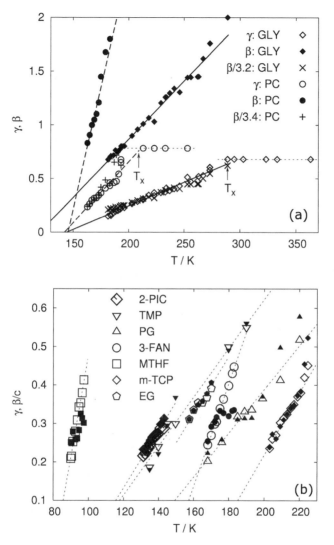

Figure 30. Temperature dependence of the parameters β and γ of the GGE distribution Eq. (36). (a) β and γ are shown for the glass formers glycerol (GLY) and propylene carbonate (PC). In addition β/c (crosses) shows that $\beta \propto \gamma$ holds in good approximation. For clarity data sets of the Augsburg and Bayreuth group are not distinguished. (b) γ (open symbols) and β/c (full symbols) for 2-picoline (2-PIC), trimethyl phosphate (TMP), propylene glycol (PG), 3-fluoro aniline (3-FAN), methyl tetrahydrofuran (MTHF), m-tricresyl phosphate (m-TCP) and ethylene glycol (EG); $c = 3.0 \pm 0.5$. Figures taken from Ref. [275].

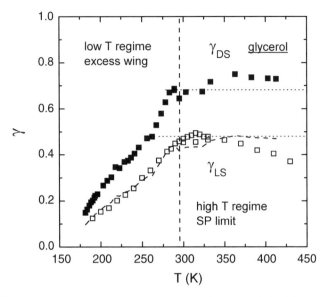

Figure 31. Comparison of the apparent exponent γ for glycerol as obtained from analyzing (cf. Eq. 30) the susceptibility minimum in the dielectric spectra (DS) and light scattering spectra (LS) including data from the high- as well as low-temperature regime; dotted lines: average values as obtained from the minimum scaling (cf. Fig. 18(b) and Fig. 22(b), dashed line: rescaled DS data to demonstrate proportionality of γ_{LS} and γ_{DS} (adapted from Ref. [64].)

As demonstrated, below T_x the spectral shape of the slow dynamics changes in a characteristic way: The power-law of the excess wing emerges with an exponent γ, that shows a linear temperature dependence, and all the parameters of the distribution $G_{GGE}(\ln \tau)$, which describe the overall shape of the susceptibility, are intimately connected with each other. Here, the question arises how $\gamma(T)$ and $\tau_\alpha(T)$ are related in detail. An often applied phenomenological interpolation of $\tau_\alpha(T)$ is provided by the Vogel-Fulcher-Tammann (VFT) law, $\tau_\alpha(T) = \tau_0 e^{\frac{D}{T-T_{VFT}}}$. The VFT behavior holds for many glass forming liquids close to T_g and up to a certain temperature, where a crossover to a different temperature dependence in the high-temperature regime is found [170,171]. From the VFT and from Eq. (38), the following relation between γ and $\lg \tau_\alpha$ can be derived [275]:

$$\lg \tau_\alpha = \lg \tau_0 + \frac{C_1}{\gamma + C_2} \tag{39}$$

and the result is seen as solid line in Fig. 27(b). To first approximation it turns out that $C_2 \approx 0$, which implies that $T_{VFT} \cong T_\gamma$. We note that while discussing a possible divergent susceptibility at $T < T_g$, Menon and Nagel [266] assumed that $T_{VFT} = T_\gamma$, whereas in the present context we find indication for the latter from a phenomenological point of view. A detailed analysis reveals that the above relation holds for all systems under study as soon as the exponent is rescaled by its value at T_g, i.e. γ/γ_g is considered [275]:

$$\lg(\tau_\alpha/\tau_0) = \frac{A_1}{\gamma/\gamma_g + A_2} \tag{40}$$

with $A_1 = C_1/\gamma_g = 25.0$, $A_2 = C_2/\gamma_g = 0.5$ and $\lg(\tau_0/s) = -15$, where the latter constant $\lg\tau_0$ was determined independently. The result is seen in Fig. 28, where γ/γ_g is plotted as a function of $\lg\tau_\alpha$ and Eq. (40) is included as a solid line. Whether or not $C_2 \neq 0$ is significant, which would imply slightly different T_{VFT} and T_γ, cannot be decided at the present stage, as γ might be slightly underestimated at low temperatures due to a certain curvature of $\varepsilon''(v)$ in the region of the excess wing, which occurs in some of the systems probably due to some weak β-process contribution. A similar scaling relation can be established for $\lg\sigma$ as a function of $\lg\tau_\alpha$ so that τ_α fully determines the line shape of the susceptibility as soon as γ_g is fixed as a system parameter [275].

Given all of this, one may expect that the crossover temperature discussed by Stickel et al., which marks the failure of the low-temperature VFT law, agrees with the crossover temperature T_x defined by the spectral changes in the dynamic susceptibility. In order to check this, Fig. 32 shows the α-relaxation time of different glass formers as a function of $T - T_g$ (Fig. 32(a)) and as a function of a variable $z = m(T/T_g - 1)$ (Fig. 32(b)), which rescales the plot in order to remove the effects of an individual fragility m (cf. Eq. (29)). Note that a similar approach was proposed in several publicatios [167,168,169]. We stress that the plot of Fig. 32(a) is somewhat different from the renowned "Angel plot". Whereas the systems clearly show a different behavior in the first case, one can show [275] that in the second plot a master curve is expected for $\tau_\alpha(z)$, if (i) the time constants indeed follow a VFT law and (ii) the pre-factor τ_0 in the VFT equation is identical or at least similar for the systems under consideration. Under these conditions $\tau_\alpha(z)$ will fall on the following master curve [275]:

$$\lg(\tau_{VFT}/\tau_0) = \frac{K_0^2}{m(T/T_g - 1) + K_0} = \frac{K_0^2}{z + K_0} \tag{41}$$

Note, that T_g contained in the variable z is assumed to be defined by the condition $\tau_\alpha(T_g) = \tau_g = 100\,s$ and the constant K_0 is given by $K_0 = \lg(\tau_g/\tau_0)$. As can be seen in Fig. 32(b), the datasets of all the systems that were considered in the

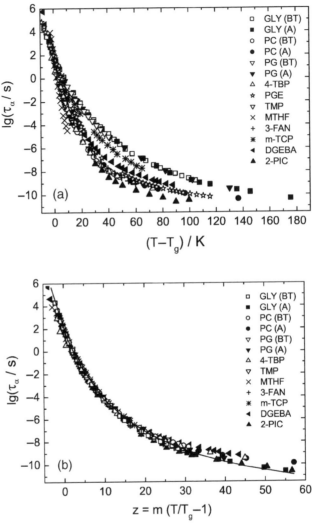

Figure 32. The α-relaxation times τ_α for the glass formers studied in the present work (cf. Fig. 27). In addition data of diglycidyl ether of bisphenol A (DGEBA) and phenyl glycidyl ether (PGE) are included; time constants as obtained from DS; data sets of m-TCP and 2-picoline were combined with τ_α from conductivity and light scattering measurements, respectively. (a) Relaxation times as a function of $T - T_g$. The systems differ by the slope of τ_α at T_g. (b) By plotting τ_α as a function of the rescaled temperature $z = m(T/T_g - 1)$ the effect of an individual fragility is removed and a master curve is obtained for systems with similar τ_0. Solid line represents Eq. (41) with $K_0 = 17$. (c) Upper part: master curve for τ_α according to Eq. (42). Deviations of the data from Eq. (42) (solid line) indicate break-down of the VFT equation. Lower part: The ratio $\lg(\tau_\alpha/\tau_{VFT})$ shows deviations from a VFT behavior most clearly. Dashed vertical lines indicate shortest and fastest τ_x, respectively, observed. All the figures taken from Ref. [275].

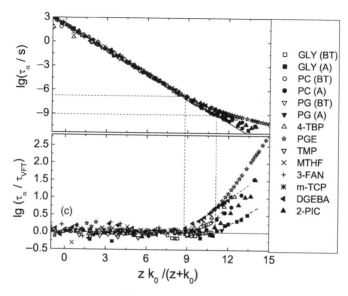

Figure 32. *Continued*

previous specral shape analysis coincide on the master curve Eq. (42) in the low-temperature regime with a constant $K_0 = 17.0$ or, equivalently, $\lg(\tau_0/s) = -15.0$, which was already used in Eq. (40). In Fig. 32(b) the time constants of the systems, for which also high-temperature data are available follow the master curve up to a certain value of z_x, which, e.g. for PC, corresponds to a relaxation time of $\lg(\tau_x/s) \approx -8.4$, being close to that earlier identified at the spectral crossover temperature T_x. At higher temperatures, the curves spread into individual traces indicating deviations from the VFT behavior. However, in the representation of Eq. (41) slight variations in $\lg\tau_0$ might still play a role and the deviations from the VFT law do not show up easily. Thus, in order to identify deviations from the VFT behavior more reliably, one may rewrite Eq. (41) to obtain [275]:

$$\lg\left(\tau_{VFT}/\tau_g\right) = \frac{-K_0 z}{z + K_0} \qquad (42)$$

Fig. 32(c) shows $\lg\tau_\alpha$ as a function of $K_0 z/(z + K_0)$. Here, K_0 is chosen such that for each system a slope of -1 is obtained in the low-temperature regime. Deviations from this slope now unambiguously indicate departure from a VFT behavior. This is best seen by considering the ratio $\lg(\tau_\alpha/\tau_{VFT})$. By comparing the upper and lower part of Fig. 32(c) τ_x is now most easily read off for each system. Indeed it turns out that for glycerol, PC and 2-picoline the crossover occurs on a rather similar time scale of $\lg(\tau_x/s) \approx -8.5$ which closely agrees with the crossover identified from changes in the shape of the DS spectra (cf. Figs. 27 and 28). For glycerol and PC similar crossover temperatures were reported by applying the "Stickel analysis" to viscosity or τ_α data [171,276]. Thus, the crossover temperatures identified from changes in the relaxation spectra and in the VFT behavior, respectively, appear to coincide for glycerol, PC and 2-picoline (cf. also table 1 in Sect. IV.E.). Here, the corresponding relaxation time $\lg\tau_x$ is identical and so is the von Schweidler exponent in the high-temperature regime. In principle, the crossover relaxation time τ_x must be conceived as a system dependent quantity, as the exponent parameter $\gamma(T)$ will follow the relation Eq. 39 up to different values of $\lg\tau_\alpha$ and thus a von Schweidler behavior with temperature independent but system- or method-dependent γ will result. To illustrate this we included the relaxation times of diglycidyl ether of bisphenol A (DGEBA) and phenyl glycidyl ether (PGE) [431] in Figs. 32. As a strong secondary process covers the high-frequency part of the α-relaxation an excess wing cannot be analyzed in these systems. However the time constants clearly show a crossover at $\lg(\tau_x/s) \approx -7$ and thus demonstrate that τ_x is a system dependent quantity. Correspondingly a crossover in the evolution of the spectral shape would be expected at longer relaxation times, though this cannot be checked directly.

The spectral analysis presented above allows for drawing a coherent picture of the evolution of the dynamic susceptibility in molecular glass formers of type A. The shape of the susceptibility spectra in the low-temperature regime is determined by three system parameters, namely T_g, fragility m and excess wing exponent γ_g at T_g. The temperature dependence of the spectral shape (given by the scaling relations for the parameters γ, β and σ) as well as τ_α can be calculated therefrom [275]. It is interesting to ask whether a similarly unique relation holds for the parameters of the LS spectra. As already discussed, first inspection shows that $\gamma_{DS}(T) \propto \gamma_{LS}(T)$ (cf. Fig. 31). Yet, currently only few data for $\gamma_{LS}(T)$ are available, which cover both the high- and low-temperature regime. Independent of these questions, the findings presented in this section strongly support the idea that the excess wing, whether part of the α-process or an independent secondary relaxation process, is a generic property of the susceptibility in the low-temperature regime at least of type A glass formers. In contrast, in the high-temperature regime no significant changes of the spectral shape are observed, and FTS applies. The temperature T_x marks the crossover between the two regimes and may be determined from monitoring $\gamma(T)$, another spectral shape parameter, or from deviations from the VFT behavior of the relaxation times τ_α. It is a challenge to incorporate in this picture the body of data provided by pressure studies (for a review, see Refs. [152,282]).

Finally, independent of the detailed analysis performed by applying the GGE distribution, the qualitative result obtained is similar to that given within the

Nagel scaling approach and also to that obtained by decomposition into separate relaxation processes [9,258]. Though different in detail, all the mentioned approaches agree that the high-frequency exponents of the α-peak and the excess wing both vary with temperature and thus that the FTS principle does not apply close to T_g.

2. Glass Formers with a β-Peak

In this chapter we consider glass formers for which a well-resolved secondary relaxation peak, the β-process, is identified on the high-frequency side of the α-peak. Most of the results compiled so far stem from DS experiments [151,153,283–292], but the β-process may also be probed by mechanical relaxation [293–297] as well as by thermal relaxation [298], neutron scattering in polymers [277,299–304], and NMR studies [96,97,305,306]; the latter will be discussed in some detail in Section V.B. Reviews have been provided by, for example, Johari [1], Murthy [307], Kudlik et al. [6], Ngai and Paluch [152], and Vogel et al. [15].

There are strong arguments that favor the view that in many cases the β-process is of intermolecular in origin and thus has to be regarded as an intrinsic property of glasses [1,6,153]. Most important, the β-process is observed in glasses formed by rigid molecules such as toluene, and therefore cannot be explained by assuming internal relaxational degrees of freedom, at least in cases like those already mentioned. Also, the β-process is observed in polymers, often among further relaxation processes, suggesting that the β-process found in polymers may be an intrinsic processes, too [15,283]. In some of these polymers, NMR investigations have demonstrated that indeed the main chain participates in the β-relaxation [308,309].

Some DS spectra of type B glass formers have been given in Fig. 14. The β-peak emerges close to but above T_g and persists below as a symmetric relaxation peak. Its most prominent property is the Arrhenius temperature-dependence of the time constants τ_β obtained from analyzing the peak position at $T < T_g$. This is shown in Fig. 15 for the two glass formers fluoroaniline and toluene. Above a certain temperature $(T > T_g)$, both α- and β-relaxation have merged, and only a single peak susceptibility survives at high temperatures. There, no difference among type B and type A glass formers has been reported (cf. Section IV.B) regarding the evolution of the susceptibility spectra. So far it is unclear whether a merging temperature can unambiguously be defined and whether the merging is yet another feature of the dynamic crossover dividing the high- and the low-temperature regime of a glass former [190,272,274].

Below T_g, the spectral width of the β-peak increases as temperature decreases, and consequently FTS does not apply. In the glass the β-process exhibits all the characteristics of a thermally activated process; and because the spectra are much broader than a Debye relaxation, the process is governed by a broad distribution of activation enthalpies. Indeed, dielectric hole burning experiments have demonstrated that a long-lived heterogeneous distributions of relaxation times exist for the β-process in sorbitol [310] as well as for the excess wing in glycerol [311]. Assuming a distribution of activation enthalpies $g(\Delta H_\beta)$, the contribution of the β-process to the complex permittivity can be written as follows:

$$\varepsilon_\beta(\omega) - \varepsilon_\infty = \Delta\varepsilon_\beta \int_0^\infty g(\Delta H_\beta) \frac{1}{1 + i\omega\tau(\Delta H_\beta)} d\Delta H_\beta, \qquad T < T_g \quad (43)$$

where ε_∞ is the permittivity at optical frequencies, $\Delta\varepsilon_\beta$ denotes the relaxation strength of the β-process, and $g(\Delta H_\beta)$ is usually assumed to be a Gaussian distribution [6,153,277,284,285,303,312,313]. Relaxation time τ_β and activation enthalpy ΔH_β are connected via the Arrhenius law, $\tau_\beta \propto \exp(\Delta H_\beta/RT)$. The mean activation enthalpy $\langle\Delta H_\beta\rangle$ can be extracted from the Arrhenius plot of Fig. 15. The Cole–Cole function is also often used to describe β-peaks; however, we emphasize that such a form for the susceptibility is incompatible with the assumption of a temperature-independent distribution of activation energies [34,142]. Sometimes the high-frequency wing of the β-peak is better interpolated by power laws. To take account of this, another distribution of relaxation times has been introduced, which is also compatible with a temperature-independent distribution of activation enthalpies—that is, the simplest case of thermally activated dynamics [142,157].

The β-process exhibits further two commonly observed features. Its relaxation strength is virtually temperature-independent below T_g, while it strongly increases with temperature above T_g, and the mean activation enthalpy $\langle\Delta H_\beta\rangle$ often correlates with T_g; for example, $\langle\Delta H_\beta\rangle = (24 \pm 3) RT_g$ was reported for a series of systems [6,153,285] though exceptions are found [314]. In some cases, values of $\langle\Delta H_\beta\rangle = 12 RT_g$ were observed, and as a consequence of such low activation enthalpies the β-process does not merge with the α-process at high-temperatures, so that this kind of secondary relaxation is expected to be observable even in the high-temperature liquid. Examples are given by ethanol [315,316], trimethyl phosphate [230], epoxy systems [317] and also some polymers [318] (cf. Fig. 33). Of course, it is not clear at all whether all these secondary relaxation processes are intrinsic processes, and up to the present no generally accepted criterion has been established although attempts have been made in that direction [152]. Experiments under pressure are expected to shed more light on this problem. So far, pressure effect studies report that some secondary processes are relatively insensitive to the pressure, while others exhibit a pronounced pressure dependence, and a definitive conclusion is not possible at the present stage [282,290,319–322].

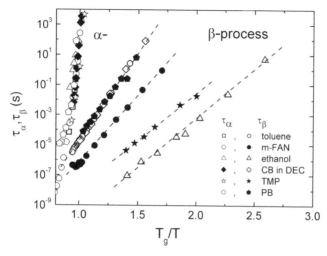

Figure 33. Time constants of the β-process obtained from dielectric spectroscopy as a function of the reduced reciprocal temperature T_g/T for the glasses toluene, fluoroaniline (*m*-FAN), ethanol, a mixture of chlorobenzene and decaline (CB/DEC), trimethyl phosphate (TMP), and polybutadiene (PB). (Compiled from Refs. 6, 137, 230, 306, 315, and 344.)

Both features, the increase of the relaxation strength with temperature and $\langle \Delta H_\beta \rangle$ being correlated with T_g, strongly suggest that the β-process is related to the glass transition phenomenon. Due to similar attempt frequencies and the approximate relation $\langle \Delta H_\beta \rangle \propto T_g$, the relaxation times τ_β of various glasses often appear similar, when plotted versus the reduced inverse temperature T_g/T. This is demonstrated in Fig. 33. Once again, we emphasize that similar β-processes are also observed in polymers. For instance, polybutadiene (PB), a polymer without side groups, shows a quite similar β-relaxation to toluene (cf. Fig. 33).

Using Eq. (43) with a suitable distribution function, time constants of the β-process can be extracted from experimental susceptibility spectra in the glassy state $(T < T_g)$. However, above T_g, where both α- and β-process are present, the spectral shape analysis becomes more involved. Taking into account that also fast (ps) relaxational and vibrational dynamics are present (cf. Section IV.B), the correlation function of a type B glass former near T_g is a three-step function, reflecting the dynamics occurring on different widely separated time scales. This is schematically shown in Fig. 34.

Following the approach for the high-temperature regime [Section IV.B, Eq. (32)], one may assume that the different relaxation processes are statistically independent, in which case the correlation functions factorize:

$$C(t) = C_{\text{fast}}(t) C_\beta(t) C_\alpha(t) \tag{44}$$

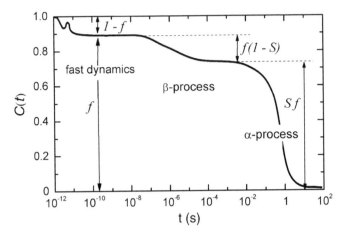

Figure 34. Schematic three-step correlation function of type B glass formers close to T_g. Here f is the nonergodicity parameter, and S is related to the β-relaxation strength [cf. Eq. (46)]. (Adapted from Ref. 65.)

where the normalized (initial value 1) functions $C_{\text{fast}}(t), C_\beta(t)$, and $C_\alpha(t)$ describe the relative partial correlation losses due to the fast dynamics, β-process and α-process, respectively [65]. The magnitude of the correlation loss in the course of the fast, picosecond dynamics is $1 - f$, where f, as before, is the generalized nonergodicity parameter. Note that in type B glass formers, f is no longer equal to the strength of the α-process but rather of α and β combined. For the β-process correlation function we write

$$C_\beta(t) = (1 - S)\phi_\beta(t) + S \qquad (45)$$

where $1 - S$ is the correlation loss brought about by the β-process, and $\phi_\beta(t)$ decays from 1 to 0. Then, at times much longer than the fast dynamics Eq. (44) reduces to

$$C(t) \cong f[(1 - S)\phi_\beta(t) + S]\phi_\alpha(t) \qquad \text{(long times)} \qquad (46)$$

We note that both f and S may be temperature-dependent (cf. Section IV.D). The approach, based on Eq. (46), is essentially identical to the so-called Williams–Watts ansatz [6,277,283,324]. Alternatively, the α- and β-processes can be assumed additive [325], implying that they occur in parallel rather than in series, as in Eq. (44). Since all the molecules participate in both the α- and β-dynamics that occur on different time rather than space scales, we believe that the Williams–Watts ansatz is more adequate for spectral analyses at $T > T_g$.

At $T > T_g$, a quantitative analysis using Eq. (46) is further complicated for the following reasons. Regarding the description of $\phi_\beta(t)$, it is not clear whether Eq. (43) still applies, since above T_g the assumption of a thermally activated process is no longer justified. Indeed, there are indications that the temperature-dependence of τ_β changes upon passing T_g [157,326–329]. In addition, it is not clear what function has to be used to interpolate $\phi_\alpha(t)$. Again, one may use the GGE function formalism; however, when a pronounced β-process is seen, an excess wing, which could still be present below the β-relaxation, will not be recognized in the spectra, and an analysis based on the GGE function becomes ambiguous. Only for weak β-process do both features become distinguishable and the spectra can be analyzed using the GGE distribution function. However, even if both features are present, the quality of the experimental data available is often insufficient to allow for a reliable extraction of parameters. Figure 14a shows sample fits of fluoroaniline, where both the excess wing and a β-process are clearly distinguishable. By contrast, the spectra of toluene with a strong β-process can be well-interpolated using Eq. (43) for the β-process and a simple model for the α-peak, such as the CD function or the generalized gamma (GG) distribution formalism, even though an additional excess wing may still be present under the β-peak [6,142,153,157,285]. Corresponding interpolations are shown in Fig. 14b. In all the analyses presented so far, a possible temperature dependence of the nonergodicity parameter $f(T)$ [cf. Eq. (46)] has been ignored, which is, however, not justified (cf. Section IV.D).

In the cases when both the excess wing and a β-process are well-resolved in the spectra, one may ask the question: Does the spectral shape of the slow dynamics including an α-peak and the excess wing still show the same behavior as in type A systems? Figure 35 shows spectra of several type B systems, rescaled so that their α-peaks overlap. For comparison, we include the data for glycerol, which is a type A glass former, rescaled to overlap with the other spectra at high frequencies. Clearly, the five lower spectra (which include glycerol) agree over a broad range of frequencies. Thus, type B systems with a weak or relatively fast β-peak (with respect to the α-process) exhibit a quite similar excess wing, which also resembles that of type A glass formers, such as glycerol in this example. It therefore appears probable that the "type A characteristics" are preserved in some type B glass formers. However, for a strong and relatively slow β-process (PB, toluene and DMP) we recognize that the high- and the low-frequency flanks of the α-peak are significantly broadened. It now appears that the presence of the β-process destroys the type A characteristics. The broadening on the low-frequency side of the α-peak can be taken into account by choosing a smaller value of α in the GG distribution. One finds the trend that the stronger the β-peak and the closer it is

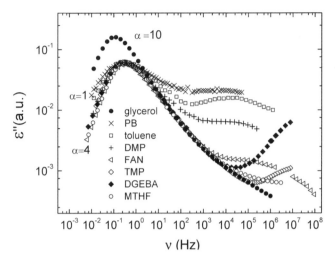

Figure 35. Dielectric spectra of several type B glass formers compared to that of glycerol (type A). The spectra were vertically shifted to coincide best at the relaxation maximum, and the data of glycerol were shifted to coincide at high frequencies. Note that, given a rather weak and comparatively fast β-process, an excess wing similar to that of glycerol is recognized; indicated is the parameter α fixing the long-time behavior of the GGE distribution; the systems included are glycerol, polybutadiene (PB), toluene, dimethyl phthalate (DMP) fluoroaniline (FAN), trimethyl phosphate (TMP), and *m*-toluidine. (Compiled from Refs. 6, 137, 142, 230, and 331.)

to the α-peak, the broader the latter or, more precisely, the smaller the parameter α in the GG distribution, cf. Fig. 35 [137,154,155,157]. These results give confidence that at least two parameters are needed to adequately describe the α-peak, such as in the GG and GGE distribution functions. The apparent broadening of the α-relaxation in the presence of a strong β-peak bears a certain similarity to the apparent stretching (lowering of the von Schweidler exponent) of the α-relaxation in systems with a strong vibrational band (see Fig. 17). We mention that only PB, toluene, and DMP belong to the group of glass formers with an activation of $\langle \Delta H_\beta \rangle \cong 24\,RT_g$; the other exhibit smaller values. Finally, we note that several reports dealing with the relation of α- and β-processes within the so-called coupling model conclude that shorter $\tau_\beta(T_g)$ are associated with smaller values of the Kohlrausch parameter β_K—that is, a more pronounced stretching of the α-relaxation [152,330]. Inspecting Fig. 35, one realizes this relation is not generally observed; thus the authors have suggested a classification scheme for β-processes based on whether or not this relation is obeyed.

While β-relaxation is commonly regarded as a genuine feature of the glassy state, there is no satisfactory understanding of its origin. Johari and Goldstein [151,332] explained it by postulating the existence of "islands of mobility," where loosely packed molecules undergo relatively unhindered motion; that is, they assumed that only a fraction of molecules takes part in the β-process, which appears as due to defects in a nonuniform glass structure (cf. also Refs. 1 and 333]. By contrast, Williams and Watts [36,283] assumed that all molecules participate in the β-dynamics, the latter being a precursor of the ultimate α-relaxation. As will be demonstrated in Section V.B, NMR experiments are able to access the β-dynamics, indicating that essentially all molecules participate in the β-relaxation [96,97]. Similar conclusions were reported from solvation dynamics experiments [334]. However, there is a broad distribution of the geometries of motion, which cannot be described by simple approaches such as the two-site jump model.

The β-process provides a further puzzle: There are almost no reports of its detection by depolarized light scattering, even though the latter is believed to probe the same molecular reorientation dynamics as DS. In part, this must be related to the fact that the relevant kilohertz-to-megahertz range is not easily accessible to LS. However, from the rare reports where LS and DS results of β-processes are compared, it appears that β-processes only weakly show up in LS, if at all [65,160,161]. As an example, Fig. 36 shows normalized correlation

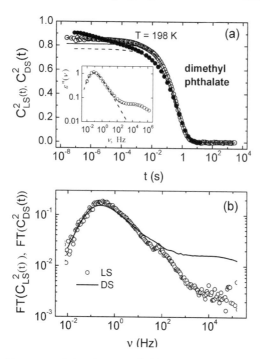

Figure 36. (a) Photon correlation function $C^2_{LS}(t)$ (unfilled circle) of dimethyl phthalate compared with the corresponding function $C^2_{DS}(t)$ (filled cicle) obtained from dielectric experiments. Lines are fits to different models. Inset: dielectric spectrum. (b) The same data compared in the frequency domain. FT, Fourier transform [65].

functions and the corresponding spectra of dimethyl phthalate, obtained by DS and PCS [65]. Obviously, the β-process is much weaker in the PCS than in the DS data of dimethyl phthalate. By contrast, in polybutadiene at high-temperatures a tertiary relaxation process was identified by DS, NMR, and LS [323,335]. In the light of these findings, it appears probable that secondary processes observed by PCS in other more complex glass formers are not intrinsic in nature and may be due to intramolecular degrees of freedom [256,336–338]. Here, again our remark from Section IV.A.1 seems applicable: Given relaxation dynamics with a small (angular) amplitude, one expects the relaxation strength $1 - f$ be *higher* in light scattering, $1 - f_{LS} \cong 3(1 - f_{DS})$; a similar relation is expected for the relaxation strength $1 - S$ of the β-process. Qualitatively, this is indeed observed for those dynamics that are faster than 10^{-8} s and whose strength determines the apparent initial values of the correlation functions in Fig. 36, but obviously not for the relaxation strength $1 - S$ of the β-process. It is a great challenge to explain these findings.

We now return to the discussion started already in Section IV.A.1 concerning the relationship of the excess wing and the β-process phenomenon—more specifically, the question as to whether the excess wing can be interpreted as a special β-process submerged under the α-relaxation [152,158,159,258]. In our view, the major achievement of Johari and co-workers was to demonstrate that there are intrinsic slow secondary relaxation processes in glasses, which first emerge already above T_g upon cooling. The excess wing also emerges upon cooling a liquid toward its T_g and persists, in the form of the nearly constant loss, in the glass. In this sense, the excess wing has the character of a secondary relaxation process. Excess wings as well as β-processes are characterized by a heterogeneous distribution of relaxation times [310,311]. Because both types of processes are known to occur in liquids composed of rigid molecules, both can be regarded as an intrinsic feature of the deeply supercooled liquid state, and in this sense both could rightly be called Johari–Goldstein (JG) processes. Of course, in cases of molecules with more complicated structure, there may be additional, nonintrinsic secondary relaxations.

At temperatures above T_g, the evolution of different kinds of secondary relaxations, or JG processes, is to varying degree related to the evolution of the α-relaxation. One extreme is the excess wing, discussed in Section IV.C.1. This secondary process manifests itself in quite a universal way: There is a unique coupling of the time scale of both processes, the main and the secondary relaxation, in type A glass formers. Moreover, the exponent γ and the position σ of the excess wing are uniquely related to the time constant τ_α. Such a strong coupling may even survive when pressure is applied [320,339]. The other extreme is a fast β-process, which does not even merge with the α-relaxation. Between these extremes, a large variety of β-processes exists. They exhibit varying degrees of separation from the α-process, along with varying relaxation strengths (cf.

Fig. 35). In some systems, it appears that a continuous change from type A to type B behavior takes place. For example, the extent of coupling to the α-process may change in a homologous series of glass formers; this may be expressed in continuous change of the attempt frequency of the β-process [154–157].

A further complication arises, because many glass formers exist exhibiting an excess wing in addition to a β-process, and it is by no means clear whether in those cases two intrinsic relaxation processes are present [6,142,152,230, 331,137,321]. In particular, for a relatively weak and well-separated β-process, it appears that the excess wing as well as α-peak still manifest themselves distinctly (i.e., the "type A characteristics"), and a simple superposition model may apply for both processes [6,284]. On the other hand, a strong β-process leads to a pronounced change of the α-peak, and then it is quite impossible to tell whether or not an additional excess wing is present. Furthermore, one may even have to face the fact that different JG processes may be probed in a different manner by the various methods. This, if true, would allow for a further distinction.

As discussed, up to now we do not fully understand the very nature of the JG processes. Regarding the underlying dynamics multidimensional NMR has provided some answers. In the case of the β-process, characteristic changes of ^{2}H NMR spectra typical of highly hindered reorientation were observed (cf. Section V.B). These spectral changes were found well below T_g. Interestingly, glass formers of type A exhibit similar spectra above T_g [15,306,340,341]—more precisely, at temperatures where the α-process is still too slow to be probed by the NMR techniques, demonstrating that near T_g molecular dynamics in the microsecond regime also still exist in type A systems. For the type A glass former polystyrene, the effect was attributed to the α-process in the context of the rotational diffusion model [72,341]. However, equally well, the typical effects of small-angle motions may be associated with the excess wing, suggesting that both excess wing and β-process involve similar and spatially highly restricted motion; of course, one may say that this is as expected for a secondary process. The important point is whether NMR is able to yield further information on details of the reorientational mechanism to specify possible differences regarding the various JG processes and in particular shed further light on the question how in some molecules two possibly intrinsic secondary relaxation processes occur coexisting on a similar time scale. All of this is a task for the future.

3. Mechanism of Molecular Reorientation—Results from NMR

In this section, we briefly review NMR results on relaxation in supercooled molecular liquids. For more details, the reader is referred to the articles by Böhmer et al. [11] and Böhmer and Kremer [12]. Various ^{2}H, ^{13}C, and ^{31}P NMR techniques can be used to access molecular dynamics related to the glass transition. In particular,

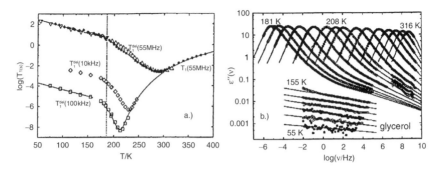

Figure 37. (a) Spin–lattice relaxation time T_1 of glycerol as a function of temperature: ^{2}H NMR experimental data at 55 MHz (filled circles) compared with calculation results for three frequencies. Dielectric loss spectra were used to estimate the spectral density in the calculations. The vertical line indicates T_g. (b) Dielectric spectra of glycerol together with interpolations (solid line) used to calculate the spin–lattice relaxation times in (a). (Adapted from Ref. 220.)

^{2}H (deuteron) NMR proved particularly powerful, since it detects solely the molecular reorientation and allows readily for isotopic labeling.

A classical NMR approach to probing molecular dynamics in liquids is to measure the spin–lattice (T_1) and spin–spin relaxation time (T_2) (cf. Section II.D) At high-temperatures, when the relaxation rate exceeds the Larmor frequency of the experiment ($\omega_L \tau_\alpha \ll 1$, "extreme narrowing" limit), T_1 and T_2 are equal and simply related to the correlation time τ_α, so that the latter can be easily obtained if the relevant coupling constant is known—for example, from analyzing low-temperature NMR line-shapes. In the moderately viscous regime, T_1 reaches a minimum when the correlation time τ_α is close to the reciprocal Larmor frequency ($\omega_L \tau_\alpha \cong 1$; cf. also Fig. 37a). The whole temperature-dependence of T_1 can be quantitatively analyzed by modeling the spectral density $S_2(\omega)$ or the associated correlation function $g_2(t)$. A combined analysis of T_1 and T_2 allows the imposition of constraints on the possible form of $S_2(\omega)$. Usually, the Cole–Davidson (CD) spectral density is adequate when the correlation times are short ($\tau_\alpha < 10^{-6}$ s, i.e., significantly above T_g) [177,342,343]. Close to T_g the CD model is insufficient and secondary relaxation processes, such as the excess wing and β-relaxation, have to be taken into account [154,155,178,220,324,345]. Examples of time constants τ_α obtained with the CD relaxation model are included in Fig. 15. They agree well with the results from other techniques.

Figure 37a presents ^{2}H T_1 data of glycerol-d_5. A typical T_1 minimum is observed at $T > T_g$, while below T_g the temperature-dependence of T_1 is weak (cf. also Fig. 52). In order to test whether NMR relaxation probes similar relaxation processes as DS, T_1 was calculated utilizing the dielectric spectra (cf. Fig. 37b), assuming that the NMR spectral density $S_2(\omega)$ can be approximated

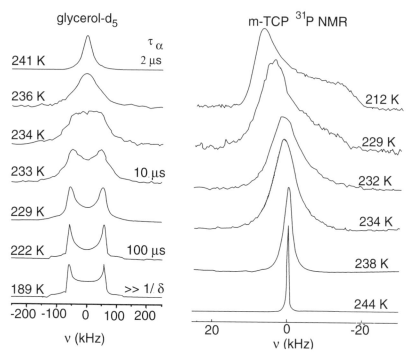

Figure 38. (left) Solid-echo ^{2}H NMR spectra of glycerol-d_5 ($T_g = 189$ K) [305]. A collapse of the solid-state spectrum is observed upon heating; the corresponding time constants of the α-process are indicated. (right) Hahn-echo ^{31}P NMR spectra of m-tricresyl phosphate (m-TCP, $T_g = 210$ K) determined by the anisotropic chemical shift interaction [324].

from the dielectric permittivity, $S_2(\omega) \approx \varepsilon''(\omega)/(\Delta\varepsilon\omega)$. Measured and calculated T_1 values agree well over the whole temperature range including temperatures below T_g. Thus, not only the α-process but also the excess wing is probed by NMR in a similar way as DS.

While cooling, when the limit $\tau_\alpha \cong 1/\delta_\lambda$ (cf. Eq. 15) is reached, the central Lorentzian NMR line, which is characteristic of a liquid ($\tau_\alpha\delta_\lambda \ll 1$), broadens, becomes structured, and eventually transforms into the solid-state spectrum, in the case of ^{2}H NMR the Pake spectrum. The breadth of the solid-state spectra makes it difficult to measure the corresponding (short) free induction decay (FID), so that it is necessary to use echo-techniques (cf. Section II.D.2). Figure 38 (left) shows solid-echo ^{2}H NMR spectra of glycerol-d_5. The crossover from a Lorentzian line to the Pake spectrum is observed some 20% above T_g. Below T_g the spectrum is independent of temperature. In Fig. 37 (right), the corresponding ^{31}P NMR spectra of m-tricresyl phosphate (m-TCP) are displayed. The characteristic spectral shape is now determined by the anisotropic chemical

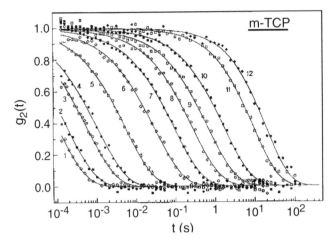

Figure 39. ^{31}P NMR stimulated-echo decays of m-tricresyl phosphate (m-TCP, $T_g = 214$ K) at short t_p, measuring the orientational correlation function $g_2(t)$. The numbers correspond to different temperatures in the range 199.5–226.4 K. The decay curves are fitted with a Kohlrausch stretched exponential (solid line). The corresponding time constants are shown in Fig. 15; at lowest temperatures, spin-diffusion (magnetization exchange via magnetic dipolar coupling of the spins) also influences the decay. (Adapted from Ref. 324.)

shift interaction. While NMR line-shape analyses of viscous liquids reliably yield characteristic relaxation times, it is in general not possible to extract unambiguous information on the shape of the spectral density $S_2(\omega)$ [341].

Most important in the current context are stimulated-echo experiments, since they allow to draw conclusions about the mechanism of the molecular reorientations related to the α-relaxation at temperatures close to T_g. A three-pulse sequence is used to create an echo signal, the amplitude of which depends on the evolution time t_p and the mixing time t_m (cf. Section II.D.3 and Fig. 7). The amplitude of the echo that forms at $t = t_p$ is recorded as a function of t_m, resulting in an echo function $C(t_m; t_p)$. In the limit $t_p \to 0$, the echo function yields the second rank orientational correlation $g_2(t)$. Using this technique, correlation functions in the time window $100\,\mu s < \tau_\alpha < T_1$ were obtained [87,88,324,346]. An example is given in Fig. 39. The nonexponential $g_2(t)$ can be interpolated by the stretched exponential function $\exp[-(t/\tau_K)^{\beta_K}]$ with a temperature-independent stretching parameter β_K, which is consistent with a virtually temperature-independent shape of the long-time tail of the α-relaxation close to T_g. Examples of time constants obtained from stimulated-echo decay measurements are included in Fig. 15. They are in good agreement with the other data. Summarizing, stimulated echo experiments in combination with T_1 and T_2 measurements allow one to determine orientational correlation times τ_2 in the range 10^{-12} through 100 s—that is, in the entire glass transition range.

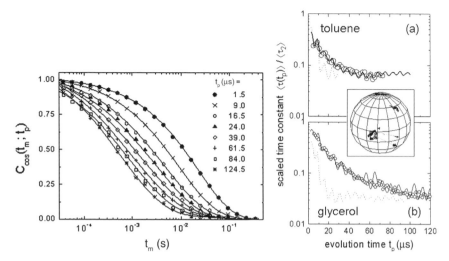

Figure 40. Results from ^{2}H NMR stimulated-echo experiments of the glass formers toluene and glycerol close to T_g. (left) Typical stimulated-echo correlation functions of glycerol for different t_p values. Note that the time constants shorten with increasing t_p (adapted from Ref. 215), (right) comparison of the $\tau(t_p)$ pattern for toluene (a) and glycerol (b). Solid lines represent random walk simulations assuming a bimodal distribution of jump angles with 80% 4° jumps and 20% 25° jumps for toluene, and 98% 2° jumps and 2% 20° jumps in the case of glycerol. Dotted line represent random walk simulations assuming a single jump angle of 10° (toluene) and 8° (glycerol); inset gives a schematic picture of an orientational trajectory. (From Ref. 12; cf. also Ref. 189).

Details of the molecular reorientation process are revealed when stimulated-echo decays $C(t_m; t_p)$ are measured for a series of t_p values. The resulting time constants and stretching parameters exhibit characteristic t_p dependences, which are fingerprints of the geometry of the motion; some examples are given in Fig. 40. Most important, such experiments allow one to determine the elementary jump time τ_{AJ}, which usually is much shorter than τ_α, thus indicating a multistep character of the α-relaxation reorientation dynamics. Several studies on supercooled liquids reported that the crossover from $C(t_m; t_p \to 0)$ to $C(t_m; t_p \to \infty)$ can be reproduced by random walk simulations with a bimodal jump angle distribution, with small angles in the range 2°–3° and large angles in the range 30°–50° [86,87,215]. The relative weight of these fractions of the distribution may however vary among the glass formers and with temperature [331,347].

While the advantages of the stimulated-echo technique lie in the resolution of small jump angles, measurement of two-dimensional (2D) NMR spectra is best suited to the study of large angular displacements, as typically found in crystalline rotator phases, for example. They lead to characteristic off-diagonal patterns [cf.

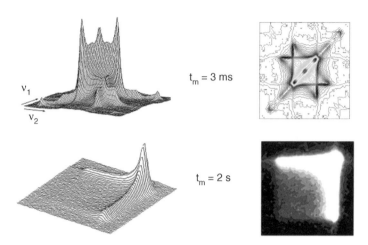

Figure 41. 2D spectra for long mixing times; oblique (left) and contour plots (right) are shown. (top) 2D ^{2}H NMR spectra of cyano-adamantane-d_{15} (CNA) in its glassy crystalline phase [344]; (bottom) 2D ^{31}P NMR spectra of the glass former m-tricresyl phosphate (m-TCP) close to $T_g = 214$ K [348]; in the case of the structural glass former, off-diagonal intensity is found all over the ν_1, ν_2 plane, in contrast to CNA, where characteristic ridges typical of sixfold jump in a cubic lattice are observed.

Fig. 41(top)], which are straightforwardly related to the jump angles [11,72]. In contrast, the dynamics involved in the α-process results in complete randomization of the molecular orientation so that the NMR frequencies at $t_m = 0$ and $t_m \gg \tau_\alpha$, respectively, are uncorrelated and a "box-like" 2D spectrum with off-diagonal intensity spread all over the frequency plane is observed [cf. Fig. 41(bottom)]. Information about the course of the reorientation process is available when 2D spectra are measured for different mixing times t_m [72,94].

A number of multidimensional NMR experiments were carried out in order to detect and explore the nature of dynamic heterogeneities in supercooled liquids. In NMR, dynamic heterogeneities reveal themselves as non-Markovian rotational dynamics. Dynamic heterogeneities and experiments to access them are discussed in several review articles [11,12,200,349–351]. These studies conclude the existence of transient, dynamically different subensembles of molecules that can be selectively probed. Most related NMR studies report that the lifetime of the heterogeneities is of the order of τ_α. Multidimensional NMR has also been used to demonstrate that the dynamics in supercooled liquids is spatially heterogeneous and to determine the length scale associated with this heterogeneity [352]. The authors reported a length scale of the order of 1 nm.

D. Temperature Dependence of the Nonergodicty Parameter

The nonergodicity parameter f denoting the relative relaxation strength of the slow relaxation may be introduced phenomenologically to describe a two-step correlation function as was done in Eq. (32). On the other hand, it plays a central role in the mode coupling theory (MCT; cf. Section III.B). Because the theory is concerned with density fluctuations, the nonergodicity parameter f_q depends on the momentum transfer q and is measured as an effective Debye–Waller (coherent scattering) or Mössbauer–Lamb factor (incoherent scattering). Most important, its temperature-dependence is expected to show a characteristic "singularity" or "cusp"—that is, a crossover from a virtually temperature-independent behavior in the high-temperature regime to a strong temperature-dependence below the critical temperature T_c. The latter is expected to be described by a square-root temperature-dependence [cf. Eq. (28)], and this behavior is also referred to as the anomaly of the nonergodicity parameter. The determination of f from experimental data can be accomplished completely independent of any theoretical model. Some of the results reported in the literature are discussed below.

Concerning the temperature-dependence of f_q at $T > T_g$, no agreement has been reached yet. Whereas in the first MCT tests applying neutron scattering (NS) techniques [147,241–243], this singularity was reported to be identified; in later NS investigations of the ionic glass former CKN, the previous finding was interpreted as an artifact of the data analysis [353]. Yet, reviewing the NS results compiled for the glass former o-terphenyl, Tölle [148] gave confidence for the presence of the square-root singularity; the corresponding results are found in Fig. 42. Also, an NS study on salol [354] and propylene carbonate [355] reported the singularity. From impulsive stimulated scattering of light, f_q at

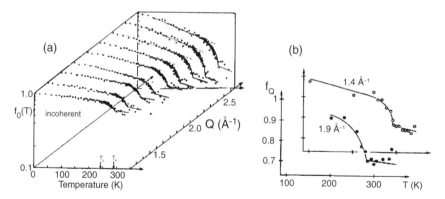

Figure 42. Nonergodicity parameter f_Q (effective Debye–Waller factor) of o-terphenyl as obtained from neutron scattering experiments for different values of the momentum transfer Q. (a) Incoherent, (b) coherent; full curves below T_c represent fits to the square root law of MCT yielding $T_c = 290$ K (From Ref. 201.)

$q = 0$ was obtained, clearly showing the singularity in propylene carbonate as well as in salol [248,249,356]; this also was found by applying Brillouin spectroscopy [357]. Finally, the anomaly was observed in a molecular dynamics simulation of o-terphenyl [358]. In contrast, for glycerol Petry and Wuttke [147] refrained from giving an estimate of T_c, and in a Brillouin study [359] the authors did not find any peculiarity concerning f_q. An impulsive thermally stimulated scattering study of glycerol did not detect any effect below 265 K [250].

Independent of whether or not a well-defined crossover temperature can be observed in NS data above T_g, it has been well known for a considerable time that on heating a glass from low temperatures a strong decrease of the Debye–Waller factor, respectively Mössbauer–Lamb factor, is observed close to T_g [360,361], and more recent studies have confirmed this observation [147,148,233]. Thus, in addition to contributions from harmonic dynamics, an anomalously strong delocalization of the molecules sets in around T_g due to some very fast precursor of the α-process and increases the mean square displacement. Regarding the free volume as probed by positron annihilation lifetime spectroscopy (PALS), for example, qualitatively similar results were reported [362–364].

Concerning the reorientational dynamics, it is obvious from the discussion in Section IV.B. that in addition to the slow α-process a fast process can be identified in the DS and LS spectra, and the interplay of both defines the manifestation of the susceptibility minimum in the gigahertz regime. Because in molecular liquids reorientational dynamics is strongly coupled to the structural relaxation, one can define a quantity analogous to the nonergodicity parameter f_q, giving the (relative) relaxation strength of the LS or DS spectrum associated with the α-process [cf. Eq. (32)]. Since this quantity f is often very close to unity, it is more appropriate to discuss the quantity $1 - f$ being the integral over the fast relaxation dynamics [64,65,136,183]. One can write

$$1 - f^{\mathrm{rel}} = \frac{\int\limits_{-\infty}^{\ln \nu_c} \chi''_{\mathrm{fast}}(\nu)\, d\ln\nu}{\int\limits_{-\infty}^{\ln \nu_c} \chi''(\nu)\, d\ln\nu} \tag{47}$$

The frequency ν_c defines an appropriately chosen cutoff to avoid contributions from the vibrational band, and $\chi''_{\mathrm{fast}}(\nu)$ may be singled out according to Eq. (31), for example. The superscript "rel" indicates that only the relaxational contributions are involved in the integration of Eq. (47). The results from LS studies can be examined in Fig. 43a. While at high-temperatures $1 - f^{\mathrm{rel}}$ is approximately temperature-independent, close to but above T_g, $1 - f^{\mathrm{rel}}$ decreases rapidly with lowering temperature until a much weaker temperature-dependence is found below T_g. This resembles what was reported in NS studies

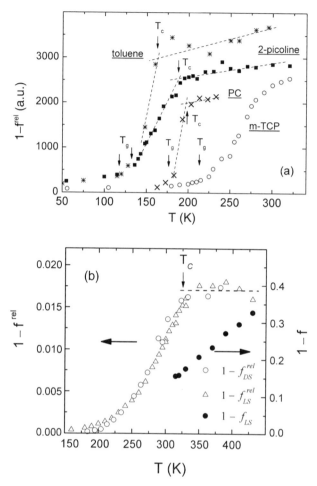

Figure 43. (a) Nonergodicity parameter $1 - f^{rel}$ for toluene, 2-picoline, propylene carbonate (PC) and m-tricresyl phosphate (m-TCP) as provided by analyzing light scattering (LS) spectra and applying Eq. (47) (compiled from [181,183,185]); (b) corresponding dielectric and LS data of $1 - f^{rel}$ for glycerol, for comparison $1 - f_{LS}$ (including vibrational contribution) directly extracted from data in Fig. 12b (adapted from Ref. 64); dashed lines: guide for the eye used to define the critical temperature T_c.

for the Debye–Waller factor (cf. Fig. 42); however, the break at T_g is more pronounced in the LS data. This may be explained by the fact that in the NS data, vibrational contributions are also included, which is not so in the LS data due to the introduction of the cutoff frequency in Eq. (47). Clearly, a crossover is observed in the nonergodicity parameter as defined by Eq. (47). Thus, from the crossover in the temperature-dependence of $1 - f^{rel}(T)$ the critical temperature T_c can be determined. However, since the crossover in $1 - f^{rel}(T)$ is often somewhat blurred, only an estimate of T_c can be given in such cases (cf. Table I in Section IV.F).

A model-independent estimate of the nonergodicity parameter is provided if one takes the inflection point of the correlation functions (cf. Fig. 13). However, here one faces another problem. The so determined $1 - f$ contains, as said, contributions originating from both relaxations and vibrations. In the case of the LS glycerol data, which are presented in Fig. 43b, clearly no crossover temperature can be identified in $1 - f(T)$, and a virtually linear increase with temperature is observed. However, when the relaxational contribution is singled out by again introducing a cutoff frequency, one rediscovers the crossover temperature around 300 K in both DS and LS data [64]. Thus, the effect in $1 - f$ is rather small and may easily be obscured by the vibrational contribution. In contrast to the LS data of glycerol the vibrational contributions are rather small in the corresponding DS spectra. Then, one may expect to see the anomaly of the nonergodicity parameter directly in the time domain data. Therefore, we Fourier-transformed the DS spectra of glycerol reported in Ref. 9 (cf. Fig. 12a); the result is seen in Fig. 44. Sufficient data including highest frequencies are only available for four temperatures. Nevertheless, although the effect is again rather small, the relaxation strength f stays virtually constant at 363 K and 295 K, whereas it increases with further cooling at 253 K and 184 K. In order to demonstrate that at the two highest temperatures the stretching is indeed very similar, we rescaled the 363 K data to coincide with those at 295 K, which works very well. Thus at these temperatures, the amplitude of the α-process was estimated by applying a fit with the same Kohlrausch stretched exponential. At lower temperatures this is not the case predominately due to the appearance of the excess wing. To our knowledge, this is one of the rare cases where the anomaly of the nonergodicity parameter can directly be distinguished in the data.

Magnetic resonance experiments—specifically, NMR and electron para-magnetic resonance (EPR)—probe predominantly slow motion. Thus, they usually collect information on the α-process in the low-temperature regime, and tests of MCT predictions are rare. Still, some important applications are found. For example, by applying ^{31}P NMR, indications of an anomalous temperature-dependence of the isotropic chemical shift was reported, indicating a dynamic crossover in a metallic glass [365]. The progress of high-field EPR offers another possibility [366–368]. EPR spectra measured at 285 GHz, for example, are extremely broad and offer a uniquely high angular resolution since, as in NMR, the EPR interaction is predominantly governed by the molecular orientation. Thus, solid-state EPR spectra are observed up to the high-temperature regime of glass formers, yielding the possibility of detecting the effect of the fast dynamics on the change of the effective coupling constant. The fast motion leads

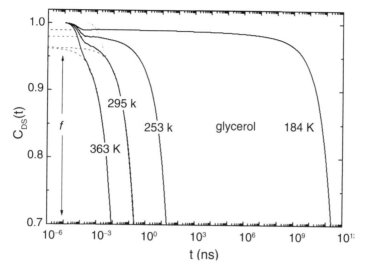

Figure 44. Correlation function as obtained from Fourier transformation of some of the dielectric spectra displayed in Fig. 12a [9]. Only the initial part is shown. Dashed lines represent extrapolation to short times providing an estimate of the nonergodicity parameter f; at $T = 363\,\text{K}$ and 295 K, extrapolations were done via a fit by a Kohlrausch decay with $\beta_K = 0.73$. Dotted line represents data at 363 K scaled to coincide with data at $T = 295\,\text{K}$ demonstrating similar stretching.

to a so-called pre-averaging of the solid-state spectrum yielding a slight reduction of the spectral width. Indeed, this effect was measured convincingly for EPR probe molecules dissolved in o-terphenyl and polybutadiene, demonstrating a quite sharp dynamic crossover well above T_g [369]. Some of the results are shown in Fig. 45. The quantity displayed characterizes the spectral width, and it is affected by the changing amplitude of the fast process.

All in all, regarding the temperature-dependence of the nonergodicity parameter, f, we think its clear change at temperatures well above T_g is by now well-documented by quite different techniques, and this is a major success of MCT. It can be singled out without referring explicitly to the theory. One should keep in mind that the change in f may be rather small (less than 5% in glycerol; cf. Fig. 44), which is why it may be difficult to extract it unambiguously in some cases. Thus, it may be more appropriate to search for the anomaly in $1 - f$ as has been done in Fig. 43. The somewhat rounded crossover displayed in most plots of f respectively $1 - f$, rendering the determination of the critical temperature T_c with some error margins, may rather be a consequence of the difficulties to extract the nonergodicity parameter from the data. Here, the EPR results provide a model-independent access to this parameter, and it appears that the crossover is indeed sharp. Note that the anomaly in $1 - f(T)$ seems to be in

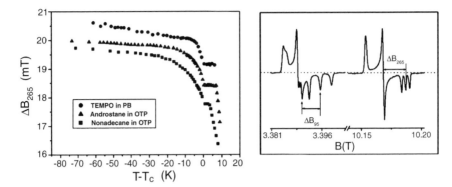

Figure 45. (left) The quantity displayed ΔB_{285} is a measure of the spectral width which is reduced by the change of the amplitude of the fast dynamics with temperature; different EPR probes (as indicated) were chosen to probe the dynamic crossover; T_c values reported are $297 \pm 2\,K$ (*o*-terphenyl, OTP) and $221 \pm 2\,K$ (polybutadiene, PB). (right) Typical EPR spectrum; indicated are two parameters characterizing the line width. (From Ref. 369.)

close proximity to T_x determined within the phenomenological analysis of the susceptibility spectra (for further discussion, see Section IV.F).

E. Tests of Mode Coupling Theory

Summarizing the phenomenological analysis presented in Sections IV.B and IV.C. there is strong evidence that two dynamical regimes, in the present context called the high- and the low-temperature regime, can be distinguished at least for the molecular glass formers of type A. The crossover from one regime to the other is marked by the temperature T_x. Whereas in the high-temperature regime the FTS principle holds, it breaks down in the low-temperature regime, accompanied by the emergence of the excess wing (or β-process). Moreover, the crossover in the spectral shape at T_x is accompanied by a clear change in the temperature dependence of the nonergodicity parameter. Since the critical temperature T_c of MCT may be determined by that change, it is very tempting to identify both temperatures, T_x and T_c.

Indeed, all the qualitative features of MCT that are supposed to appear at $T \geq T_c$, are observed in the high-temperature regime; that is, a two-step correlation function, the FTS principle for a stretched relaxation (α-peak scaling), a temperature-independent shape of the susceptibility minimum (minimum scaling), and the time scale of the α-process probed by the various techniques are proportional to each other. Beyond this qualitative scenario, the theory, being the most advanced attempt at describing the glass transition phenomenon, also makes quantitative predictions concerning the evolution of the susceptibility spectra (cf. Section III.B). Therefore, in Section IV.E.1, we shall review tests of the validity of the asymptotic scaling laws of the theory reflecting the universal properties of the dynamics. In Section IV.E.2 we shall briefly discuss more elaborated approaches of the theory to describe molecular glass formers. Again, we restrict the discussion to reorientational dynamics. For other tests of MCT the reader is referred to Yip and Nelson [3], Kob and Andersen [111–113], Götze [19], Tölle [148], and Das [21].

1. The Asymptotic Scaling Laws

The asymptotic scaling laws of MCT (cf. Section III.B) describe the crossover from the fast relaxation to the onset of the slow relaxation (α-process). In the frequency domain, this defines the shape of the susceptibility minimum in the gigahertz regime. They are derived as leading-order results from the solutions of the MCT equations and are expected to hold when the liquid approaches the critical temperature T_c from above. However, close to T_c, which in molecular liquids turns out to lie well above T_g, a further relaxation channel (so-called phonon-assisted hopping) is expected by the theory to become effective, which leads to a breakdown of the asymptotic scaling laws close to T_c. Thus, regarding that temperature range, no clear-cut prediction for the validity of the scaling laws can be given. Assuming that the reorientational dynamics is strongly coupled to the density fluctuations, the asymptotic laws were tested by several groups analyzing DS as well as LS susceptibility spectra.

Within MCT, the two exponents interpolating the susceptibility minimum in the gigahertz range of the LS and DS spectra, namely $\gamma = b_{MCT}$ (von Schweidler exponent) and a_{MCT} (fast dynamics), are not independent of each other but rather fixed by the exponent parameter λ [cf. Eq. (24), Section III.B]. In Figs. 18a, 18b, and 20 the corresponding MCT interpolations of the minimum are included. We note that interpolating the full relaxation spectrum by applying the Cole–Davidson function in addition to a power-law contribution for the fast dynamics is mathematically identical with the MCT interpolation of the minimum, provided that $\beta_{CD} = \gamma = b_{MCT}$ is fixed according to the exponent parameter λ. This approach was taken in Fig. 18b for 2-picoline. Usually, the parameter λ is determined by fitting the low-frequency side of the minimum. For the LS spectra of toluene and 2-picoline as well as DS spectra for glycerol the experimental data around the susceptibility minimum are compatible with this choice of the exponents. At high frequencies some deviations are observed which are attributed to the presence of the boson peak, and at high-temperatures the fit fails since no minimum is observed any longer, because both effects are not taken into account by the asymptotic laws.

Providing a satisfying interpolation of the susceptibility minimum, the temperature-dependence of the parameters $\chi''_{min}(T)$ and $\nu_{min}(T)$ (i.e., the locus of the minimum), as well as that of $\tau_\alpha(T)$, is expected to yield the critical temperature T_c [cf. Eq. (27)]. Again, all the exponents determining the respective temperature-dependence are fixed by the exponent parameter λ being temperature-independent. The corresponding rectification graphs are shown in Fig. 46 [136,183]. Indeed, the three parameters consistently lead to a critical temperature T_c. The scaling laws appear to work well up to highest temperatures, and we stress that only by including high-temperature data one is one able to clearly identify the deviations of the scaling laws appearing close

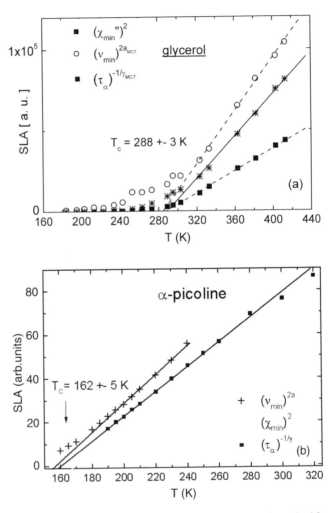

Figure 46. Linearized plots of the three parameters τ_α, ν_{min}, and χ_{min}, determined from the MCT analyses of the relaxation spectrum in the high-temperature regime. Plotted are the scaling law amplitude (SLA) as indicated: (a) from the dielectric spectra of glycerol (cf. Fig. 18a) (adapted from Ref. 136); (b) from the light scattering spectra of 2-picoline (cf. Fig. 18b) (from Ref. 183).

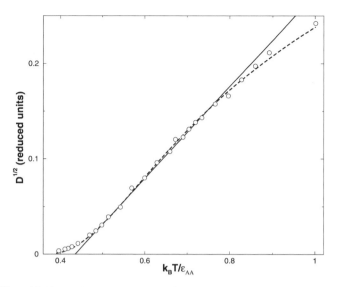

Figure 47.   Diffusion coefficient D as obtained from a molecular dynamics simulation study of a binary Lennard-Jones system reaching temperatures below the crossover temperature of mode coupling theory (MCT). Solid line represents interpolation by MCT power law; note the large temperature range covered by the power law. (From Ref. 371.)

to T_c. This result should be compared with a molecular dynamics simulation of Lennard-Jones systems [371], attaining very low-temperatures (cf. Fig. 47). Here, regarding the temperature-dependence of the diffusion coefficient D, the MCT power law is found to be appropriate also over a rather large temperature interval. Of course, at highest temperatures in the fluid liquid the MCT description is not expected to hold.

Similarly successful interpolations of LS spectra by applying the asymptotic laws of MCT were reported by Cummins and co-workers for salol [133], propylene carbonate [372], o-terphenyl [373,374], and trimethyl heptane [375,376]. Actually, it was Cummins' group that introduced the tandem-Fabry–Perot interferometry into the field to test MCT. The glass former o-terphenyl was also studied by Steffen et al. [377]. LS spectra analyses being compatible with MCT predictions were carried out also for m-tricresyl phosphate [217,218], toluene [370], 2-picoline [136,230], o-toluidine [247], and m-toluidine [229]. Lunkenheimer et al. [9,10] applied the universal laws for analyzing the DS spectra of glycerol and propylene carbonate. In most of the cases, the various studies for a given system reported similar crossover temperatures. In contrast, in the case of the nonfragile glass former glycerol, quite different T_c values were published, and no agreement was achieved on how to determine T_c from experimental spectra in this case ([64, and references therein). Also the more advanced schematic models of MCT were successfully applied to molecular glass formers [106,374,378,379]. In the so-called extended theory, which incorporates the hopping mechanism to restore ergodicity below T_c, spectral analyses were carried out also including spectra below T_c [374,380–384]. LS and NS spectra of several liquids reaching in some cases temperatures close to the boiling point were analyzed in the same spirit [385]. Even for benzene, traces of glassy dynamics in terms of a two-step correlation with a more or less stretched long time decay could be observed and analyzed within a schematic MCT model [149]. Also, the dynamics of water was interpreted in terms of MCT [386]. This once again demonstrates that "glassy dynamics" is a generic property of fluid liquids, so no supercooling is necessary.

For glycerol and propylene carbonate, DS as well as LS spectra were measured and an analysis within the idealized MCT was carried out [9,10,372,387], finding fair agreement. However, inspecting more recent LS data in Fig. 20, it is obvious that the interpolation by the asymptotic scaling laws of MCT does not provide a consistent interpolation of both DS and LS spectra. Most important, as discussed (cf. Section IV.B.1), the von Schweidler exponent is not method independent, and the broader α-peak in the LS spectra is accompanied by a much stronger boson peak. Similar results were observed for PC (cf. Fig. 20 and also Refs. 9). Thus, for two glass formers, at least, for which both LS and DS spectra were measured so far, application of the asymptotic laws of MCT clearly fails, and the various MCT analyses performed on single data sets (from LS or DS) have to be regarded with some care. It appears that for molecular reorientation, the predictions of the idealized theory are obscured by corrections depending, for example, on the particular shape of the molecule or on the magnitude of the boson peak contribution. As reviewed in Section IV.A.2, the following trend was observed: The stronger the boson peak, the smaller the apparent von Schweidler exponent γ_{LS}. Thus, it could be possible that LS data are less well suited to testing the asymptotic laws. A pronounced boson peak always hampers the analysis of the relaxational contribution at the susceptibility minimum. As will be discussed in Section IV.E.2, MCT models exist which allow one to explicitly take into account rotational dynamics.

Concerning reorientational dynamics, the asymptotic scaling laws of MCT were also tested in the time domain by analyzing optical Kerr effect (OKE) data for salol [69,70], m-toluidine [388–390], o-terphenyl [390], dibutyl phthalate [391], an epoxy compound [392], benzene [150], and water [393]. The OKE signal is assumed to be proportional to the negative time derivative of the LS correlation function [394]. In this representation of the data the different power-law regimes of MCT are expected to show up directly. For example, the von Schweidler law can be well distinguished from the final decay due to the α-process. Indeed, this is seen in Fig. 48 (top). In Fig. 48 (bottom), the asymptotic laws of MCT are tested. For the data at high-temperatures, a very good interpolation is provided with a single fitting parameter, λ, and the corresponding rectification plot consistently yields the crossover temperature T_c [inset in Fig. 48 (bottom)]. At lowest temperatures, actually below T_c, deviations from the MCT laws are observed. The crossover temperature agrees well with what was reported by depolarized LS experiments [133].

Studying recently a series of other glass-forming liquids, Fayer and co-workers claimed that the short-time dynamics actually cannot be ascribed to the dynamics described by the universal laws of MCT [395,396]. Rather than a critical decay, a power law t^{-z} with a temperature-independent exponent z is found in the OKE signal. The values of z are equal to or somewhat less than 1, implying that this so-called intermediate power law reflects a logarithmic decay in the reorientational correlation function $C_{LS}(t)$, corresponding to a nearly constant loss in the susceptibility spectrum. It falls in between the critical decay of MCT and the von Schweidler law. In the case of salol, this is recognized at lowest temperatures measured [dashed line in Fig. 48 (top)]. Similar decay curves are found in the isotropic phase of liquids forming a liquid crystal at lower temperature [397]. The authors are led to the assumption that glass-forming liquids contain a similar pseudo-domain structure to liquid crystals in their isotropic phase approaching the isotropic–nematic phase transition. Addressing this phenomenon, nearly logarithmic correlation decays were also reported and interpreted as a mani-festation of the β-peak scenario of MCT [398,399]. We mention that such a phenomenon is not observed when the LS as well as the DS susceptibility data of glycerol or propylene carbonate (Figs. 18 and 20) are converted to the cor-responding derivative decay curves. Of course, any more or less broad suscepti-bility minimum may also produce a decay, which may appear similar to a logarithmic decay in a limited time interval. For example, the appearance of the excess wing below T_x leads to a rather flat minimum.

2. Beyond the Asymptotic Scaling Laws

Although in some cases a consistent analysis of LS or DS spectra was carried out by applying the asymptotic laws of MCT, there are strong indications that these features are not completely appropriate to quantitatively describe, for example, DS as well as LS spectra. As discussed above, this is by now well known for PC and glycerol, at least. In order to tackle the problem of different experimental probes in a more realistic fashion, several MCT approaches have been published [265,380,400]. In a two-correlator schematic model, in which the dynamics of some probe (e.g., molecular reorientation in a dielectric experiment) is coupled to the overall structural relaxation in a simple manner, a simultaneous description of LS, DS, and NS spectra was possible even below T_c. Some of the results are

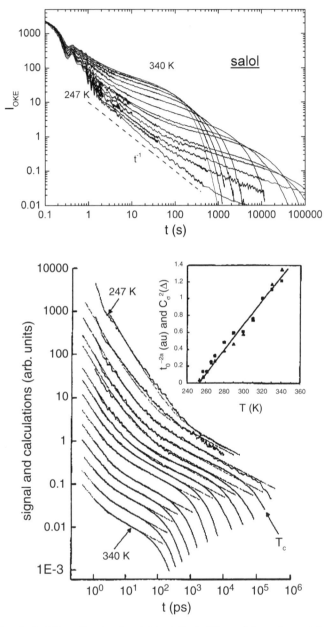

Figure 48. Optical Kerr effect data for salol (top). Dashed line: t^{-1} decay (bottom). Interpolation of the data by the universal β-correlator of MCT fixed by the exponent parameter $\lambda = 0.73$. Inset: Corresponding rectification plot yielding a crossover temperature 252 K (compiled from Ref. 70.)

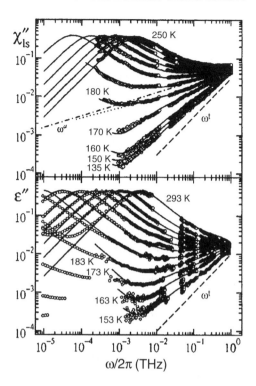

Figure 49. Susceptibility spectra for propylene carbonate $(T_g = 160\,\text{K})$ as measured by depolarized light scattering (top, data from Ref. 372) and dielectric spectroscopy (bottom, data from Ref. 9), each normalized by a temperature-independent static susceptibility. The full lines are fits from solutions of a two-component schematic MCT model. The dashed lines indicate a white noise spectrum. The dash–dotted line in the upper panel exhibits the asymptote of the critical spectrum. The dotted line shows the solution of the model at $T = T_c$ with hopping terms being neglected. (From Ref. 380.)

seen in Fig. 49, describing the experimental DS (rank $l = 1$ correlation function) and LS $(l = 2)$ spectra of propylene carbonate. Similar results were observed using the even more elaborate models, and they can be summarized as follows. (Indeed within such more advanced descriptions, MCT allows for a complete interpolation of the high-frequency spectra including, as said, also temperatures below T_c). For a given data set the FTS principle is rediscovered and the time scales of the α-process for each probe are proportional to each other. These are two features of the universal properties of MCT, which are thus expected to survive also in molecular systems. Nonuniversal features are found regarding the relaxation strength $1 - f$ of the fast dynamics. For $l = 2$ the contribution of the fast dynamics is stronger as compared to that of the $l = 1$

data. Also, the stretching of the α-process is more pronounced in the correlation function with $l = 2$ than for $l = 1$. In other words, the apparent von Schweidler exponent is not independent of the probe, a result also experimentally confirmed for glycerol and propylene carbonate (cf. Fig. 20).

F. Summary: The Evolution of the Dynamic Susceptibility above T_g

In this section, we shall attempt to summarize the results concerning the evolution of the dynamic susceptibility above the glass transition as probed by various techniques while cooling a simple molecular liquid. In previous sections, regarding the spectral shape analysis, arguments were given that two qualitatively different dynamic regimes can be distinguished, at high and low-temperatures, separated by a crossover temperature T_x. In the high-temperature regime, a two-step correlation function with a stretched long-time decay is observed. The shape of the correlation function appears to be virtually independent of the temperature; only the time scale of the α-process shifts. From this it naturally follows that the FTS principle holds, or, equivalently, that the α-peak scaling and the minimum scaling are applicable in this regime. Moreover, it turns out that the relaxation features characterizing the high-temperature regime, which are typical features of "glassy dynamics," are observed up to highest temperatures in most real liquids, while the transition to low-density fluid dynamics is so far only observed in computer experiments.

The low-temperature regime in type A systems is characterized by the emergence of the excess wing, which marks quite a sharp transition at T_x associated with the failure of the high-temperature scaling properties. In type B glass formers the situation is somewhat less clear from an experimental point of view; however, it may be possible that the emergence of the β-process equivalently marks the crossover. Although the excess wing, like the β-process, has to be seen as an intrinsic secondary relaxation process, it differs from the latter in that it exhibits quite a generic coupling to the α-relaxation, which is expressed most prominently in the relation of the excess wing exponent $\gamma(T)$ with the time scale τ_α. In Table I, we compiled data for T_x as obtained (i) from the crossover in the exponent $\gamma(T)$ (cf. Figs. 29 and 30), and (ii) from scaling the time constant $\tau_\alpha = \tau_\alpha(z = m(T - T_g)/T_g)$ (cf. Fig. 30b). For a given glass former the so-determined values of T_x agree well and demonstrate that phenomenologically the crossover temperature can be extracted unambiguously. The emergence of the excess wing and the change of the temperature-dependence of $\tau_\alpha(T)$ happen to occur at the same temperature. One should keep in mind that still not many data sets are available, which cover a sufficiently large temperature range in order to allow for a reliable determination of T_x. So far, the most extensive data analysis of DS and LS spectra is possible for only two glass formers glycerol and propylene carbonate.

TABLE I
Crossover Temperature for Various Glass Formers as Reported by the Different Methods: From the Temperature Dependence of the Stretching Parameter $\gamma(T)$, Scaling the Time Constant $\tau_\alpha = \tau_\alpha(T)$ [cf. Eq. (42)], Non-ergodicity Parameter $1 - f(T)$ Obtained from Spectra Analysis, Electron Paramagnetic Resonance (EPR), and from Tests of the Asymptotic Laws of Mode Coupling Theory

	T_x $\gamma(T)$	T_x $\tau_\alpha = \tau_\alpha(T)$	T_c $1 - f$	T_c EPR	T_c MCT asymptotic laws
glycerol	290 ± 8 (Fig. 30)	295 ± 4 (Fig. 32c)	325 ± 10 (Fig. 43b)	—	288 ± 3 (Fig. 46)
propylene carbonate	200 ± 5 (Fig. 30)	206 ± 3 (Fig. 32c)	197 ± 5 (Fig. 43a)	—	187 ± 5 [102,372]
2-picoline	175 ± 5 (Fig. 21)	174 ± 3 (Fig. 32c)	188 ± 5 (Fig. 43a)	—	162 ± 5 (Fig. 46)
o-terphenyl	—	—	290 (Fig. 42)	298.5 (Fig. 45)	290 ± 2 [148]
toluene	—	—	162 ± 5 (Fig. 43a)	—	155 ± 5 [185]
polybutadiene	—	—	—	221.5 (Fig. 45)	216 [147]

Concerning MCT, all the qualitative features of the theory are rediscovered in the experimental spectra in the high-temperature regime. However, the validity of the asymptotic laws, which could indicate universal quantitative features of the shape of the susceptibility spectra, is rather limited. For example, the apparent von Schweidler exponent γ turns out to be probe-dependent; a related difference in α-peak stretching was also reproduced in the framework of so-called schematic models of MCT. Nevertheless, we included in Table I such T_c values reported for cases where the asymptotic laws were confirmed to hold over a large temperature interval. The asymptotic laws appear to work well for relaxation spectra with a weak vibrational band. The reported values of T_c are close to T_x. Usually, T_c as obtained from applying the asymptotic laws is found somewhat below T_x. Anticipating that T_x and T_c are related to the same crossover phenomenon, this is not surprising, because in the linearized plots one extrapolates to low-temperatures in order to determine T_c, but deviations from the asymptotic laws are observed as expected already somewhat *above* T_c. Since empirically the relaxation features of the high-temperature regime (α-peak and minimum scaling) are observed up to the highest attainable temperatures, it seems to be justified that also the asymptotic laws, if applicable, have to be tested in a temperature range as large as possible, although this is not *a priori* expected by the theory.

Given that the asymptotic laws may fail, the only quantity expected to signal the crossover forecast by MCT is the anomaly of the nonergodicity parameter f.

Here indeed, a quite subtle change of the dynamics was revealed well above T_g. Regarding reorientational dynamics, this subtle effect seems to be always present; however, it is often difficult to unambiguously assign a distinct crossover temperature, because the crossover may be somewhat blurred, while the actual change in f is below 10%; in addition, one has to single out the spectral contribution originating from the fast relaxation. In most cases the latter is obscured by the vibrational contributions in the spectra. Moreover, *a priori*, T_c and T_x are not necessarily connected. It could be possible that that the asymptotic laws simply fail because some secondary process (e.g., the excess wing) appears actually not to be taken into account by the theory. Now, one would expect for a T_c determined by some extrapolation procedure that $T_c < T_x$. However, when inspecting the tentatively extracted values for T_c indicated by the crossover in the nonergodicity parameter, from analyzing the relaxation spectra as well as from EPR experiments, one realizes that T_c appears close to or even slightly above T_x (cf. Table I). Thus, it may be possible that the discrepancy between T_c and T_x results from uncertainties in the determination of $f(T)$. All in all, taking into account the general difficulties involved in determining T_x and T_c from experimental data, we are inclined to draw the conclusion that all the crossover temperatures are associated with the same phenomenon, that is, $T_x \approx T_c$.

Concerning the slow dynamics below the crossover temperature T_c, the predictive power of the theory seems to be rather limited. In particular, the emergence of *intrinsic* slow secondary processes, which seems to be associated with the dynamic crossover in the experimental spectra, is not contained even in the extended versions of the theory; consequently, the slow dynamics spectrum is not reproduced correctly. In this respect, the extended theory introducing the hopping mechanism for describing the susceptibility minimum below T_c is misleading. On the other hand, the most prominent prediction of MCT *below* T_c is the anomaly of the nonergodicity parameter, which, as discussed, is found by different model-independent approaches. However, within the framework of MCT, this anomaly is closely connected with the appearance of a so-called "knee" feature in the spectral shape of the fast dynamics spectrum below T_c. This feature, however, has not been identified experimentally in molecular liquids, and only indications for its existence are observed in colloidal systems [19]. In molecular systems, merely a more or less smooth crossover to a white noise spectrum has been reported in some cases [183,231,401]. Thus, it may be possible that the knee phenomenon is also smeared out.

Since α-relaxation stretching and minimum scaling can be observed up to highest temperatures, even well above the melting point, and since T_c is at a rather elevated temperature, one may conclude that MCT describes the slowing down of the dynamics in a rather dense liquid and may fail at some point due to the emergence of further slow secondary relaxation processes, which are precursors for dynamic properties in the glassy state and as such not contained in the theory. Still, the nontrivial temperature-dependence of the nonergodicity parameter below T_c is a remarkable prediction, and its change appears to be associated with the emergence of the secondary relaxation processes. This helps to sustain the hope that the theory may be extended to cover the slow dynamics also below T_c.

V. EXPERIMENTAL RESULTS OF MOLECULAR GLASSES $(T < T_g)$

When the temperature is decreased below T_g, the structural relaxation freezes, so that all the relaxation processes persisting in the glassy state occur without changes of the structure—that is, isostructurally. Here we do not consider aging processes that occur below but close to T_g on long-time scales (e.g., hours, days, or even weeks) and that lead to further structural relaxation even in the glassy state [7,402]. As will be shown, the dielectric loss of type A glasses (as well as NMR relaxation; cf. Fig. 37) is dominated by the nearly constant loss (NCL) phenomenon, whereas in type B glasses one observes instead a β-relaxation peak. As mentioned, it is not clear how β-processes manifest themselves in light scattering, with the scarce available data suggesting only a weak manifestation as compared to the dielectric spectroscopy [65,160,161]. In contrast, β-processes are detected by multidimensional NMR techniques, which, however, became obvious only recently [96,97,305,306]. Small-angle reorientational motion in the glassy state can also be sensitively probed by electron paramagnetic resonance (EPR) techniques, especially in high fields [403,404,405,406]. There are also numerous publications concerning LS below T_g, often in inorganic glasses, and dealing with vibrational as well as fast relaxational ($>$ GHz) properties (e.g., Refs. 407–411). We refrain from further reviewing these works here, and we restrict ourselves to discussing slow secondary relaxation processes. In Section V.A we briefly summarize dielectric relaxation phenomena below T_g for type A systems, and then in Section V.B we discuss NMR results of type B glass formers.

A. Nearly Constant Loss and Low-Temperature Properties of Molecular Glasses

Upon cooling into the glass, the α-relaxation peak moves out of the accessible frequency window, leaving behind secondary relaxation features. In type A glass formers, the excess wing then degenerates into the nearly constant loss (NCL)—that is, almost flat, frequency-independent spectrum. More specifically, one observes a power-law spectrum with a small exponent $\gamma = 0.1$–0.2, extending over many decades in frequency [6,412]. Utilizing a high-precision bridge, it is

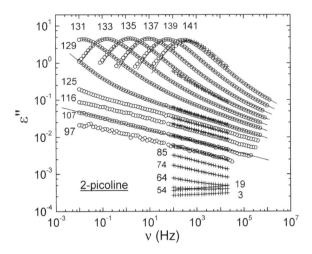

Figure 50. Dielectric spectra of the glass former 2-picoline ($T_g = 133$ K), measured with conventional broadband equipment (unfilled circles) and a high-precision bridge (crosses); the excess wing emerges at high-temperatures and degenerates into a nearly constant loss below T_g; note that at lowest temperatures (3 K and 19 K) the exponent of the power law spectrum changes from negative to positive. Solid lines represent fit of α-process including excess wing by the GGE distribution [cf. Eq. (36)]. (Compiled from Refs. 137 and 183.)

now possible to monitor the dielectric loss over a limited frequency range down to cryogenic temperatures [50,137]. Figures 12a, 14, and 50 present some examples. The corresponding temperature-dependence of the apparent exponent γ, shown in Fig. 51b as a function of the reduced temperature T/T_g, clearly displays a temperature range where γ is small and virtually temperature-independent (i.e., a NCL). The temperature-dependence of NCL was found to be exponential (see Fig. 51a) [6,412]. At lowest temperatures a crossover to a power-law spectrum with a small but positive exponent is observed in $\varepsilon''(\nu)$. The crossover is accompanied also by a change of the temperature-dependence. For the NCL contribution, we thus write

$$\varepsilon''(\nu) \propto \nu^{-\gamma} \exp(T/T_{\mathrm{NCL}}), \qquad T < T_g \tag{48}$$

where the constant T_{NCL} is introduced, in many cases found to be $T_{\mathrm{NCL}} \cong 30$ K. We note that a power law appears to be a reasonable but not perfect approximation of the spectra below T_g, which in some cases exhibit a small curvature in a double logarithmic plot of $\varepsilon''(\nu)$. As a note of caution, we emphasize that in such cases it is difficult to decide whether a weak β-peak is present. The NCL contribution is probably identical with what was previously referred to as "background loss" in molecular glasses [1]. We note that the NCL phenomenon is widely discussed for crystalline and amorphous ionic conductors

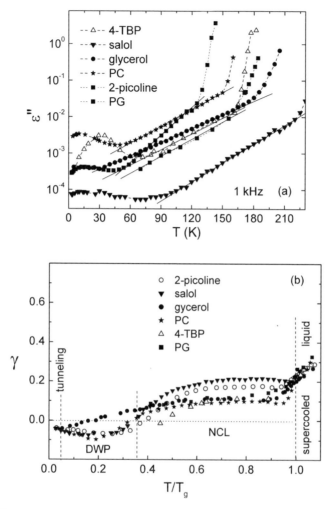

Figure 51. (a) Dielectric loss as a function of temperature for molecular glass formers (type A) measured with a high-precision bridge at 1 kHz. The solid lines are interpolations by Eq. (48). (b) Exponent of the power-law spectrum observed below T_g; a reduced temperature scale T/T_g is chosen, and the relevant relaxation mechanism is indicated. DWP, activated dynamics in double-well potentials; NCL, nearly constant loss. (Adapted from Ref. 50.)

[222–225], where it is, however, related to the dynamics of ions dynamics rather than to molecular reorientations. In these studies, NCL exhibits a nearly exponential temperature-dependence, resembling that of the Debye–Waller factor. It was accordingly proposed that NCL originates from vibrational dynamics in anharmonic potentials and may be present in molecular glasses, too, actually in addition to the excess wing [8,224,225]. The analysis of the DS spectra presented in Section IV.C.1 demonstrated that no additional constant loss contribution is needed to interpolate the spectra below T_x. Moreover, one should keep in mind that the NCL contribution in molecular glass formers covers an extremely wide range of frequencies, at least down to 10^{-2} Hz. In our view, the NCL phenomenon is reminiscent of the excess wing emerging close to T_x, which in the glass survives as a separate secondary relaxation process [6]. We note that, in addition to DS, the excess wing as well as NCL phenomenon were detected in NMR spin–lattice relaxation and in internal friction experiments [6,220,414] (cf. Fig. 37). In the spectral density, the NCL contribution approximately corresponds to $1/f$ noise. The latter was directly observed in the spectrum of spontaneous polarization fluctuations [415].

Inspecting Fig. 51, at temperatures below $T/T_g \cong 0.3$–0.5, further relaxation processes are identified in the $\varepsilon''(T)$ data. At the same time, the frequency dependence of $\varepsilon''(v)$ changes from a negative to a positive slope. Relaxational features emerging in this temperature range are likely associated with the low-temperature anomalies of glasses [50,416,417]. For example, plotting the dielectric loss as a function of temperature on a logarithmic scale (cf. Fig. 52), one sees that $\tan \delta$ reaches a plateau below about 3 K. This is similar to the behavior of amorphous silica (SiO_2), which is a paradigmatic inorganic glass, well-studied for

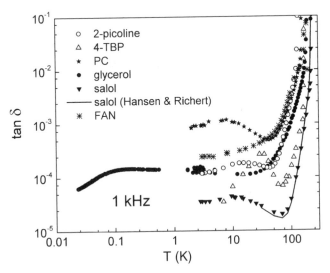

Figure 52. Double logarithmic plot of the dielectric loss data ($\tan \delta$) of several type A glass formers as a function of temperature. Below 3 K the tunneling regime with the "tunneling plateau" is recognized; the temperature range $3 < T < 30$ K corresponds to thermally activated dynamics in double-well potentials; at higher temperatures, nearly constant loss is found in the corresponding spectra. (Adapted from Ref. 50.)

its low-temperature properties. This plateau is called the "tunneling plateau," and it is also observed in internal friction data [418,419]. The plateau can be explained within the standard tunneling model assuming a distribution of two-level systems originating from the presence of double-well potentials (DWP). Inspecting Fig. 51b, the frequency dependence of the plateau (i.e., at lowest temperatures) exhibits a very weak but rather universal frequency dependence [50]. At temperatures slightly above the plateau, molecular glasses show a small absorption maximum also found in silica. These peaks can be explained by assuming thermally activated transitions in the double-well potentials [50,410,420,421].

B. The β-Process—Results from NMR

Relevant dielectric results of type B glasses were already discussed in Section IV.C.2. The spectra below T_g exhibit a broad symmetric secondary relaxation peak that can be interpolated assuming a Gaussian distribution of activation enthalpies. Only recently it became clear that also NMR is able to identify secondary relaxation processes in glasses, moreover providing information on the mechanisms of molecular reorientation that is not easily accessible to most of the other methods. For detailed reports the reader is referred to the reviews by Böhmer et al. [11] and Vogel et al. [15]. Here, we summarize the major results.

We first discuss the ^{2}H NMR spin–lattice relaxation results of molecular glass formers at $T < T_g$. In Fig. 53, we present the mean relaxation time $\langle T_1 \rangle$, equal to the integral of the corresponding (nonexponential) relaxation function, for several glasses including a polymer (polybutadiene-d_6). The temperature-

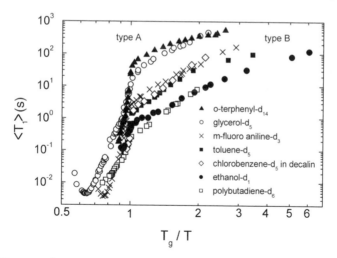

Figure 53. Mean ^{2}H spin–lattice relaxation times for type A and type B glass formers as a function of reduced reciprocal temperature in a double logarithmic plot. (Compiled from Refs. 177, 178, 323, 422, and 423.)

dependence of $\langle T_1 \rangle$ allows one to discriminate between type A and type B glasses. The relaxation of type A glass formers is significantly slower than that of type B glass formers, and the temperature-dependence is different. Due to the presence of the β-process, the spectral density of type B systems is enhanced, resulting in a faster relaxation. The temperature-dependences within A and B groups are, however, rather similar. For type B glass formers, this is a consequence of the fact that the activation enthalpy ΔH_β, the relaxation strength $1 - S$, and the shape of the β-process are often similar for diverse systems, leading to a comparable spectral density $S_\beta(\omega)$ [306]. Exceptions are polybutadiene and ethanol exhibiting further relaxation processes [315,316,323]. It is important to note that the temperature-dependence of $\langle T_1 \rangle$ is significantly weaker than that expected based on the activation energy determined from DS spectra. Furthermore, a power-law behavior rather than Arrhenius temperature-dependence is found [167]. The reason is that in addition to the activation energy, the broadening of the distribution function $G_\beta(\ln\tau)$ affects the temperature-dependence of $\langle T_1 \rangle$.

Following the reasoning in Section IV.C.2 and using Eq. (46), it follows that for the spin–lattice relaxation time [15] we obtain

$$1/\langle T_1 \rangle \propto f(1 - S)[S_\beta(\omega_L) + 4S_\beta(2\omega_L)] \qquad (49)$$

Well below T_g, one can neglect the temperature-dependence of the nonergodicity parameter f. The spectral density $S_\beta(\omega_L)$ at the Larmor frequency ω_L can be calculated assuming, for example, a logarithmic Gaussian distribution $G_\beta(\ln\tau)$ of relaxation times corresponding to a Gaussian distribution of activation enthalpies (and entropies), as is often done in dielectric spectroscopy [6,153,277,284,285,303,312,313]. One can then determine $1 - S$, from which an estimate of the mean angular displacement follows. It was found that, in all the cases, small-angle dynamics of essentially *all* the molecules govern the spin–lattice relaxation of molecular glasses at $T/T_g > 0.5$, which rules out the assumption of "islands of mobility" [154,155,167,345,424] (cf. also below).

Qualitatively comparing NMR relaxation data with those from DS in different systems, larger magnitudes of the β-peak in dielectric spectra correspond to faster spin–lattice relaxation times near T_g. For example, the glass former *m*-fluoro aniline, showing the slowest spin–lattice relaxation among the type B glass formers considered in Fig. 53, exhibits also the smallest dielectric β-peak. The existence of glasses exhibiting an even smaller relaxation strength of the β-process [cf. *m*-toluidine in Fig. (35)] suggests that there may be a continuous crossover from type B to type A relaxation behavior.

It has been usually argued that the NMR line-shape reaches the solid-state, rigid-lattice limit at temperatures around T_g (cf. Fig. 38), and consequently no dynamical effects are expected when measuring solid-echo spectra. While such

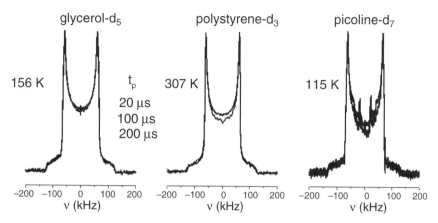

Figure 54. ^{2}H NMR spectra for the type A glasses glycerol-d_5, polystyrene-d_3, and picoline-d_7 at $T/T_g \approx 0.85$. Results for solid-echo delays $t_p = 20$, 100, and 200 μs are shown. In the case of picoline-d_7, the subspectrum of the methyl group was removed. (From Ref. 306.)

behavior is indeed observed for type A glasses, the situation is different for type B systems. Figure 54 shows spectra of several type A glasses measured at the same reduced temperature $T/T_g \approx 0.85$ and with different solid-echo delay time t_p. The spectra are independent of t_p and thus give no indication of dynamics on the μs-timescale, which is consistent with the absence of a secondary relaxation peak in the DS spectra. In contrast, the β-process of type B glasses manifests itself in a pronounced dependence of the ^{2}H NMR spectrum on t_p (cf. Fig. 55) [306]. These changes of the line-shape are typical of a small-amplitude rotational motion, as becomes clear when comparing the spectra with random walk simulations in Fig. 56a [93].

Figure 56b displays ^{2}H NMR spectra of chlorobenzene-d_5 in a mixture with *cis*-decalin at $T < T_g$, obtained with a long echo delay $t_p = 300$ μs. Strong deviations from the rigid-lattice spectrum are seen at higher temperatures, when the β-peak in the DS spectrum is located in the microsecond-range [306]. Upon cooling, these deviations are reduced until the rigid-lattice spectrum is reached near $T \approx 70$ K, where the β-process exits the NMR time window. These findings confirm that the ^{2}H NMR spectrum is indeed sensitive to the β-relaxation dynamics. For the long delay t_p, we see in Fig. 56b that the intensity in the center of the spectrum almost vanishes, but this only happens unless all the C–^{2}H bonds move on this time scale. Thus, the results imply that basically all chlorobenzene molecules participate in the β-process. A similar trend was reported for toluene [96]. Therefore, the NMR results provide strong evidence against a model proposed by Goldstein and Johari [1,151,332,333], where only a fraction of molecules moves in "islands of mobility." Also, the data do not support the conjecture that the β-process results from a reorientation about the

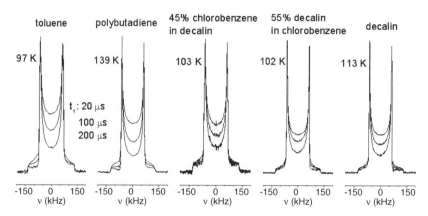

Figure 55. ^{2}H NMR spectra for type B glasses at $T/T_g \approx 0.85$. Results for the neat systems toluene-d_5, polybutadiene-d_6, and decaline-d_{18} are shown together with those for the binary mixtures 45% chlorobenzene-d_5 in decaline and 55% decaline-d_{18} in chlorobenzene. Solid-echo delays $t_p = t_1 = 20$, 100, and 200 μs were used. (From Ref. 306.)

molecular symmetry axis; essentially, an isotropic but spatially restricted motion is revealed. Moreover, from the DS spectra it follows that the molecules in toluene exhibit $\tau < 1$ ms at $T/T_g = 0.83$. Assuming, for example, a two-site jump, these molecules would contribute to a rigid-lattice spectrum because $\tau_\beta \gg 1/\delta_Q$ (cf. Section II.D.1) holds, which is not observed; that is, molecules

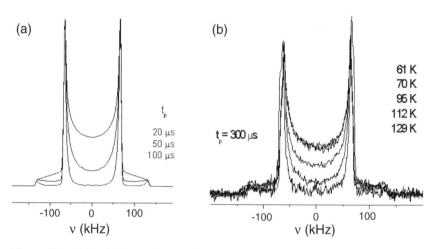

Figure 56. (a) Results from random walk simulations for different solid-echo delay times t_p; the C–^{2}H bond performs rotational random jumps on the surface of a cone with a full opening angle $\chi = 6°$ (from Ref. 93). (b) Experimental ^{2}H NMR spectra of chlorobenzene-d_5 in a mixture with *cis*-decalin at various temperatures $T < T_g$ and at a long solid-echo delay t_p [429].

characterized by correlation times $\tau_\beta \gg 1/\delta_Q$ do show some motion already in the microsecond regime. Thus, the β-process is a multistep process (as is the α-process), so that the overall loss of correlation is not achieved until a number of elementary steps on a time scale $\tau_{AJ} \ll \tau_\beta$ are performed. We also note that inspecting the spectra of chlorobenzene-d_5 and decalin-d_{18} in the mixture in Fig. 55, quite similar line-shape effects are recognized. Thus, independent of size and shape, both molecules participate in the similar manner in the β-process of the binary glass.

The stimulated-echo technique is well-suited not only to investigate the primary relaxation of glass-forming liquids above T_g (cf. Section IV.C.3) but also secondary relaxations [96,97]. Figure 57 shows correlation functions $C_{\cos}(t_m; t_p)$ for three type B glasses at $T/T_g \approx 0.83$. It was demonstrated that the β-process manifests itself in a stretched decay at $t_m < 100$ ms. We see that these short-time decays of $C_{\cos}(t_m; t_p)$ are nearly identical for the glasses studied, in agreement with results in DS, where the compounds exhibit very similar β-peaks at a given value of T/T_g. The amplitude of the decays up to $t_m \approx 100$ ms strongly increases as the evolution time t_p increases, indicating again the presence of spatially restricted rotational jumps. At times $t_m > 100$ ms, nuclear spin diffusion further reduces the correlation functions.

In the context of the ^{2}H NMR data, it has been possible to design a model of the β-process in molecular glasses [97]. Although the rotational jumps are

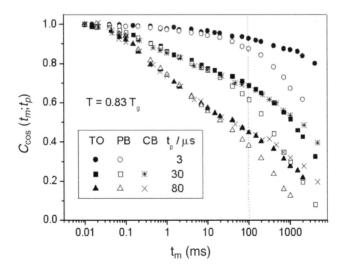

Figure 57. Correlation functions $C_{\cos}(t_m; t_p)$ for the type B glasses toluene-d_5 (TO), polybutadiene-d_6 (PB), and 45% chlorobenzene-d_5 (CB) in *cis*-decalin. Results for various evolution times t_p at $T/T_g \approx 0.83$ are compared. (From Ref. 96.)

restricted to small angular displacement of the majority of the molecules, a distribution of geometries of the motion may exist due to the disorder of the glassy state. Furthermore, the experimental data of the β-process are incompatible with simple models, such as a two-site jump model, but rather suggest a multistep process. Finally, there is evidence that at $T < T_g$ the geometry of motion is basically independent of temperature, consistent with an essentially constant dielectric relaxation strength. In the case of toluene for example, it has been shown that a random barrier model with a temperature-independent distribution of cone opening angles $V(\chi)$ was able to account for all experimental NMR findings. The barriers were taken from the distribution of activation energies determined in DS. While the values of the larger angles are somewhat ambiguous, it was argued that a maximum at $\chi = 5 \pm 1°$ is necessary for the small-angle contribution. Despite the success of this energy-landscape model, it is evidently not a unique way of describing the experimental data.

A brief discussion of NMR results of the β-process in a broader context now follows. ^{2}H NMR data suggest that essentially all molecules participate in the β-process, albeit with different angular amplitudes. The mean angular amplitude of the β-process exhibits a weak temperature-dependence below T_g, whereas it increases strongly with the temperature above T_g. The latter finding correlates with a strong increase of the dielectric relaxation strength with temperature at $T > T_g$. Thus, in aging experiments at $T < T_g$, one may expect that the behavior of the β-process approaches the behavior obtained from extrapolation of the strong temperature-dependence at $T > T_g$, suggesting that the amplitude of the secondary relaxation decreases when the structural relaxation proceeds, as was indeed observed by dielectric spectroscopy [286,333,425]. The molecular motions associated with the β-relaxation contribute to the configurational entropy, and it was argued that the experimental values of the excess entropy of glasses are smaller than the values expected for a situation where all molecules take part in the β-process [333]. Commenting on this conclusion, it relies on a simple calculation for molecules that move independently between sites of equal energy. One can imagine that concerted motions reduce the contribution of the β-process to the configurational entropy. In addition, disordered materials likely possess a distribution of site energies.

The origin of the β-process can be discussed in terms of the topologies of the potential energy landscapes of supercooled liquids obtained in molecular dynamics simulations [126,426]. The β-process may thus be related to transitions between neighboring local minima of the energy surface, or else to exploration of metabasins (cf. Section III.A and Fig. 58). In the latter picture, the β-dynamics is associated with jumps of the system among several local minima belonging to the same metabasin, and the time-scale separation of the α- and β-process upon cooling is a consequence of the increasingly important trapping of the system in a metabasin. The β-process is thus seen as a multistep

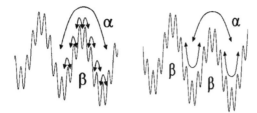

Figure 58. Schematic picture of the potential energy landscape of a fragile glass former suggesting that the local minima are grouped into "metabasins." Transitions between these metabasins lead to the α-process. Concerning the β-process, one can imagine that it results from either transitions between neighboring local minima (left) or the attainment of metabasins; that is, the system jumps among local minima grouped into a metabasin (right). (From Ref. 15, also Ref. 426.)

process, which is consistent with experimental results discussed above. In a similar approach to the free-energy landscape model, the α-process is associated with transitions between free-energy minima, while the β-process is attributed to intra-minima relaxation [427]. Finally, the evolution of the dielectric β-peak during cooling and heating cycles has been qualitatively described using a double-well potential model [428]. In conclusion, it appears an interesting challenge to refine the energy-landscape model of the β-process—which uses the distribution of activation energies determined by dielectric spectroscopy as an external input—so as to enable not only a description of NMR experiments, but also of results from other experimental techniques.

VI. CONCLUSIONS

The present review has discussed the results reported for the reorientational dynamics in molecular glass formers. Experimental data that cover the full relevant frequency range of, say, 18 decades are now available. Important progress has been achieved by introducing the tandem-Fabry–Perot interfero-metry as well as pushing dielectric spectroscopy into the gigahertz-to-terahertz regime. Also, the development of the optical Kerr effect technique represents important progress. Only by compiling spectra in the gigahertz range it has been possible to test MCT and to establish the features of the high-temperature regime of glass formers. Still, the experimental situation is somewhat unsatisfactory. Some of the conclusions drawn so far rely on the comparison of the LS and DS data of only a few glass formers—in particular, glycerol and propylene carbonate. Thus, more data are needed, for example, to allow for a systematic comparison between the spectra of different probes, like LS and DS. We note that in the near future, high-frequency dielectric spectra may also be made available by applying nonlinear optical techniques for generating and detecting far-infrared signals [430]. Comparing DS and LS spectra, it appears that the apparent von Schweidler exponent is probe-dependent. Moreover, only a few data sets are available that cover a sufficiently wide range of temperatures below as well as above the crossover temperature T_c. Nevertheless, we think, a consistent picture is starting to emerge.

Without doubt, MCT provides the most detailed and most convincing approach to the glass transition phenomenon, allowing one to quantitatively describe the high-temperature dynamics. The latter regime is demarcated by a dynamic crossover well above T_g, the presence of which is observed by different experimental techniques and which appears to be model-independent. Most prominently, the crossover is seen in the temperature-dependence of the nonergodicity parameter, which facilitates the determination of the critical temperature T_c. Although the crossover to a white noise spectrum, which is anticipated by the theory, is not as such observed experimentally, specific spectral changes do occur around T_c. In particular, the emergence of the excess wing in type A systems, the breakdown of the FTS principle, and a crossover in the temperature-dependence of the α-relaxation times are found, within experimental accuracy, to coincide at the same temperature.

In reality, it appears that glassy dynamics persist up to temperatures well above the melting point and that the asymptotic scaling laws of MCT, though not applicable to every system or experimental probe, often describe the dynamic susceptibility up to the highest available temperatures. Thus, all efforts to model normal liquid dynamics in sophisticated phenomenological approaches have to face the fact that the cage effect may well be the dominating mechanism even at the highest temperatures which are experimentally accessible.

In the low-temperature regime, clearly, there is a need for further theoretical development because, for example, the excess wing as well as the β-process phenomenon should be included in a theory of the glass transition phenomenon, since they seem to be intrinsic features of glassy dynamics which determine the shape of the susceptibility spectra starting in the low-temperature regime of supercooled liquids down to the glassy state. Although NMR has revealed important details such as the fact that secondary relaxation dynamics involves highly restricted motion of all molecules, the complex nature of this process still makes its description a strong challenge for theory. In order to possibly distinguish excess wing and β-process, it is of importance to clarify their manifestation in different experiments. For example, no neutron scattering study is known to have identified a β-process in molecular glass formers, although it was studied in polymers. Applying X-ray photon correlation spectroscopy may provide more insights. Also, further NMR studies would certainly be useful due to their capacity for disclosing details of the mechanism of molecular motion as already demonstrated for the α-process. Another challenge is to integrate the findings reported here with the body of results of pressure-induced effects. What is the role of temperature and density on the various relaxation phenomena?

Most astonishing, after several decades of intensive research efforts, we do not fully understand the temperature-dependence of the α-relaxation time.

In the present work, we have reviewed results on molecular glass formers, but it is not clear to what extent the observed relaxation features are also found in more complex systems, such as polymers and inorganic network glasses. Susceptibility spectra covering, say, more than 10 decades in frequency are still rare for the case of more complex systems. For example, how does the manifestation of the dynamic crossover change when going from a molecular liquid to a polymer?

Below T_g, and down to the temperature range of the low-temperature anomalies of molecular glasses, experimental data are scarce and understanding of the phenomena is rather incomplete. Only recently, by applying dielectric high-precision bridges a wealth of new data has been reported and further results are yet to be expected. The role of the nearly constant loss of the thermally activated dynamics in double-well potentials as well as the crossover to the tunneling regime still have to be understood. Thus, due to the low amplitude of the remaining relaxation processes below T_g, the glassy state still is an experimental challenge, waiting to be explored by the combined effort of different techniques, and one may hope that similar progress can be made as in the case of supercooled liquids above T_g.

Acknowledgments

The authors thank R. Böhmer, C. Gainaru, P. Medick, D. Leporini, F. Kremer, S. A. Lusceac, V. N. Novikov, and A. Rivera for helpful advice and discussions. We thank G. Hinze and P. Lunkenheimer for kindly sending us some of their optical Kerr effect and dielectric spectra, respectively. The financial support of the Deutsche Forschungsgemeinschaft (DFG) through grants Ro 907/2-6 and Ro907/8 as well as through SFB 279 is gratefully acknowledged.

References

1. G. P. Johari, *Ann. New York Acad. Sci.* **279**, 117 (1976).

2. E. Rössler and H. Sillescu, in Materials Science and Technology, Vol. 12, R. W. Cahn, P. Haasen, and E. J. Kramer, eds., Verlag Chemie, Weinheim, 1991, p. 574.

3. S. Yip and P. Nelson, eds., *Transport Theory Stat. Phys.* **24**, (1995).

4. M. D. Ediger, C. A. Angell and S. R. Nagel, *J. Phys. Chem.* **100**, 13200 (1996).

5. H. Z. Cummins, G. Li, Y. H. Hwang, G. Q. Shen, W. M. Du, J. Hernandez and N. J. Tao, *Z. Phys. B* **103**, 501 (1997).

6. A. Kudlik, S. Benkhof, T. Blochowicz, C. Tschirwitz and E. Rössler, *J. Mol. Struct.* **479**, 201 (1999).

7. C. A. Angell, K. L. Ngai, G. B. McKenna, P. F. McMillan and S. W. Martin, *J. Appl. Phys.* **88**, 3113 (2000).

8. K. L. Ngai, *J. Non-Cryst. Solids* **275**, 7 (2000).

9. P. Lunkenheimer, U. Schneider, R. Brand and A. Loidl, *Contemp. Phys.* **41**, 15 (2000).

10. P. Lunkenheimer and A. Loidl, in *Broadband Dielectric Spectroscopy*, F. Kremer, A. Schönhals, eds., Springer, Berlin, 2003, p. 130.

11. R. Böhmer, G. Diezemann, G. Hinze and E. Rössler, *Prog. NMR Spectrosc.* **39**, 191 (2001).

12. R. Böhmer and F. Kremer, in *Broadband Dielectric Spectroscopy*, F. Kremer and A. Schönhals, eds., Springer, Berlin, 2003, p. 265.

13. E. Donth, *Relaxation and Thermodynamics in Polymers: Glass Transition*, Akademie, Berlin, 1992.

14. E. Donth, *The Glass Transition*, Springer, Berlin 2001.

15. M. Vogel, P. Medick and E. A. Rössler, *Annual Reports on NMR Spectroscopy*, 2005, p. 231.

16. W. Götze, in Liquids, *Freezing and The Glass Transition*, Proceedings of the Les Houches Summer School, J. P. Hansen, D. Levesque, and J. Zinn-Justin, eds., Elsevier, New York, 1991, p. 287.

17. W. Götze and L. Sjögren, *Rep. Prog. Phys.* **55**, 241 (1992).

18. R. Schilling, in *Disorder Effects on Relaxational Processes*, R. Richert and A. Blumen, eds., Springer, Berlin, 1994

19. W. Götze, *J. Phys. Condens. Matter* **11**, A1 (1999).

20. R. Schilling, in *Collective Dynamics of Nonlinear and Disordered Systems*, G. Radons, W. Just, and P. Häussler, eds., Springer, Berlin, 2003, p. 171.

21. S. P. Das, *Rev. Mod. Phys.* **76**, 785 (2004).

22. U. Balucani and M. Zoppi, *Dynamics of the Liquid State*, Clarendon Press, Oxford, 1994.

23. L. Fabian, A. Latz, R. Schilling, F. Sciortino, P. Tartaglia and C. Theis, *Phys. Rev. E* **60**, 5768 (1999).

24. P. A. Madden, in *Liquids, Freezing and The Glass Transition*, Proceedings of the Les Houches Summer School, J. P. Hansen, D. Levesque and J. Zinn-Justin, eds., Elsevier, New York, 1991, p. 547.

25. B. J. Berne and R. Pecora, *Dynamic Light Scattering*, John Wiley & Sons, New York, 1976.

26. J. P. Hansen and I. R. McDonald, *Theory of Simple Liquids*, Academic Press, London, 1986.

27. W. M. Evans, *Adv. Chem. Phys.* **81**, p. 361 (1992).

28. G. Diezemann and H. Sillescu, *J. Chem. Phys.* **111**, 1126 (1999).

29. P. Madden and D. Kivelson, *Adv. Chem. Phys.*, **56**, 467 (1975).

30. S. Kämmerer, W. Kob and R. Schilling, *Phys. Rev. E* **56**, 5450 (1997)

31. H. B. Callen and T. A. Welton, *Phys. Rev.* **83**, 34 (1951).

32. R. Kubo, *J. Phys. Soc. Japan* **12**, 570 (1957).

33. R. Kubo, *Rep. Prog. Phys.* **29**, 255 (1966).

34. C. J. F. Böttcher and P. Bordewijk, *Theory of Electric Polarization*, Vol. 2, Elsevier, Amsterdam, 1973.

35. G. Williams, *J. Non-Cryst. Solids* **131–133**, 1 (1991).

36. G. Williams, in *Materials Science and Technology*, Vol. 12, R. W. Cahn, P. Haasen and E. J. Kramer, eds., Verlag Chemie, Weinheim, 1993, p. 471.

37. F. Kremer and A. Schönhals, eds., *Broadband Dielectric Spectroscopy*, Springer, Berlin, 2003.

38. E. V. Russell, N. E. Israeloff, L. E. Walther, and H. Alvarez-Gomariz, *Phys. Rev Lett.* **81**, 1461 (1998).

39. E. V. Russell and N. E. Israeloff, *Nature* **408**, 695 (2000).
40. G. Williams, *Chem. Rev.* **72**, 55 (1972).
41. S. H. Glarum, *J. Chem. Phys.* **33**, 1371 (1960).
42. R. H. Cole, *J. Chem. Phys.* **42**, 637 (1965).
43. E. Fatuzzo and P. R. Mason, *Proc. Phys. Soc. London* **90**, 741 (1967).
44. H. Fröhlich, *Theory of Dielectrics*, Clarendon Press, Oxford, 1949; 2nd ed., 1958.
45. P. Lunkenheimer, *Habilitationsschrift*, Augsburg, 1999.
46. F. I. Mopsik, *Rev. Sci. Instrum.* **55**, 79 (1984).
47. R. Richert and H. Wagner, *J. Phys. Chem.* **99**, 10948 (1995).
48. R. H. Cole, *J. Phys. Chem.* **79**, 1459 (1975).
49. H. W. Starkweather, Jr., P. Avakian and R. R. Matheson, *Macromolecules* **25**, 6871 (1992).
50. C. Gainaru, A. Rivera, S. Putselyk, G. Eska and E. A. Rössler, *Phys. Rev. B.* **72**, 174203 (2005).
51. A. A. Volkov, Y. G. Goncharov, G. V. Kozlov, S. P. Lebedev and A.M. Prokhorov, *Infrared Phys.* **25**, 369 (1985).
52. U. Schneider, P. Lunkenheimer, A. Pimenov, R. Brand and A. Loidl, *Ferroelectrics* **249**, 89 (2001).
53. H. Z. Cummins, G. Li, W. Du, R. M. Pick and C. Dreyfus, *Phys. Rev. E* **53**, 896 (1996).
54. A. Patkowski, W. Steffen, H. Nilgens, E. W. Fischer and R. Pecora, *J. Chem. Phys.* **106**, 8401 (1997).
55. G. Li, W. M. Du, X. K. Chen and H. Z. Cummins, *Phys. Rev. A* **45**, 3867 (1992).
56. R. Lobo, J. E. Robinson and S. Rodrigues, *J. Chem. Phys.* **59**, 5992 (1973).
57. E. L. Pollock and B. J. Alder, *Phys. Rev. Lett.* **46**, 950 (1981).
58. V. Mazzacurati, P. Benassi and G. Ruocco, *J. Phys. E: Sci. Instrum.* **21**, 798 (1988).
59. N. Surovtsev, J. Wiedersich, V. N. Novikov, E. Rössler and A. P. Sokolov, *Phys. Rev. B* **58**, 14888 (1998).
60. J. Gapinski, W. Steffen, A. Patkowski, A. P. Sokolov, A. Kisliuk, U. Buchenau, M. Russina, F. Mezei and H. Schober, *J. Chem. Phys.* **110**, 2312 (1999).
61. H. C. Barshila, G. Li, G. Q. Shen and H. Z. Cummins, *Phys. Rev. E* **59**, 5625 (1999).
62. H. Z. Cummins, in *Photon Correlation and Light Beating Spectroscopy*, H. Z. Cummins and E. R. Pike, eds., Plenum Press, New York, 1974, p. 225.
63. T. Shibata, T. Matsuoka, S. Koda and H. Nomura, *J. Chem. Phys.* **109**, 2038 (1998).
64. A. Brodin and E. A. Rössler, *Eur. Phys. J. B* **44**, 3 (2005).
65. A. Brodin, E. A. Rössler, R. Bergman and J. Mattsson, *Eur. Phys. J. B* **36**, 349 (2003).
66. M. S. Beevers, J. Crossley D. C. Garrington, and G. Williams, *J. Chem. Soc. Faraday Trans. II* **73**, 458 (1977).
67. A. D. Buckingham, *Proc. Phys. Soc. B* **69**, 344 (1956).
68. M. Cho, M. Du, N. F. Scherer, G. R. Fleming, and S. Mukamel, *J. Phys. Chem.* **99**, 2410 (1993).
69. R. Torre, P. Bartolini and R. M. Pick, *Phys. Rev. E* **57**, 1912 (1998).
70. G. Hinze, D. D. Brace, S. D. Gottke and M. D. Fayer, *J. Chem. Phys.* **113**, 3723 (2000).
71. V. J. McBrierty and K. J. Packer, *Nuclear Magnetic Resonance in Solid Polymers*, Cambridge University Press, Cambridge, 1993.

72. K. Schmidt-Rohr and H. W. Spiess, *Multidimensional Solid State NMR and Polymers*, Academic, London, 1997.

73. K. Schmidt-Rohr and H. W. Spiess, *Ann. Rep. NMR Spectrosc.* **48**, 1 (2002).

74. R. Kimmich, *NMR Tomography, Diffusometry, Relaxometry*, Springer, Berlin, 1997.

75. R. Kimmich and E. Anoardo, *Prog. Nucl. Magn. Reson.* **44**, 257 (2004).

76. J. F. Stebbins, *Chem. Rev.* **91**, 1353 (1991).

77. D. Brinkmann, *Prog. NMR Spectrosc.* **24**, 527 (1992).

78. H. Eckert, *Prog. NMR Spectrosc.* **24**, 159 (1992).

79. M. J. Duer, ed., *Solid State NMR Spectroscopy*, Blackwell Science, Oxford, 2002.

80. H. W. Spiess, in P. Diehl, E. Fluck, H. Günther, R. Kosfeld and J. Seelig, eds., *NMR Basic Principles and Progress*, Vol. 15, Springer, Heidelberg, 1978, p. 55.

81. A. Abragam, *The Principles of Nuclear Magnetism*, Clarendon Press, Oxford, 1961.

82. N. Bloembergen, E. M. Purcell and R. V. Pound, *Phys. Rev.* **73**, 679 (1948).

83. W. Schnauss, F. Fujara, K. Hartmann and H. Sillescu, *Chem. Phys. Lett.* **166**, 381 (1990).

84. E. Rössler and M. Taupitz, in *Disorder Effects on Relaxation Processes*, R. Richert and A. Blumen, eds., Springer, Berlin, 1994.

85. H. W. Spiess and H. Sillescu, *J. Magn. Reson.* **42**, 381 (1981).

86. G. Hinze, *Phys. Rev. E* **57**, 2010 (1998).

87. B. Geil, F. Fujara and H. Sillescu, *J. Magn. Reson.* **130**, 18 (1998).

88. F. Fujara, S. Wefing and H. W. Spiess, *J. Chem. Phys.* **84**, 4579 (1986).

89. G. Fleischer and F. Fujara, in *NMR—Basic Priciples and Progress*, Springer, Berlin, 1994, p. 159.

90. E. N. Ivanov, *Sov. Phys. JETP* **18**, 1041 (1964).

91. E. N. Ivanov and K. A. Valiev, *Opt. Spektrosk.* **35**, 289 (1973).

92. J. E. Anderson, *Faraday Symp. Chem. Soc.* **6**, 82 (1972).

93. M. Vogel and E. Rössler, *J. Magn. Reson.* **147**, 43 (2000).

94. P. Medick, M. Vogel and E. Rössler, *J. Magn. Reson.* **159**, 126 (2002).

95. P. Debye, *Polare Molekeln*, S. Hirzel, Leipzig, 1929; also *Chemical Catalog*, New York, 1929, reprinted by Dover, New York, 1954.

96. M. Vogel and E. Rössler, *J. Chem. Phys.* **114**, 5802 (2001).

97. M. Vogel and E. Rössler, *J. Chem. Phys.* **115**, 10883 (2001).

98. W. Kauzmann, *Chem. Rev.*, **43**, 219 (1948).

99. K. Binder, *Comp. Phys. Commun.* **121–122**, 168 (1999).

100. G. Adams and J. H. Gibbs, *J. Chem. Phys.* **43**, 139 (1965).

101. M. H. Cohen and D. Turnbull, *J. Chem. Phys.* **31**, 1164 (1959).

102. M. H. Cohen and G. S. Grest, *Phys. Rev. B* **20**, 1077 (1979).

103. K. Kawasaki, in *Phase Transitions and Critical Phenomena*, C. Domb and M. S. Green, eds., Academic Press, London, 1979.

104. U. Bengtzelius, W. Götze, and A. Sjölander, *J. Phys. Solid State Phys. C* **17**, 5915 (1984).

105. E. Leutheusser, *Phys. Rev. A* **29**, 2765 (1984).

106. T. Franosch, W. Götze, M. R. Mayr and A. P. Singh, *Phys. Rev. E* **55**, 3183 (1997).

107. R. Schilling and T. Scheidsteger, *Phys. Rev. E* **56**, 2932 (1997).

108. S.-H. Chong and W. Götze, *Phys. Rev. E* **65**, 041503 (2002).

109. H. Z. Cummins, *J. Phys. Condens. Matter* **11**, A95 (1999).

110. J.-L. Barrat, J. N. Roux and J.-P. Hansen, *Chem. Phys.* **149**, 197 (1990).

111. W. Kob and H. C. Andersen, *Phys. Rev. E* **47**, 3281 (1993).

112. W. Kob and H. C. Andersen, *Phys. Rev. E* **48**, 4364 (1993).

113. W. Kob and H. C. Andersen, *Transport Theory and Statistical Physics* **24**, 1179–1198 (1995).

114. W. Kob, *J. Phys. Condens. Matt.* **11**, R85 (1999).

115. F. Sciortino, L. Fabbian, S. H. Chen and P. Tartaglia, *Phys. Rev. E* **56**, 5397 (1997).

116. S. Mossa, G. Ruocco, R. Di Leonardo and M. Sampoli, *Phys. Rev. E* **62**, 612 (2000).

117. S. Mossa, G. Ruocco and M. Sampoli, *Phys. Rev. E* **64**, 021511 (2001).

118. E. Bartsch, *Transport Theory Stat. Phys.* **24**, 1125 (1995).

119. W. Van Megen *Transport Theory Stat. Phys.* **24**, 1017 (1995).

120. M. Mezard and G. Parisi, *J. Phys. A* **29**, 6515 (1996).

121. M. Mezard and G. Parisi, *J. Chem. Phys.* **111**, 1076 (1999).

122. T. R. Kirkpatrick and D. Thirumalai, *Transport Theory Stat. Phys.* **24**, 927 (1995).

123. D. Kivelson, S. A. Kivelson, G. Tarjus, X. L. Zhao, and G. Tarjus, *Physica A* **219**, 27 (1995).

124. C. A. Angell, *J. Non-Cryst. Solids* **131–133**, 13 (1991).

125. C. A. Angell, *Science* **267**, 1924 (1995).

126. F. H. Stillinger, *Science* **267**, 1935 (1995).

127. P. G. Debenedetti and F. H. Stillinger, *Nature (London)* **410**, 259 (2001).

128. F. H. Stillinger and P. G. Debenedetti, *J. Chem. Phys.* **116**, 3353 (2002).

129. B. Doliwa and A. Heuer, *Phys. Rev. Lett.* **91**, 235501 (2003).

130. W. Götze, *J. Phys. C* **21**, 3407 (1988).

131. S. P. Das and G. F. Mazenko, *Phys. Rev. A* **34**, 2265 (1986).

132. W. Götze and L. Sjögren, *Z. Phys. B: Condens. Matter* **65**, 415 (1987).

133. G. Li, W. M. Du, A. Sakai and H. Z. Cummins, *Phys. Rev. A* **46**, 3343 (1992).

134. H. Z. Cummins, G. Li, W. M. Du, J. Hernandez and N. J. Tao, *Transport Theory Stat. Phys.* **24**, 981 (1995).

135. S. M. Lindsay, M. W. Anderson and J. R. Sandercock, *Rev. Sci. Instrum.* **52**, 1478 (1981).

136. S. Adichtchev, T. Blochowicz, C. Tschirwitz, V. N. Novikov, and E. A. Rössler, *Phys. Rev. E* **68**, 011504 (2003).

137. C. Gainaru, Universität Bayreuth PhD thesis, 2006.

138. A. Kisliuk, V. N. Novikov, and A. P. Sokolov, *J. Polym. Sci. B* **40**, 201 (2002).

139. D. W. Davidson and R. H. Cole, *J. Chem. Phys.* **19**, 1484 (1951).

140. P. K. Dixon, L. Wu, S. R. Nagel, B. D. Williams and J. P. Carini, *Phys. Rev. Lett.* **65**, 1108 (1990).

141. A. Hoffmann, F. Kremer, E. W. Fischer and A. Schönhals, in *Disorder Effects on Relaxation Processes*, R. Richert and A. Blumen, eds., Springer, Berlin, 1994, p. 309.

142. T. Blochowicz, C. Tschirwitz, S. Benkhof and E. A. Rössler, *J. Chem. Phys.* **118**, 7544 (2003).

143. R. L. Leheny and S. R. Nagel, *Europhys. Lett.* **39**, 447 (1997).

144. R. L. Leheny and S. R. Nagel, *J. Non-Cryst. Solids* **235–237**, 278 (1998).

145. A. P. Sokolov, A. Kisliuk, D. Quitmann and E. Duval, *Phys. Rev. B* **48**, 7692 (1993).

146. A. P. Sokolov, E. Rössler, A. Kisliuk and D. Quitmann, *Phys. Rev. Lett.* **71**, 2062 (1993).

147. W. Petry and J. Wuttke, *Transport Theory Stat. Phys.* **24**, 1075 (1995).

148. A. Tölle, *Rep. Prog. Phys.* **64**, 1473 (2001).

149. S. Wiebel and J. Wuttke, *New J. Phys.* **4**, 56 (2002).

150. M. Ricci, S. Wiebel, P. Bartolino, A. Taschin and R. Torre, *Philos. Mag.* **84**, 1491 (2004).

151. G. P. Johari and M. Goldstein, *J. Chem. Phys.* **53**, 2372 (2000).

152. K. L. Ngai and M. Paluch, *J. Chem. Phys.* **120**, 857 (2004).

153. A. Kudlik, C. Tschirwitz, S. Benkhof, T. Blochowicz, and E. Rössler, *Europhys. Lett.* **40**, 649 (1997).

154. A. Döß, M. Paluch, H. Sillescu and G. Hinze, *Phys. Rev. Lett.* **88**, 095701 (2002).

155. A. Döß, M. Paluch, H. Sillescu and G. Hinze, *J. Chem. Phys.*, 2002, **117**, 6582 (2002).

156. J. Mattsson, R. Bergman, P. Jacobsson and L. Börjesson, *Phys. Rev. Lett.* **90**, 075702 (2003).

157. T. Blochowicz and E. A. Rössler, *Phys. Rev. Lett.* **92**, 225701 (2004).

158. U. Schneider, R. Brand, P. Lunkenheimer and A. Loidl, *Phys. Rev. Lett.* **84**, 5560 (2000).

159. P. Lunkenheimer, R. When, T. Riegger and A. Loidl, *J. Non-Cryst. Solids* **307–310**, 336 (2002).

160. L. Comez, D. Fioretto, L. Palmieri, L. Verdini, P. A. Rolla, J. Gapinski, T. Pakula, A. Patkoski, W. Steffen and E. W. Fischer, *Phys. Rev. E* **60**, 3086 (1999).

161. S. Kahle, J. Gapinski, G. Hinze, A. Patkowski and G. Meier, *J. Chem. Phys.* **122**, 074506 (2005).

162. A. Napolitano and P. B. Macedo, *J. Res. Natl. Bur. Stand.* A**72**A, 425 (1968).

163. H. Tweer, J. H. Simmons and P. B. Macedo, *J. Chem. Phys.* **54**, 1952 (1971).

164. W. T. Laughlin and D. R. Uhlmann, *J. Phys. Chem.* **76**, 2317 (1972).

165. P. K. Dixon, *Phys. Rev. B* **42**, 8179 (1990).

166. E. Rössler, *J. Chem. Phys.* **92**, 3725 (1990).

167. E. Rössler, A. P. Sokolov, P. Eiermann and U. Warschewske, *Physica A* **201**, 237 (1993).

168. E. Rössler, V. N. Novikov and A. P. Sokolov, Phase Transitions **63**, 201 (1997).

169. E. Rössler, K. U. Hess and V. N. Novikov, *J. Non-Cryst. Solids* **223**, 207 (1998).

170. F. Stickel, E. W. Fischer and R. Richert, *J. Chem. Phys.* **102**, 6251 (1995).

171. F. Stickel, E. W. Fischer and R. Richert, *J. Chem. Phys.* **104**, 2043 (1996).

172. C. Hansen, F. Stickel, R. Richert and E. W. Fischer, *J. Chem. Phys.* **108**, 6408 (1998).

173. S. V. Nemilov, *Thermodynamic and Kinetic Aspects of the Vitrous State*, CRC Press, Boca Raton, FL, 1995.

174. J. Dyre, N. B. Olsen and T. Christensen, *Phys. Rev. B* **53**, 2171 (1996).

175. J. Dyre and N. B. Olsen, *Phys. Rev. E* **69**, 042501 (2004).

176. D. Kivelson and G. Tarjus, *J. Non-Cryst. Solids* **235–237**, 86 (1998).

177. T. Dries, F. Fujara, M. Kiebel, E. Rössler and H. Sillescu, *J. Chem. Phys.* **88**, 2139 (1988).

178. W. Schnauss, F. Fujara and H. Sillescu, *J. Chem. Phys.* **97**, 1378 (1992).

179. B. Geil, PhD thesis, Universität Dortmund, 1997.

180. A. Kudlik, PhD thesis, Universität Bayreuth, 1998.

181. S. Adichtchev and E. A. Rössler unpublished results, (2005).

182. D. J. Plazek, C. A. Bero, and I. C. Chay, *J. Non-Cryst. Solids* **172–174**, 181 (1994).

183. S. Adichtchev, S. Benkhof, T. Blochowicz, V. N. Novikov, E. Rössler, C. Tschirwitz and J. Wiedersich, *Phys. Rev. Lett.* **88**, 055703 (2002).

184. C. Alba-Simionesco, private communication, 2005.

185. J. Wiedersich, N. Surovtsev and E. Rössler, *J. Chem. Phys.* **113**, 1143 (2000).

186. G. Hinze, H. Sillescu and F. Fujara, *Chem. Phys. Lett.* **232**, 154 (1995).

187. E. Rössler, *Ber. Bunsenges. Phys. Chem.* **94**, 392 (1990).

188. R. Diehl, F. Fujara and H. Sillescu, *Europhys. Lett.* **13**, 257 (1990).

189. G. Diezemann, R. Böhmer, G. Hinze and H. Sillescu, *Phys. Rev. E* **57**, 4398 (1998).

190. E. Rössler, *Phys. Rev. Lett.* **65**, 1595 (1990).

191. D. Ehlich and H. Sillescu, *Macromolecules* **23**, 1600 (1990).

192. F. Fujara, B. Geil, H. Sillescu and G. Fleischer, *Z. Phys. B* **88**, 195 (1992).

193. I. Chang, F. Fujara, B. Geil, G. Heuberger, T. Mangel and H. Sillescu, *J. Non-Cryst. Solids* **172–174**, 248 (1994).

194. G. Heuberger and H. Sillescu, *J. Phys. Chem.* **100**, 15255 (1996).

195. M. T. Cicerone and M. D. Ediger, *J. Chem. Phys.* **104**, 7210 (1996).

196. D. B. Hall, D. D. Deppe, K. E. Hamilton, A. Dhinojwala and J. M. Torkelson, *J. Non-Cryst. Solids* **235–237**, 48 (1998).

197. S. F. Swallen, P. A. Bonvallet, R.J. McMahon and M. D. Ediger, *Phys. Rev. Lett.* **90**, 015901 (2003).

198. I. Chang and H. Sillescu, *J. Phys. Chem. B* **101**, 8794 (1997).

199. G. Diezemann, R. Böhmer, G. Hinze and H. Sillescu, *J. Non-Cryst. Solids* **235–237**, 121 (1998).

200. M. D. Ediger, *Annu. Rev. Phys. Chem.* **51**, 99 (2000).

201. E. Bartsch, F. Fujara, B. Geil, M. Kiebel, W. Petry, W. Schnauss, H. Sillescu and J. Wuttke, *Physica A* **201**, 223 (1993).

202. M. Cukiermann, J. W. Lane and D. R. Uhlmann, *J. Chem. Phys.* **59**, 3639 (1973).

203. M. Lofink and H. Sillescu, *Slow Dynamics in Condensed Matter*, AIP Conference Proceedings 256, 1992, p. 30.

204. W. Oldekop, *Glastech. Ber.* **30**, 8 (1957).

205. C. A. Angell and J. C. Tucker, in *Chemistry of Process Metallurgy*, Richardson Conference (Imperial College of Science, London 1973), J. H. E. Jeffes and R. J. Tait, eds., Institute of Mining Metallurgy Publication, London, 1973, p. 207.

206. R. Böhmer and C. A. Angell, *Phys. Rev. B* **45**, 10091 (1992).

207. R. Böhmer and C. A. Angell, in *Disorder Effects and Relaxational Processes*, R. Richert and A. Blumen, eds., Springer, Berlin 1994.

208. V. N. Novikov and A. P. Sokolov, *Nature* **431**, 961 (2004).

209. T. Scopigno, G. Ruocco, F. Sette and G. Monaco, *Science* **302**, 849 (2003).

210. P. Bordat, F. Affouard and M. Descamps, *Phys. Rev. Lett.* **93**, 105502 (2004).

211. S. Yannopoulos and G. Papatheodorou, *Phys. Rev. E* **62**, 3728 (2000).

212. N. V. Surovtsev, A. M. Pugachev, B. G. Nenashev, V. K. Malinovsky, *J. Phys. Condens. Matter* **15**, 7651 (2003).

213. V. N. Novikov, Y. Ding and A. P. Sokolov, *Phys. Rev. E* **71**, 1 (2005).

214. R. Böhmer, K. L. Ngai, C. A. Angell and D. J. Plazek, *J. Chem. Phys.* **99**, 4201 (1993).

215. R. Böhmer and G. Hinze, *J. Chem. Phys.* **109**, 241 (1998).

216. A. Pugachev, A. Brodin and E. A. Rössler, unpublished data, 2005.

217. E. Rössler, A. P. Sokolov, A. Kisliuk and D. Quitmann, *Phys. Rev. B* **49**, 14967 (1994).

218. A. P. Sokolov, W. Steffen, and E. Rössler, *Phys. Rev. E* **52**, 5105 (1995).

219. D. Prevosto, P. Bartolini, R. Torre, S. Capaccioli, M. Ricci, A. Taschin, D. Pisignano and M. Lucchesi, *Philos. Mag. B* **82**, 553 (2002).

220. T. Blochowicz, A. Kudlik, S. Benkhof, J. Senker, E. Rössler and G. Hinze, *J. Chem. Phys.* **110**, 12011 (1999).

221. M. J. Lebon, C. Dreyfus, Y. Guissani, R. M. Pick and H. Z. Cummins, *Phys. B* **103**, 433 (1997).

222. D. L. Sidebottom, P. F. Green, and R. K. Brow, *Phys. Rev. Lett.* **74**, 5068 (1995).

223. C. León, A. Rivera, A. Várez, J. Sanz, and J. SantamarÀa, *Phys. Rev. Lett.* **86**, 1279 (2001).

224. R. Casalini, K. L. Ngai, and C. M. Roland, *J. Chem. Phys.* **112**, 5181 (2000).

225. K. L. Ngai, *J. Non-Cryst. Solids* **274**, 155 (2000).

226. R. Hilfer, *Phys. Rev. E* **65**, 061510 (2002).

227. A. Brodin, M. Frank, S. Wiebel. G. Shen, J. Wuttke and H. Z. Cummins, *Phys. Rev. E* **65**, 051503 (2002).

228. X. C. Zeng, D. Kivelson and G. Tarjus, *Phys. Rev. E* **50**, 1711 (1994).

229. A. Aouadi, C. Dreyfus, W. Massot, R. Pick, T. Berger, W. Steffen, A. Patkowski and C. Alba-Simionesco, *J. Chem. Phys.* **112**, 9860 (2000).

230. S. Adichtchev, T. Blochowicz, C. Gainaru, V. N. Novikov, E. A. Rössler and C. Tschirwitz, *J. Phys. Condens. Matter* **15**, S835 (2003).

231. H. P. Zhang, A. Brodin, H. C. Barshilia, G. Q. Shen, H. Z. Cummins and R. M. Pick, *Phys. Rev. E* **70**, 011502 (2004).

232. T. Nicolai, J. C. Gimel and R. Johnsen, *J. Phys. II France* **6**, 697 (1996).

233. M. Kiebel, E. Bartsch, O. Debus, F. Fujara, W. Petry and H. Sillescu, *Phys. Rev. B* **45**, 10301 (1992).

234. J. Colmenero, A. Arbe and A. Alegria, *Phys. Rev. Lett.* **71**, 2603 (1993).

235. E. Bartsch, F. Fujara, J. F. Legrand, W. Petry, H. Sillescu and J. Wuttke, *Phys. Rev. E* **52**, 738 (1995).

236. V. Z. Gochiyaev, Malinovsky, V. N. Novikov and A. P. Sokolov, *Philos. Mag. B* **63**, 777 (1991).

237. A. P. Sokolov, A. Kisliuk, D. Quitmann, A. Kudlik and E. Rössler, *J. Non-Cryst. Solids* **172–174**, 138 (1994).

238. Y. P. Kalmykov and S. V. Titov, *J. Mol. Struct.* **479**, 123 (1999).

239. W. T. Coffey, *J. Mol. Liq.* **114**, 5 (2004).

240. Y. P. Kalmykov, W. T. Coffey, D. S. F. Crothers and S. V. Titov, *Phys. Rev. E.* **70**, 041103 (2004).

241. F. Mezei, W. Knaak, and B. Farago, *Phys. Scr.* **19**, 363 (1987).

242. F. Fujara and W. Petry, *Europhys. Lett.* **4**, 921 (1987).

243. D. Richter, B. Frick and B. Farago, *Phys. Rev. Lett.* **61**, 2465 (1988).

244. N. J. Tao, G. Li and H. Z. Cummins, *Phys. Rev. Lett.* **66**, 1334 (1991).

245. W. Steffen, A. Patkowski, G. Meier and E. W. Fischer, *J. Chem. Phys.* **96**, 4171 (1992).

246. W. Steffen, A. Patkowski, H. Gläser, G. Meier, and E. W. Fischer, *Phys. Rev. E* **49**, 2992 (1994).

247. G. Pratesi, P. Bartolini, D. Senatra, M. Ricci, R. Righini, F. Barocchi, and R. Torre, *Phys. Rev. E* **67**, 021505 (1981).

248. I. C. Halalay, Y. Yang and K. A. Nelson, *Transport Theory Stat. Phys.*, **24**, 1153 (1995).

249. Y. Yang and K. A. Nelson, *J. Chem. Phys.* **104**, 5429 (1996).

250. D. M. Paolucci and K. A. Nelson, *J. Chem. Phys.* **112**, 6725 (2000).

251. H. Dux and T. Dorfmüller, *Chem. Phys.* **40**, 219 (1979).

252. G. Fytas, C. H. Wang, D. Lilge and T. Dorfmüller, *J. Chem. Phys.* **75**, 4247 (1981).

253. C. H. Wang, R. J. Ma, G. Fytas and T. Dorfmüller, *J. Chem. Phys.* **78**, 5863 (1983).

254. G. Fytas, T. Dorfmüller and C. H. Wang, *J. Phys. Chem.* **87**, 5045 (1983).

255. G. Meier, B. Gerharz, D. Boese and E. W. Fischer, *J. Chem. Phys.* **94**, 3050 (1991).

256. R. Bergman, L. Börjesson, L. M. Torell and A. Fontana, *Phys. B* **56**, 11619 (1997).

257. C. León, K. L. Ngai and C. M. Roland, *J. Chem. Phys.* **110**, 11585 (1999).

258. K. L. Ngai, P. Lunkenheimer, C. Leon, U. Schneider, R. Brand and A. Loidl, *J. Chem. Phys.* **115**, 1405 (2001).

259. A. Kudlik, S. Benkhof, R. Lenk and E. Rössler, *Europhys. Lett.* **32**, 511 (1995).

260. A. Kudlik, T. Blochowicz, S. Benkhof and E. Rössler, *Europhys. Lett.* **36**, 475 (1996).

261. R. L. Leheny, N. Menon, and S. R. Nagel, *Europhys. Lett.* **36**, 473 (1996).

262. U. Schneider, R. Brand, P. Lunkenheimer, and A. Loidl, *Eur. Phys. J. E* **2**, 67 (2000).

263. M. Fuchs, I. Hofacker, and A. Latz, *Phys. Rev. A* **45**, 898 (1992).

264. B. Kim and G. F. Mazenko, *Phys. Rev. A* **46**, 1992 (1992).

265. S.-H. Chong, W. Götze, and A. P. Singh, *Phys. Rev. E* **63**, 011206 (2000).

266. N. Menon and S. R. Nagel, *Phys. Rev. Lett.* **74**, 1230 (1995).

267. R. V. Chamberlin, *Phys. Rev. B* **48**, 15638 (1993).

268. C. Hansen, R. Richert and E. W. Fischer, *J. Non-Cryst. Solids* **215**, 293 (1997).

269. R. Viot, G. Tarjus and D. Kivelson, *J. Chem. Phys.* **112**, 10368 (2000).

270. H. Z. Cummins, *J. Phys. Condens. Matter* **17**, 1457 (2005).

271. T. Blochowicz, PhD thesis, Universität Bayreuth (2003).

272. V. N. Novikov and A. P. Sokolov, *Phys. Rev. E* **67**, 031507 (2003).

273. C. Tschirwitz, T. Blochowicz and E. A. Rössler, unpublished data.

274. M. Beiner, H. Huth, and K. Schröter, *J. Non-Cryst. Solids* **279**, 126 (2001).

275. T. Blochowicz, C. Gainaru, P. Medick, and E. A. Rössler, *J. Chem. Phys.* (2006), in press.

276. K. Schröter and E. Donth, *J. Chem. Phys.* **113**, 9101 (2000).

277. A. Arbe, D. Richter, J. Colmenero and B. Farago, *Phys. Rev. E* **54**, 3853 (1996).

278. N. B. Olsen, T. Christensen and J. C. Dyre, *Phys. Rev. Lett.* **86**, 1271 (2001).

279. P. K. Dixon, N. Menon and S. R. Nagel, *Phys. Rev. E* **50**, 1717 (1994).

280. H. Z. Cummins, Y. H. Hwang, G. Li, W. M. Du, W. Losert and G. Q. Shen, *J. Non-Cryst. Solids* **235–237**, 254 (1997).

281. Y. H. Hwang and G. Q. Shen, *J. Phys. Condens Matter* **11**, 1453 (1999).

282. C. M. Roland, S. Hensel-Bielowka, M. Paluch and R. Casalini, *Rep. Prog. Phys.* (2005).

283. G. Williams and D. C. Watts, *Trans. Faraday Soc.* **66**, 80 (1970).

284. L. Wu, *Phys. Rev. B* **43**, 9906 (1991).

285. A. Kudlik, C. Tschirwitz, S. Benkhof, T. Blochowicz and E. Rössler, *J. Non-Crystal. Solids* **235–237**, 406 (1998).

286. H. Wagner and R. Richert, **103**, 4071 (1999).

287. S. Capaccioli, R. Casaini, M. Lucchesi, G. Lovicu, D. Prevosto, D. Pisigano, G. Romano, P. A. Rolla, *J. Non-Crystal. Solids* **307–310**, 238 (2002).

288. R. Nozaki, H. Zenitani, A. Minoguchi and K. Kitai, *J. Non-Crystal. Solids* **307–310**, 349 (2002).

289. S. Sudo, M. Shimomura, S. Shinyashiki and S. Yagihara, *J. Non-Cryst. Solids* **307–310**, 356 (2002).

290. A. Reiser, G. Kasper, and S. Hunklinger, *Phys. Rev. Lett.* **92**, 125701 (2004).

291. Md. Shahin and S. S. N. Murthy, *J. Chem. Phys.* **122**, 014507 (2005).

292. S. Maslanka, M. Paluch, W. W. Sulkowski and C. M. Roland, *J. Chem. Phys.* **122**, 084511 (2005).

293. N.G. McCrum, B. E. Read and G. Williams, *Anelastic and Dielectric Effects in Polymeric Solids*, Dover, New York, 1967.

294. V. M. Bershtein, V. M. Egorov, Y. A. Emelyanov, A. Stepanov, *Polymer Bull.* **9**, 98 (1983).

295. G. P. Johari, *J. Mol. Liq.* **36**, 255 (1987).

296. A. Faivre, G. Niquet, M. Maglione, J. Fornazero, J. F. Jal and L. David, *Eur. Phys. J. B* **10**, 277 (1999).

297. J. Perez, J. Y. Cavaille and L. David, *J. Mol. Struct.* **479**, 183 (1999).

298. H. Fujimori and M. Oguni, *Solid State Commun.* **94**, 157 (1995).

299. D. Richter, R. Zorn, B. Farago, B. Frick and L. J. Fetters, *Phys. Rev. Lett.* **68**, 71 (1991).

300. E. Rössler, *Phys. Rev. Lett.* **69**, 1620 (1992).

301. D. Richter, R. Zorn, B. Farago, B. Frick, and L. J. Fetters, *Phys. Rev. Lett.* **69**, 1621 (1992).

302. D. Richter, A. Arbe, J. Comenero, M. Monkenbusch, B. Farago and R. Faust, *Macromolecules* **31**, 1133 (1998).

303. A. Arbe, J. Colmenero, B. Frick, M. Monkenbusch and D. Richter, *Macromolecules* **31**, 4926 (1998).

304. S. Kahle, L. Willner, M. Monkenbusch, D. Richter, A. Arbe, J. Colmenero and B. Frick, *Appl. Phys. A* **74**, S371 (2002).

305. M. Vogel and E. Rössler, *J. Phys. Chem. B* **104**, 4285 (2000).

306. M. Vogel, C. Tschirwitz, G. Schneider, C. Koplin, P. Medick and E. Rössler, *J. Non-Cryst. Solids* **307–310**, 326 (2002).

307. S. S. N. Murthy, *J. Chem. Soc., Faraday Trans. II* **85**, 581 (1989).

308. K. Schmidt-Rohr, A. S. Kudlik, H. W. Beckham, A. Ohlemacher, U. Pawelzik, C. Boeffel and H. W. Spiess, *Macromolecules* **27**, 4733 (1994).

309. A. S. Kulik, H. W. Beckham, K. Schmidt-Rohr, U. Radloff, U. Pawelzik, C. Boeffel and H. W. Spiess, *Macromolecules* **27**, 4746 (1994).

310. R. Richert, *Europhys. Lett.* **54**, 767 (2001).

311. K. Duvvuri and R. Richert, *J. Chem. Phys.* **118**, 1356 (2003).

312. L. Wu and S. R. Nagel, *Phys. Rev. B* **46**, 11198 (1992).

313. R. D. Deegan and S. R. Nagel, *Phys. Rev. B* **52**, 5653 (1995).

314. K. L. Ngai and S. Capaccioli, *Phys. Rev. E* **69**, 031501 (2004).

315. S. Benkhof, A. Kudlik, T. Blochowicz and E. Rössler, *J. Phys. C, Condens. Mattter* **10**, 8155 (1998).

316. R. Brand, P. Lunkenheimer, and A. Loidl, *J. Chem. Phys.* **116**, 10386 (2002).

317. D. Pisignano, S. Capaccioli, R. Casalini, M. Lucchesi, P. A. Rolla, A. Justl and E. Rössler, *J. Phys. Condensed Matter* **13**, 4405 (2001).

318. D. Gomez, A. Alegria, A. Arbe and J. Colmenero, *Macromolecules* **34**, 503 (2001).

319. S. Hensel-Bielowka, M. Paluch, J. Ziolo, and C. M. Roland, *J. Phys. Chem. B* **106**, 12459 (2002).

320. S. Hensel-Bielowka and M. Paluch, *Phys. Rev. Lett.* **89**, 025704 (2002).

321. R. Casalini and C. M. Roland, *Phys. Rev. Lett.* **91**, 015702 (2003).

322. M. Sekula, S. Pawlus, S. Hensel-Bielowka, J. Ziolo, M. Paluch, and C. M. Roland, *J. Phys. Chem. B* **108**, 4997 (2004).

323. S. A. Lusceac, C. Gainaru, M. Vogel, C. Koplin, P. Medick, and E. A. Rössler, *Macromolecules* **38**, 5625 (2005).

324. E. Rössler and P. Eiermann, *J. Chem. Phys.* **100**, 5237 (1994).

325. E. Garwe, A. Schönhals, H. Lockwenz, M. Beiner, K. Schröter, and E. Donth, *Macromolecules* **29**, 247 (1996).

326. N. B. Olsen, *J. Non-Cryst. Solids* **235–237**, 399 (1998).

327. N. B. Olsen, T. Christensen and J. C. Dyre, *Phys. Rev. E* **62**, 4435 (2000).

328. T. Fujima, H. Frusawa and K. Ito, *Phys. Rev. E* **66**, 031503 (2002).

329. M. Paluch, C. M. Roland, S. Pawlus, J. Ziolo and K. L. Ngai, *Phys. Rev. Lett.* **91**, 115701 (2003).

330. K. L. Ngai, *Physica A* **261**, 36 (1998).

331. F. Qi, T. El Goresy, R. Böhmer, A. Döß, G. Diezemann, G. Hinze, H. Sillescu, T. Blochowicz, C. Gainaru, E. A. Rössler and H. Zimmermann, *J. Chem. Phys.* **118**, 7431 (2003).

332. M. Goldstein, *J. Chem. Phys.* **51**, 3728 (1968).

333. G. P. Johari, *J. Non-Cryst. Solids* **307–310**, 317 (2002).

334. H. Wagner and R. Richert, *J. Non-Cryst. Solids* **242**, 19 (1998).

335. Y. Ding, V. N. Novikov and A. P. Sokolov, *J. Polym. Sci. B* **42**, 994 (2003).

336. G. Fytas, *Macromolecules* **22**, 21 (1989).

337. G. D. Patterson, P. K. Jue, D. J. Ramsay and J. R. Stevens, *J. Polym. Sci. B* **32**, 1137 (1994).

338. D. Sidebottom, R. Bergman, L. Börjesson and L. M. Torell, *Phys. Rev. Lett.* **71**, 2260 (1993).

339. C. M. Roland, R. Casalini and M. Paluch, *Chem. Phys. Lett.* **367**, 259 (2003).

340. E. Rössler, H. Sillescu and H. W. Spiess, *Polymer* **26**, 203 (1985).

341. U. Pschorn, E. Rössler, H. Sillescu, H. W. Spiess, S. Kaufmann and D. Schaefer, *Macromolecules* **24**, 398 (1991).

342. E. Rössler and H. Sillescu, *Chem. Phys. Lett.* **112**, 94 (1984).

343. O. Kircher, R. Böhmer and C. Alba-Simionesco, *J. Mol. Struct.* **479**, 195 (1999).

344. S. A. Lusceac, I. Roggatz, P. Medick, J. Gmeiner, and E. A. Rössler, *J. Chem. Phys.* **121**, 4770 (2004).

345. G. Hinze and H. Sillescu, *J. Chem. Phys.* **104**, 314 (1996).

346. S. Dvinskikh, G. Benini, J. Senker, M. Vogel, J. Wiedersich, A. Kudlik and E. Rössler, *J. Phys. Chem. B* **103**, 1727 (1999).

347. T. Jörg, R. Böhmer, H. Sillescu and H. Zimmermann, *Europhys. Lett.* **49**, 748 (2000).

348. O. Baldus, PhD thesis, Universität Bayreuth, 2005.

349. R. Böhmer, G. Diezemann, G. Hinze and H. Sillescu, *J. Chem. Phys.* **108**, 890 (1998).

350. H. Sillescu, *J. Non-Cryst. Solids* **243**, 81 (1999).

351. R. Richert, *J. Phys. Condens. Matter* **14**, R703 (2002).

352. S. A. Reinsberg, X. H. Qiu, M. Wilhelm, H. W. Spiess and M. D. Ediger, *J. Chem. Phys.* **114**, 7299 (2001).

353. F. Mezei and M. Russina, *J. Phys. Condens Matter* **11**, A341 (1999).

354. J. Toulouse, G. Coddens and R. Pattnaik, *Physica A* **201**, 305 (1993).

355. L. Börjesson and W. S. Howells, *J. Non-Cryst. Solids* **131–133**, 53 (1991).

356. C. Glorieux, K. A. Nelson, G. Hinze and M. D. Fayer, *J. Chem. Phys.* **116**, 3384 (2002).

357. C. Dreyfus, M. J. Lebon, H. Z. Cummins and J. Toulouse, *Physica A* **201**, 270 (1993).

358. L. J. Lewis and G. Wahnström, *Phys. Rev. E* **50**, 3865 (1994).

359. L. Comez, D. Fioretto and F. Scarpini, *J. Chem. Phys.* **119**, 6032 (2003).

360. A. Vasquez and P. A. Finn, *J. Chem. Phys.* **72**, 1958 (1972).

361. S. L. Ruby, B. J. Zabransky and P. A. Flinn, *J. Phys. Coll.* **37**, C6 745 (1976).

362. A. Mermet, E. Duval, N. V. Surovtsev, J. F. Jal, A. J. Dianoux and A. F. Yee, *Europhys. Lett.* **38**, 515 (1997).

363. J. Bartos and J. Kristiak, *J. Phys. Condens. Matter* **11**, A371 (1999).

364. K. L. Ngai, L. R. Bao, A. F. Yee and C. L. Soles, *Phys. Rev. Lett.* **87**, 215901 (2001).

365. L. Li, J. Schroers and Y. Wu, *Phys. Rev. Lett.* **91**, 265502 (2003).

366. M. A. Ondar, O. Y. Grinberg, L. G. Oranskii, V. I. Kurochkin and Y. L. Lebedev, *J. Struct. Chem.* **22**, 626 (1981).

367. D. E. Budil, K. A. Earle and J. H. Freed, *J. Phys. Chem.* **97**, 1294 (1993).

368. A. Schweiger and G. Jeschke, *Principles of Pulse Electron Paramagnetic Resonance*, Oxford University Press, Oxford, 2001.

369. V. Bercu, M. Martinelli, C. A. Massa, L. A. Pardi, E. A. Rössler, and D. Leporini, manuscript in preparation, 2006.

370. J. Wiedersich, N. Surovtsev, and E. Rössler *Chem. Phys.* **113**, 1143 (2000).

371. S. S. Ashwin and S. Sastry, *J. Phys. Condens. Matter* **15**, S1253 (2003).

372. W. M. Du, G. Li, G. H. Z. Cummins, M. Fuchs, J. Toulouse and L. A. Knauss, *Phys. Rev. E* **49**, 2192 (1994).

373. H. Z. Cummins, Y. H. Hwang, Gen Li, W. M. Du, W. Losert, G. Q. Shen, *J. Non-Cryst. Solids* **235–237**, 254 (1998).

374. A. P. Singh, G. Li. W. Götze, M. Fuchs, T. Franosch and H. Z. Cummins, *J. Non-Cryst. Solids* **235–237**, 66 (1998).

375. Y. H. Hwang, G. Q. Shen and H. Z. Cummins, *J. Non-Cryst. Solids* **235–237**, 180 (1998).

376. G. Q. Shen, J. Toulouse, S. Beuafils, B. Bonello, Y. H. Hwang, P. Finkel, J. Hernandez, M. Bertault, N. Maglione, C. Ecolivet and H. Z. Cummins, *Phys. Rev. E* **62**, 783 (2000).

377. W. Steffen, G. Meier, A. Patkowski and E. W. Fischer, *Physica A* **201**, 300 (1993).

378. V. Krakoviak, C. Alba-Simionesco and M. Krauzmann, *J. Chem. Phys.* **107**, 3417 (1997).

379. V. Krakoviak and C. Alba-Simionesco, *J. Chem. Phys.* **117**, 2161 (2002).

380. W. Götze and T. Voigtmann, *Phys. Rev. E* **61**, 4133 (2000).

381. H. Z. Cummins, W. M. Du, M. Fuchs, W. Götze, S. Hildebrand, A. Latz, G. Li and N. J. Tao, *Phys. Rev. E* **47**, 4223 (1993).

382. H. Z. Cummins, W. M. Du, M. Fuchs, W. Götze, A. Latz, G. Li and N. J. Tao, *Physica A* **201**, 207 (1993).

383. G. Li, M. Fuchs, W. M. Du, A. Latz, N. J. Tao, J. Hernandez, W. Goetze and H. Z. Cummins, *J. Non-Cryst. Solids* **172–174**, 43 (1994).

384. H. Z. Cummins, G. Li, W. Du, Y. H. Hwang and G. Q. Shen, *Prog. Theor. Phys. S* **126**, 21 (1997).

385. S. Wiebel, PhD thesis, Technische Universität München, 2003.

386. A. P. Sokolov, J. Hurts and D. Quitmann, *Phys. Rev. B* **51**, 12865 (1995).

387. J. Wuttke, J. Hernandez, G. Li, G. Coddens, H. Z. Cummins, F. Fujara, W. Petry, and H. Sillescu, *Phys. Rev. Lett.* **72**, 3052 (1994).

388. R. Torre, M. Ricci, P. Bartolini, C. Dreyfus and R. M. Pick, *Philos. Mag.* **79**, 1897 (1999).

389. R. Torre, P. Bartolini, M. Ricci and R. M. Pick, *Europhys. Lett.* **52**, 324 (2000).

390. S. D. Gottke, D. D. Brace, G. Hinze and M. D. Fayer, *J. Phys. Chem. B* **105**, 238 (2001).

391. D. D. Brace, S. D. Gottke, H. Cang and M. D. Fayer, *J. Chem. Phys.* **116**, 1598 (2002).

392. D. Prevosto, P. Bartolini, R. Torre, M. Ricci, A. Taschin, S. Cappaccioli, M. Lucchesi and P. Rolla, *Phys. Rev. E* **66**, 011502 (2002).

393. R. Torre, P. Bartolini and R. Righini, *Nature* **428**, 296 (2004).

394. S. Kinoshita, Y. Kai, M. Yamaguchi and T. Yagi, *Phys. Rev. Lett.* **75**, 148 (1995).

395. H. Cang, V. N. Novikov and M. D. Fayer, *J. Chem. Phys.* **118**, 2800 (2003).

396. H. Cang, J. Li, V. N. Novikov and M. D. Fayer, *J. Chem. Phys.* **118**, 9303 (2003).

397. H. Cang, V. V. Novikov and M. D. Fayer, *J. Chem. Phys.* **119**, 10421 (2003).

398. W. Götze and M. Sperl, *Phys. Rev. E* **66**, 011405 (2002).

399. W. Götze and M. Sperl, *Phys. Rev. Lett.* **92**, 105701 (2004).

400. W. Götze, A. P. Singh and T. Voigtmann, *Phys. Rev. E* **61**, 6934 (2000).

401. J. Wiedersich, N. Surovtsev, T. Blochowicz, C. Tschirwitz, A. Kudlik, V. Novikov and E. Rössler, *J. Phys. Condense Matter* **11**, A147 (1999).

402. P. Lunkenheimer, R. Wehn, U. Schneider and A. Loidl, *Phys. Rev. Lett.* **95**, 055702 (2005).

403. G. G. Maresch, M. Weber, A. A. Dubinskii and H. W. Spiess, *Chem. Phys. Lett.* **193**, 134 (1992).

404. D. Leporini, V. Schädler, U. Wiesner, H. W. Spiess and G. Jeschke, *J. Chem. Phys.* **119**, 11829 (2003).

405. V. Bercu, M. Martenelli, C. A. Massa, L. A. Pardi and D. Leporini, *J. Phys. Condens. Matter* **16**, L1–L10 (2004).

406. S. A. Dzuba, E. P. Kirilina, E. S. Salnikov and L. V. Kulik, *J. Chem. Phys.* **122**, 094702 (2005).

407. C. Levelut, N. Gaimes, F. Terki, G. Cohen-Solal, J. Pelous, and R. Vacher, *Phys. Rev. B* **51**, 8606 (1995).

408. A. P. Sokolov, V. N. Novikov and B. Strube, *Phys. Rev. B* **56**, 5042 (1997).

409. G. Monaco, D. Fioretto, C. Masciovecchio, G. Ruocco and F. Sette, *Phys. Rev. Lett.* **82**, 1776 (1999).

410. J. Wiedersich, S. Adichtchev and E. Rössler, *Phys. Rev. Lett.* **84**, 2718 (2000).

411. J. A. H. Wiedersich, N. V. Surovtsev, V. N. Novikov and E. Rössler, *Phys. Rev. B* **64**, 064207 (2001).

412. C. Hansen and R. Richert, *J. Phys. Condens. Matter* **9**, 9661 (1997).

413. M. Crissman, J. A. Sauer and A. E. Woodward, *J. Polym. Sci. A* **2**, 5075 (1964).

414. Y. Akagi and N. Nakamura, *J. Phys. Condens. Matter* **12**, 5155 (2000).

415. N. E. Isrealoff and T. S. Grigera, *Europhys. Lett.* **43**, 308 (1998).

416. W. A. Phillips, ed., *Amorphous Solids, Low Temperature Properties*, Springer, Berlin, 1981.

417. P. Esquinazi, ed., *Tunneling Systems in Amorphous and Crystalline Solids*, Springer, Berlin, 1998.

418. K. A. Topp and D. G. Cahill, *Z. Phys. B* **101**, 235–245 (1996).

419. R. O. Pohl, X. Liu and E. Thompson, *Rev. Mod. Phys.* **74**, 991 (2002).

420. N. Theodorakopoulos and J. Jäckle, *Phys. Rev. B* **14**, 2637–2641 (1976).

421. K. S. Gilroy and W. A. Phillips, *Philos. Mag. B* **43**, 735 (1986).
422. G. Schneider and E. A. Rössler, unpublished results, 2005.
423. S. A. Lusceac, PhD thesis, Universität Bayreuth, 2005.
424. R. Böhmer, G. Hinze, T. Jörg, F. Qi and H. Sillescu, *J. Phys. Condens. Matter* **12**, A383 (2000).
425. G. P. Johari, *J. Chem. Phys.* **77**, 4619 (1982).
426. M. Vogel, B. Doliwa, A. Heuer, and S. C. Glotzer, *J. Chem. Phys.* **120**, 4404 (2004).
427. G. Diezemann, U. Mohanty, and I. Oppenheim, *Phys. Rev. E* **59**, 2067 (1999).
428. J. Dyre and N. B. Olsen, *Phys. Rev. Lett.* **91**, 155703 (2003).
429. C. Koplin and E. A. Rössler, unpublished data.
430. T. F. Crimmins, M. J. Gleason, D. W. Ward, and K. A. Nelson, in *Ultrafast Phenomena*, Vol. XII, T. Elsaesser, S. Mukamel, M. M. Murnane and N. F. Scherer, eds., Springer Series in Chemical Physics Vol. 66, Springer, Berlin, 2001, p. 221.
431. A. Justl, P. A. Rolla, E. A. Rössler, unpublished data.

CHAPTER 3

SLOW RELAXATION, ANOMALOUS DIFFUSION, AND AGING IN EQUILIBRATED OR NONEQUILIBRATED ENVIRONMENTS

NOËLLE POTTIER

Matière et Systèmes Complexes, UMR 7057 CNRS and Université Paris 7—Denis Diderot, 75251 Paris Cedex 05, France

CONTENTS

Fractals, Diffusion, and Relaxation in Disordered Complex Systems: A Special Volume of Advances in Chemical Physics, Volume 133, Part A, edited by William T. Coffey and Yuri P. Kalmykov. Series editor Stuart A Rice.

I. INTRODUCTION

This chapter relates to some recent developments concerning the physics of out-of-equilibrium, slowly relaxing systems. In many complex systems such as glasses, polymers, proteins, and so on, temporal evolutions differ from standard laws and are often much slower. Very slowly relaxing systems display aging effects [1]. This means in particular that the time scale of the response to an external perturbation, and/or of the associated correlation function, increases with the age of the system (i.e., the waiting time, which is the time elapsed since the preparation). In such situations, time-invariance properties are lost, and the fluctuation–dissipation theorem (FDT) does not hold.

For instance, the correlation function $\tilde{C}_{00}(t, t')$ of an aging variable O at two subsequent times t' and t $(t' \leq t)$ is not invariant under time translation even in the limit of large age t', in contrast with the correlation function of a dynamic variable in equilibrium. Otherwise stated, the correlation function $\tilde{C}_{00}(t, t')$ depends both on the observation time $\tau = t - t'$ and on the waiting time (or age) $t_w = t'$. This feature may be shared by the response function $\tilde{\chi}_{OO}(t, t')$. As far as an aging variable is concerned, the fluctuation–dissipation theorem (FDT) is violated. Classically, the out-of-equilibrium violation of the FDT is characterized by a fluctuation–dissipation ratio, which is a function of both τ and t_w [2–4]. One may associate to this ratio an effective temperature, governing the slow dynamics, and sharing some properties of a temperature in the thermodynamic sense [5,6]. These concepts are currently under experimental investigation in glassy systems [7–12] and in granular materials [13,14].

A system well-adapted to the analysis of these concepts is a diffusing particle in contact with an environment, which itself may be in equilibrium (thermal bath) or out of equilibrium (aging medium).

Actually, a (possibly anomalously) diffusing particle is, first, an interesting system *per se*. Roughly speaking, its velocity thermalizes, but not its displacement with respect to a given initial position, which is out of equilibrium at any time. The violation of the corresponding FDT can be characterized by a fluctuation–dissipation ratio and an effective temperature. Besides, such a particle constitutes also a convenient tool for investigating the properties of its surrounding medium. In particular, in a nonequilibrated aging environment, itself characterized by an effective temperature, a diffusing particle acts as a thermometer: Independent measurements (at the same age of the medium) of the particle mobility and mean-square displacement give access to the effective temperature characterizing the out-of-equilibrium medium.

The aim of this chapter is to show how the concepts of FDT violation and effective temperature can be illustrated in the framework of the above quoted system, as done experimentally in Ref. 12 and theoretically in Refs. 15–19. We do not discuss here the vast general domain of aging effects in glassy systems, which are reviewed in Refs. 2–4. Since the present contribution should be understood by beginners in the field, some relevant fundamental topics of equilibrium statistical physics—namely, on the one hand, the statistical description of a system coupled to an environment and, on the other hand, the fluctuation–dissipation theorem (in a time domain formulation)—are first recalled. Then, questions specifically related to out-of-equilibrium dynamics, such as the description of aging effects by means of an effective temperature, are taken up in the framework of the above-quoted model system.

The chapter is organized as follows.

In Section II, motivated by the fact that in typical experiments an aging system is not isolated, but coupled to an environment which acts as a source of dissipation, we recall the general features of the widely used Caldeira–Leggett model of dissipative classical or quantum systems. In this description, the system of interest is coupled linearly to an environment constituted by an infinite ensemble of harmonic oscillators in thermal equilibrium. The resulting equation of motion of the system can be derived exactly. It can be given, under suitable conditions, the form of a generalized classical or quantal Langevin equation.

In Section III, we recall the fluctuation–dissipation theorem (FDT), valid for dynamical variables evolving in equilibrium. We provide its formulation in the time domain (in the whole range of temperatures). The choice of the time-domain formulation is motivated by the fact that, when out-of-equilibrium variables are concerned, the necessary modification of the equilibrium FDT can conveniently be carried out by introducing a fluctuation–dissipation ratio defined in terms of time-dependent quantities.

In Section IV, we turn to our model system, namely a classical or quantal dissipative free particle. We focus the study on the so-called Ohmic dissipation case, in which the noise is white and the particle equation of motion can be given the form of a classical nonretarded Langevin equation in the high-temperature regime (the Brownian particle then undergoes normal diffusive motion). We show how, as far as the particle displacement is concerned, the deviation from FDT can, for any temperature of the bath, be described by means of an effective temperature expressed in terms of the time-dependent diffusion coefficients $D(\tau)$ and $D(t_w)$.

In Section V, we turn to non-Ohmic dissipation. The classical or quantal dissipative particle then obeys a generalized Langevin equation with a frequency-dependent noise and the associated friction. In this problem, the complexity stems from the thermal noise, which is not white. As a result (except for a particular case of zero temperature localization), the particle undergoes anomalous diffusion. While its velocity thermalizes at large times (slowly, in contrast to standard Brownian motion), its displacement never attains equilibrium: It ages. This feature of normal diffusion is thus shared by a subdiffusive or superdiffusive motion. The associated fluctuation–dissipation ratio, which is shown to allow for a proper description of finite temperature aging effects, is a self-similar scaling function of the ratio of the observation time to the waiting time.

In Section VI, we consider a classical particle diffusing in an out-of-equilibrium environment. In this case, all the dynamical variables attached to the particle, even its velocity, are aging variables. We analyze how the drift and diffusion properties of the particle can be interpreted in terms of an effective temperature of the medium. From an experimental point of view, independent measurements of the mean-square displacement and of the mobility of a particle immersed in an aging medium such as a colloidal glass give access to an out-of-equilibrium generalized Stokes–Einstein relation, from which the effective temperature of the medium can eventually be deduced.

II. STATISTICAL MECHANICS OF SYSTEMS IN CONTACT WITH ENVIRONMENTS

The simplest example of a classical or quantum dissipative system is a particle evolving in a potential $\mathcal{V}(x)$ and coupled linearly to a fluctuating dynamical reservoir or bath. If the bath is only weakly perturbed by the system, it can be considered as linear, described by an ensemble of harmonic oscillators. Starting from the corresponding system-plus-bath Hamiltonian and using some convenient approximations, it is possible to get a description of the dissipative dynamics of the system.

The resulting dissipation model, first introduced in Refs. 20–22, has later been popularized by A.O. Caldeira and A.J. Leggett, who applied it to the quantum mechanical tunneling of a macroscopic variable [23,24]. This model (also put forward at the same time in Refs. 25–27 for the description of quantum Brownian motion) is now referred to as the Caldeira–Leggett dissipation model. It is widely used for the description of the dynamics of classical as well as quantal dissipative systems (see, for instance, Refs. 28 and 29). Since our subsequent analysis of aging effects will rely upon equations of motion deduced from the Caldeira–Leggett model, we find it useful, to begin with, to briefly recall its main features.

A. The Caldeira–Leggett Model

One considers a particle interacting linearly with an environment constituted by an infinite number of independent harmonic oscillators in thermal equilibrium. The particle equation of motion, which can be derived exactly, takes the form of a generalized Langevin equation, in which the memory kernel and the correlation function of the random force are assigned well-defined microscopic expressions in terms of the bath operators.

1.   The Caldeira–Leggett Hamiltonian

Let us consider a particle of mass m, as described by its coordinate x and the conjugate momentum p, evolving in a potential $\mathcal{V}(x)$. The particle is coupled with a bath of independent harmonic oscillators of masses m_n, described by the coordinates $\{x_n\}$ and the conjugate momenta $\{p_n\}$ $(n = 1, \ldots, N)$. The coupling between the particle and each one of the bath oscillators is taken as bilinear. The Hamiltonian of the global system constituted by the particle and the oscillators is

$$H_{\text{C.-L.}} = \frac{p^2}{2m} + \mathcal{V}(x) + \frac{1}{2}\sum_{n=1}^{N}\left[\frac{p_n^2}{m_n} + m_n\omega_n^2\left(x_n - \frac{c_n}{m_n\omega_n^2}x\right)^2\right] \tag{1}$$

In Eq. (1), the constants c_n measure the strength of the coupling.

In the case of a free particle $(\mathcal{V}(x) = 0)$, the model described by the Caldeira–Leggett Hamiltonian (1) is exactly solvable.[1]

2.   The Coupled Particle Equation of Motion

From the Hamiltonian (1) with $\mathcal{V}(x) = 0$, one deduces the Hamilton equations for all the degrees of freedom of the global system; that is, for the particle, we have

$$\frac{dx}{dt} = \frac{p}{m}, \qquad \frac{dp}{dt} = \sum_n c_n\left(x_n - \frac{c_n}{m_n\omega_n^2}x\right) \tag{2}$$

and for the bath oscillators we obtain

$$\frac{dx_n}{dt} = \frac{p_n}{m_n}, \qquad \frac{dp_n}{dt} = -m_n\omega_n^2 x_n + c_n x \tag{3}$$

---

[1]The same is true for a particle evolving in a harmonic potential.

Equations (3) can be formally solved by considering the particle position $x(t)$ as given. One thus gets

$$x_n(t) = x_n(t_i) \cos \omega_n(t - t_i) + \frac{p_n(t_i)}{m_n \omega_n} \sin \omega_n(t - t_i) + c_n \int_{t_i}^{t} \frac{\sin \omega_n(t - t')}{m_n \omega_n} x(t') \, dt'$$

(4)

where t_i denotes the initial time at which the coupling is switched on. The equation of motion of the coupled particle, as deduced from the particle Hamilton equations (2), is of the form

$$m\ddot{x}(t) = \sum_n c_n \left[x_n(t) - \frac{c_n^2}{m_n \omega_n^2} x(t) \right]$$

(5)

With the help of Eq. (4) (after an integration by parts of the last term of its right-hand side), Eq. (5) can be reformulated as a closed integrodifferential equation for $x(t)$:

$$m\ddot{x}(t) + m \int_{t_i}^{t} \gamma(t - t') \dot{x}(t') \, dt' = -m x(t_i) \gamma(t - t_i) + F(t)$$

(6)

In Eq. (6), the functions $\gamma(t)$ and $F(t)$ have the following microscopic expressions:

$$\gamma(t) = \frac{1}{m} \sum_n \frac{c_n^2}{m_n \omega_n^2} \cos \omega_n t$$

(7)

and

$$F(t) = \sum_n c_n \left[x_n(t_i) \cos \omega_n(t - t_i) + \frac{p_n(t_i)}{m_n \omega_n} \sin \omega_n(t - t_i) \right]$$

(8)

Equation (8) can be rewritten equivalently, in terms of the bath oscillators annihilation and creation operators, as

$$F(t) = \sum_n \lambda_n [b_n(t_i) e^{-i\omega_n(t - t_i)} + b_n^\dagger(t_i) e^{i\omega_n(t - t_i)}]$$

(9)

with $\lambda_n = c_n (\hbar / 2m_n \omega_n)^{1/2}$.

Except for the term $-mx(t_i)\gamma(t - t_i)$, which we will discuss later, Eq. (6) is similar to a generalized Langevin equation, in which $\gamma(t)$ acts as a memory kernel and $F(t)$ acts as a random force. By using, instead of $\gamma(t)$, the retarded memory kernel[2] $\tilde{\gamma}(t) = \Theta(t)\gamma(t)$, the upper integration bound of the integral in the left-hand side of Eq. (6) can be set equal to $+\infty$. This latter equation can then be rewritten in the following equivalent form:

$$m\ddot{x}(t) + m \int_{t_i}^{\infty} \tilde{\gamma}(t - t') \dot{x}(t') \, dt' = -m x(t_i) \tilde{\gamma}(t - t_i) + F(t) \qquad (10)$$

Equations (6) and (10) have been deduced without approximation from the Hamilton equations (2) and (3). The memory kernel and the random force are expressed in terms of the parameters of the microscopic Caldeira–Leggett Hamiltonian (1).

3. The Statistical Properties of the Random Force

Let us assume that, at time $t = t_i$, the density operator of the global particle-plus-bath system is factorized in the form $\rho_{\text{part.}}$ where ρ_{bath} denotes the thermal equilibrium density operator of the unperturbed bath and $\rho_{\text{part.}}$, the particle density operator.

The random force $F(t)$ can be viewed as corresponding to a stationary Gaussian stochastic process [the Gaussian character ensuing from the statistical properties of the bath operators involved in Eqs. (8) and (9)]. It is fully characterized by its average value and its correlation function.

The average value of the random force $F(t)$ is then equal to zero:

$$\langle F(t) \rangle = 0 \qquad (11)$$

The symmetrized correlation function of the random force, namely the function $\tilde{C}_{FF}(t, t')$ as defined by

$$\tilde{C}_{FF}(t, t') = \frac{1}{2} \langle [F(t), F(t')]_+ \rangle \qquad (12)$$

depends only on the time difference $t - t'$. In Eqs. (11) and (12) (and in the following), the symbol $\langle \ldots \rangle$ denotes the average value with respect to the bath density operator.

---

[2] The notation $\Theta(t)$ stands for the Heaviside function:

$$\Theta(t) = \begin{cases} 0, & t < 0 \\ 1, & t > 0 \end{cases}$$

B. Phenomenological Modeling of Dissipation

1. The Spectral Density of the Coupling

The equations of motion (6) or (10) do not *per se* describe irreversible dynamics. This is only possible if the number N of the bath oscillators tends toward infinity, with their angular frequencies forming a continuum in this limit.

A central ingredient in the model is the generalized friction coefficient $\gamma(\omega)$, which is the Fourier transform of the retarded memory kernel $\tilde{\gamma}(t)$. To compute $\gamma(\omega)$, following a standard procedure, one first attributes a small imaginary part $\epsilon > 0$ to ω. One thus defines

$$\gamma(\omega + i\epsilon) = \int_0^\infty \tilde{\gamma}(t)\, e^{i\omega t}\, e^{-\epsilon t}\, dt, \qquad \epsilon > 0 \tag{13}$$

The generalized friction coefficient $\gamma(\omega)$ is obtained by taking the limit $\epsilon \to 0^+$:

$$\gamma(\omega) = \lim_{\epsilon \to 0^+} \gamma(\omega + i\epsilon) \tag{14}$$

Thus, from the microscopic expression of $\tilde{\gamma}(t)$, namely

$$\tilde{\gamma}(t) = \Theta(t)\, \frac{1}{m} \sum_n \frac{c_n^2}{m_n \omega_n^2} \cos \omega_n t \tag{15}$$

one derives the microscopic expression of $\gamma(\omega)$:

$$\gamma(\omega) = \frac{i}{2m} \sum_n \frac{c_n^2}{m_n \omega_n^2} \lim_{\epsilon \to 0^+} \left(\frac{1}{\omega - \omega_n + i\epsilon} + \frac{1}{\omega + \omega_n + i\epsilon} \right) \tag{16}$$

In particular, one has

$$\Re e\gamma(\omega) = \frac{\pi}{2m} \sum_n \frac{c_n^2}{m_n \omega_n^2} \left[\delta(\omega - \omega_n) + \delta(\omega + \omega_n) \right] \tag{17}$$

Note that $\Re e\gamma(\omega)$ is an even function of ω.

At this stage, one generally introduces the spectral density $J(\omega)$ of the coupling with the environment, which is a function of ω defined for $\omega > 0$ by the following formula:

$$J(\omega) = \frac{\pi}{2} \sum_n \frac{c_n^2}{m_n \omega_n} \delta(\omega - \omega_n), \qquad \omega > 0 \tag{18}$$

For $\omega > 0$, one has between $\Re e\gamma(\omega)$ and $J(\omega)$ the relation

$$\Re e\gamma(\omega) = \frac{J(\omega)}{m\omega} \tag{19}$$

In the continuum limit, both $\Re e\gamma(\omega)$ and $J(\omega)$ can be considered as continuous functions of ω.

As will be seen below, the spectral density is a central quantity in the model, since it determines the dynamics of the coupled particle. In particular, the large-time particle dynamics is controlled by the behavior of $J(\omega)$ at small angular frequencies. In numerous situations, this behavior can be described by a power law of the type $J(\omega) \sim \omega^\delta$. The exponent $\delta > 0$ is most often an integer. Its value depends on the dimensionality of the space corresponding to the considered environment. Noninteger values of δ correspond to fractal environments. The general Caldeira–Leggett dissipation model is versatile enough to generate various damped equations of motion, according to the value chosen for δ.

2. The Fluctuation–Dissipation Theorem of the Second Kind

The correlation function of the random force $F(t)$ (Eq. (12)) only depends on the observation time. It is given by

$$\tilde{C}_{FF}(t - t') = m \int_{-\infty}^{\infty} \frac{d\omega}{2\pi} \Re e\gamma(\omega)\, \hbar\omega\, \coth\frac{\beta\hbar\omega}{2}\, e^{-i\omega(t-t')} \tag{20}$$

In the classical (high-temperature) limit, Eq. (20) simplifies into

$$\tilde{C}_{FF}(t - t') = mkT\,\gamma(t - t') \tag{21}$$

The detailed analysis of the behavior of $\tilde{C}_{FF}(t)$ as a function of t at any bath temperature can be found in Ref. 25.

Equation (20) (or Eq. (21) in the classical limit) constitute the formulation of the fluctuation–dissipation theorem of the second kind or second FDT (using the Kubo terminology [30,31]). This theorem applies to the random force $F(t)$, which is a bath dynamical variable. It expresses the fact that the bath is in equilibrium.

3. The Generalized Classical or Quantal Langevin Equation

Let us now come back to the equation of motion (10). Its right-hand side contains, besides the random force $F(t)$, the term $-mx(t_i)\tilde{\gamma}(t - t_i)$, which depends on the particle initial position.

As will be seen below, the kernel $\tilde{\gamma}(t)$ decreases on a characteristic time of the order of ω_c^{-1}, the angular frequency ω_c characterizing the bandwidth of the bath oscillators effectively coupled to the particle. The quantity $\tilde{\gamma}(t - t_i)$ is thus negligible if $\omega_c(t - t_i) \gg 1$. Mathematically, this condition can be realized at any time t by referring the initial time t_i to $-\infty$. Then, the initial particle position becomes irrelevant in Eq. (10), which takes the form of the generalized Langevin equation, namely,

$$m\frac{dv}{dt} + m \int_{-\infty}^{+\infty} \tilde{\gamma}(t - t') v(t') \, dt' = F(t), \qquad v = \frac{dx}{dt} \tag{22}$$

In Eq. (22), the Langevin force $F(t)$ may be considered as a Gaussian stationary random process of zero mean with correlation function given by Eq. (20).

C. Ohmic Dissipation

1. The Ohmic Spectral Density

Let us consider again the spectral density $J(\omega)$, assumed to vary proportionally to ω^δ at small angular frequencies. The value $\delta = 1$ is especially important since it corresponds to frequency-independent damping. The corresponding dissipation model, as defined by the relations

$$J(\omega) = \eta\omega = m\gamma\omega \quad (\omega > 0), \qquad \Re e\gamma(\omega) = \gamma \tag{23}$$

where η denotes the viscosity coefficient, is referred to as the Ohmic model (the reason for this denomination will be made clear later). Formulas (23) are only valid for small angular frequencies: Actually, the spectral density $J(\omega)$ does not increase without bounds, but decreases toward zero when $\omega \to \infty$. One thus writes, instead of the relations (23), the formulas

$$J(\omega) = m\gamma\omega f_c\left(\frac{\omega}{\omega_c}\right), \qquad \Re e\gamma(\omega) = \gamma f_c\left(\frac{\omega}{\omega_c}\right), \qquad \omega > 0 \tag{24}$$

The definition of $\Re e\gamma(\omega)$ is then extended to $\omega < 0$ by imposing that it must be an even function of ω. In Eq. (24), the function $f_c(\omega/\omega_c)$ is a cutoff function tending toward zero when $\omega \to \infty$. One often chooses a Lorenzian cutoff function:

$$f_c\left(\frac{\omega}{\omega_c}\right) = \frac{\omega_c^2}{\omega_c^2 + \omega^2} \tag{25}$$

With this choice, one has

$$J(\omega) = m\gamma\omega \frac{\omega_c^2}{\omega_c^2 + \omega^2} \quad (\omega > 0), \qquad \Re e\gamma(\omega) = \gamma \frac{\omega_c^2}{\omega_c^2 + \omega^2} \qquad (26)$$

Thus, in the Ohmic model, the retarded memory kernel is represented by an exponentially decreasing function of time constant ω_c^{-1}:

$$\tilde{\gamma}(t) = \Theta(t)\gamma\omega_c \, e^{-\omega_c t} \qquad (27)$$

Its Fourier transform is

$$\gamma(\omega) = \gamma \frac{\omega_c}{\omega_c - i\omega} \qquad (28)$$

As for the random force correlation function, it is given by

$$\tilde{C}_{FF}(t) = m\hbar\omega_c^2\gamma \int_{-\infty}^{\infty} \frac{d\omega}{2\pi} \frac{\omega}{\omega^2 + \omega_c^2} \coth\frac{\beta\hbar\omega}{2} \, e^{-i\omega t} \qquad (29)$$

2. The Infinitely Short Memory Limit

The Ohmic model memory kernel admits an infinitely short memory limit $\gamma(t) = 2\gamma\delta(t)$, which is obtained by taking the limit $\omega_c \to \infty$ in the memory kernel $\gamma(t) = \gamma\omega_c e^{-\omega_c|t|}$ [this amounts to the use of the dissipation model as defined by Eq. (23) for any value of ω]. Note that the corresponding limit must also be taken in the Langevin force correlation function (29). In this limit, Eq. (22) reduces to the nonretarded Langevin equation:

$$m\frac{dv}{dt} + m\gamma v(t) = F(t), \qquad v = \frac{dx}{dt} \qquad (30)$$

The formal similitude between Eq. (30) and Ohm's law in an electrical circuit justifies the term of Ohmic model given to the dissipation model as defined by the spectral density (23).

III. TIME-DOMAIN FORMULATION OF THE FLUCTUATION–DISSIPATION THEOREM

When out-of-equilibrium dynamic variables are concerned, as will be the case in the following sections of this chapter, the equilibrium fluctuation–dissipation theorem is not applicable. In order to discuss properties such as the aging effects which manifest themselves by the loss of time translational invariance in response and/or correlation functions, it is convenient to have at hand a formulation of the equilibrium fluctuation–dissipation theorem in the time domain.

The equilibrium FDT is usually written in a form that involves frequency-dependent quantities such as generalized susceptibilities and spectral densities [30–35]. We show below how this theorem can be formulated in the time domain. Our arguments do not reduce to a simple Fourier transformation of the usual frequency-domain formulation. Instead they are developed from the very beginning in the time domain, and they use only the various time-dependent quantities entering into play. The corresponding formulations of the FDT, which are established in the whole range of temperatures, allow in particular for a discussion of both the classical limit and the zero-temperature case [36].

A. General Time-Domain Formulation

Let us consider a system in equilibrium, described in the absence of external perturbations by a time-independent Hamiltonian H_0. We will be concerned with equilibrium average values which we will denote as $\langle \ldots \rangle$, where the symbol $\langle \ldots \rangle$ stands for $\text{Tr}\,\rho_0 \ldots$ with $\rho_0 = e^{-\beta H_0}/\text{Tr}\,e^{-\beta H_0}$ the canonical density operator. Since we intend to discuss linear response functions and symmetrized equilibrium correlation functions generically denoted as $\tilde{\chi}_{BA}(t, t')$ and $\tilde{C}_{BA}(t, t')$, we shall assume that the observables of interest A and B do not commute with H_0 (were it the case, the response function $\tilde{\chi}_{BA}(t, t')$ would indeed be zero). This hypothesis implies in particular that A and B are centered: $\langle A \rangle = 0$, $\langle B \rangle = 0$.

Generally speaking, the equilibrium FDT establishes a link between the dissipative part $\tilde{\xi}_{BA}(t, t')$ of the linear response function $\tilde{\chi}_{BA}(t, t')$ and the symmetrized equilibrium correlation function $\tilde{C}_{BA}(t, t') = \frac{1}{2}\langle [A(t'), B(t)]_+ \rangle$ (or the derivative $\partial \tilde{C}_{BA}(t, t')/\partial t'$ with $t' < t$ the earlier time).

1. The Dissipative Part of the Response Function in Terms of the Symmetrized Correlation Function

Consider two quantum-mechanical observables A and B with thermal equilibrium correlation functions verifying the Kubo–Martin–Schwinger condition [35], that is,

$$\langle A(t' - i\hbar\beta)B(t) \rangle = \langle B(t)A(t') \rangle, \qquad \beta = 1/kT \tag{31}$$

and compute the contour integral

$$I = \oint_\Gamma \langle A(\tau)B(t) \rangle \, \frac{\pi}{\beta\hbar} \, \frac{1}{\sinh\frac{\pi(\tau-t')}{\beta\hbar}} \, d\tau \tag{32}$$

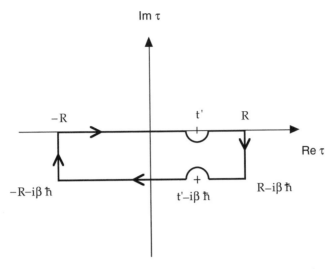

Figure 1. Integration contour for the calculation of the integrals I [Eq. (32)] and J [Eq. (39)].

where t is a real time and Γ is the closed contour in the complex τ-plane represented in Fig. 1. One easily sees that the integrand in I has no singularities inside Γ. Actually, denoting by $|\lambda\rangle$ and E_λ the eigenstates and eigenenergies of H_0, one has

$$\langle A(\tau)B(t)\rangle = \frac{1}{Z}\sum_{\lambda,\lambda'} A_{\lambda\lambda'}B_{\lambda'\lambda}\, e^{-\beta E_\lambda + i(\tau-t)(E_\lambda - E_{\lambda'})}, \qquad Z = \mathrm{Tr}\, e^{-\beta H_0} \qquad (33)$$

which shows that the continuation to the region $-\beta \leq \Im m\tau \leq 0$ of the complex τ-plane of the correlation function $\langle A(\tau)B(t)\rangle$ is analytic [34,37].

According to the Cauchy theorem, one thus has $I = 0$. In the limit $R \to \infty$, one obtains the relation[3]

$$i\pi\langle [B(t),A(t')]\rangle = \mathrm{vp}\int_{-\infty}^{\infty} dt''\,\langle [A(t''),B(t)]_+\rangle\,\frac{\pi}{\beta\hbar}\,\frac{1}{\sinh\frac{\pi(t''-t')}{\beta\hbar}} \qquad (34)$$

Taking into account the Kubo formula $\tilde{\chi}_{BA}(t,t') = \Theta(t-t')(i/\hbar)\langle [B(t),A(t')]\rangle$, one gets from Eq. (34) an expression of the response function in terms of the

---

[3]The symbol vp denotes the Cauchy principal value.

symmetrized correlation function:

$$\tilde{\chi}_{BA}(t, t') = \frac{2}{\pi\hbar} \Theta(t - t') \, \mathrm{vp} \int_{-\infty}^{\infty} dt'' \, \tilde{C}_{BA}(t, t'') \frac{\pi}{\beta\hbar} \frac{1}{\sinh \frac{\pi(t'' - t')}{\beta\hbar}} \tag{35}$$

Then, introducing the dissipative part $\tilde{\xi}_{BA}$ of $\tilde{\chi}_{BA}$, as defined by the formula

$$\tilde{\chi}_{BA}(t, t') = 2i \, \Theta(t - t') \, \tilde{\xi}_{BA}(t, t') \tag{36}$$

one gets from Eq. (35) an expression of the dissipative part of the response function in terms of the symmetrized correlation function:

$$i\hbar \, \tilde{\xi}_{BA}(t, t') = \frac{1}{\pi} \, \mathrm{vp} \int_{-\infty}^{\infty} dt'' \, \tilde{C}_{BA}(t, t'') \frac{\pi}{\beta\hbar} \frac{1}{\sinh \frac{\pi(t'' - t')}{\beta\hbar}} \tag{37}$$

Since the two-time equilibrium averages involved in $\tilde{\xi}_{BA}$ and $\tilde{C}_{BA}$ only depend on the time differences involved, Eq. (37) can be rewritten as a convolution equation:

$$i\hbar \, \tilde{\xi}_{BA}(t) = \frac{1}{\pi} \, \tilde{C}_{BA}(t) * \mathrm{vp} \frac{\pi}{\beta\hbar} \frac{1}{\sinh \frac{\pi t}{\beta\hbar}} \tag{38}$$

The convolution in the right-hand side of Eq. (38) is taken with respect to t.

2. The Symmetrized Correlation Function in Terms of the Dissipative Part of the Response Function

The expression of the symmetrized correlation function in terms of the dissipative part of the response function can be derived, in like manner (i.e., by using contour integration), or by inverting the convolution equation (38).

Thus we compute on the contour Γ (Fig. 1) the integral

$$J = \oint_{\Gamma} \langle A(\tau)B(t) \rangle \frac{\pi}{\beta\hbar} \coth \frac{\pi(\tau - t')}{\beta\hbar} \, d\tau \tag{39}$$

Using similar arguments as above (i.e., noticing that $J = 0$ since there are no singularities of the integrand of J inside Γ), one obtains the relation

$$\frac{1}{2} \langle [A(t'), B(t)]_+ \rangle = -\frac{i}{2\pi} \, \mathrm{vp} \int_{-\infty}^{\infty} dt'' \, \langle [B(t), A(t'')] \rangle \frac{\pi}{\beta\hbar} \coth \frac{\pi(t'' - t')}{\beta\hbar} \tag{40}$$

that is,

$$\tilde{C}_{BA}(t,t') = -\frac{1}{\pi}\,\mathrm{vp}\int_{-\infty}^{\infty} dt''\, i\hbar\,\tilde{\xi}_{BA}(t,t')\,\frac{\pi}{\beta\hbar}\,\coth\frac{\pi(t''-t')}{\beta\hbar} \tag{41}$$

Before going further, let us add a comment. The arguments previously used in the computation of I must be refined in order to show, first, that the integrals in the right-hand side of Eqs. (40)–(41) are properly defined and, second, that the contributions to J [Eq. (39)] of the two vertical segments of abscissae R and $-R$ of the contour Γ (Fig. 1) do vanish in the limit $R \to \infty$. The question stems from the fact that the function $\coth(\pi t/\beta\hbar)$ tends toward a finite limit for large values of its argument, contrary to the function $1/\sinh(\pi t/\beta\hbar)$, involved in the computation of I, which tends toward zero [this latter property ensures that the integrals in the right-hand side of Eqs. (34), (35), and (37) are properly defined and that the contribution to I of the above-mentioned segments is actually zero in the limit $R \to \infty$].

A sufficient condition is $\lim_{R\to\infty}\langle A(\pm R - i\hbar y)B(t)\rangle = \langle A\rangle\langle B\rangle$ $(= 0$ since A and B are centered).[4] In systems with a finite number of degrees of freedom, and thus with oscillating response and correlation functions, this limit does not even exist. Nevertheless, Eqs. (40) and (41) still apply, provided we introduce a small damping which will be eventually let equal to zero, or, which amounts to the same, we treat the response and correlation functions as distributions.

Equation (41) can be viewed as the reciprocal of Eq. (37). It can be rewritten, using convolution product notations, as

$$\tilde{C}_{BA}(t) = -\frac{1}{\pi}\,i\hbar\,\tilde{\xi}_{BA}(t) * \frac{\pi}{\beta\hbar}\,\mathrm{vp}\,\coth\frac{\pi t}{\beta\hbar} \tag{42}$$

Note that this latter expression can also be obtained directly by inverting the convolution equation (38), which is easily done by using the identity (demonstrated in Appendix A):

$$\frac{\pi}{\beta\hbar}\,\mathrm{vp}\,\frac{1}{\sinh\frac{\pi t}{\beta\hbar}} * \frac{\pi}{\beta\hbar}\,\mathrm{vp}\,\coth\frac{\pi t}{\beta\hbar} = -\pi^2\,\delta(t) \tag{43}$$

---

[4]Note that for $y = 0$ this condition amounts to $\lim_{R\to\infty}\langle A(\pm R)B(t)\rangle = 0$, while for $y = \beta$ it amounts to $\lim_{R\to\infty}\langle B(t)A(\pm R)\rangle = 0$.

Equation (38), together with the inverse relation (42), both being reproduced below together for clarity,

$$ i\hbar\, \tilde{\xi}_{BA}(t) = \frac{1}{\pi}\, \tilde{C}_{BA}(t) * \frac{\pi}{\beta\hbar}\, \text{vp}\, \frac{1}{\sinh\frac{\pi t}{\beta\hbar}}, \quad \tilde{C}_{BA}(t) = -\frac{1}{\pi}\, i\hbar\, \tilde{\xi}_{BA}(t) * \frac{\pi}{\beta\hbar}\, \text{vp}\, \coth\frac{\pi t}{\beta\hbar} $$

$$(44)$$

constitute the formulation of the equilibrium FDT in the time domain. As it should, these relations correspond by Fourier transformation to the usual fluctuation–dissipation relations between $\xi_{BA}(\omega)$ and $C_{BA}(\omega)$ [30–35], namely,

$$ \xi_{BA}(\omega) = \frac{1}{\hbar}\tanh\frac{\beta\hbar\omega}{2}\, C_{BA}(\omega), \qquad C_{BA}(\omega) = \hbar\,\coth\frac{\beta\hbar\omega}{2}\, \xi_{BA}(\omega) \qquad (45) $$

This can be checked by making use of the Fourier formulas (A5) and (A10), established in Appendix A.

The time-domain formulation constitutes a proper framework for the necessary modifications of the fluctuation–dissipation formulas when the frequency-dependent quantities cannot be properly defined, as may be the case when out-of-equilibrium dynamic variables are concerned.

B.   The Classical Limit

Before entering the discussion of the classical limit, we will propose another formulation of the FDT as given by Eq. (37) or Eq. (38) in the time-domain. It provides the expression of the dissipative part of the linear response function in terms of the derivative of the symmetrized correlation function, an expression which becomes particularly simple in the classical limit [34].

1.   The Dissipative Part of the Response Function in Terms of the Derivative of the Symmetrized Correlation Function

Integrating by parts, one can recast Eq. (37) into the equivalent form

$$ i\hbar\, \tilde{\xi}_{BA}(t, t') = \frac{1}{\pi}\int_{-\infty}^{\infty} dt''\, \frac{\partial \tilde{C}_{BA}(t, t'')}{\partial t''}\, \log\coth\frac{\pi|t'' - t'|}{2\beta\hbar} \qquad (46) $$

that is, making use of convolution product notations,

$$ i\hbar\, \tilde{\xi}_{BA}(t) = -\frac{1}{\pi}\frac{d\tilde{C}_{BA}(t)}{dt} * \log\coth\frac{\pi|t|}{2\beta\hbar} \qquad (47) $$

an expression equivalent to Eq. (38). Thus, at any temperature, the dissipative part of the response function [i.e., the function $\tilde{\xi}_{BA}(t)$] is proportional to the convolution product, taken with respect to t, of the functions $d\tilde{C}_{BA}(t)/dt$ and $\log \coth(\pi|t|/2\beta\hbar)$. This latter function is very peaked around $t = 0$ at high temperature, while it spreads out more and more around this value as the temperature decreases.

2. The Classical FDT

In the classical limit, making use of the property (A14), that is,

$$\log \coth |ax| \sim_{|a| \to \infty} \frac{\pi^2}{4|a|} \delta(x) \tag{48}$$

with $a = \pi/2\beta\hbar$, one shows that Eq. (47) reduces to

$$\tilde{\xi}_{BA}(t) = \frac{i\beta}{2} \frac{d\tilde{C}_{BA}(t)}{dt} \tag{49}$$

One deduces from Eq. (49) that the expression of the response function $\tilde{\chi}_{BA}(t, t')$ involves the derivative of the correlation function $\tilde{C}_{BA}(t, t')$ with respect to the earlier time t':

$$\tilde{\chi}_{BA}(t, t') = \beta\, \Theta(t - t') \frac{\partial \tilde{C}_{BA}(t, t')}{\partial t'} \tag{50}$$

Equation (49) for the dissipative part of the response function, or Eq. (50) for the response function, constitutes time-domain formulations of the classical FDT.

C. The Extreme Quantum Case

1. The Zero-Temperature FDT

At $T = 0$, coming back to the formulation (44) of the FDT, one gets

$$i\hbar\, \tilde{\xi}_{BA}(t) = \frac{1}{\pi} \tilde{C}_{BA}(t) * \mathrm{vp}\, \frac{1}{t}, \qquad \tilde{C}_{BA}(t) = -\frac{1}{\pi} i\hbar\, \tilde{\xi}_{BA}(t) * \mathrm{vp}\, \frac{1}{t} \tag{51}$$

In other words, Hilbert transformation relations between $-i\hbar\tilde{\xi}_{BA}(t)$ and $\tilde{C}_{BA}(t)$ exist, that is,

$$-i\hbar\, \tilde{\xi}_{BA}(t) = \frac{1}{\pi} \mathrm{vp} \int_{-\infty}^{\infty} \tilde{C}_{BA}(t') \frac{1}{t' - t}\, dt' \tag{52}$$

and

$$\tilde{C}_{BA}(t) = -\frac{1}{\pi} \, \mathrm{vp} \int\limits_{-\infty}^{\infty} \left(-i\hbar\,\tilde{\xi}_{BA}(t')\right) \frac{1}{t'-t} \, dt' \tag{53}$$

Interestingly enough, these relations are formally similar to the usual Kramers–Kronig relations between the real and imaginary parts of the generalized susceptibility (except for the evident fact that they hold in the time-domain, and not in the frequency domain). Otherwise stated, at $T = 0$, the quantities $\tilde{C}_{BA}(t)$ and $-i\hbar\,\tilde{\xi}_{BA}(t)$ must constitute, respectively, the real and imaginary parts of an analytic signal $\tilde{Z}_{BA}(t)$ with only positive frequency Fourier components [38,39].

2. The Zero-Temperature Analytic Signal

Following these lines, consider the signal $\tilde{Z}_{BA}(t) = \langle B(t)A \rangle$. One has

$$\tilde{Z}_{BA}(t) = \tilde{C}_{BA}(t) + \hbar\,\tilde{\xi}_{BA}(t) \tag{54}$$

By Fourier transformation, one gets from Eq. (54)

$$Z_{BA}(\omega) = C_{BA}(\omega) + \hbar\,\xi_{BA}(\omega) \tag{55}$$

At $T = 0$, the fluctuation–dissipation relations (45) reduce to[5]

$$\xi_{BA}(\omega) = \frac{1}{\hbar} \, \mathrm{sgn}(\omega) \, C_{BA}(\omega) \tag{56}$$

so that one gets, as expected,

$$Z_{BA}(\omega) = \begin{cases} 2\,C_{BA}(\omega), & \omega > 0 \\ 0, & \omega < 0 \end{cases} \tag{57}$$

At $T = 0$, the function $\tilde{Z}_{BA}(t) = \langle B(t)A \rangle$ has only positive frequency Fourier components and thus possesses the characteristics of an analytic signal [38,39]. This implies that the integral definition $\tilde{Z}_{BA}(\tau) = \int_0^\infty Z_{BA}(\omega)e^{-i\omega\tau}\,d\omega/2\pi$ can then be extended into the whole lower half of the complex τ-plane[6] (i.e., $\Im m\,\tau \leq 0$).

---

[5]One defines:

$$\mathrm{sgn}(\omega) = \begin{cases} +1, & \omega > 0, \\ -1, & \omega < 0. \end{cases}$$

[6]This is in accordance with the above noted fact that at finite temperature the extension to the region $-\beta \leq \Im m\,\tau \leq 0$ of the complex τ-plane of the correlation function $\langle B(\tau)A \rangle$ is analytic.

Similar considerations can be applied to the function $\tilde{Y}_{BA}(t) = \langle AB(t) \rangle$, which possesses only negative frequency Fourier components.

3. Other Representations of the Analytic Signal

Let us here focus on the analytic signal $\tilde{Z}_{BA}(t)$ of positive frequency Fourier components [similar remarks can be made for $\tilde{Y}_{BA}(t)$].

For $\Im m\tau \leq 0$, one can write, taking advantage of Eqs. (55) and (57),

$$\tilde{Z}_{BA}(\tau) = \frac{1}{2\pi} \int_{0}^{\infty} d\omega \, 2\hbar \, \xi_{BA}(\omega) \, e^{-i\omega\tau} \tag{58}$$

This yields the following representation of $\tilde{Z}_{BA}(\tau)$ for $\Im m\tau \leq 0$ in terms of $\tilde{\xi}_{BA}(t)$:

$$\tilde{Z}_{BA}(\tau) = \frac{1}{\pi} \int_{-\infty}^{\infty} (-i\hbar \, \tilde{\xi}_{BA}(t')) \, \frac{1}{\tau - t'} \, dt' \tag{59}$$

Note that, when $\Im m\tau = 0$, the integral in Eq. (59) must be understood as a principal value.

Equation (59) can in turn be used as a definition of $\tilde{Z}_{BA}(\tau)$ in the upper half of the complex τ-plane (i.e., $\Im m\,\tau > 0$), where the integral definition $\tilde{Z}_{BA}(\tau) = \int_{0}^{\infty} Z_{BA}(\omega)e^{-i\omega\tau}d\omega/2\pi$ cannot be used. It verifies the property

$$Z_{BA}^{*}(\tau) = Z_{BA}(\tau^{*}) \tag{60}$$

IV. AGING EFFECTS IN CLASSICAL OR QUANTAL BROWNIAN MOTION

Out–of–equilibrium dynamics of slowly relaxing systems is characterized by aging effects. Aging phenomena have been intensively studied, both experimentally and theoretically, mostly in complex systems such as spin glasses and other types of glasses [1–4]. Interestingly enough, aging can also be encountered in simpler, nondisordered systems [40–45]. For instance, as first noticed in Ref. 40, the simplest example of a dynamic variable which does not reach equilibrium and exhibits aging is the random walk: Indeed the correlation function of the displacement of a free Brownian particle at two different times depends explicitly on the two times involved, and not simply on their difference.

Up to now, aging effects have mostly been discussed in a classical framework (see, however Refs. 46–48). In order to investigate quantum aging effects, it is interesting, to begin with, to study the displacement of a free Brownian particle, since, as quoted above, this variable displays aging in the classical case. Free quantum Brownian motion can conveniently be described in the framework of the Ohmic dissipation model.

We compute below the velocity and displacement correlation functions, first, of a classical, then, of a quantal, Brownian particle. In contrast to its velocity, which thermalizes, the displacement $x(t) - x(t_0)$ of the particle with respect to its position at a given time never attains equilibrium (whatever the temperature, and even at $T = 0$). The model allows for a discussion of the corresponding modifications of the fluctuation–dissipation theorem.

A. Aging Effects in Overdamped Classical Brownian Motion

The one-dimensional dynamics of an overdamped classical free Brownian particle of viscosity η is described by the first-order differential equation

$$\eta \frac{dx}{dt} = F(t) \tag{61}$$

in which the random force $F(t)$ is modeled by a stationary Gaussian random process of zero mean and of correlation function:

$$\langle F(t)F(t') \rangle = 2\eta \, kT \, \delta(t - t') \tag{62}$$

In Eq. (62), the symbol $\langle \ldots \rangle$ denotes the average over the realizations of the noise.

Let us choose some initial time t_0 and consider the displacement of the particle at any later time $t > t_0$, as defined by

$$x(t) - x(t_0) = \frac{1}{\eta} \int_{t_0}^{t} F(t_1) \, dt_1 \tag{63}$$

1. The Displacement Response and Correlation Functions

The displacement response function $\tilde{\chi}_{xx}(t, t')$ characterizes the average displacement $\langle x(t) - x(t_0) \rangle$ at time t, due to a unit impulse of force taking place at a previous time $t' < t$. Is is easily deduced from the equation of motion (61), in which one adds to the random force $F(t)$ a nonrandom force proportional to $\delta(t - t')$. One thus gets

$$\tilde{\chi}_{xx}(t, t') = \Theta(t - t') \frac{1}{\eta} \tag{64}$$

Let us introduce the two-time displacement correlation function $\tilde{C}_{xx}(t, t'; t_0)$ as defined by

$$\tilde{C}_{xx}(t, t'; t_0) = \langle [x(t) - x(t_0)][x(t') - x(t_0)] \rangle \tag{65}$$

It is easy to check that, for $t_0 \leq t' \leq t$,

$$\tilde{C}_{xx}(t, t'; t_0) = 2 \frac{kT}{\eta} (t' - t_0) \qquad (66)$$

or, in terms of the observation time $\tau = t - t'$ and of the waiting time $t_w = t' - t_0$:

$$\tilde{C}_{xx}(\tau, t_w) = 2 \frac{kT}{\eta} t_w \qquad (67)$$

The displacement correlation function of the overdamped free Brownian particle is proportional to the waiting time [40,41].

2. The Fluctuation–Dissipation Ratio

As first noticed in Ref. 40, the FDT has to be generalized in order to handle such a case. Equation (50), valid for a classical dynamic variable in equilibrium, is not applicable to $x(t) - x(t_0)$. One can write the actual relation between $\tilde{\chi}_{xx}(t, t')$ and $\tilde{C}_{xx}(t, t'; t_0)$ with $t_0 \leq t' \leq t$ as

$$\tilde{\chi}_{xx}(t, t') = \beta \Theta(t - t') X(t, t'; t_0) \frac{\partial \tilde{C}_{xx}(t, t'; t_0)}{\partial t'} \qquad (68)$$

where the fluctuation–dissipation ratio $X(t, t'; t_0)$ acts as a violation factor of the FDT [40]. Defining for the dynamic variable under study an inverse effective temperature $\beta_{\mathrm{eff}}^{\mathrm{cl}}(t, t'; t_0)$ by

$$\beta_{\mathrm{eff}}^{\mathrm{cl}}(t, t'; t_0) = \beta X(t, t'; t_0) \qquad (69)$$

one can rewrite Eq. (68) as

$$\tilde{\chi}_{xx}(t, t') = \beta_{\mathrm{eff}}^{\mathrm{cl}} \Theta(t - t') \frac{\partial \tilde{C}_{xx}(t, t'; t_0)}{\partial t'}, \qquad t_0 \leq t' < t \qquad (70)$$

In the present model, using Eqs. (64) and (66), one gets, for any t and $t'(t' < t)$,

$$X(t, t'; t_0) = \frac{1}{2} \qquad (71)$$

In this simple description, the fluctuation–dissipation ratio associated to the particle displacement is a constant. Since it does not depend on T, it can be viewed as rescaling the temperature (as far as the variable of interest, namely the particle displacement, is concerned):

$$\beta_{\text{eff}}^{\text{cl}} = \frac{\beta}{2} \qquad (72)$$

B. Aging Effects in the Langevin Model

In the more refined Langevin description, the free Brownian particle equation of motion contains an inertial term and reads

$$m\frac{dv}{dt} + m\gamma v = F(t), \qquad v = \frac{dx}{dt} \qquad (73)$$

where $F(t)$ is the above-defined Gaussian delta-correlated random force, m the mass of the particle, and $\gamma = \eta/m$ the friction coefficient.

1. The Velocity Correlation Function

If, at a given initial time t_i, the velocity takes the value $v(t_i) = v_i$, at any later time $t \geq t_i$, it is given by

$$v(t) = v_i\, e^{-\gamma(t-t_i)} + \frac{1}{m}\int_{t_i}^{t} F(t_1)\, e^{-\gamma(t-t_1)}\, dt_1 \qquad (74)$$

The velocity correlation function, as defined by

$$\tilde{C}_{vv}(t_1, t_2) = \langle v(t_1)v(t_2)\rangle \qquad (75)$$

is equal to

$$\tilde{C}_{vv}(t_1; t_2) = \left(v_i^2 - \frac{kT}{m}\right) e^{-\gamma(t_1+t_2-2t_i)} + \frac{kT}{m}\, e^{-\gamma|t_1-t_2|} \qquad (76)$$

Letting $t_i \to -\infty$, the velocity $v(t)$ reduces to its stationary part:

$$v(t) = \frac{1}{m}\int_{-\infty}^{t} F(t_1)\, e^{-\gamma(t-t_1)}\, dt_1 \qquad (77)$$

Correspondingly, the correlation function $\tilde{C}_{vv}(t_1, t_2)$ becomes the equilibrium correlation function

$$\tilde{C}_{vv}(t_1 - t_2) = \frac{kT}{m}\, e^{-\gamma|t_1-t_2|} \qquad (78)$$

which only depends on the observation time. Note that formula (78) can also be obtained by taking as initial velocity a thermally distributed random variable v_i.

The velocity, which equilibrates at large times, is not an aging variable. Thus, Fourier analysis and the Wiener–Khintchine theorem can be used equivalently to obtain the equilibrium correlation function $\tilde{C}_{vv}(t_1 - t_2)$.

2. The Displacement Response and Correlation Functions

Adding to the Langevin force in Eq. (73) a nonrandom force proportional to $\delta(t - t')$, one gets the expression of the displacement response function:

$$\tilde{\chi}_{xx}(t, t') = \Theta(t - t') \frac{1}{\eta} [1 - e^{-\gamma(t-t')}] \tag{79}$$

This quantity only depends on the observation time $\tau = t - t'$.

From now on, we assume that the limit $t_i \to -\infty$ has been taken. The correlation function $\tilde{C}_{xx}(t, t'; t_0)$ as defined by Eq. (65) can be deduced from the equilibrium velocity correlation function *via* a double integration over time:

$$\tilde{C}_{xx}(t, t'; t_0) = \int_{t_0}^{t} dt_1 \int_{t_0}^{t'} \tilde{C}_{vv}(t_1 - t_2) \, dt_2 \tag{80}$$

Using Eq. (78), one gets, for $t_0 \le t' \le t$,

$$\tilde{C}_{xx}(t, t'; t_0) = \frac{kT}{\eta} \left[2(t' - t_0) - \frac{1 + e^{-\gamma(t-t')} - e^{-\gamma(t-t_0)} - e^{-\gamma(t'-t_0)}}{\gamma} \right] \tag{81}$$

or, in terms of the observation time $\tau = t - t'$ and of the waiting time $t_w = t' - t_0$:

$$\tilde{C}_{xx}(\tau, t_w) = \frac{kT}{\eta} \left[2t_w - \frac{1 + e^{-\gamma\tau} - e^{-\gamma(\tau+t_w)} - e^{-\gamma t_w}}{\gamma} \right] \tag{82}$$

Thus, as in the overdamped limit, the displacement $x(t) - x(t_0)$ is not a stationary random process.

3. The Fluctuation–Dissipation Ratio in the Langevin Model

According to Eqs. (79) and (81), the fluctuation–dissipation ratio in Eq. (68) is given by

$$X(t, t'; t_0) = \frac{1 - e^{-\gamma(t-t')}}{2 - e^{-\gamma(t-t')} - e^{-\gamma(t'-t_0)}} \tag{83}$$

that is, in terms of τ and t_w:

$$X(\tau, t_w) = \frac{1 - e^{-\gamma\tau}}{2 - e^{-\gamma\tau} - e^{-\gamma t_w}} \tag{84}$$

Since $X(\tau, t_w)$ does not depend on T, it can still be viewed as rescaling the temperature. The temperature rescaling factor now depends on both times τ and t_w. In the limit of large τ and $t_w (\gamma\tau \gg 1, \gamma t_w \gg 1)$, it tends toward the constant value $1/2$ which corresponds to the viscous limit of the Langevin model [Eq. (71)]. As for the associated inverse effective temperature, since one has

$$\beta_{\text{eff}}^{\text{cl}}(\tau, t_w) = \beta X(\tau, t_w) \tag{85}$$

the discussion of its behavior as a function of τ and t_w can be reduced to the one concerning the fluctuation–dissipation ratio.

4. Analysis in Terms of the Time-Dependent Diffusion Coefficient

In order to analyze the behavior of $X^{\text{cl}}(\tau, t_w)$ or $\beta_{\text{eff}}^{\text{cl}}(\tau, t_w)$ as a function of τ and t_w, it is convenient to introduce the time-dependent diffusion coefficient $D(t)$, which is an odd function of time defined for $t > 0$ by

$$D(t) = \frac{1}{2}\frac{d}{dt}\langle[x(t) - x]^2\rangle, \qquad t > 0 \tag{86}$$

that is,

$$D(t) = \frac{1}{2}\frac{d}{dt}\tilde{C}_{xx}(t, t; 0), \qquad t > 0 \tag{87}$$

At any time t, one can write

$$D(t) = \int_0^t \tilde{C}_{vv}(t')\, dt' \tag{88}$$

In the Langevin model, using Eq. (78), one gets

$$D(t) = \frac{kT}{\eta}(1 - e^{-\gamma t}), \qquad t > 0 \tag{89}$$

For $t > 0, D(t)$ increases montonously with t. In the limit $\gamma t \gg 1$, it tends toward the Einstein diffusion coefficient $D = kT/\eta$.

The fluctuation–dissipation ratio (84) can be rewritten in terms of $D(\tau)$ and $D(t_w)$ as

$$X(\tau, t_w) = \frac{D(\tau)}{D(\tau) + D(t_w)} \tag{90}$$

This quantity is obviously smaller than 1 and, as in the overdamped case, does not depend on T. The associated inverse effective temperature is

$$\beta_{\text{eff}}^{\text{cl}}(\tau, t_w) = \beta \frac{D(\tau)}{D(\tau) + D(t_w)} \tag{91}$$

Interestingly enough, X and $\beta_{\text{eff}}^{\text{cl}}$ depend on τ and t_w through the corresponding time-dependent diffusion coefficients[7] $D(\tau)$ and $D(t_w)$.

Returning to Eq. (68), one sees that the numerator of the fluctuation–dissipation ratio $X(t, t'; t_0)$ is proportional to the response function $\tilde{\chi}_{xx}(t, t')$ while its denominator is proportional to the derivative $\partial \tilde{C}_{xx}(t, t'; t_0)/\partial t'$ of the correlation function $\tilde{C}_{xx}(t, t'; t_0)$. Between t_0 and t' (i.e., during the waiting time t_w), the diffusing particle does not move on the average: $\langle x(t') \rangle = \langle x(t_0) \rangle$. Clearly, in order to compute the response of the displacement to an impulse of force at time t', one cannot apply the FDT in its standard form during the full time interval (t_0, t) (i.e., to the variable $x(t) - x(t_0)$), but only during the restricted time interval (t', t) (i.e., to the variable $x(t) - x(t')$) since in this latter case there is no waiting time. Indeed, proceeding in this way, that is, writing

$$\tilde{\chi}_{xx}(t, t') = \beta \, \Theta(t - t') \left. \frac{\partial \tilde{C}_{xx}(t, t'; t_0)}{\partial t'} \right|_{t_0 = t'} \tag{92}$$

one correctly gets for the displacement response function the expression

$$\tilde{\chi}_{xx}(t, t') = \beta \, \Theta(t - t') \int_{t'}^{t} \langle v(t_1) v(t') \rangle \, dt_1 \tag{93}$$

that is,

$$\tilde{\chi}_{xx}(t, t') = \beta \, \Theta(t - t') \, D(t - t') \tag{94}$$

[7]In fact, the validity of Eqs. (90) and (91) is not restricted to the simple (i.e., nonretarded) Langevin model as defined by Eq. (73). These formulas can be applied in other classical descriptions of Brownian motion in which a time-dependent diffusion coefficient can be defined. This is for instance, the case in the presence of non-Ohmic dissipation, in which case the motion of the Brownian particle is described by a retarded Langevin equation (see Section V).

In the Langevin model, the formula (94) for $\tilde{\chi}_{xx}(t,t')$ is identical to the expression (79), as it should. As for the denominator of $X(t,t';t_0)$, it involves the quantity

$$\frac{\partial \tilde{C}_{xx}(t,t';t_0)}{\partial t'} = \int_{-t_w}^{\tau} \tilde{C}_{vv}(u)\,du \tag{95}$$

that is, since $\tilde{C}_{vv}(u)$ is an even function of u [Eq. (78)]:

$$\frac{\partial \tilde{C}_{xx}(t,t';t_0)}{\partial t'} = D(\tau) + D(t_w) \tag{96}$$

This result can be easily understood because fluctuations in the displacement $x(t) - x(t_0)$ take place even during the waiting time $t_w = t' - t_0$.

Summing up, the fluctuation–dissipation ratio $X(t,t';t_0)$ and the associated inverse effective temperature $\beta_{\text{eff}}^{\text{cl}}(t,t';t_0)$ allow one to write a modified FDT relating $\tilde{\chi}_{xx}(t,t')$ to $\partial \tilde{C}_{xx}(t,t';t_0)/\partial t'$ with $t_0 \leq t' < t$, this latter quantity taking into account even those fluctuations of the displacement which take place during the waiting time.

Note, however, that the parameter $T_{\text{eff}}^{\text{cl}} = 1/\beta_{\text{eff}}^{\text{cl}}$, if it allows one to write a modified FDT for the displacement of the free Brownian particle [Eq. (70)], cannot be given the full significance of a physical temperature. In particular, it does not control the thermalization: in the absence of potential, the thermalization, which is solely linked with the behavior of the velocity, is effective once the limit $t_i \to -\infty$ has been taken.

5. Behavior of the Fluctuation–Dissipation Ratio and of the Effective Temperature in the Langevin Model

Let us now return to the expression (84) of $X(\tau,t_w)$ in the Langevin model. Clearly, for all values of τ, X is equal to 1 when $t_w = 0$, and it differs from 1 when $t_w \neq 0$. In that sense, it can be said that aging is always present as far as the displacement of the free Brownian particle is concerned. The curves representing $X(\tau,t_w) = \beta_{\text{eff}}^{\text{cl}}(\tau,t_w)/\beta$ as a function of τ for various nonzero values of t_w are plotted on Fig. 2.

For a given value of t_w, $X(\tau,t_w)$ and $\beta_{\text{eff}}^{\text{cl}}(\tau,t_w)$ increase monotonously with τ towards a finite limit $X_\infty(t_w)$. One has

$$X_\infty(t_w) \equiv \lim_{\gamma\tau\gg 1} X(\tau,t_w) = \frac{1}{2 - e^{-\gamma t_w}} \tag{97}$$

For small values of $t_w\,(\gamma t_w \ll 1)$, $X_\infty(t_w)$ does not depart very much from 1. For large values of $t_w\,(\gamma t_w \gg 1)$, $X_\infty(t_w)$ approaches the value $1/2$.

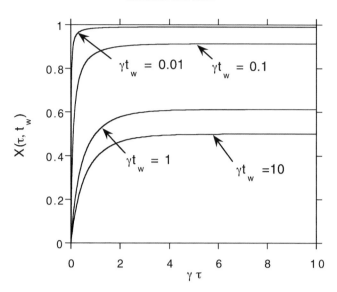

Figure 2. The violation factor $X(\tau, t_w) = \beta_{\mathrm{eff}}(\tau, t_w)/\beta$ of the classical FDT in the Langevin model plotted as a function of $\gamma\tau$ for various values of $\gamma t_w : \gamma t_w = 0.01$, $\gamma t_w = 0.1$, $\gamma t_w = 1$, $\gamma t_w = 10$.

In the opposite limit—that is, for small values of $\tau(\gamma\tau \ll 1)$—the motion is ballistic, and one gets

$$X \simeq \frac{\gamma\tau}{1 + \gamma\tau - e^{-\gamma t_w}} \tag{98}$$

For small values of $t_w(\gamma t_w \ll 1)$, one has, in this regime,

$$X \simeq \frac{\tau}{\tau + t_w} \tag{99}$$

and for large values of t_w one has

$$X \simeq \gamma\tau \tag{100}$$

C. Aging Effects in Quantum Brownian Motion

We now turn to quantum Brownian motion as described by the generalized Langevin equation (22), with the symmetrized correlation function of the random force as given by Eq. (20).

1. The Velocity Correlation Function

The symmetrized velocity correlation function is defined by

$$\tilde{C}_{vv}(t_1, t_2) = \frac{1}{2} \langle [v(t_1), v(t_2)]_+ \rangle \tag{101}$$

As stated above, the Langevin force $F(t)$ can be viewed as corresponding to a stationary random process. Clearly, the same is true of the solution $v(t)$ of the generalized Langevin equation (22), an equation which is valid once the limit $t_i \to -\infty$ has been taken. Thus, Fourier analysis and the Wiener–Khintchine theorem can be used to obtain the velocity correlation function, which only depends on the observation time: $\tilde{C}_{vv}(t_1, t_2) = \tilde{C}_{vv}(t_1 - t_2)$. As in the classical case, the velocity does not age.

As a result, one gets

$$\tilde{C}_{vv}(t) = \frac{1}{m} \int_{-\infty}^{\infty} \frac{d\omega}{2\pi} \frac{\Re e \gamma(\omega)}{|\gamma(\omega) - i\omega|^2} \hbar\omega \coth \frac{\beta\hbar\omega}{2} e^{-i\omega t} \tag{102}$$

In the Ohmic dissipation model with a Lorentzian cutoff function, $\gamma(\omega)$ is given by Eq. (28), and Eq. (102) reads

$$\tilde{C}_{vv}(t) = \frac{\gamma}{m} \int_{-\infty}^{\infty} \frac{d\omega}{2\pi} \frac{\omega_c^2}{(\gamma\omega_c - \omega^2)^2 + \omega^2\omega_c^2} \hbar\omega \coth \frac{\beta\hbar\omega}{2} e^{-i\omega t} \tag{103}$$

In the case $\omega_c/\gamma > 4$, which corresponds to a weak coupling between the particle and the bath,[8] one can write

$$\tilde{C}_{vv}(t) = \frac{\gamma}{m} \frac{1}{\alpha} \int_{-\infty}^{\infty} \frac{d\omega}{2\pi} \left(\frac{1}{\omega^2 + \omega_-^2} - \frac{1}{\omega^2 + \omega_+^2} \right) \hbar\omega \coth \frac{\beta\hbar\omega}{2} e^{-i\omega t} \tag{104}$$

where for brevity α stands for $(1 - 4\gamma\omega_c^{-1})^{1/2}$, and the angular frequencies $\omega_\pm$ are defined by

$$\omega_\pm = \frac{\omega_c}{2}(1 \pm \alpha) \tag{105}$$

---

[8]The following expressions remain valid even when $\omega_c/\gamma < 4$, provided that the appropriate analytical continuations are done. In this case, $\omega_\pm$ has an imaginary part, and therefore oscillations in time appear in the correlation functions. We shall only here consider the case $\omega_c/\gamma > 4$, since the weak-coupling assumption is already contained in the Caldeira–Leggett Hamiltonian (1).

In the infinitely short memory limit $\omega_c \to \infty$, one simply has

$$\tilde{C}_{vv}^{\omega_c \to \infty}(t) = \frac{\gamma}{m} \int_{-\infty}^{\infty} \frac{d\omega}{2\pi} \frac{1}{\omega^2 + \gamma^2} \hbar\omega \coth\frac{\beta\hbar\omega}{2} e^{-i\omega t} \qquad (106)$$

At any nonzero temperature, formulas (104) and (106) can, respectively, be recast as

$$\tilde{C}_{vv}(t) = \frac{kT}{m} \frac{1}{\alpha} \left[\frac{\gamma}{\omega_-} f_T(t, \omega_-) - \frac{\gamma}{\omega_+} f_T(t, \omega_+) \right] \qquad (107)$$

and

$$\tilde{C}_{vv}^{\omega_c \to \infty}(t) = \frac{kT}{m} f_T(t, \gamma) \qquad (108)$$

where $f_T(t, \gamma)$ stands for the series expansion

$$f_T(t, \gamma) = e^{-\gamma|t|} + 2 \sum_{n=1}^{\infty} \frac{e^{-\gamma|t|} - \frac{n}{\gamma t_{th}} e^{-\frac{n|t|}{t_{th}}}}{1 - \left(\frac{n}{\gamma t_{th}}\right)^2} \qquad (109)$$

and t_{th} denotes a thermal time as defined by $t_{th} = \hbar/2\pi kT$.

The detailed analysis of the behavior of $\tilde{C}_{vv}(t)$ as a function of t can be found in Ref. 25. Let us here just quote the following important feature: There exists a crossover temperature T_c, which is given by $T_c = \hbar\omega_-/\pi k$ (or $T_c = \hbar\gamma/\pi k$ in the infinitely short memory limit), above and below which the velocity correlation function exhibits two markedly different behaviors. Namely, for $T > T_c$, $\tilde{C}_{vv}(t)$ is positive at any time, and therefore, despite the existence of quantum corrections, this regime may be roughly qualified as classical. For $T < T_c$, $\tilde{C}_{vv}(t)$ is positive at small times, passes through a zero, and is negative at large times, a regime which may be qualified as quantal.

2.   The Quantum Time-Dependent Diffusion Coefficient

The quantum time-dependent diffusion coefficient $D(t) = \int_0^t \tilde{C}_{vv}(t') \, dt'$ is an odd function of time given by:

$$D(t) = \frac{\gamma}{m} \int_{-\infty}^{\infty} \frac{d\omega}{2\pi} \frac{\omega_c^2}{(\gamma\omega_c - \omega^2)^2 + \omega^2\omega_c^2} \hbar \coth\frac{\beta\hbar\omega}{2} \sin\omega t \qquad (110)$$

For simplicity, we restrict ourselves from now on to the short-memory limit $\omega_c \to \infty$. Eq. (110) then simplifies to

$$D^{\omega_c \to \infty}(t) = \frac{\gamma}{m} \int_{-\infty}^{\infty} \frac{d\omega}{2\pi} \frac{1}{\omega^2 + \gamma^2} \hbar \coth \frac{\beta \hbar \omega}{2} \sin \omega t \qquad (111)$$

For $t > 0$, formula (111) can (at any nonzero temperature) be recast as the following series expansion:

$$D^{\omega_c \to \infty}(t) = \frac{kT}{\eta} \left\{ 1 - e^{-\gamma t} + 2 \sum_{n=1}^{\infty} \frac{e^{-\frac{nt}{t_{\text{th}}}} - e^{-\gamma t}}{1 - \left(\frac{n}{\gamma t_{\text{th}}}\right)^2} \right\}, \qquad t > 0 \qquad (112)$$

The quantum time-dependent diffusion coefficient for $t > 0$ thus appears as the sum of the classical time-dependent diffusion coefficient $(kT/\eta)(1 - e^{-\gamma t})$, which from now on we shall denote as $D^{\text{cl}}(t)$, and of a supplementary contribution due to quantum effects, which can be shown to be always positive.

Let us now briefly recall the main features of the behavior of $D^{\omega_c \to \infty}(t)$ as a function of t for $t > 0$ [25]. First, the limiting value at infinite time of $D^{\omega_c \to \infty}(t)$ is, at any nonzero temperature, the usual Einstein diffusion coefficient kT/η. Above the crossover temperature T_c as defined above, $D^{\omega_c \to \infty}(t)$ increases monotonously toward its limiting value. Below the crossover temperature (i.e., $T < T_c$), $D^{\omega_c \to \infty}(t)$ first increases, then passes through a maximum and finally slowly decreases toward its limiting value. Thus, in the quantum regime, $D^{\omega_c \to \infty}(t)$ can exceed its stationary value, and the diffusive regime is only attained very slowly, namely after times $t \gg t_{\text{th}}$. At $T = 0$, $D^{\omega_c \to \infty}(t)$ can be expressed in terms of exponential integral functions:

$$D_{T=0}^{\omega_c \to \infty}(t) = \frac{\hbar}{2\pi m \gamma} [e^{-\gamma t} \overline{\text{Ei}}(\gamma t) - e^{\gamma t} \text{Ei}(-\gamma t)], \qquad t > 0 \qquad (113)$$

$D_{T=0}^{\omega_c \to \infty}(t)$ passes through a maximum at a time $t_m(T = 0) \sim \gamma^{-1}$. For $\gamma t \gg 1$, one has

$$D_{T=0}^{\omega_c \to \infty}(t) \sim \frac{\hbar}{\pi m} \frac{1}{\gamma t}, \qquad \gamma t \gg 1 \qquad (114)$$

The behavior of $D^{\omega_c \to \infty}(t)$ for several different temperatures on both sides of the crossover temperature T_c is illustrated in Fig. 3. Interestingly enough, these curves never intersect. One can indeed easily check that the slopes at the origin $dD^{\omega_c \to \infty}(t)/dt|_{t=0} = \tilde{C}_{vv}^{\omega_c \to \infty}(t = 0)$ increase with T, as well as the asymptotic values at infinite time kT/η. One can also show that, at any fixed time t,

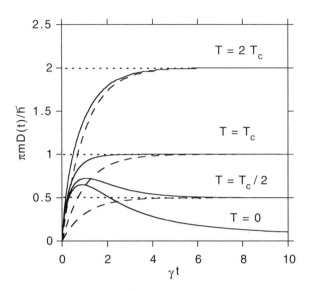

Figure 3. In the short-memory limit ($\omega_c \to \infty$), the quantum time-dependent diffusion coefficient $D^{\omega_c \to \infty}(t)$ plotted as a function of γt for several different bath temperatures on both sides of T_c (full lines): $\gamma t_{\text{th}} = 0.25 (T = 2T_c$, classical regime); $\gamma t_{\text{th}} = 0.5$ ($T = T_c$, crossover); $\gamma t_{\text{th}} = 1$ ($T = T_c/2$, quantum regime); $\gamma t_{\text{th}} = +\infty$ ($T = 0$). The corresponding curves for the classical diffusion coefficient $D^{\text{cl}}(t)$ are plotted in dotted lines in the same figure.

$D^{\omega_c \to \infty}(t)$ is a monotonic increasing function of the temperature (Fig. 4). This physically relevant property, which is valid in the whole temperature range, will play a crucial role in the following.

3. The Displacement Response and Correlation Functions

From the generalized Langevin equation (22) with $\tilde{\gamma}(t)$ as given by Eq. (27), one deduces, in the case $\omega_c/\gamma > 4$,

$$\tilde{\chi}_{xx}(t,t') = \Theta(t-t') \frac{1}{\eta} \left[1 - \frac{1}{\alpha} \frac{\omega_+^2}{\omega_c^2} e^{-\omega_-(t-t')} + \frac{1}{\alpha} \frac{\omega_-^2}{\omega_c^2} e^{-\omega_+(t-t')} \right] \qquad (115)$$

For well-separated time scales ω_c^{-1} and $\gamma^{-1}(\omega_c^{-1} \ll \gamma^{-1})$, one has simply

$$\tilde{\chi}_{xx}(t,t') = \Theta(t-t') \frac{1}{\eta} \left[1 - e^{-\gamma(t-t')} + \gamma^2 \omega_c^{-2} e^{-\omega_c(t-t')} \right] \qquad (116)$$

In the infinitely short memory limit $\omega_c \to \infty$, one recovers as expected for $\tilde{\chi}_{xx}(t,t')$ the expression (79) corresponding to the nonretarded Langevin model.

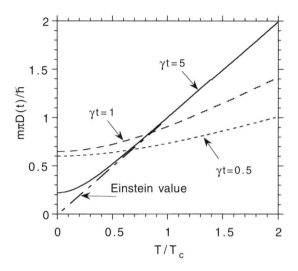

Figure 4. Quantum time-dependent diffusion coefficient $D^{\omega_c \to \infty}(t)$ plotted as a function of T/T_c for several different values of γt : $\gamma t = 0.5$; $\gamma t = 1$; $\gamma t = 5$; classical limit ($\gamma t \gg 1$).

The symmetrized displacement correlation function $\tilde{C}_{xx}(t, t'; t_0)$ is defined by the formula:

$$\tilde{C}_{xx}(t, t'; t_0) = \frac{1}{2} \left\langle [(x(t) - x(t_0)), (x(t') - x(t_0))]_+ \right\rangle \qquad (117)$$

Still assuming that the limit $t_i \to -\infty$ has been taken from the beginning, one can compute $\tilde{C}_{xx}(t, t'; t_0) = \int_{t_0}^{t} dt_1 \int_{t_0}^{t'} \tilde{C}_{vv}(t_1 - t_2) dt_2$. However, the full expression of $\tilde{C}_{xx}(t, t'; t_0)$ is of no use in the analysis of aging effects, which will be seen to rely only on the expression for the time-dependent diffusion coefficient $D(t) = (1/2)d\tilde{C}_{xx}(t, t; 0)/dt$.

4. Relation Between the Displacement Response Function and the Time-Dependent Diffusion Coefficient

A modified form of the quantum FDT in an out-of-equilibrium situation, allowing for the description of aging effects, has been proposed in [46,47] for mean-field spin-glass models. Here, we propose a modified form of the quantum FDT which can conveniently be applied to the displacement of the free quantum Brownian particle—that is, to the problem of diffusion.[9]

---

[9]Interestingly, this modified quantum FDT can also be used in non-Ohmic models with $0 < \delta < 2$, that is, when anomalous diffusion is taking place (see Section V).

Since the limit $t_i \to -\infty$ has been taken from the beginning, the particle velocity has reached equilibrium. The equilibrium FDT (46) can be applied to this dynamic variable, yielding, with obvious notations, an expression for the velocity response function,

$$\tilde{\chi}_{vx}(t_1, t') = \frac{2}{\pi\hbar} \Theta(t_1 - t') \int_{-\infty}^{\infty} dt'' \frac{\partial \tilde{C}_{vx}(t_1 - t'')}{\partial t''} \log \coth \frac{\pi|t'' - t'|}{2\beta\hbar} \quad (118)$$

that is,

$$\tilde{\chi}_{vx}(t_1, t') = \frac{2}{\pi\hbar} \Theta(t_1 - t') \int_{-\infty}^{\infty} dt'' \, \tilde{C}_{vv}(t_1 - t'') \log \coth \frac{\pi|t'' - t'|}{2\beta\hbar} \quad (119)$$

Since the displacement response function $\tilde{\chi}_{xx}(t, t')$ can be expressed as $\tilde{\chi}_{xx}(t, t') = \int_{t'}^{t} \tilde{\chi}_{vx}(t_1, t') \, dt_1$, one gets for this quantity, making use of Eq. (119) and of the identity

$$\frac{\partial \tilde{C}_{xx}(t, t''; t_0)}{\partial t''}\bigg|_{t_0 = t'} = \int_{t'-t''}^{t-t''} \tilde{C}_{vv}(u) \, du \quad (120)$$

the expression

$$\tilde{\chi}_{xx}(t, t') = \frac{2}{\pi\hbar} \Theta(t - t') \int_{-\infty}^{\infty} dt'' \frac{\partial \tilde{C}_{xx}(t, t''; t_0)}{\partial t''}\bigg|_{t_0 = t'} \log \coth \frac{\pi|t'' - t'|}{2\beta\hbar} \quad (121)$$

Note that writing formula (121) amounts to making use of the same argument as in the classical case, namely to apply the FDT as expressed by Eq. (46) to the variable $x(t) - x(t')$. Since one has, in terms of the time-dependent diffusion coefficient,

$$\frac{\partial \tilde{C}_{xx}(t, t''; t_0)}{\partial t''}\bigg|_{t_0 = t'} = D(t - t'') + D(t'' - t') \quad (122)$$

one can rewrite Eq. (121) as

$$\tilde{\chi}_{xx}(t, t') = \frac{2}{\pi\hbar} \Theta(t - t') \int_{-\infty}^{\infty} dt'' \, [D(t - t'') + D(t'' - t')] \log \coth \frac{\pi|t'' - t'|}{2\beta\hbar}$$

$$(123)$$

that is, the time-dependent diffusion coefficient being an odd function of its time argument:

$$\tilde{\chi}_{xx}(t,t') = \frac{2}{\pi\hbar}\,\Theta(t-t')\int\limits_{-\infty}^{\infty}dt''D(t-t'')\log\coth\frac{\pi|t''-t'|}{2\beta\hbar} \tag{124}$$

One can also write equivalently

$$\tilde{\chi}_{xx}(t,t') = \frac{2}{\pi\hbar}\,\Theta(t-t')\int\limits_{-\infty}^{\infty}dt_1\,D(t-t'-t_1)\log\coth\frac{\pi|t_1|}{2\beta\hbar} \tag{125}$$

Exactly like its classical analog Eq. (94), Eq. (125) allows one to express the displacement response function in terms of the time-dependent diffusion coefficient. However, contrary to the classical case in which $\tilde{\chi}_{xx}(t,t')$ is directly proportional to $D(t-t')$, in the quantum formulation $\tilde{\chi}_{xx}(t,t')$ is a convolution product, for the value $t-t'$ of the argument, of the functions $D(t_1)$ and $\log\coth(\pi|t_1|/2\beta\hbar)$. Inverting the convolution equation (125) yields an expression for $D(t)$ in terms of the dissipative part of the displacement response function:

$$D(t) = -\frac{1}{\pi}\,\mathrm{vp}\int\limits_{-\infty}^{\infty}dt''\,i\hbar\,\frac{\partial\tilde{\xi}_{xx}(t,t'')}{\partial t''}\,\frac{\pi}{\beta\hbar}\coth\frac{\pi t''}{\beta\hbar} \tag{126}$$

Since the displacement response function does not depend on the bath temperature [Eq. (115) or (116)], Eq. (126) displays the above quoted property that, at any fixed time $t, D(t)$ is a monotonic increasing function of the temperature. In the infinitely short memory limit, taking into account the corresponding expression (79) of $\tilde{\chi}_{xx}$, one gets from Eq. (126)

$$D^{\omega_c\to\infty}(t) = \frac{\hbar}{2\pi m}\,\mathrm{vp}\int\limits_{-\infty}^{\infty}dt_1[\Theta(t-t_1)e^{-\gamma(t-t_1)}+\Theta(t_1-t)e^{\gamma(t-t_1)}]\,\frac{\pi}{\beta\hbar}\coth\frac{\pi t_1}{\beta\hbar}$$

$$\tag{127}$$

5.   The Modified Quantum Fluctuation–Dissipation Theorem

Let us write a modified quantum FDT as

$$\frac{\partial\tilde{C}_{xx}(t,t';t_0)}{\partial t'} = -\frac{1}{\pi}\,\mathrm{vp}\int\limits_{-\infty}^{\infty}dt''\,i\hbar\,\frac{\partial\tilde{\xi}_{xx}(t,t'')}{\partial t''}\,\frac{\pi}{\beta_{\mathrm{eff}}\hbar}\coth\frac{\pi(t''-t')}{\beta_{\mathrm{eff}}\hbar} \tag{128}$$

where, for short, β_{eff} stands for $\beta_{\mathrm{eff}}(t, t'; t_0)$, or, equivalently,

$$D(t - t') + D(t' - t_0) = -\frac{1}{\pi} \mathrm{vp} \int_{-\infty}^{\infty} dt'' \, i\hbar \, \frac{\partial \tilde{\xi}_{xx}(t, t'')}{\partial t''} \frac{\pi}{\beta_{\mathrm{eff}} \hbar} \coth \frac{\pi(t'' - t')}{\beta_{\mathrm{eff}} \hbar}$$

(129)

The quantity β_{eff} introduced in Eqs. (128)–(129) can be interpreted as an inverse effective temperature, defined for $\tau = t - t' > 0$ and $t_w = t' - t_0 > 0$.

Like its classical analog, the parameter $T_{\mathrm{eff}} = 1/\beta_{\mathrm{eff}}$, if it allows one to write the modified quantum FDT (129), cannot however be considered as a *bona fide* physical temperature. In particular, it does not control the thermalization of the particle (which is effective here since the limit $t_i \to -\infty$ has been taken from the very beginning).

6. Determination of the Effective Temperature

When compared with the right-hand side of Eq. (126), the right-hand side of Eq. (129) appears to be formally identical to the time-dependent diffusion coefficient $D_{T_{\mathrm{eff}}}(t - t')$ at the effective temperature $T_{\mathrm{eff}} = 1/k\beta_{\mathrm{eff}}$. One can thus rewrite Eq. (129) in the following more compact form:

$$D(t - t') + D(t' - t_0) = D_{T_{\mathrm{eff}}}(t - t')$$

(130)

Equation (130), in turn, allows the determination of the effective temperature as a function of $t - t'$ and $t' - t_0$. Since the time-dependent diffusion coefficient is, at any positive time, a monotonic increasing function of the temperature, Eq. (130) yields for $T_{\mathrm{eff}}(t - t', t' - t_0)$ a uniquely defined value.

In the classical limit in which the time-dependent diffusion coefficient is proportional to the temperature, the inverse effective temperature $\beta_{\mathrm{eff}}^{\mathrm{cl}}$ deduced in this limit from Eq. (130) satisfies the equation

$$\beta_{\mathrm{eff}}^{\mathrm{cl}}(t, t'; t_0) = \beta \frac{D(t - t')}{D(t - t') + D(t' - t_0)}$$

(131)

in accordance with Eq. (91) with $\tau = t - t'$ and $t_w = t' - t_0$.

Note that, except in the classical limit, β_{eff} is not simply proportional to β, so that the quantum FDT violation cannot be described by a simple rescaling of the bath temperature.

7. The Effective Temperature in the Ohmic Model

In contrast with its classical counterpart, the inverse effective temperature is not given by a simple ratio of time-dependent diffusion coefficients like Eq. (131), but has to be deduced from the implicit equation (130), which in general can only

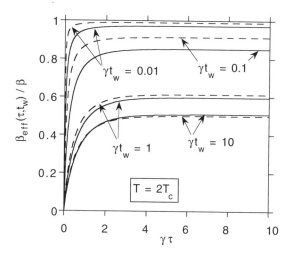

Figure 5. In the short-memory limit $(\omega_c \to \infty)$, the ratio $\beta_{\text{eff}}(\tau, t_w)/\beta$, at bath temperature $T = 2T_c$ (classical regime), plotted as a function of $\gamma\tau$ for various values of γt_w (full lines): $\gamma t_w = 0.01$; $\gamma t_w = 0.1$; $\gamma t_w = 1$; $\gamma t_w = 10$. The corresponding curves for the classical violation factor $X(\tau, t_w) = \beta_{\text{eff}}^{\text{cl}}(\tau, t_w)/\beta$ are plotted in dotted lines on the same figure.

be solved numerically. The solution $\beta_{\text{eff}}(t, t'; t_0)$ of Eq. (130) is a function of the observation time $\tau = t - t'$ and of the waiting time $t_w = t' - t_0$.

Let us now restrict the study to the particular case of the Ohmic model in the short memory limit $\omega_c \to \infty$. Equation (130) then reads

$$\int_{-\infty}^{\infty} \frac{d\omega}{2\pi} \frac{1}{\omega^2 + \gamma^2} \hbar \coth \frac{\beta_{\text{eff}}(t, t'; t_0)\hbar\omega}{2} \sin \omega(t - t')$$

$$= \int_{-\infty}^{\infty} \frac{d\omega}{2\pi} \frac{1}{\omega^2 + \gamma^2} \hbar \coth \frac{\beta\hbar\omega}{2} [\sin \omega(t - t') + \sin \omega(t' - t_0)] \quad (132)$$

The curves representing $\beta_{\text{eff}}(\tau, t_w)/\beta$ as a function of τ for various nonzero values of t_w are plotted on Figs. 5 to 7 for several nonzero bath temperatures on both sides of T_c. Note that Eqs. (130) and (132) also allow to define the effective temperature when $T = 0$. The corresponding curves are plotted on Fig. 8.

Let us now comment upon the results obtained (Figs. 5 to 8). For all values of τ, and at any nonzero temperature, the ratio $\beta_{\text{eff}}(\tau, t_w)/\beta$ is equal to 1 when $t_w = 0$ and it differs from 1 when $t_w \neq 0$. Thus, as in the classical case, aging is always present as far as the displacement of the quantum Brownian particle is concerned. For a given value of t_w, the ratio $\beta_{\text{eff}}(\tau, t_w)/\beta$ increases monotonously with τ toward a finite limit $X_\infty^Q(t_w)$ (Figs. 5 to 7). At $T = 0, T_{\text{eff}}(\tau, t_w)$ decreases monotonously with τ toward zero (Fig. 8).

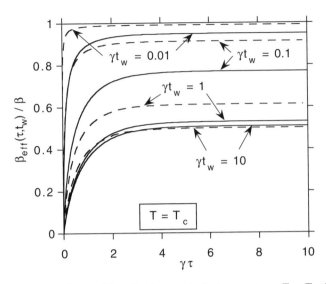

Figure 6. Same as above (Fig. 5), but at bath temperature $T = T_c$ (crossover temperature).

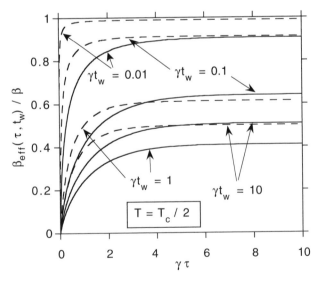

Figure 7. Same as above (Fig. 5 or Fig. 6), but at bath temperature $T = T_c/2$ (quantum regime).

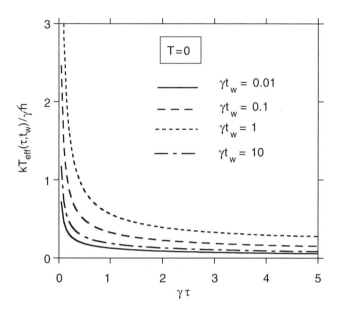

Figure 8. Zero temperature case. The effective temperature $T_{\text{eff}}(\tau, t_w)$ plotted as a function of $\gamma\tau$ for various values of γt_w : $\gamma t_w = 0.01$, $\gamma t_w = 0.1$, $\gamma t_w = 1$, $\gamma t_w = 10$.

Let us discuss the limit value of $\beta_{\text{eff}}(\tau, t_w)/\beta$ for large values of τ (i.e., $\tau \gg \gamma^{-1}, t_{\text{th}}$). For a given t_w, one has

$$X_\infty^Q(t_w) = \lim_{\tau \gg \gamma^{-1}, t_{\text{th}}} \beta_{\text{eff}}(\tau, t_w)/\beta = \frac{1}{1 + \dfrac{\eta}{kT} D(t_w)} \tag{133}$$

Above the crossover temperature T_c, $D(t_w)$ is a monotonously increasing function of t_w, which is reflected in the fact that $X_\infty^Q(t_w)$ decreases monotonously with t_w (see Fig. 5 for $T = 2T_c$). Below the crossover temperature (i.e., $T < T_c$), $D(t_w)$ is not a monotonous function of t_w. Consequently, the same is true of $X_\infty^Q(t_w)$ (see Fig. 7 for $T = T_c/2$). A similar phenenomenon takes place at $T = 0$, in which case, for any given value of τ, $T_{\text{eff}}(\tau, t_w)$ first increases with t_w, passes through a maximum for a value $t_{wm} \sim \gamma^{-1}$, then decreases toward zero (Fig. 8).

8. Discussion

The parameter $\beta_{\text{eff}}(t, t'; t_0)$ having been determined, one can write, using Eqs. (129) and (130),

$$D_{T_{\text{eff}}}(t - t') = -\frac{1}{\pi} \text{vp} \int_{-\infty}^{\infty} dt'' \, i\hbar \, \frac{\partial \tilde{\xi}_{xx}(t, t'')}{\partial t''} \, \frac{\pi}{\beta_{\text{eff}}\hbar} \coth \frac{\pi(t'' - t')}{\beta_{\text{eff}}\hbar} \tag{134}$$

On another hand, returning to the expression (124) of $\tilde{\chi}_{xx}(t,t')$ and taking advantage of the fact that the displacement response function does not actually depend on β (Eq. (115) or (116)), one may replace β by $\beta_{\mathrm{eff}}(t - t', t' - t_0)$ in this latter expression, yielding

$$\tilde{\chi}_{xx}(t,t') = \frac{2}{\pi\hbar}\,\Theta(t - t')\int\limits_{-\infty}^{\infty} dt''D_{T_{\mathrm{eff}}}(t - t'')\log\coth\frac{\pi|t'' - t'|}{2\beta_{\mathrm{eff}}\hbar} \qquad (135)$$

Equations (134) (respectively (135)) are formally similar to Eqs. (126) (respectively (124)), except for the fact that, due to waiting time effects, the diffusing particle is considered as being in contact with a bath at T_{eff}, an effective temperature depending on both time arguments $t - t'$ and $t' - t_0$.

Let us emphasize, however, that, in contrast with the set of equations [(124), (126)] that are reciprocal convolution relations, Eq. (135) cannot be viewed as the reciprocal relation of Eq. (134), since these two equations do not involve the same functions. Actually, the convolution product structure is lost in Eqs. (134) and (135), because of the dependence of T_{eff} and β_{eff} on $t - t'$ and $t' - t_0$.

V. AGING EFFECTS IN CLASSICAL OR QUANTAL ANOMALOUS DIFFUSION

In this section, we extend the previous study to the case of non-Ohmic dissipation, in the presence of which the particle damped motion is described by a truly retarded equation even in the classical limit, and either localization or anomalous diffusion phenomena are taking place. Such situations are encountered in various problems of condensed matter physics [28].

A. Non-Ohmic Noise and Friction

Non-Ohmic dissipation models are defined by a generalized friction coefficient varying at small angular frequencies like a power-law characterized by the exponent $\delta - 1$ (with $\delta \neq 1$):

$$\Re e\gamma(\omega) = \gamma_\delta\left(\frac{|\omega|}{\tilde{\omega}}\right)^{\delta-1}, \qquad |\omega| \ll \omega_c \qquad (136)$$

In Eq. (136), ω_c is the previously defined cutoff angular frequency typical of the environment, and $\tilde{\omega} \ll \omega_c$ is a reference angular frequency allowing for the coupling constant $\eta_\delta = m\gamma_\delta$ to have the dimension of a viscosity for any δ [28].

Correspondingly, according to the second FDT (20), the noise spectral density, which is the inverse Fourier transform of the random force correlation function $\tilde{C}_{FF}(t)$, is given by

$$C_{FF}(\omega) = \hbar\omega \coth\frac{\beta\hbar\omega}{2} \, \eta_\delta \left(\frac{|\omega|}{\tilde{\omega}}\right)^{\delta-1} \qquad |\omega| \ll \omega_c \qquad (137)$$

For $\delta = 1$, the noise spectral density is a constant (white noise), at least in the angular frequency range $|\omega| \ll \omega_c$. In the limit $\omega_c \to \infty$, the Langevin force $F(t)$ is delta-correlated, and the Langevin equation is nonretarded. The white noise case corresponds to Ohmic friction. The cases $0 < \delta < 1$ and $\delta > 1$ are known respectively as the sub-Ohmic and super-Ohmic models. Here we will assume that $0 < \delta < 2$, for reasons to be developed below [28,49–51].

At this stage, following Ref. 28, it is convenient to introduce the δ-dependent angular frequency ω_δ as defined by

$$\omega_\delta^{2-\delta} = \gamma_\delta \, \frac{1}{\tilde{\omega}^{\delta-1}} \, \frac{1}{\sin\frac{\delta\pi}{2}} \qquad (138)$$

Using this notation, Eq. (136) reads

$$\Re e\gamma(\omega) = \omega_\delta \left(\frac{|\omega|}{\omega_\delta}\right)^{\delta-1} \sin\frac{\delta\pi}{2}, \qquad |\omega| \ll \omega_c \qquad (139)$$

Note for further purpose that the function $\gamma(\omega)$ can be defined for any complex ω (except for a cut on the real negative axis). It is given by

$$\gamma(\omega) = \omega_\delta \left(\frac{-i\omega}{\omega_\delta}\right)^{\delta-1}, \qquad |\omega| \ll \omega_c, \qquad -\pi < \arg\omega \leq \pi \qquad (140)$$

B. Non-Ohmic One- and Two-Time Dynamics

Let us return to the Caldeira–Leggett microscopic model. The motion of the particle can then be described by the generalized Langevin equation (22), which we reproduce below for practical convenience:

$$m\frac{dv}{dt} = -m \int_{-\infty}^{\infty} \tilde{\gamma}(t-t') \, v(t') \, dt' + F(t), \qquad v = \frac{dx}{dt} \qquad (141)$$

In Eq. (141), it is assumed that the diffusing particle and the surrounding medium have been put in contact in an infinitely remote past, as pictured by the lower integration bound $-\infty$ in the retarded friction term. Both $F(t)$ and $v(t)$ can be viewed as stationary random processes. Their spectral densities are linked by

$$C_{vv}(\omega) = |\mu(\omega)|^2 C_{FF}(\omega) \qquad (142)$$

where

$$\mu(\omega) = \frac{1}{m[\gamma(\omega) - i\omega]} \tag{143}$$

denotes the frequency-dependent particle mobility. It has been demonstrated that, for $0 < \delta < 2$, the total mass of the particle and of the bath oscillators diverges, while, for $\delta > 2$, it remains finite and can be considered as a renormalized mass (in this latter case, the dynamics is governed at large times by a kinematical term involving the renormalized mass, and the initial expected value of the velocity is never forgotten) [28,49–51].

In the following, we limit ourselves to the case $0 < \delta < 2$. In this range of values of δ, the expected initial value of the velocity is forgotten at large times, a situation which in this sense may be qualified as ergodic.

1. Mittag–Leffler Relaxation of an Initial Velocity Fluctuation

The function $\gamma(\omega)$ characterizes the relaxation of the average particle velocity. For instance, with $\gamma(\omega)$ as given by Eq. (140), one can show that the average velocity at time t corresponding to an initial velocity $v(t_i) = v_i$ is given by [52]

$$\langle v(t) \rangle = v_i\, E_{2-\delta}(-[\omega_\delta(t - t_i)]^{2-\delta}), \qquad t \geq t_i \tag{144}$$

In Eq. (144), $E_\alpha(x)$ denotes the Mittag–Leffler function[10] of index α [53,54], and the δ-dependent angular frequency ω_δ acts as an inverse relaxation time. For $\delta = 1$, the function $E_{2-\delta}[-(\omega_\delta t)^{2-\delta}]$ reduces to a decreasing exponential $(E_1(-\omega_1 t) = e^{-\omega_1 t})$, while, for other values of δ in the range $0 < \delta < 2$, it decays algebraically at large times, as pictured by [52,53]

$$E_{2-\delta}[-(\omega_\delta t)^{\delta-2}] \simeq \frac{1}{\Gamma(\delta - 1)}\,(\omega_\delta t)^{\delta-2} \tag{145}$$

---

[10]The Mittag–Leffler function is defined by the series expansion

$$E_\alpha(x) = \sum_{n=0}^{\infty} \frac{x^n}{\Gamma(\alpha n + 1)}, \qquad \alpha > 0$$

where Γ is the Euler Gamma function. The Mittag–Leffler function $E_\alpha(x)$ reduces to the exponential e^x when $\alpha = 1$. The asymptotic behavior at large x of the Mittag–Leffler function $E_\alpha(x)$ is as follows:

$$E_\alpha(x) \simeq -\frac{1}{x}\frac{1}{\Gamma(1 - \alpha)}, \qquad x \gg 1$$

Equations (144) and (145) demonstrate that, in the range of values $0 < \delta < 2$, the initial value of the velocity is forgotten at large times.

2. The Velocity Correlation Function

Applying the Wiener–Khintchine theorem, one obtains the velocity correlation function as the inverse Fourier transform of $C_{vv}(\omega)$, that is, in terms of the noise spectral density $C_{FF}(\omega)$:

$$\tilde{C}_{vv}(t) = \frac{1}{m^2} \int\limits_{-\infty}^{\infty} \frac{d\omega}{2\pi} e^{-i\omega t} C_{FF}(\omega) \frac{1}{\gamma(\omega) - i\omega} \frac{1}{\gamma^*(\omega) + i\omega} \tag{146}$$

Equation (146) can be rewritten as

$$\tilde{C}_{vv}(t) = \frac{1}{m^2} \int\limits_{-\infty}^{\infty} \frac{d\omega}{2\pi} e^{-i\omega t} \frac{C_{FF}(\omega)}{2\Re e\gamma(\omega)} \left\{ \frac{1}{\gamma(\omega) - i\omega} + \frac{1}{\gamma^*(\omega) + i\omega} \right\} \tag{147}$$

The detailed analysis of the behavior of $\tilde{C}_{vv}(t)$ as a function of t can be found in Ref. 51.

3. The Particle Coordinate and Displacement

Let us now turn to the study of the particle coordinate. One may attempt to define its spectral density as $C_{xx}(\omega) = C_{vv}(\omega)/\omega^2$. If convergent, the integral $\int_{-\infty}^{\infty} (d\omega/2\pi) C_{xx}(\omega)$ represents $\langle x^2(t) \rangle$, an equilibrium average value which must be independent of t. Checking the small-ω behavior of the integrand with the chosen model for $\gamma(\omega)$ [Eq. (140)], one sees that this is only possible at $T = 0$ and for $0 < \delta < 1$. In this case, the particle is localized, and it makes sense to define its position in an absolute way as $x(t) = \int_{-\infty}^{t} v(t') dt'$. The two-time symmetrized position correlation function, as defined by

$$\tilde{C}_{xx}(t, t') = \frac{1}{2} \langle [x(t), x(t')]_+ \rangle \tag{148}$$

only depends on the observation time: it does not age.

In other cases, that is at $T = 0$ for $1 \leq \delta < 2$, and at finite T for $0 < \delta < 2$, the integral $\int_{-\infty}^{\infty} (d\omega/2\pi) C_{xx}(\omega)$ diverges. Then $\langle x^2(t) \rangle$ and $\tilde{C}_{xx}(t, t')$ as defined by Eq. (148) are infinite. The particle diffuses (possibly anomalously). The integrated velocity correlation function $\int_0^t \tilde{C}_{vv}(t') dt'$ represents the time-dependent diffusion coefficient $D(t)$ (in an extended sense when diffusion is anomalous, that is at $T = 0$ and for $1 < \delta < 2$, and at finite T for $0 < \delta < 1$ and $1 < \delta < 2$). It is then no longer possible to define an absolute position. We thus focus our interest on the displacement $x(t) - x(t_0)(t \geq t_0)$. This quantity does not equilibrate with the bath, even at large times. The displacement correlation function $\tilde{C}_{xx}(t, t'; t_0)$ as defined by Eq. (117), depends on both $\tau = t - t'$ and $t_w = t' - t_0$: it ages. As in the Ohmic case, the aging properties of $\tilde{C}_{xx}(t, t'; t_0)$ can be described in terms of the time-dependent quantum generalized diffusion coefficients $D(\tau)$ and $D(t_w)$.

4. The Time-Dependent Quantum Generalized Diffusion Coefficient

Therefore, before describing the modification of the equilibrium FDT, we need to study in details the behavior of $D(t)$. Note, however, that the integrated velocity correlation function $\int_0^t C_{vv}(t') \, dt'$ takes on the meaning of a time-dependent diffusion coefficient only when the mean-square displacement increases without bounds (when the particle is localized, this quantity characterizes the relaxation of the mean square displacement $\Delta x^2(t)$ toward its finite limit $\Delta x^2(\infty)$).

Since $0 < \delta < 2$, one can restrict the study to the infinite bath bandwidth limit $\omega_c \to \infty$. One has

$$D^{\omega_c \to \infty}(t) = \frac{\hbar}{m\pi} \gamma_\delta^{\delta-2}$$

$$\times \int_0^\infty d\omega \left[\omega^{\delta-1} + \omega^{3-\delta} \left(\omega^{\delta-2} \cot \frac{\delta\pi}{2} - \gamma_\delta^{\delta-2} \right)^2 \right]^{-1} \coth \frac{\beta\hbar\omega}{2} \sin \omega t \quad (149)$$

At finite T, it is interesting to discuss on the same footing the classical counterpart of $D(t)$, namely $D^{cl}(t)$ deduced from $D(t)$ by replacing $\coth(\beta\hbar\omega/2)$ by $2/\beta\hbar\omega$ in Eq. (148). Several important features of $D(t)$ and $D^{cl}(t)$ can be obtained by contour integration.

At $T = 0, D(t)$ is found to be the sum of a pole contribution, which exists only for $0 < \delta < 1$, given by the oscillating function

$$D(t)_{\text{pole}} \sim \frac{\hbar}{m} \frac{1}{2 - \delta} e^{-\Lambda t} \sin \Omega t \quad (150)$$

where Ω and Λ are known functions of δ and γ, and of a cut contribution behaving at large times as a power law,

$$D(t)_{\text{cut}} \sim \frac{\hbar}{m\pi} (\gamma t)^{\delta-2} \sin^3 \frac{\delta\pi}{2} \Gamma(2 - \delta) \quad (151)$$

where Γ denotes the Euler Gamma function.

At finite $T, D^{cl}(t)$ is also found to be the sum with an oscillating function, which exists only for $0 < \delta < 1$, namely

$$D^{cl}(t)_{\text{pole}} \sim \frac{kT}{m\gamma} \frac{2}{2 - \delta} \left(\sin \frac{\delta\pi}{2} \right)^{\frac{1}{2-\delta}} e^{-\Lambda t} \sin(\Omega t - \phi) \quad (152)$$

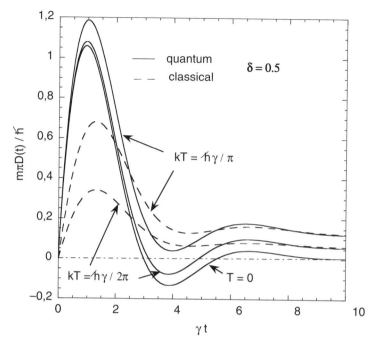

Figure 9. Full lines: $D(t)$ plotted as a function of γt for $\delta = 0.5$ and for bath temperatures $T = 0, kT = \hbar\gamma/2\pi$, $kT = \hbar\gamma/\pi$ [at $T = 0$, $D(t)$ is not an extended diffusion coefficient, but characterizes the relaxation of $\Delta x^2(t)$ toward its finite limit value $\Delta x^2(\infty)$]. Dashed lines: The corresponding $D^{\mathrm{cl}}(t)$.

with $\phi = \pi\delta/2(2 - \delta)$, and of a cut contribution behaving at large times as a power law:

$$D^{\mathrm{cl}}(t)_{\mathrm{cut}} \sim \frac{kT}{m\gamma} (\gamma t)^{\delta - 1} \frac{\sin\frac{\delta\pi}{2}}{\Gamma(\delta)} \tag{153}$$

The behaviors of $D(t)$ and $D^{\mathrm{cl}}(t)$ at several different temperatures are illustrated on Fig. 9 for $\delta = 0.5$ and on Fig. 10 for $\delta = 1.5$. Interestingly enough, for any given δ, the curves corresponding to different bath temperatures do not intersect. Actually, it can be shown that, at any fixed time $t, D(t)$, like $D^{\mathrm{cl}}(t)$, is a monotonously increasing function of T.

For times $t \ll t_{\mathrm{th}}(t_{\mathrm{th}} = \hbar/2\pi kT)$ and for any value of δ, the curves for $D(t)$ at finite T nearly coincide with those at $T = 0$, as it should.

At intermediate times and for $0 < \delta < 1$, an oscillation due to the pole contribution takes place in $D(t)$ (and also in $D^{\mathrm{cl}}(t)$ but with a smaller

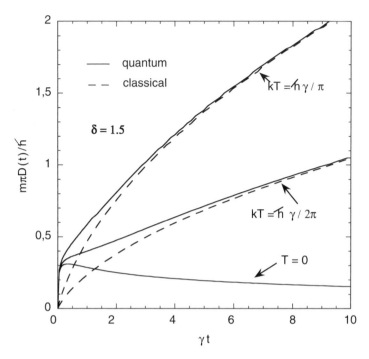

Figure 10. Full lines: The quantum extended diffusion coefficient $D(t)$ plotted as a function of γt for $\delta = 1.5$ and for bath temperatures $T = 0$, $kT = \hbar\gamma/2\pi$, $kT = \hbar\gamma/\pi$. Dashed lines: The corresponding $D^{cl}(t)$.

amplitude). For certain values of δ and T, this oscillation may even result in negative values of $D(t)$ during a finite time interval.

At large times $t \gg \gamma^{-1}, t_{th}$ and for any finite T, the curves of $D(t)$ and $D^{cl}(t)$ join together: $D(t)$ describes a subdiffusive regime when $0 < \delta < 1$, and a superdiffusive one when $\delta > 1$. At $T = 0, D(t)$ describes the relaxation of $\Delta x^2(t)$ toward $\Delta x^2(\infty)$ for $0 < \delta < 1$ and describes a subdiffusive regime for $1 < \delta < 2$.

C. Non-Ohmic Classical Aging Effects

In the classical case, a modified FDT associated with the particle displacement can be written in the form (68). For a diffusing particle, the fluctuation–dissipation ratio $X(\tau, t_w)$ is given by Eq. (90), rewritten here for convenience as

$$X(\tau, t_w) = \frac{D^{cl}(\tau)}{D^{cl}(\tau) + D^{cl}(t_w)} \tag{154}$$

For any τ and t_w, one can define an effective inverse temperature as $\beta_{\text{eff}}^{\text{cl}}(\tau, t_w) = \beta X(\tau, t_w)$. Since X does not depend on T, the bath temperature is simply rescaled by a factor $1/X$ larger than 1, due to those fluctuations of the particle displacement which take place during the waiting time.

At large times $(\tau, t_w \gg \gamma^{-1}, t_{\text{th}})$, one can use in Eq. (154) the asymptotic expressions of $D^{\text{cl}}(\tau)$ and $D^{\text{cl}}(t_w)$ as given by Eq. (153). Equation (154) then displays the fact that, in a sub-Ohmic or super-Ohmic model of exponent $\delta (0 < \delta < 1$ or $1 < \delta < 2)$, a self-similar aging regime takes place at large times, as pictured by the fluctuation–dissipation ratio:

$$X(\tau, t_w) \simeq \frac{1}{1 + \left(\dfrac{t_w}{\tau}\right)^{\delta-1}} \qquad (155)$$

Interestingly enough, X and $T_{\text{eff}} = 1/k\beta_{\text{eff}}$ are functions of the ratio t_w/τ, solely parameterized by δ. They do not depend on the other parameters of the model. For $\delta = 1$, one retrieves the results of the Ohmic model, namely $X = 1/2$ and $T_{\text{eff}} = 2T$. For any other value of δ, X and T_{eff} are algebraic functions of the ratio t_w/τ. Note that the limits $\tau \to \infty$ and $t_w \to \infty$ do not commute.

D. Non-Ohmic Quantum Aging Effects

In the quantum case, the effective temperature $T_{\text{eff}} = (k\beta_{\text{eff}})^{-1}$ can be obtained from Eq. (130), an equation which also allows one to define T_{eff} at $T = 0$ for $1 \leq \delta < 2$. Since $D(t)$ is a monotonic increasing function of T, Eq. (130) yields for $T_{\text{eff}}(\tau, t_w)$ a uniquely defined value, as in the Ohmic dissipation case.

The curves representing $\beta_{\text{eff}}(\tau, t_w)$ as a function of τ for $\delta = 0.5$ and $\delta = 1.5$ at a given finite temperature and for a given $t_w \gg \gamma^{-1}, t_{\text{th}}$ are plotted on Fig. 11. Quantum effects do not persist beyond times $\tau \sim t_{\text{th}}$. Thus, for times $\tau \gg \gamma^{-1}, t_{\text{th}}$, the classical fluctuation–dissipation ratio $X(\tau, t_w)$ as given by Eq. (155) allows for a proper description of finite temperature aging.

VI.   ANOMALOUS DIFFUSION IN OUT-OF-EQUILIBRIUM ENVIRONMENTS

In all that follows, the discussion will be purely classical. We now want to address the question whether the study of the (possibly anomalous) diffusion of a particle in an out-of-equilibrium medium is likely to provide information about the out-of-equilibrium properties of the latter. Generally speaking, the medium in which a diffusing particle evolves may be, or not, in a state of thermal equilibrium. For instance, when it is composed of an aging medium such as a glassy colloidal suspension of Laponite [8,12,55,56], the environment of a

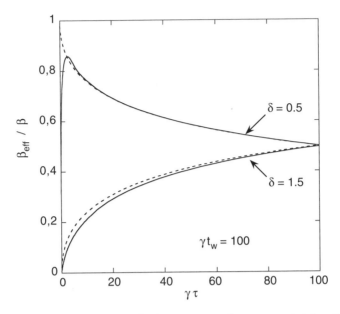

Figure 11. Full lines: The inverse effective temperature β_{eff}, as computed from Eq. (130), plotted as a function of $\gamma\tau$ for a bath temperature $kT = \hbar\gamma/2\pi$ and two different values of δ. Dashed lines: The corresponding classical effective inverse temperature β_{eff}^{cl}, as deduced from Eq. (155).

diffusing particle is not in thermal equilibrium. This feature renders the analysis of the particle motion much more involved, in comparison to the thermal equilibrium case.

A. Diffusion in a Thermal Bath

To begin with, we summarize some results about the (possibly anomalous) diffusion of a particle in a thermal bath.

1. The Fluctuation–Dissipation Theorems

One usually studies diffusion in a thermal bath by writing two fluctuation–dissipation theorems, generally referred to as the first and second FDTs (using the Kubo terminology [30,31]). As recalled for instance in Ref. 57, the first FDT expresses a necessary condition for a thermometer in contact solely with the system to register the temperature of the bath. As for the second FDT, it expresses the fact that the bath itself is in equilibrium.

Interestingly, each one of the two FDTs can be formulated in two equivalent ways, depending on whether one is primarily interested in writing a Kubo formula for a generalized susceptibility $\chi(\omega)$ [namely, in the present case, $\mu(\omega)$ or $\gamma(\omega)$], or an expression for its dissipative part [namely, the Einstein relation for $\Re e\mu(\omega)$ or the Nyquist formula for $\Re e\gamma(\omega)$]. This feature is far from being academic: Indeed the equivalence between these two types of formulations, which holds at equilibrium, has to be carefully reconsidered out-of-equilibrium, when one attempts to extend the FDTs with the help of a frequency-dependent effective temperature [5,6].

Linear response theory, applied to the particle velocity, considered as a dynamic variable of the isolated particle-plus-bath system, allows to express the mobility in terms of the equilibrium velocity correlation function. Since the mobility $\mu(\omega)$ is simply the generalized susceptibility $\chi_{vx}(\omega)$, one has the Kubo formula

$$\mu(\omega) = \frac{1}{kT} \int_0^\infty \langle v(t)v \rangle \, e^{i\omega t} \, dt \tag{156}$$

where T is the bath temperature [29,30]. Accordingly, the velocity spectral density is related to the real part of the mobility:

$$C_{vv}(\omega) = \int_{-\infty}^\infty \langle v(t)v \rangle \, e^{i\omega t} \, dt = kT \, 2\Re e\mu(\omega) \tag{157}$$

Introducing the frequency-dependent diffusion coefficient $D(\omega)$ as defined by

$$D(\omega) = \int_0^\infty \langle v(t)v \rangle \, \cos \omega t \, dt \tag{158}$$

one can rewrite Eq. (157) as the celebrated Einstein relation:

$$\frac{D(\omega)}{\Re e\mu(\omega)} = kT \tag{159}$$

The Einstein relation (159) or the expression (157) of the dissipative part $\Re e\mu(\omega)$ of the mobility constitute another formulation of the first FDT. Indeed they contain the same information as the Kubo formula (156) for the mobility, since $\mu(\omega)$ can be deduced from $\Re e\mu(\omega)$ with the help of the usual Kramers–Kronig relations valid for real ω [29,30]. Equation (156) on the one hand, and Eq. (157) or Eq. (159) on the other hand, are thus fully equivalent, and they all involve the thermodynamic bath temperature T. Note, however, that while $\mu(\omega)$ as given by Eq. (156) can be extended into an analytic function in the upper complex half-plane, the same property does not hold for $D(\omega)$.

Then, using the expression of $C_{vv}(\omega)$ as given by Eq. (157), that is, in view of the relation (143) between $\mu(\omega)$ and $\gamma(\omega)$,

$$C_{vv}(\omega) = \frac{kT}{m} \frac{2\Re e\gamma(\omega)}{|\gamma(\omega) - i\omega|^2} \tag{160}$$

one gets from Eq. (142) the Nyquist formula yielding the noise spectral density in terms of the dissipative part $\Re e\gamma(\omega)$ of the generalized friction coefficient:

$$C_{FF}(\omega) = mkT\, 2\Re e\gamma(\omega) \tag{161}$$

Correspondingly, one can write a Kubo formula relating the generalized friction coefficient $\gamma(\omega)$ to the random force correlation function:

$$\gamma(\omega) = \frac{1}{mkT} \int_0^\infty \langle F(t)F \rangle\, e^{i\omega t}\, dt \tag{162}$$

Equations (161) and (162) are two equivalent formulations of the second FDT [30,31]. The Kubo formula (162) for the generalized friction coefficient can also be established directly by applying linear response theory to the force exerted by the bath on the particle, this force being considered as a dynamical variable of the isolated particle-plus-bath system. We will come back to this point in Section VI.B.

Summing up, when the particle environment is a thermal bath, the two fluctuation–dissipation theorems are valid. In both theorems the bath temperature T plays an essential role. In its form (157) or (159) (Einstein relation), the first FDT involves the spectral density of a dynamical variable linked to the particle (namely its velocity), while, in its form (161) (Nyquist formula), the second FDT involves the spectral density of the random force, which is a dynamical variable of the bath.

2. Regression Theorem

When the environment is a thermal bath, we may introduce in the expression (147) of $\tilde{C}_{vv}(t)$ as a Fourier integral the value of the ratio $C_{FF}(\omega)/2\Re e\gamma(\omega)$ as given by the second FDT (161). One gets

$$\tilde{C}_{vv}(t) = \frac{kT}{m} \int_{-\infty}^{\infty} \frac{d\omega}{2\pi} e^{-i\omega t} \left\{ \frac{1}{\gamma(\omega) - i\omega} + \frac{1}{\gamma^*(\omega) + i\omega} \right\} \tag{163}$$

In the above integral, the function $1/[\gamma(\omega) - i\omega]$ is analytic for $\Im m\omega > 0$, while the function $1/[\gamma^*(\omega) + i\omega]$ is analytic for $\Im m\omega < 0$ [31]. Computing $\tilde{C}_{vv}(t)$ for $t > 0$, one chooses as an integration contour the semicircle of large radius in the lower complex half-plane. The second term does not contribute. One thus has

$$\tilde{C}_{vv}(t) = \frac{kT}{m} \int_{-\infty}^{\infty} \frac{d\omega}{2\pi} e^{-i\omega t} \frac{1}{\gamma(\omega) - i\omega}, \qquad t > 0 \tag{164}$$

Formula (164) shows that, when diffusion takes place in a thermal bath, the velocity correlation function is characterized by the same law as is the average velocity. This result constitutes the regression theorem, valid at equilibrium for any $\gamma(\omega)$.

In particular, with the previously introduced model for $\gamma(\omega)$ [namely, $\Re e\gamma(\omega) \propto |\omega|^{\delta-1}$ with $0 < \delta < 2$] [Eq. (140)], one has

$$\tilde{C}_{vv}(t) = \frac{kT}{m} E_{2-\delta}[-(\omega_\delta t)^{2-\delta}], \qquad t > 0 \tag{165}$$

Setting $t = 0$, one gets from Eq. (165) the equipartition result:

$$\langle v^2 \rangle = \frac{kT}{m} \tag{166}$$

B. Diffusion in an Out-of-Equilibrium Environment

The fully general situation of a particle diffusing in an out-of-equilibrium environment is much more difficult to describe. Except for the particular case of a stationary environment, the motion of the diffusing particle cannot be described by the generalized Langevin equation (22). A more general equation of motion has to be used. The fluctuation–dissipation theorems are *a fortiori* not valid. However, one can try to extend these relations with the help of an age- and frequency-dependent effective temperature, such as proposed and discussed, for instance, in Refs. 5 and 6.

An equation of motion is derived in Section VI.B for the case of a bilinear coupling between the particle and its non equilibrated environment. As for the out-of-equilibrium extension of the linear response theory, it will be discussed in Section VI.C.

1. Equation of Motion of a Particle Linearly Coupled to an Out-of-Equilibrium Environment

The validity of the generalized Langevin equation (22) is restricted to a stationary medium. In other situations, for instance when the diffusing particle evolves in an aging medium such as a glassy colloidal suspension of Laponite [8,12,55,56], another equation of motion has to be used.

We shall derive such an equation by assuming a bilinear coupling of the form $-\Phi x$ between the particle and its environment. In the presence of such a coupling, the particle equation of motion is of the form

$$m \frac{dv}{dt} = \Phi(t) \tag{167}$$

where $\Phi(t)$ denotes the global force exerted by the environment. In turn, the latter is perturbed by the coupling with the particle. The linear response relation yielding the average force exerted by the environment is of the form

$$\langle\Phi(t)\rangle = \int_{t_i}^{t} \tilde{\chi}_{\Phi\Phi}(t,t')x(t')dt' \tag{168}$$

where the causal function $\tilde{\chi}_{\phi\phi}(t,t')(t > t')$ is a linear response function of the surrounding medium [in Eq. (168), it is assumed that the particle and the environment have been put in contact at time t_i]. The equation of motion (167) takes the form

$$m\frac{dv}{dt} = \int_{t_i}^{t} \tilde{\chi}_{\Phi\Phi}(t,t')x(t')dt' + \Phi_E(t) \tag{169}$$

where $\Phi_E(t)$ is a fluctuating force with zero mean generated by the environment. When the latter is out-of-equilibrium, there is no specific relation between the response function $\tilde{\chi}_{\Phi\Phi}(t,t')$ and the correlation function $\langle\Phi_E(t)\Phi_E(t')\rangle$. In the most general situations, neither $\Phi_E(t)$ nor $v(t)$ can be viewed as stationary random processes. Before considering such situations, let us come back briefly on the particular cases—first, of a stationary medium, second, of a thermal bath—in order to see how the previously described properties can be retrieved.

2. Stationary Medium Case

Let us consider again the particular case of a particle diffusing in a stationary medium, in order to see how the generalized Langevin equation (22) can be deduced from the more general equation (169). When the medium is stationary, the response function $\tilde{\chi}_{\Phi\Phi}(t,t')$ reduces to a function of $t - t'(\tilde{\chi}_{\Phi\Phi}(t,t') = \tilde{\chi}_{\Phi\Phi}(t-t'))$. Introducing then the causal function $\tilde{\gamma}(t)$ as defined by

$$\tilde{\chi}_{\Phi\Phi}(t) = -m\frac{d\tilde{\gamma}(t)}{dt} \tag{170}$$

and integrating by parts, one gets from Eq. (169)

$$m\frac{dv}{dt} = -m\int_{t_i}^{t} \tilde{\gamma}(t-t')v(t')dt' + \Phi_E(t) + m\tilde{\gamma}(t_i)x(t-t_i) - m\tilde{\gamma}(t-t_i)x(t_i)$$

$$\tag{171}$$

The equation of motion (171) can be identified with the generalized Langevin equation (22) by letting $t_i \rightarrow -\infty$ (one has then $\tilde{\gamma}(t_i) = 0$), provided that the term $-m\tilde{\gamma}(t - t_i)x(t_i)$ is negligible in this limit. The fluctuating force $\Phi_E(t)$ then identifies with the Langevin force $F(t)$. The fluctuating force and the particle velocity can be viewed as stationary random processes. When equilibrium is not realized, neither the Kubo formulas (156) and (162) nor the regression theorem (165) are valid.

3. Thermal Bath Case

When equilibrium is realized, linear response theory, applied to the bath dynamical variable $\Phi_E(t)$, yields

$$\tilde{\chi}_{\Phi\Phi}(t - t') = \beta\Theta(t - t')\frac{\partial}{\partial t'}\langle F(t)F(t')\rangle, \qquad \beta = (kT)^{-1} \qquad (172)$$

that is, using Eq. (170) and the causality property, and the fact that at equilibrium $\Phi_E(t)$ is identical to the Langevin force $F(t)$:

$$\tilde{\gamma}(t - t') = \frac{1}{mkT}\,\Theta(t - t')\,\langle F(t)F(t')\rangle \qquad (173)$$

C. Out-of-Equilibrium Linear Response Theory

When the environment is not stationary, response functions such as $\tilde{\chi}_{\Phi\Phi}(t, t')$ and $\tilde{\chi}_{vx}(t, t')$ depend separately on the two times t and t' entering into play, and not only on the time difference or observation time $\tau = t - t'$. However, the observation time continues to play an essential role in the description. Hence, it has been proposed to define time- and frequency-dependent response functions as Fourier transforms with respect to τ of the corresponding two-time quantities [5,6,58]. The time t, which represents the waiting time or the age of the system, then plays the role of a parameter.

Let us first briefly recall these definitions, and examine under which conditions the age—and frequency-dependent response functions share the analytic properties of the corresponding stationary quantities.

1. Age- and Frequency-Dependent Response Functions

Considering the response function $\tilde{\chi}(t, t')$ as a function of t and of the observation time $\tau = t - t'$ [i.e., writing $\tilde{\chi}(t, t') = \tilde{\chi}_1(t, \tau)$], one introduces the Fourier transform of $\tilde{\chi}_1$ with respect to τ (for a fixed t):

$$\chi_1(\omega, t) = \int \tilde{\chi}_1(t, \tau)\,e^{i\omega\tau}\,d\tau \qquad (174)$$

Due to the causality of $\tilde{\chi}(t, t')$, the lower integration bound in formula (174) is 0. As for the upper integration bound, it is equal to t minus the lower bound over t'.

If one assumes that the perturbation is applied at $t' = 0$, the upper integration bound is equal to t. One then writes

$$\chi_1(\omega, t) = \int_0^t \tilde{\chi}_1(t, \tau) e^{i\omega\tau} d\tau \qquad (175)$$

that is,

$$\chi_1(\omega, t) = \int_0^t \tilde{\chi}(t, t') e^{i\omega(t-t')} dt' \qquad (176)$$

The Fourier relation (176) can be inverted, which yields

$$\int_{-\infty}^{\infty} \frac{d\omega}{2\pi} e^{-i\omega(t-t')} \chi_1(\omega, t) = \Theta(t') \Theta(t - t') \tilde{\chi}(t, t') \qquad (177)$$

The time t can be interpreted as the waiting time or the age t_w of the system under study. In other words, one has introduced [5,6,58]:

$$\chi_1(\omega, t_w) = \int_0^{t_w} \tilde{\chi}(t_w, t_w - \tau) e^{i\omega\tau} d\tau \qquad (178)$$

Note that the stationary regime is obtained, not only by assuming that $\tilde{\chi}(t, t')$ is invariant by time translation $(\tilde{\chi}(t, t') = \tilde{\chi}_1(\tau))$, but also by assuming that the age of the system tends toward infinity:

$$\chi(\omega) = \lim_{t_w \to \infty} \int_0^{t_w} \tilde{\chi}_1(\tau) e^{i\omega\tau} d\tau \qquad (179)$$

However, from the point of view of linear response theory, the definitions (174) or (178) suffer from several drawbacks. Actually, the function $\chi_1(\omega, t_w)$ as defined by Eq. (174) is not the Fourier transform of the function $\tilde{\chi}_1(t_w, \tau)$, but a partial Fourier transform computed in the restricted time interval $0 < \tau < t_w$. As a consequence, it does not possess the same analyticity properties as the generalized susceptibility $\chi(\omega)$ defined by Eq. (179). While the latter, extended to complex values of ω, is analytic in the upper complex half-plane $(\Im m\omega > 0)$, the function $\chi_1(\omega, t_w)$ is analytic in the whole complex plane. As a very simple example, consider the exponentially decreasing response function

$$\tilde{\chi}(t, t') = \Theta(t - t') e^{-\gamma(t-t')} \qquad (180)$$

One has[11]

$$\chi_1(\omega, t_w) = \frac{1}{\gamma - i\omega} \left[1 - e^{-(\gamma - i\omega)t_w} \right] \tag{181}$$

The corresponding generalized susceptibility is obtained by taking the limit $t_w \to \infty$ in Eq. (181):

$$\chi(\omega) = \frac{1}{\gamma - i\omega} \tag{182}$$

In this example, $\chi(\omega)$ as given by Eq. (182) has a pole in the lower complex half-plane ($\omega = -i\gamma$). This pole can be traced back to a removable singularity on the right-hand side of Eq. (181), so that $\chi_1(\omega, t_w)$ is analytic everywhere. This is an important drawback, since the information about the nature of the modes of the unperturbed system, which is contained in the poles of the generalized susceptibility, is not contained in the partial Fourier transform $\chi_1(\omega, t_w)$ as defined by Eq. (178).

2. Quasi-Stationary Regime: Introduction of an Effective Temperature

However, if $\omega t_w \gg 1$, one can assume that $\chi_1(\omega, t_w)$ varies very slowly with t_w in the range of values of ω of interest [5,6]. Two well-separated time scales then exist, respectively, characterizing the times pertinent for the measuring process, and the waiting time or the age of the system. Here, $\chi_1(\omega, t_w)$ describes a quasi-stationary regime taking place for a given waiting time. Considered as a function of ω, $\chi_1(\omega, t_w)$ is assumed to have the analytic properties required from a generalized susceptibility (i.e., analyticity in a complex half-plane, namely, the upper one with our definition of the Fourier transformation). From now, we will omit the parameter t_w, it being understood for simplicity that the analysis is valid for a given t_w.

Let us now come back to the specific problem of the diffusion of a particle in an out-of-equilibrium environment. In a quasi-stationary regime, the particle velocity obeys the generalized Langevin equation (22). The generalized susceptibilities of interest are the particle mobility $\mu(\omega) = \chi_{vx}(\omega)$ and the generalized friction coefficient $\gamma(\omega) = -(1/m\omega)\chi_{\Phi\Phi}(\omega)$ [the latter formula deriving from the relation (170) between $\tilde{\gamma}(t)$ and $\tilde{\chi}_{\Phi\Phi}(t)$). The results of linear response theory as applied to the particle velocity, namely the Kubo formula (156) and the Einstein relation (159), are not valid out-of-equilibrium. The same

---

[11]Despite the time translational invariance of $\tilde{\chi}(t, t')$, the partial Fourier transform $\chi_1(\omega, t_w)$ effectively depends on t_w, because the perturbation is applied at time 0 and not $-\infty$.

is true of the Kubo formula for the generalized friction coefficient [Eq. (162)] and of the Nyquist formula (161).

One can then try to extend the linear response theory with the help of an effective temperature [5,6].

D. Effective Temperature in an Out-of-Equilibrium Medium: The Link with the Kubo Formulas for the Generalized Susceptibilities

In an out-of-equilibrium environment, no well-defined thermodynamical temperature does exist. Since, in an out-of-equilibrium regime, even if stationary, the FDTs are not satisfied, one can try to rewrite them in a modified way, and thus to extend linear response theory, with the help of a (frequency-dependent) effective temperature.

Such a quantity, denoted as $T_{\mathrm{eff}}(\omega)$, and parameterized by the age of the system, has been defined, for real ω, via an extension of both the Einstein relation and the Nyquist formula. It has been argued in Refs. 5 and 6 that the effective temperature defined in this way plays in out-of-equilibrium systems the same role as does the thermodynamic temperature in systems at equilibrium (namely, the effective temperature controls the direction of heat flow and acts as a criterion for thermalization).

The effective temperature can, in principle, be deduced from independent measurements, for instance, of $\Re e\mu(\omega)$ and $D(\omega)$ [or of $\Re e\gamma(\omega)$ and $C_{FF}(\omega)$]. However, experimentally it may be preferable to make use of the modified Kubo formulas for the corresponding generalized susceptibilities. The Kubo formula for $\mu(\omega)$ [and also the one for $\gamma(\omega)$] cannot be extended to an out-of-equilibrium situation by simply replacing T by $T_{\mathrm{eff}}(\omega)$ in Eq. (156) [and in Eq. (162)]. In the following, we will show in details how the Kubo formulas have then to be rewritten.

1. Definition of the Effective Temperature

To begin with, following Refs. 5 and 6, we write the relation between the dissipative part of the mobility and the velocity spectral density in an out-of-equilibrium, quasi-stationary, regime as

$$C_{vv}(\omega) = kT_{\mathrm{eff}}(\omega)\, 2\Re e\mu(\omega) \qquad (183)$$

This amounts to assuming a modified Einstein relation:

$$\frac{D(\omega)}{\Re e\mu(\omega)} = kT_{\mathrm{eff}}(\omega) \qquad (184)$$

Equation (183) [or Eq. (184)] defines a frequency-dependent effective temperature $T_{\mathrm{eff}}(\omega)$.

Interestingly, due to the linearity of the generalized Langevin equation (22), the same effective temperature $T_{\text{eff}}(\omega)$ can consistently be used in the modified Nyquist formula linking the noise spectral density $C_{FF}(\omega) = \int_{-\infty}^{\infty} \langle F(t)F \rangle e^{i\omega t} dt$ and the dissipative part $\Re e\gamma(\omega)$ of the generalized friction coefficient. One has

$$C_{FF}(\omega) = mkT_{\text{eff}}(\omega)\, 2\Re e\gamma(\omega) \qquad (185)$$

Let us note for future purposes that $T_{\text{eff}}(\omega)$ is only defined for ω real. Equations (183)–(185) cannot be extended to complex values of ω.

2. The Modified Kubo Formula for the Mobility

Out–of–equilibrium, it is not possible to rewrite the Kubo formula for $\mu(\omega)$ in a form similar to Eq. (156) with $T_{\text{eff}}(\omega)$ in place of T:

$$\mu(\omega) \neq \frac{1}{kT_{\text{eff}}(\omega)} \int_0^{\infty} \langle v(t)v \rangle\, e^{i\omega t}\, dt \qquad (186)$$

Let us instead write a modified Kubo formula as

$$\mu(\omega) = \frac{1}{kT_{\text{eff}}^{(1)}(\omega)} \int_0^{\infty} \langle v(t)v \rangle\, e^{i\omega t}\, dt \qquad (187)$$

Equation (187) defines a complex frequency-dependent function $T_{\text{eff}}^{(1)}(\omega)$, in which terms is the velocity spectral density $C_{vv}(\omega) = \int_{-\infty}^{\infty} \langle v(t)v \rangle e^{i\omega t} dt$

$$C_{vv}(\omega) = 2\Re e[\mu(\omega)kT_{\text{eff}}^{(1)}(\omega)] \qquad (188)$$

According to Eq. (187), the quantity $\mu(\omega)kT_{\text{eff}}^{(1)}(\omega) = \int_0^{\infty} \langle v(t)v \rangle e^{i\omega t} dt$ can be extended into an analytic function in the upper complex half-plane. Since this analyticity property holds for $\mu(\omega)$, it also holds for $T_{\text{eff}}^{(1)}(\omega)$.

For real ω, one has the identity

$$T_{\text{eff}}(\omega)\, \Re e\mu(\omega) = \Re e[\mu(\omega)T_{\text{eff}}^{(1)}(\omega)] \qquad (189)$$

Equation (189) allows to derive the function $T_{\text{eff}}^{(1)}(\omega)$ from a given effective temperature $T_{\text{eff}}(\omega)$. Conversely, it provides the link allowing one to deduce $T_{\text{eff}}(\omega)$ from the modified Kubo formula (187) for $\mu(\omega)$.

Thus, the velocity correlation function can be expressed in two equivalent forms, either in terms of the effective temperature, as

$$\tilde{C}_{vv}(t) = \int_{-\infty}^{\infty} \frac{d\omega}{\pi}\, e^{-i\omega t}\, kT_{\text{eff}}(\omega)\, \Re e\mu(\omega) \qquad (190)$$

or, in terms of the function $T_{\text{eff}}^{(1)}(\omega)$, as

$$\tilde{C}_{vv}(t) = \int\limits_{-\infty}^{\infty} \frac{d\omega}{\pi}\, e^{-i\omega t}\, \Re e[\mu(\omega)kT_{\text{eff}}^{(1)}(\omega)] \tag{191}$$

Let us note for future purposes that since $C_{vv}(\omega)$ is an even function of ω, Eq. (191) can be rewritten as a Fourier cosine integral,

$$\tilde{C}_{vv}(t) = \int\limits_{-\infty}^{\infty} \frac{d\omega}{\pi}\, \cos \omega t\, \Re e[\mu(\omega)kT_{\text{eff}}^{(1)}(\omega)] \tag{192}$$

that is,[12]

$$\tilde{C}_{vv}(t) = \int\limits_{-\infty}^{\infty} \frac{d\omega}{\pi}\, \cos \omega t\, \mu(\omega)kT_{\text{eff}}^{(1)}(\omega) \tag{193}$$

3. The Modified Kubo Formula for the Generalized Friction Coefficient

Similarly, out-of-equilibrium, it is not possible to rewrite the Kubo formula for $\gamma(\omega)$ in a form similar to Eq. (162) with $T_{\text{eff}}(\omega)$ in place of T:

$$\gamma(\omega) \neq \frac{1}{mkT_{\text{eff}}(\omega)} \int\limits_{0}^{\infty} \langle F(t)F\rangle\, e^{i\omega t}\, dt \tag{194}$$

Let us instead write a modified Kubo formula as

$$\gamma(\omega) = \frac{1}{mkT_{\text{eff}}^{(2)}(\omega)} \int\limits_{0}^{\infty} \langle F(t)F\rangle\, e^{i\omega t}\, dt \tag{195}$$

One has

$$C_{FF}(\omega) = 2m\Re e[\gamma(\omega)kT_{\text{eff}}^{(2)}(\omega)] \tag{196}$$

According to Eq. (195), the quantity $m\gamma(\omega)kT_{\text{eff}}^{(2)}(\omega) = \int_0^\infty \langle F(t)F\rangle\, e^{i\omega t}\, dt$ can be extended into an analytic function in the upper complex half-plane. The same property holds for $T_{\text{eff}}^{(2)}(\omega)$.

For real ω, one has the identity

$$T_{\text{eff}}(\omega)\,\Re e\gamma(\omega) = \Re e[\gamma(\omega)T_{\text{eff}}^{(2)}(\omega)] \tag{197}$$

---

[12]Since $\Re e[\mu(\omega)kT_{\text{eff}}^{(1)}(\omega)]$ is even, $\mathcal{I}m[\mu(\omega)kT_{\text{eff}}^{(1)}(\omega)]$ is odd, as it can be deduced from the Kramers–Kronig relations linking these two quantities. For this reason, Eqs. (192) and (193) are equivalent.

Equation (197) allows to derive the function $T_{\text{eff}}^{(2)}(\omega)$ from a given effective temperature $T_{\text{eff}}(\omega)$, and conversely.

Note that although the same effective temperature $T_{\text{eff}}(\omega)$ can be consistently used in both the modified Einstein relation (184) and the modified Nyquist formula (185), the quantities $T_{\text{eff}}^{(1)}(\omega)$ and $T_{\text{eff}}^{(2)}(\omega)$ involved in the corresponding Kubo formulas [i.e., Eq. (187) for $\mu(\omega)$ and Eq. (195) for $\gamma(\omega)$] are not identical.

E. The Out-of-Equilibrium Generalized Stokes–Einstein Relation and the Determination of the Effective Temperature

Let us consider a particle evolving in an out-of-equilibrium environment (for instance, an aging colloidal glass). In a quasi-stationary regime, as pictured by the condition $\omega t_w \gg 1$, there exist two well-separated time scales: one characterizing the times pertinent for the measuring process (of order ω^{-1}), and the other the aging time t_w. At a given t_w, the particle velocity can then be assumed to obey the generalized Langevin equation (22).

Independent measurements of the particle mean-square displacement and mobility in the aging medium allow for the determination of its effective temperature. The aim of this section is to present a simple and convenient way of deducing $T_{\text{eff}}(\omega)$ from the experimental results.

Experimentally, the effective temperature of a colloidal glass can be determined by studying the anomalous drift and diffusion properties of an immersed probe particle. More precisely, one measures, at the same age of the medium, on the one hand, the particle mean-square displacement as a function of time, and, on the other hand, its frequency-dependent mobility. This program has recently been achieved for a micrometric bead immersed in a glassy colloidal suspension of Laponite. As a result, both $\Delta x^2(t)$ and $\mu(\omega)$ are found to display power-law behaviors in the experimental range of measurements [12].

1. The Out-of-Equilibrium Generalized Stokes–Einstein Relation

Here we show how the modified Kubo formula (187) for $\mu(\omega)$ leads to a relation between the (Laplace transformed) mean-square displacement and the z-dependent mobility (z denotes the Laplace variable). This out-of-equilibrium generalized Stokes–Einstein relation makes explicit use of the function $T_{\text{eff}}^{(1)}(\omega)$ involved in the modified Kubo formula (187), a quantity which is not identical to the effective temperature $T_{\text{eff}}(\omega)$; however $T_{\text{eff}}(\omega)$ can be deduced from this using the identity (189). Interestingly, this way of obtaining the effective temperature is completely general (i.e., it is not restricted to large times and small frequencies). It is therefore well adapted to the analysis of the experimental results [12].

The mean-square displacement of the diffusing particle, as defined by

$$\Delta x^2(t) = \langle [x(t) - x(t = 0)]^2 \rangle, \qquad t > 0 \tag{198}$$

can be deduced from the velocity correlation function *via* a double integration over time:

$$\Delta x^2(t) = 2 \int_0^t dt_1 \int_0^{t_1} dt_2 \, \langle v(t_1) v(t_2) \rangle \tag{199}$$

Introducing the Laplace transformed quantities $\hat{v}(z) = \int_0^\infty v(t) e^{-zt} \, dt$ and $\widehat{\Delta x^2}(z) = \int_0^\infty \Delta x^2(t) e^{-zt} \, dt$, one gets, by Laplace transforming Eq. (199),

$$z^2 \, \widehat{\Delta x^2}(z) = 2 \, \langle \hat{v}(z) v \rangle \tag{200}$$

By Laplace transforming the expression of $\tilde{C}_{vv}(t)$ as a Fourier cosine transform of $C_{vv}(\omega)$, one obtains

$$\langle \hat{v}(z) v \rangle = \int_{-\infty}^{\infty} \frac{d\omega}{2\pi} \frac{z}{z^2 + \omega^2} \, C_{vv}(\omega) \tag{201}$$

Introducing in Eq. (201) the expression (188) of $C_{vv}(\omega)$ (and using the fact that $\Im m[\mu(\omega) k T_{\text{eff}}^{(1)}(\omega)]$ is an odd function of ω), one gets

$$z^2 \, \widehat{\Delta x^2}(z) = 2 \int_{-\infty}^{\infty} \frac{d\omega}{\pi} \frac{z}{z^2 + \omega^2} \, \mu(\omega) k T_{\text{eff}}^{(1)}(\omega) \tag{202}$$

Using standard contour integration, we obtain

$$z^2 \, \widehat{\Delta x^2}(z) = 2\hat{\mu}(z) k \, \hat{T}_{\text{eff}}^{(1)}(z) \tag{203}$$

In Eq. (203), one has introduced the z-dependent quantities:

$$\hat{\mu}(z) = \mu(\omega = iz), \qquad \hat{T}_{\text{eff}}^{(1)}(z) = T_{\text{eff}}^{(1)}(\omega = iz) \tag{204}$$

Equation (203) is the extension to an out-of-equilibrium medium of the generalized Stokes–Einstein relation $z^2 \, \widehat{\Delta x^2}(z) = 2\hat{\mu}(z) kT$, valid for a medium in equilibrium [59].

Let us emphasize that the function $T_{\text{eff}}^{(1)}(\omega)$ is not the effective temperature involved in both the modified Einstein relation (184) and the modified Nyquist formula (185). Instead it is the quantity which appears in the modified Kubo formula for the mobility [Eq. (187)]. Let us add that one could also have obtained Eq. (203) from the modified Kubo formula (187) by setting $z = -i\omega$ in this latter relation [which is allowed since the function $\mu(\omega)T_{\text{eff}}^{(1)}(\omega)$ is analytic in the upper complex half-plane]:

$$\langle \hat{v}(z)v \rangle = \hat{\mu}(z)k\,\hat{T}_{\text{eff}}^{(1)}(z) \tag{205}$$

Equation (203) then directly follows [using Eq. (200)].

2. The Effective Temperature

As displayed by the out-of-equilibrium generalized Stokes–Einstein relation (203), independent measurements of the particle mean-square displacement and frequency-dependent mobility in an aging medium give access, once $\widehat{\Delta x^2}(z)$ and $\hat{\mu}(z) = \mu(\omega = iz)$ are determined, to $\hat{T}_{\text{eff}}^{(1)}(z)$ and to $T_{\text{eff}}^{(1)}(\omega) = \hat{T}_{\text{eff}}^{(1)}(z = -i\omega)$. Then, the identity (189) yields the effective temperature:

$$T_{\text{eff}}(\omega) = \frac{\Re e[\mu(\omega)T_{\text{eff}}^{(1)}(\omega)]}{\Re e\mu(\omega)} \tag{206}$$

Formula (206), together with the out-of-equilibrium generalized Stokes–Einstein relation (203), is the central result of the present section.

The above-described method has recently been used in Ref. 12 to obtain the effective temperature of a glassy colloidal suspension of Laponite through an immersed micrometric bead diffusion and mobility measurements.

F. The Effective Temperature in a Model with Power-Law Behaviors

To illustrate the whole procedure, let us consider a model in which both $\Delta x^2(t)$ and $\mu(\omega)$ are assumed to display power-law behaviors.

1. The Generalized Friction Coefficient and the Mobility

We take for $\Re e\gamma(\omega)$ the same function we used in our previous study of aging effects in anomalous diffusion—that is, a function behaving like a power-law of exponent $\delta - 1$ with $0 < \delta < 2$:

$$\Re e\gamma(\omega) = \omega_\delta \left(\frac{|\omega|}{\omega_\delta}\right)^{\delta-1} \sin\frac{\delta\pi}{2}, \qquad |\omega| \ll \omega_c \tag{207}$$

The function $\gamma(\omega)$ can be defined for any complex ω (except for a cut on the real negative axis). It is given by

$$\gamma(\omega) = \omega_\delta \left(\frac{-i\omega}{\omega_\delta}\right)^{\delta-1}, \qquad |\omega| \ll \omega_c, \qquad -\pi < \arg\omega \le \pi \tag{208}$$

The mobility $\mu(\omega)$ is related to $\gamma(\omega)$ by Eq. (143). In the frequency range $|\omega| \ll \omega_\delta$, inertia can be neglected. Then, we may use the overdamped mobility:

$$\mu(\omega) \simeq \frac{1}{m\gamma(\omega)}, \qquad |\omega| \ll \omega_\delta \qquad (209)$$

One has, for ω real:

$$\Re e\mu(\omega) \simeq \frac{1}{m\omega_\delta} \left(\frac{|\omega|}{\omega_\delta}\right)^{1-\delta} \sin\frac{\delta\pi}{2}, \qquad |\omega| \ll \omega_\delta \qquad (210)$$

For any complex ω (except for a cut on the real negative axis), one has

$$\mu(\omega) \simeq \frac{1}{m\omega_\delta} \left(\frac{-i\omega}{\omega_\delta}\right)^{1-\delta}, \qquad -\pi < \arg\omega \le \pi, \qquad |\omega| \ll \omega_\delta \qquad (211)$$

From Eq. (211), one deduces the function $\hat{\mu}(z)$:

$$\hat{\mu}(z) = \frac{1}{m\omega_\delta} \left(\frac{z}{\omega_\delta}\right)^{1-\delta}, \qquad |z| \ll \omega_\delta \qquad (212)$$

2.   The Mean-Square Displacement

Let us set

$$\Delta x^2(t) \sim 2 \frac{kT}{m\omega_\nu^2} \frac{(\omega_\nu t)^\nu}{\Gamma(\nu+1)}, \qquad \omega_\nu t \gg 1 \qquad (213)$$

where ω_ν^{-1} denotes some characteristic time after which the behavior (213) is well-settled. The behavior of the mean-square displacement is a (possibly anomalously) diffusive one, as characterized by the exponent $\nu(0 < \nu < 2)$. Correspondingly, one has

$$\widehat{\Delta x^2}(z) = 2 \frac{kT}{m\omega_\nu^2} \omega_\nu^\nu z^{-\nu-1}, \qquad |z| \ll \omega_\nu \qquad (214)$$

3.   Determination of the Effective Temperature

From the out-of-equilibrium generalized Stokes–Einstein relation (203), together with the expressions of $\hat{\mu}(z)$ [Eq. (212)] and of $\widehat{\Delta x^2}(z)$ [Eq. (214)], one deduces

$$\hat{\mathcal{T}}_{\text{eff}}^{(1)}(z) = T \frac{\omega_\delta^{2-\delta}}{\omega_v^{2-v}} z^{\delta-v}, \qquad |z| \ll \omega_\delta, \omega_v \tag{215}$$

and

$$\mathcal{T}_{\text{eff}}^{(1)}(\omega) = T \frac{\omega_\delta^{2-\delta}}{\omega_v^{2-v}} (-i\omega)^{\delta-v}, \qquad |\omega| \ll \omega_\delta, \omega_v \tag{216}$$

Then, using Eq. (206) and the expressions (211) of $\mu(\omega)$ and (216) of $\mathcal{T}_{\text{eff}}^{(1)}(\omega)$, one obtains the effective temperature of the aging medium:

$$T_{\text{eff}}(\omega) = T \frac{\omega_\delta^{2-\delta}}{\omega_v^{2-v}} \frac{\sin \frac{v\pi}{2}}{\sin \frac{\delta\pi}{2}} |\omega|^{\delta-v}, \qquad |\omega| \ll \omega_\delta, \omega_v \tag{217}$$

Setting $\alpha = \delta - v$ and introducing the characteristic frequency ω_0 as defined by

$$\omega_0^\alpha = \frac{\omega_\delta^{\delta-2}}{\omega_v^{v-2}} \frac{\sin \frac{\delta\pi}{2}}{\sin \frac{v\pi}{2}} \tag{218}$$

one can write

$$T_{\text{eff}}(\omega) = T \left(\frac{|\omega|}{\omega_0} \right)^\alpha, \qquad |\omega| \ll \omega_0 \tag{219}$$

The effective temperature behaves like a power-law of ω with an exponent $\alpha = \delta - v$.

G. Discussion

The main results of this section are the out-of-equilibrium generalized Stokes–Einstein relation (203) between $\widehat{\Delta x^2}(z)$ and $\hat{\mu}(z)$, together with the formula (206) linking $T_{\text{eff}}(\omega)$ and the quantity, denoted as $\mathcal{T}_{\text{eff}}^{(1)}(\omega)$, involved in the Stokes–Einstein relation. One thus has at hand an efficient way of deducing the effective temperature from the experimental results [12]. Indeed, the present method, which avoids completely the use of correlation functions and makes use only of one-time quantities (via their Laplace transforms), is particularly well-suited to the interpretation of numerical data.

VII. SUMMARY

In this chapter, we have showed that a particle undergoing normal or anomalous diffusion constitutes a system conveniently allowing one to illustrate and to discuss the concepts of FDT violation and effective temperature. Our study was carried out using the Caldeira–Leggett dissipation model. Actually this model, which is sufficiently versatile to give rise to various normal or anomalous diffusion behaviors, constitutes an appropriate framework for such a study, in quantum as well as in classical situations.

For a particle evolving in a thermal bath, we focused our interest on the particle displacement, a dynamic variable which does not equilibrate with the bath, even at large times. As far as this variable is concerned, the equilibrium FDT does not hold. We showed how one can instead write a modified FDT relating the displacement response and correlation functions, provided that one introduces an effective temperature, associated with this dynamical variable. Except in the classical limit, the effective temperature is not simply proportional to the bath temperature, so that the FDT violation cannot be reduced to a simple rescaling of the latter. In the classical limit and at large times, the fluctuation–dissipation ratio T/T_{eff}, which is equal to $1/2$ for standard Brownian motion, is a self-similar function of the ratio of the observation time to the waiting time when the diffusion is anomalous.

When the environment of the particle is itself out-of-equilibrium, as is the case for a particle evolving in an aging medium, we showed how the study of both the mobility and the diffusion of the particle allows one to obtain the effective temperature of the medium. We derived an out-of-equilibrium generalized Stokes–Einstein relation linking the Laplace transform of the mean-square displacement and the z-dependent mobility. This relation provides an efficient way of deducing the effective temperature from the experimental results.

Acknowledgments

Special thanks are due to A. Mauger for his collaboration, as well as to B. Abou and F. Gallet for fruitful discussions.

References

1. L. C. E. Struik, *Physical Aging in Amorphous Polymers and Other Materials*, Elsevier, Amsterdam, 1978.
2. J.-P. Bouchaud, L. F. Cugliandolo, J. Kurchan, and M. Mézard, Out of equilibrium dynamics in spin-glasses and other glassy systems, in *Spin-Glasses and Random Fields*, A. P. Young ed., World Scientific, Singapore, 1997.
3. L. F. Cugliandolo, Dynamics of glassy systems, in *Slow Relaxations and Nonequilibrium Dynamics in Condensed Matter*, Les Houches—École d'Été de Physique Théorique, Vol.

77. J.-L. Barrat, M. V. Feigelman, J. Kurchan, and J. Dalibard, eds., Springer, Berlin, 2003.

4. A. Crisanti and F. Ritort, Violations of the fluctuation–dissipation theorem in glassy systems: Basic notions and the numerical evidence, *J. Phys. A: Math. Gen.* **36**, R181 (2003).

5. L. F. Cugliandolo, J. Kurchan, and L. Peliti, Energy flow, partial equilibration and effective temperatures in systems with slow dynamics. *Phys. Rev. E* **55**, 3898 (1997).

6. L. F. Cugliandolo and J. Kurchan, A scenario for the dynamics in the small entropy production limit. *Physica A* **263**, 242 (1999) and *J. Phys. Soc. Japan* **69**, 247 (2000).

7. T. S. Grigera and N. E. Israeloff, Observation of fluctuation–dissipation theorem violations in a structural glass. *Phys. Rev. Lett.* **83**, 5038 (1999).

8. L. Bellon and S. Ciliberto, Experimental study of the fluctuation–dissipation relation during an aging process. *Physica D* **168–169**, 325 (2002).

9. L. Buisson, S. Ciliberto, and A. Garcimartín, Intermittent origin of the large violations of the fluctuation–dissipation relations in an aging polymer glass. *Europhys. Lett.* **63**, 603 (2003).

10. D. Hérisson and M. Ocio, Fluctuation–dissipation ratio of a spin glass in the aging regime. *Phys. Rev. Lett.* **88**, 257202 (2002).

11. D. Hérisson and M. Ocio, Off-equilibrium fluctuation–dissipation relation in a spin glass. *Eur. Phys. J. B* **40**, 283 (2004).

12. B. Abou and F. Gallet, Probing a nonequilibrium Einstein relation in an aging colloidal glass. *Phys. Rev. Lett.* **93**, 160603 (2004).

13. H. A. Makse and J. Kurchan, Thermodynamic approach to dense granular matter: A numerical realization of a decisive experiment. *Nature* **415**, 614 (2002).

14. G. D'Anna, P. Mayor, A. Barrat, V. Loreto, and F. Nori, Observing the Brownian motion in vibro-fluidized granular matter. *Nature* **424**, 909 (2003).

15. N. Pottier and A. Mauger, Aging effects in free quantum Brownian motion. *Physica A* **282**, 77 (2000).

16. A. Mauger and N. Pottier, Aging effects in the quantum dynamics of a dissipative free particle: Non-Ohmic case. *Phys. Rev. E* **65**, 056107 (2002).

17. N. Pottier, Aging properties of an anomalously diffusing particle. *Physica A* **317**, 371 (2003).

18. N. Pottier and A. Mauger, Anomalous diffusion of a particle in an aging medium. *Physica A* **332**, 15 (2004).

19. N. Pottier, Out of equilibrium generalized Stokes–Einstein relation: Determination of the effective temperature of an aging medium. *Physica A* **345**, 472 (2005).

20. I. R. Senitzky, Dissipation in quantum mechanics. The harmonic oscillator. *Phys. Rev.* **119**, 670 (1960).

21. G. W. Ford, M. Kac, and P. Mazur, Statistical mechanics of assemblies of coupled oscillators. *J. Math. Phys.* **6**, 504 (1965).

22. P. Ullersma, An exactly solvable model for Brownian motion. *Physica* **32**, 27, 56, 74, 90 (1966).

23. A. O. Caldeira and A. J. Leggett, Quantum tunnelling in dissipative systems. *Ann. Phys.* **149**, 374 (1983).

24. A. J. Leggett, S. Chakravarty, A. T. Dorsey, M. P. A. Fisher, A. Garg, and W. Zwerger, Dynamics of the dissipative two-state system. *Rev. Mod. Phys.* **59**, 1 (1987).

25. C. Aslangul, N. Pottier, and D. Saint-James, Time behavior of the correlation functions in a simple dissipative quantum model. *J. Stat. Phys.* **40**, 167 (1985).

26. V. Hakim and V. Ambegaokar, Quantum theory of a free particle interacting with a linearly dissipative environment. *Phys. Rev. A* **32**, 423 (1985).

27. G. W. Ford and M. Kac, On the quantum Langevin equation. *J. Stat. Phys.* **46**, 803 (1987).

28. U. Weiss, *Quantum Dissipative Systems*, 2nd ed., World Scientific, Singapore, 1999.

29. R. Zwanzig, *Nonequilibrium Statistical Mechanics*, Oxford University Press, Oxford, 2001.

30. R. Kubo, The fluctuation–dissipation theorem. *Rep. Prog. Phys.* **29**, 255 (1966).

31. R. Kubo, M. Toda, and N. Hashitsume, *Statistical Physics II: Nonequilibrium Statistical Mechanics*, 2nd ed., Springer-Verlag, Berlin, 1991.

32. H. B. Callen and T. A. Welton, Irreversibility and generalized noise. *Phys. Rev.* **83**, 34 (1951).

33. H. B. Callen and R. F. Greene, On a theorem of irreversible thermodynamics. *Phys. Rev.* **86**, 702 (1952).

34. R. Kubo, Statistical–mechanical theory of irreversible processes. I. General theory and simple applications to magnetic and conduction problems. *J. Phys. Soc. Japan* **12**, 570 (1957).

35. G. Parisi, *Statistical Field Theory*, Perseus Books, Reading, MA, 1998.

36. N. Pottier and A. Mauger, Quantum fluctuation–dissipation theorem: A time-domain formulation. *Physica A* **291**, 327 (2001).

37. S. W. Lovesey, *Condensed Matter Physics: Dynamic Correlations*, The Benjamin/Cummings Publishing Company, 1980.

38. J. W. Goodman, *Statistical Optics*, Wiley, New York, 1985.

39. L. Mandel and E. Wolf, *Optical Coherence and Quantum Optics*, Cambridge University Press, New York, 1995.

40. L. F. Cugliandolo, J. Kurchan, and G. Parisi, Off-equilibrium dynamics and aging in unfrustrated systems. *J. Physique* **4**, 1641 (1994).

41. R. Mélin and P. Butaud, Glauber dynamics and aging. *J. Physique I* **7**, 691 (1997).

42. W. Krauth and M. Mézard, Aging without disorder on long time scales. *Z. Phys. B* **97**, 127 (1995).

43. F. Ritort, Glassiness in a model without energy barriers. *Phys. Rev. Lett.* **75**, 1190 (1995).

44. C. Godrèche and J. M. Luck, Long-time regime and scaling of correlations in a simple model with glassy behavior. *J. Phys. A: Math. Gen.* **29**, 1915 (1995).

45. S. Franz and F. Ritort, Dynamical solution of a model without energy barriers. *Europhys. Lett.* **31**, 507 (1995).

46. L. F. Cugliandolo and G. Lozano, Quantum aging in mean-field models. *Phys. Rev. Lett.* **80**, 4979 (1998).

47. L. F. Cugliandolo and G. Lozano, Real-time nonequilibrium dynamics of quantum glassy systems. *Phys. Rev. B* **59**, 915 (1999).

48. G. M. Schütz and S. Trimper, Relaxation and aging in quantum spin systems. *Europhys. Lett.* **47**, 164 (1999).

49. P. Schramm and H. Grabert, Low-temperature and long-time anomalies of a damped quantum particle. *J. Stat. Phys.* **49**, 767 (1987).

50. H. Grabert, P. Schramm, and G. L. Ingold, Quantum Brownian motion: The functional integral approach. *Phys. Rep.* **168**, 115 (1988).

51. C. Aslangul, N. Pottier, and D. Saint-James, Quantum dynamics of a damped free particle, *J. Physique* **48**, 1871 (1987).

52. E. Lutz, Fractional Langevin equation. *Phys. Rev. E* **64**, 051106 (2001).

53. A. Erdélyi, *Higher Transcendental Functions*, Vol. 3, McGraw-Hill, New York, 1955.

54. R. Gorenflo and F. Mainardi, Fractional oscillations and Mittag–Leffler functions, International Workshop on the Recent Advances in Applied Mathematics, State of Kuwait, May 4–7, 1996. Proceedings, Kuwait University, Department of Mathematics and Computer Science, 1996, p. 193.

55. A. Knaebel, M. Bellour, J.-P. Munch, V. Viasnoff, F. Lequeux, and J. L. Harden, Aging behavior of Laponite clay particle suspensions, *Europhys. Lett.* **52**, 73 (2000).

56. B. Abou, D. Bonn, and J. Meunier, Aging dynamics in a colloidal glass. *Phys. Rev. E* **64**, 021510 (2001).

57. L. F. Cugliandolo, D. S. Dean, and J. Kurchan, Fluctuation–dissipation theorems and entropy production in relaxational systems. *Phys. Rev. Lett.* **79**, 2168 (1997).

58. G. J. M. Koper and H. J. Hilhorst, Nonequilibrium and aging in a one-dimensional Ising spin glass. *Physica A* **155**, 431 (1989).

59. T. G. Mason and D. A. Weitz, Optical measurements of frequency-dependent linear viscoelastic moduli of complex fluids. *Phys. Rev. Lett.* **74**, 1250 (1995).

APPENDIX: SOME USEFUL FOURIER TRANSFORMS AND CONVOLUTION RELATIONS

A.1   Fourier Transforms

Consider

$$I_1 = \int_0^\infty d\omega \, \sin \omega t \, \tanh \frac{\beta \hbar \omega}{2} \tag{A.1}$$

Using the expansion

$$\tanh \frac{\pi x}{2} = \frac{4x}{\pi} \sum_{k=1}^\infty \frac{1}{(2k-1)^2 + x^2} \tag{A.2}$$

with $\omega = (\pi/\beta\hbar)x$, one gets

$$I_1 = \frac{2\pi}{\beta\hbar} \sum_{k=1}^\infty e^{-(2k-1)\pi t/\beta\hbar}, \qquad t > 0 \tag{A.3}$$

which yields the final result, valid whatever the sign of t:

$$I_1 = \begin{cases} \dfrac{\pi}{\beta\hbar} \dfrac{1}{\sinh \frac{\pi t}{\beta\hbar}}, & t \neq 0 \\ 0, & t = 0 \end{cases} \tag{A.4}$$

One thus has

$$\frac{1}{2\pi} \int_{-\infty}^{\infty} d\omega\, e^{-i\omega t}\, \tanh\frac{\beta\hbar\omega}{2} = -\frac{i}{\pi}\, I_1 \tag{A.5}$$

Similarly, let us set

$$I_2 = \int_{0}^{\infty} d\omega\, \sin\omega t\, \coth\frac{\beta\hbar\omega}{2} \tag{A.6}$$

Using the expansion

$$\coth\frac{\pi x}{2} = \frac{2}{\pi x} + \frac{4x}{\pi}\sum_{k=1}^{\infty}\frac{1}{(2k)^2 + x^2} \tag{A.7}$$

with ω defined as above, one gets

$$I_2 = \frac{\pi}{\beta\hbar} + \frac{2\pi}{\beta\hbar}\sum_{k=1}^{\infty} e^{-2k\pi t/\beta\hbar}, \qquad t > 0 \tag{A.8}$$

which yields the final result, valid whatever the sign of t:

$$I_2 = \begin{cases} \dfrac{\pi}{\beta\hbar}\coth\dfrac{\pi t}{\beta\hbar}, & t \neq 0, \\[2mm] 0, & t = 0. \end{cases} \tag{A.9}$$

One thus has

$$\frac{1}{2\pi} \int_{-\infty}^{\infty} d\omega\, e^{-i\omega t}\, \coth\frac{\beta\hbar\omega}{2} = -\frac{i}{\pi}\, I_2 \tag{A.10}$$

A.2   Convolution Relations

From the relation

$$\tanh\frac{\beta\hbar\omega}{2}\coth\frac{\beta\hbar\omega}{2} = 1 \tag{A.11}$$

one deduces, by Fourier transformation and use of the Fourier formulas (A5) and (A10), the identity

$$\frac{\pi}{\beta\hbar}\,\mathrm{vp}\,\frac{1}{\sinh\frac{\pi t}{\beta\hbar}} * \frac{\pi}{\beta\hbar}\,\mathrm{vp}\,\coth\frac{\pi t}{\beta\hbar} = -\pi^2\,\delta(t) \tag{A.12}$$

At $T = 0$, Eq. (A.12) reduces to

$$\mathrm{vp}\,\frac{1}{t} * \mathrm{vp}\,\frac{1}{t} = -\pi^2\,\delta(t) \qquad (A.13)$$

In order to study the classical high temperature limit of Eq. (A.12), let us note that

$$\log \coth |ax| \underset{|a| \to \infty}{\sim} \frac{\pi^2}{4|a|}\,\delta(x) \qquad (A.14)$$

and

$$\frac{a}{\sinh ax} \underset{|a| \to \infty}{\sim} -\frac{\pi^2}{2|a|}\,\delta'(x) \qquad (A.15)$$

Formula (A.14) can be demonstrated in a standard fashion by considering the quantity $\log|\coth ax|$ as a distribution.[13]

Formulas (A.14) and (A.15) can be used to study the classical limit of the convolution relation (A.12). Indeed, in the classical limit, using Eq. (A.15) with $a = \pi/2\beta\hbar$, the left-hand side of Eq. (A.12) is seen to reduce to $-(\pi^2/2)\,\delta'(t) * \mathrm{sgn}(t)$—that is, to $-\pi^2\delta(t)$, as it should.

---

[13]Indeed, for any well-behaved function $\phi(x)$, the integral $\int_{-\infty}^{\infty} \log \coth |ax|\,\phi(x)\,dx$ tends toward $(\pi^2/4|a|)\,\phi(0)$ in the limit $|a| \to \infty$.

CHAPTER 4

POWER-LAW BLINKING QUANTUM DOTS: STOCHASTIC AND PHYSICAL MODELS

GENNADY MARGOLIN, VLADIMIR PROTASENKO, MASARU KUNO

Department of Chemistry and Biochemistry, Notre Dame University, Notre Dame, Indiana 46556, USA

ELI BARKAI

Department of Chemistry and Biochemistry, Notre Dame University, Notre Dame, Indiana 46556, USA; and Department of Physics, Bar Ilan University, Ramat Gan 52900, Israel

CONTENTS

Fractals, Diffusion, and Relaxation in Disordered Complex Systems: A Special Volume of Advances in Chemical Physics, Volume 133, Part A, edited by William T. Coffey and Yuri P. Kalmykov. Series editor Stuart A Rice.

I. INTRODUCTION

Single quantum dots when interacting with a continuous wave laser field blink; that is, at random times the dot turns from a state on, in which many photons are emitted, to a state off, in which no photons are emitted. While stochastic intensity trails are found today in a vast number of single-molecule experiments, the dots exhibit statistical behavior that seems unique. In particular, the dots exhibit power law statistics, aging, and ergodicity breaking. While our understanding of the physical origin of the blinking behavior of the dots is not complete, several physical pictures have emerged in recent years, which explain the blinking in terms of simple physics. Here we will review a diffusion model that may explain some of the observations made so far. Then we analyze the properties of the dots, using a stochastic approach. In particular we review the behaviors of the time and ensemble average intensity correlation functions. Usually it is assumed that these two quantities are identical in the limit of long times; however, this is not so for dots. In the final part of the paper, we analyze experimental data of blinking CdSe–ZnS dots and show that they exhibit ergodicity breaking in agreement with theory. We also show that statistically all the dots in our sample are identical, and find a surprising and new symmetry in the behavior of the on and off times.

II. PHYSICAL MODELS

A typical fluorescence intensity trace of a CdSe quantum dot, or nanocrystal (NC), overcoated with ZnS (in short, CdSe–ZnS NC) under continuous laser illumination is shown in Fig. 1. Thence we learn that roughly speaking, the intensity jumps between two states: on and off. Some of the deviations from this digital behavior can be attributed to fluctuating nonradiative decay channels due to coupling to the environment and also due to time binning procedure [1–4]. Data analysis of such a time trace is often based on a distribution of on and off times. Defining a threshold above which the NC is considered in state on and below which it is in state off, one can extract the probability density functions $\psi_+(\tau)$ of on and $\psi_-(\tau)$ of off times. Surprisingly these exhibit a power-law decay $\psi_\pm(\tau) \propto \tau^{-1-\alpha_\pm}$, as shown in Fig. 2. A summary of different experimental exponents is presented in Table I, indicating such a power-law decay in most cases. In some cases $\alpha_+ \approx \alpha_-$ and the exponents are close to 1/2. In particular, Brokmann et al. [5] measured 215 CdSe–ZnS NCs and found that all are statistically identical with $\alpha_+ = 0.58 \pm 0.17$, $\alpha_- \approx 0.48 \pm 0.15$ so that $\alpha_+ \approx \alpha_- \approx 0.5$. Note that most of the uncertainty in the values of the exponents can be attributed simply to the statistical limitations of data analysis [6] (see also Section VI below). Shimizu et al. [7] found that in the limit of low temperature and weak laser fields $\alpha_+ \approx \alpha_- \approx 0.5$. The fact that in many cases we have $\alpha_\pm < 1$ leads to interesting statistical behavior—for example, ergodicity

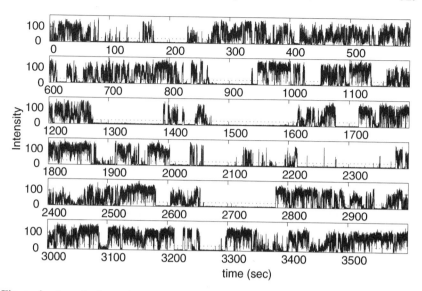

Figure 1. Intensity fluctuations in a CdSe–ZnS NC under continuous laser illumination at room temperature. Dotted horizontal line was selected as a threshold to divide off and on states.

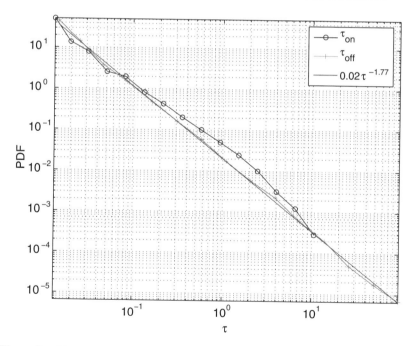

Figure 2. Distributions $\psi_\pm(\tau)$ of on and off times for the NC, whose intensity trajectory is shown in Fig. 1. The straight line is the fit to the off time distribution.

TABLE I

Summary of Experimental Exponents for On (α_+) and Off (α_-) Time Distributions for Various Single NCs Under Different Experimental Conditions.[a]

Group	Material	No.	Radii (nm)	Temp. (K)	Laser Intensity (kW/cm²)	α_+	α_-
Verberk et al. [4]	CdS	1	2.5	1.2		e^{-at}	0.65(0.2)
Brokmann et al. [5]	CdSe–ZnS	215		300		0.58(0.17)	0.48(0.15)
Shimizu et al. [7]	CdSe–ZnS, CdSe, CdTe	>200	1.5, 2.5	300, 10	0.1–0.7	0.5(0.1), cutoff	0.5(0.1)
Kuno et al. [41]	CdSe–ZnS	~200	1.7–2.9	300–394	0.24–2.4		0.5–0.75
Kuno et al. [9]	CdSe–ZnS	>300	1.7–2.7	300	0.1–100	0.9(0.05)	0.54(0.03)
Kuno et al. [42]	InP	~30	1.5	300	0.24	1.0(0.2)	0.5(0.1)
Cichos et al. [43]	Si	~1000			1.8, 6.5	1.2(0.1)	0.3, 0.7
Hohng and Ha [44]	CdSe–ZnS						0.94–1.10
Müller et al. [45]	CdSe–ZnS		4.4 (core)	300	0.025	0.55	0.05, 0.25
van Sark et al. [46]	CdSe–ZnS		~3.7	300	20	~1.2, ~0.7	~0.2, ~0.4
Kobitski et al. [47]	CdSe	41	3.6		0.04–0.38	0.97–0.66	0.42–0.64

[a]Notice that Verberk et al. use uncapped NCs, while other measurement consider capped NCs, hence exponential distribution of on times is found only for uncapped dots. Hohng and Ha used CdSe–ZnS NCs coated with streptavidin which might alter the exponent α_-.

breaking and aging. We will discuss these behaviors in Section III. A physical model for blinking was suggested by Efros and Rosen [8]. Briefly the on and off periods correspond to neutral and charged NCs respectively. Thus the on/off trace teaches us something about the elementary charging mechanism of the dot. The difficulty is to explain the power-law distributions of the on and off times, or, in other words, why should the time the charge occupies the NCs follow power law behavior?

Two types of models have been suggested, namely, a diffusion approach and a random trap model. The measurements of Dahan's and Bawendi's groups [5,7], which show the universal power law $\alpha_{\pm} = 0.5$, are consistent with the diffusion model (see details below). The fact that all dots are found to be similar [5] seems not to be consistent with models of quenched disorder [4,9,10] since these support the idea of a distribution of $\alpha_{\pm}$. However, some experiments show deviations from the $\alpha_{+} \approx \alpha_{-} \approx 0.5$ and may support a distribution of $\alpha_{\pm}$. It is possible that preparation methods and environments lead to different mechanisms of power-law blinking, along with different exponents [6]. More experimental work in this direction is needed; in particular, experimentalists still have to investigate the distribution of $\alpha_{\pm}$ and need to show whether and under what conditions are all the dots statistically identical. We discuss the diffusion model below; different aspects of the tunneling and trapping model can be found in Refs. 4, 6, and 10.

As discussed at length by Shimizu et al. [7], the on time distributions show temperature and laser power dependencies—for example, exponential cutoffs of power-law behavior. Although no direct observations of cutoffs in the off time distribution have been reported, ensemble measurements by Chung and Bawendi [3] demonstrate that such a cutoff should also exist, but at times of the order of tens of minutes to hours. Our analysis here, employing power-law decaying distributions, is of course applicable in time windows where power-law statistics holds.

A. Diffusion Model

We note that the simplest diffusion controlled chemical reaction $A + B \rightleftharpoons AB$, where A is fixed in space, can be used to explain some of the observed behavior on the uncapped NCs. As shown by the group of Orrit [4], such dots exhibit exponential distribution of *on* times and power-law distribution of *off* times. The *on* times follow standard exponential kinetics corresponding to an ionization of a neutral NC (denoted as AB). A model for exponential behavior has already been given in Ref. 8. Clearly the experiments of the group of Orrit show that the capping plays an important part in the the blinking, since capped NCs exhibit power-law behavior both for the on and off times. We will return to capped dots later.

Once the uncapped NC is ionized ($A + B$ state), we assume the ejected charge carrier undergoes a random walk on the surface of the NC or in the bulk.

This part of the problem is similar to Onsager's classical problem of an ion pair escaping neutralization (see, e.g., Refs. 11 and 12). The survival probability in the *off* state for time t, $S_-(t)$ is related to the *off* time distribution via $S_-(t) = 1 - \int_0^t \psi_-(\tau)d\tau$, or

$$\psi_-(t) = -\frac{dS_-(t)}{dt} \tag{1}$$

It is well known that in three dimensions the survival probability decays like $t^{-1/2}$, the exponent $1/2$ is close to the exponent often measured in the experiments. The $1/2$ appears in many situations, for finite and infinite systems, in completely and partially diffusion-controlled recombination, and in different dimensions, and it can govern the leading behavior of the survival probability for orders of magnitude in time [12–14]. In this picture the exponent $1/2$ does not depend on temperature, similar to what is observed in experiment. We note that it is possible that instead of the charge carrier executing the random walk, diffusing lattice defects which serve as a trap for the charge carrier are responsible for the blinking behavior of the NCs.

A long time ago, Hong, Noolandi, and Street [16] investigated geminate electron–hole recombination in amorphous semiconductors. In their model they included the effects of tunneling, Coulomb interaction, and diffusion. Combination of tunneling and diffusion leads to an $S(t) \propto t^{-1/2}$ behavior. However, when the Coulomb interactions are included in the theory, deviations from the universal $t^{-1/2}$ law are observed—for example, in the analysis of photoluminescence decay in amorphous Si:H, as a function of temperature.

Coulomb interaction between the charged NCs and the ejected electron seems to be an important factor in the Physics of NCs. The Onsager radius is a measure of the strength of the interaction

$$r_{Ons} = \frac{e^2}{k_b T \epsilon} \tag{2}$$

Krauss and Brus [15] measured the dielectric constant of CdSe dots, and found a value of 8. Hence, at room temperature we find $r_{Ons} \simeq 70 \,\text{Å}$ (angstrom) (however, note that the dielectric constant of the matrix is not identical to that of the dot). Since the length scale of the dots is of the order of a few nanometers, the Coulomb interaction seems to be an important part of the problem. This according to the theory in Ref. 16 is an indication of possible deviations from the universal $1/2$ power-law behavior. It is also an indication that an ejected electron is likely to return to the dot and not escape to the bulk (since the force is attractive). In contrast, if the Onsager radius is small, an ejected electron would most likely escape to the bulk, leaving the dot in state off forever (i.e., Polya

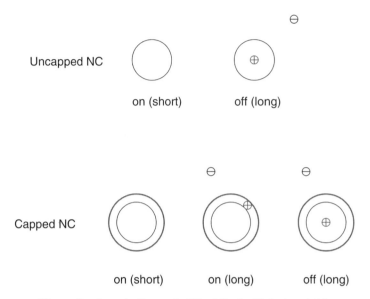

Figure 3.   On and off states for NCs, following Verberk et al. [4].

theorem in three dimensions). Unfortunately, currently there is not sufficient experimental data to determine in more qualitative ways if an Onsager-like model can be used to explain the observed data. As in standard geminate recombination processes, the dependence of blinking on temperature, on dielectric constant of the dot and of the matrix [6], and on external driving field may yield more microscopic information on the precise physical mechanism governing the fascinating blinking behavior.

One of the possible physical pictures explaining blinking of capped NCs can be based on a diffusion process, using a variation of a three state model of Verberk et al. [4]. As mentioned above, a power-law distribution of *on* and *off* times are now observed. In particular, neutral capped NC will correspond to state *on* (as for uncapped NCs). However, a capped NC can remain *on* even in the ionized state (see Fig.3) Verberk et al. assume that the ionized capped NC can be found in two states: (i) the charge remaining in the NC can be found in the center of the NC (possibly a delocalized state), (ii) charge remaining in the NC can be trapped in vicinity of capping. For case (i) the NC will be in state *off*, for case (ii) the NC will be in state *on*. Depending on exact location of this charge, the fluorescence intensity can vary. The main idea is that the rate of Auger nonradiative recombination [8] of consecutively formed electron–hole pairs will drop for case (ii) but not for case (i). We note that capping may increase effective radius of the NC, or provide trapping sites for the hole (e.g., recent studies by Lifshitz et al. [17] demonstrate that coating of NCs creates trapping sites in the interface). Thus the *off* times occur when the NC is ionized and the hole is close to the center, these *off* times are slaved to the diffusion of the electron. While *on* times occur for both a neutral NC and for a charged NC with the charge in vicinity of capping, the latter *on* times are slaved to the diffusion of the electron. In the case of power law *off* time statistics this model predicts the same power-law exponent for the *on* times, because both of them are governed by the return time of the ejected electron.

Proceeding beyond the context of nanocrystals, we note that fluorescence of single molecules [19] and of nanoparticles diffusing through a laser focus [18], switching on and off of vibrational modes of a molecule [20], opening-closing behavior of certain single ion channels [13,21,22], motion of bacteria [23], deterministic diffusion in chaotic systems [24], the sign of magnetization of spin systems at criticality [25], and others exhibit power-law intermittency behavior [26]. More generally, the time trace of the NCs is similar to the well-known Lévy walk model [27]. Hence the stochastic theory which we consider in the following section is very general. In particular, we do not restrict our attention to the exponent 1/2, because there are indications that α takes on the other values lying between 0 and 1 as well, and the analysis hardly changes.

III. STOCHASTIC MODEL AND DEFINITIONS

The random process considered in this manuscript is shown in Fig. 4. The intensity $I(t)$ jumps between two states $I(t) = +1$ and $I(t) = 0$. At the start of the measurement $t = 0$ the NC is in state *on*: $I(0) = 1$. The sojourn time τ_i is an *off* time if i is even, it is an *on* time if i is odd (see Fig. 4). The times τ_i for odd [even] i are drawn at random from the probability density function (PDF) $\psi_+(t)$, $[\psi_-(t)]$, respectively. These sojourn times are mutually independent, identically distributed random variables. The times t_i are cumulative times from the starting point of the process at time zero till the end of the ith transition. The time T' in

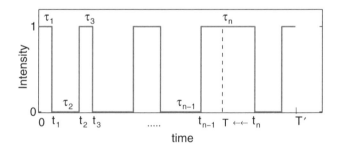

Figure 4. Schematic temporal evolution of the dichotomous intensity process.

Fig. 4 is the time of an observation. We denote the Laplace transform of $\psi_\pm(t)$ using

$$\hat{\psi}_\pm(s) = \int_0^\infty \psi_\pm(t)e^{-st}dt \tag{3}$$

In what follows we will investigate statistical properties of this seemingly simple stochastic process. In particular we will investigate its correlation function. In experiment, correlation functions are commonly used to characterize intensity trajectories. The main advantage of the use of correlation functions, as compared with PDFs of on and off times, is that in these there is no need to introduce the intensity cutoff. Correlation functions are more general than on and off time distributions. Besides, correlation functions exhibit aging and ergodicity breaking, which are in our opinion interesting.

We will consider several classes of on/off PDFs and will classify generic behaviors based on the small s expansion of $\psi_\pm(s)$. We will consider:

(i) **Case 1.** PDFs with finite mean *on* and *off* times, whose Laplace transform in the limit $s \to 0$ satisfies

$$\hat{\psi}_\pm(s) = 1 - s\tau_\pm + \cdots \tag{4}$$

Here $\tau_+(\tau_-)$ is the average *on* (*off*) time. For example, exponentially distributed *on* and *off* times,

$$\hat{\psi}_\pm(s) = \frac{1}{1 + s\tau_\pm} \tag{5}$$

belong to this class of PDFs.

(ii) **Case 2.** PDFs with infinite mean *on* and *off* times, namely PDFs with power-law behavior satisfying

$$\psi_\pm \propto t^{-1-\alpha_\pm}, \qquad \alpha_- < \alpha_+ \leq 1 \tag{6}$$

in the limit of long times. The small s behavior of these functions satisfies

$$\hat{\psi}_\pm(s) = 1 - A_\pm s^{\alpha_\pm} + \cdots \tag{7}$$

where $A_\pm$ are parameters which have units of time$^\alpha$. We will also consider cases where *on* times have finite mean ($\alpha_+ = 1$) while the *off* mean time diverges ($\alpha_- < 1$) since this situation describes the behavior of an uncapped NC [4; see also Ref. 28].

(iii) **Case 3.** PDFs with infinite mean with $\alpha_+ = \alpha_- = \alpha$

$$\hat{\psi}_\pm(s) = 1 - A_\pm s^\alpha + \cdots \tag{8}$$

As already mentioned, Brokmann et al. [5] report that for CdSe dots, $\alpha_+ = 0.58 \pm 0.17$, and $\alpha_- = 0.48 \pm 0.15$; hence within the limits of experimenal accuracy, $\alpha \simeq 0.5$.

Standard theories of data analysis commonly use the ergodic hypothesis where a time average of a process is replaced with an average over an ensemble. The simplest time average in our case is the time average intensity

$$\bar{I} = \frac{\int\limits_0^{T'} I(t)dt}{T'} \tag{9}$$

In the limit of long times and if the ergodic assumption holds, then we have $\bar{I} = \langle I \rangle$, where $\langle I \rangle$ is the ensemble average. As usual, we may generate many intensity trajectories one at a time, to obtain the ensemble-averaged correlation function

$$C(t, t') = \langle I(t)I(t + t') \rangle \tag{10}$$

and the normalized ensemble-averaged correlation function

$$g^{(2)}(t, t') \equiv \frac{\langle I(t)I(t + t') \rangle}{\langle I(t) \rangle \langle I(t + t') \rangle} = \frac{C(t, t')}{\langle I(t) \rangle \langle I(t + t') \rangle} \tag{11}$$

We may construct from a single trajectory of $I(t)$, recorded in a time interval $(0, T')$, the time average (TA) correlation function

$$C_{TA}(T', t') = \frac{\int\limits_0^{T'-t'} I(t)I(t + t')dt}{T' - t'} \tag{12}$$

In single-molecule experiments, the time averaged correlation function is considered, not the ensemble average. However, it is often assumed that the ensemble average and the time average correlation functions are identical. For nonergodic processes $C_{TA}(T', t') \neq C(t, t')$ even in the limit of large t and T'. Moreover for nonergodic processes, even in the limit of $T' \to \infty$, $C_{TA}(T', t')$ is a random function which varies from one sample of $I(t)$ to another. The ensemble-averaged function $C(t, t')$ of the considered process is nonstationary; that is, it retains its dependence on t even when $t \to \infty$. This is known as *aging*. It follows then from Eq. (12) that $\langle C_{TA}(T', t') \rangle = \int_0^{T'-t'} C(t, t')dt/(T' - t') \neq C(t, t')$.

IV. AGING

Consider the ensemble averaged correlation function $C(t, t') = \langle I(t + t')I(t) \rangle$. For processes with a finite microscopic time scale, which exhibit stationary behavior, one has $C(t, t') = f(t')$. Namely the correlation function does not depend on the observation time t. Aging means that $C(t, t')$ depends on both t and t' even in the limit when both are large [48,30]. Simple aging behavior means that at the scaling limit $C(t, t') = f(t'/t)$, which is indeed the scaling in our Case 3; in Case 2 below we find such a scaling for $g^{(2)}(t, t')$, while $C(t, t')$ will scale differently. Aging and nonergodicity are related. In our models, when single-particle trajectories become nonergodic, the ensemble average exhibits aging. Both behaviors are related to the fact that no characteristic time scale for the underlying process exists.

A.   Mean Intensity of *On–Off* Process

The ensemble-averaged intensity $\langle I(t) \rangle$ for a process switching between 1 and 0 and starting at 1 is now considered and will be used later. In the Laplace $t \rightarrow s$ domain it is easy to show that

$$\langle \hat{I}(s) \rangle = \frac{1 - \hat{\psi}_+(s)}{s} \cdot \frac{1}{1 - \hat{\psi}_+(s)\hat{\psi}_-(s)} \tag{13}$$

The Laplace $s \rightarrow t$ inversion of Eq. (13) yields the mean intensity $\langle I(t) \rangle$. Using small s expansions of Eq. (13), we find in the limit of long times

$$\langle I(t) \rangle \sim \begin{cases} \frac{\tau_+}{\tau_+ + \tau_-} & \text{case 1} \\ \frac{A_+ t^{\alpha_- - \alpha_+}}{A_- \Gamma(1 + \alpha_- - \alpha_+)} & \text{case 2} \\ \frac{A_+}{A_+ + A_-} & \text{case 3} \end{cases} \tag{14}$$

If the *on* times are exponential, as in Eq. (5), then

$$\langle \hat{I}(s) \rangle = \frac{\tau_+}{1 + s\tau_+ - \psi_-(s)} \tag{15}$$

This case corresponds to the behavior of the uncapped NCs. The expression in Eq. (15), and more generally, the case $\alpha_- < \alpha_+ = 1$ leads for long times t to

$$\langle I(t) \rangle \sim \frac{\tau_+ t^{\alpha_- - 1}}{A_- \Gamma(\alpha_-)} \tag{16}$$

For exponential *on* and *off* time distributions Eq. (5), we obtain the exact solution

$$\langle I(t) \rangle = \frac{\tau_- \exp\left[-t\left(\frac{1}{\tau_-} + \frac{1}{\tau_+}\right)\right] + \tau_+}{\tau_- + \tau_+} \tag{17}$$

The average intensity does not yield direct evidence for aging, because it depends only on one time variable, and one has to consider a correlation function in order to explore aging in its usual sense.

Remark. For the case $\alpha_+ < \alpha_- < 1$, corresponding to a situation where *on* times are in the statistical sense much longer then *off* times, $\langle I(t) \rangle \sim 1$.

B. Aging Correlation Function of *On–Off* Process

The ensemble-averaged correlation function $C(t, t') = \langle I(t)I(t + t') \rangle$ was calculated in [31]:

$$\hat{C}(t, u) = \frac{\hat{f}_t(u = 0, +) - \hat{f}_t(u, +)}{u}$$

$$+ \hat{f}_t(u, +) \frac{\hat{\psi}_-(u)[1 - \hat{\psi}_+(u)]}{u[1 - \hat{\psi}_-(u)\hat{\psi}_+(u)]} \tag{18}$$

where u is the Laplace conjugate of t' and

$$\hat{f}_s(u, +) = \frac{\hat{\psi}_+(s) - \hat{\psi}_+(u)}{(u - s)[1 - \hat{\psi}_+(s)\hat{\psi}_-(s)]} \tag{19}$$

where s is the Laplace conjugate of t. We note that $\hat{f}_s(u, +)$ is the double Laplace transform of the PDF of the so-called forward recurrence time indicating that after the aging of the process in the time interval t, the statistics of the first jump event after time t will generally depend on the age t. However, a process is said to exhibit aging, only if the statistics of this first jump depend on t even when this age is long. In particular if the microscopic time scale of the problem is infinite, no matter how large t the correlation function still depends on the age (see details below). The first term in Eq. (18) is due to trajectories which were in state on at time t and did not make any transitions (i.e., the concept of persistence), while the second term includes all the contributions from the trajectories being in state on at time t and making an even number of transitions [31].

C. Case 1

For case 1 with finite τ_+ and τ_-, and in the limit of long times t, we find

$$\lim_{t\to\infty} \hat{C}(t,u) = \frac{1}{u}\frac{\tau_+}{\tau_+ + \tau_-}\left\{ 1 - \frac{[1 - \hat{\psi}_+(u)][1 - \hat{\psi}_-(u)]}{\tau_+ u[1 - \hat{\psi}_-(u)\hat{\psi}_+(u)]} \right\} \quad (20)$$

This result was obtained by Verberk and Orrit [32], and it is seen that the correlation function depends asymptotically only on t' (since u is the Laplace pair of t')—namely, when the average *on* and *off* times are finite the system does not exhibit aging. If both $\psi_+(t)$ and $\psi_-(t)$ are exponential, then the *exact* result is

$$C(t,t') = \frac{\tau_- \exp\left[-t\left(\frac{1}{\tau_-} + \frac{1}{\tau_+}\right)\right] + \tau_+}{\tau_- + \tau_+}$$

$$\times \frac{\tau_- \exp\left[-t'\left(\frac{1}{\tau_-} + \frac{1}{\tau_+}\right)\right] + \tau_+}{\tau_- + \tau_+}$$

and $C(t,t')$ becomes independent of t exponentially fast as t increases.

D. Case 2

We consider case 2; however, we limit our discussion to $\alpha_+ = 1$ and $\alpha_- < 1$. As mentioned, this corresponds to uncapped NCs where *on* times are exponentially distributed, while *off* times are described by power-law statistics. Using the exact solution Eq. (18), we find asymptotically, when both t and t' are large,

$$C(t,t') \sim \left(\frac{\tau_+}{A_-}\right)^2 \frac{(tt')^{\alpha_- - 1}}{\Gamma^2(\alpha_-)} \quad (21)$$

Unlike case 1, the correlation function approaches zero when $t \to \infty$; this is because if t is large, we expect to find the process in state *off*. Using Eq. (16), the asymptotic behavior of the normalized correlation function Eq. (11) is

$$g^{(2)}(t,t') \sim \left(1 + \frac{t}{t'}\right)^{1-\alpha_-} \quad (22)$$

We see that the correlation functions (21) and (22) exhibit aging, since they depend on the age of the process t.

Considering the asymptotic behavior of $C(t,t')$ for large t,

$$\hat{C}(t,u) \approx \frac{1}{u}\frac{\tau_+}{A_-\Gamma(\alpha_-)t^{1-\alpha_-}}\left\{ 1 - \frac{[1 - \hat{\psi}_+(u)][1 - \hat{\psi}_-(u)]}{\tau_+ u[1 - \hat{\psi}_-(u)\hat{\psi}_+(u)]} \right\} \quad (23)$$

This equation is similar to Eq. (20), especially if we notice that the "effective mean" time of state *off* until total time t scales as $A_- t^{1-\alpha_-}$.

For the special case, where *on* times are exponentially distributed, the correlation function C is a product of two identical expressions *for all t and t'*:

$$\hat{C}(s,u) = \frac{\tau_+}{1 + s\tau_+ - \psi_-(s)} \cdot \frac{\tau_+}{1 + u\tau_+ - \psi_-(u)} \tag{24}$$

where $s(u)$ is the Laplace conjugate of $t(t')$ respectively. Comparing to Eq. (15), we obtain

$$C(t,t') = \langle I(t) \rangle \langle I(t') \rangle \tag{25}$$

and for the normalized correlation function we get

$$g^{(2)}(t,t') = \frac{\langle I(t') \rangle}{\langle I(t+t') \rangle} \tag{26}$$

Equations (26) and (25) are important since they show that measurement of mean intensity $\langle I(t) \rangle$ yields the correlation functions, for this case. While our derivation of Eqs. (26) and (25) is based on the assumption of exponential on times, it is valid more generally for any $\psi_+(t)$ with finite moments, in the asymptotic limit of large t and t'. To see this, note that Eqs. (21) and (16) yield $C(t,t') \sim \langle I(t) \rangle \langle I(t') \rangle$.

In Fig. 5 we compare the asymptotic result (21) with exact numerical double Laplace inversion of the correlation function. We use an exponential PDF of *on* times $\psi_+(s) = 1/(1+s)$, and power law distributed *off* times: $\hat{\psi}_-(s) = 1/(1+s^{0.4})$ corresponding to $\alpha_- = 0.4$. Convergence to asymptotic behavior is observed.

Remark. For fixed t the correlation function in Eq. (21) exhibits a $(t')^{\alpha_- - 1}$ decay. A $(t')^{\alpha_- - 1}$ decay of an intensity correlation function was reported in experiments of Orrit's group [4] for uncapped NCs (for that case $\alpha_- = 0.65 \pm 0.2$). However, the measured correlation function is a time-averaged correlation function [Eq. (12)] obtained from a single trajectory. Here the correlation function is independent of t; thus no comparison between theory and experiment can be made yet.

E. Case 3

We now consider case 3 and find [31]

$$C(t,t') = P_+ - P_+ P_- \frac{\sin \pi\alpha}{\pi} B\left(\frac{1}{1+t/t'}; 1-\alpha, \alpha\right) \tag{27}$$

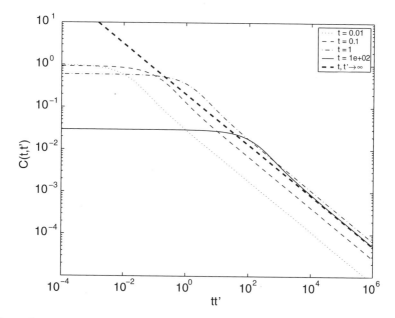

Figure 5. Exact $C(t,t')$ for case 2: exponential *on* times and power-law *off* times with $\alpha_- = 0.4$. We use $\hat{\psi}_+(s) = 1/(1+s)$ and $\hat{\psi}_-(s) = 1/(1+s^{0.4})$ and numerically obtain the correlation function. For each curve in the figure we fix the time t. The process starts in the state *on*. Thick dashed straight line shows the asymptotic behavior Eq. (21). For short times ($t' < 1$ for our example) we observe the behavior $C(t,t') \sim C(t,0) = \langle I(t) \rangle$, the correlation function is flat.

where

$$P_{\pm} = \frac{A_{\pm}}{A_+ + A_-}$$

following from Eq. (14) and where

$$B(z; a, b) = \int_0^z x^{a-1}(1-x)^{b-1} dx$$

is the incomplete beta function. The behavior in this limit does not depend on the detailed shape of the PDFs of the *on* and *off* times, besides the parameters A_+/A_- and α. We note that both terms in Eq. (18) contribute to Eq. (27). The appearance of the incomplete beta function in Eq. (27) is related to the concept of persistence. The probability of not switching from state *on* to state *off* in a time interval $(t, t + t')$, assuming the process is in state *on* at time t, is called the persistence probability. In the scaling limit this probability is

$$P_0(t, t + t') \sim 1 - \frac{\sin \pi \alpha}{\pi} B\left(\frac{1}{1 + t/t'}; 1 - \alpha, \alpha\right) \qquad (28)$$

The persistence implies that long time intervals in which the process does not jump between states *on* and *off*, control the asymptotic behavior of the correlation function. The factor P_+, which is controlled by the amplitude ratio A_+/A_-, determines the expected short and long time t' behaviors of the correlation function, namely $C(\infty, 0) = \lim_{t \to \infty}\langle I(t)I(t + 0)\rangle = P_+$ and $C(\infty, \infty) = \lim_{t \to \infty}\langle I(t)I(t + \infty)\rangle = (P_+)^2$. In slightly more detail the two limiting behaviors are

$$C(t, t') \sim \begin{cases} P_+, & \frac{t'}{t} \ll 1 \\ (P_+)^2 + P_+ P_- \frac{\sin(\pi\alpha)}{\pi\alpha}\left(\frac{t'}{t}\right)^{-\alpha}, & \frac{t'}{t} \gg 1 \end{cases} \qquad (29)$$

Using Eq. (14) the normalized intensity correlation function is $g^{(2)}(t, t') \sim C(t, t')/(P_+)^2$.

In Fig. 6 we compare the asymptotic result (27) with exact numerical double Laplace inversion of the correlation function for PDFs $\hat{\psi}_+(s) = \hat{\psi}_-(s) = 1/(1 + s^{0.4})$. Convergence to Eq. (27) is seen.

Remark. For small t'/t we get flat correlation functions. Flat correlation functions were observed by Dahan's group [33] for capped NCs. However, the measured correlation function is a single trajectory correlation function Eq. (12); thus no comparison between theory and experiment can be made yet.

V. NONERGODICITY

Nonergodicity of blinking quantum dots was first pointed out in the experiments of the group of Dahan [33]. We begin the discussion of nonergodicity in blinking NCs by plotting 100 time-averaged correlation functions from 100 NCs in Fig. 7. Clearly, the correlation functions obtained are different. The simplest explanation would be that the NCs have different statistical properties. However, similar variability is also observed for a given NC, when we calculate correlation functions for different T' (e.g., Ref. 33). To further illustrate this point, we generate on a computer a two-state process, with power-law waiting time of on and off times following $\psi_-(\tau) = \psi_+(\tau) = \alpha\tau^{-1-\alpha}$ for $\tau > 1$ (and zero

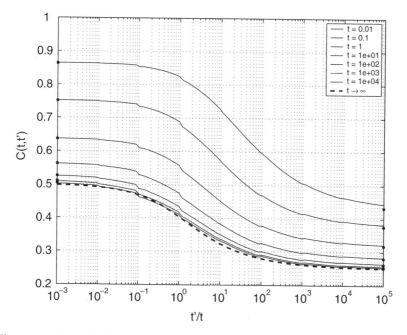

Figure 6. Exact $C(t, t')$ for case 3, when both *on* and *off* times are power-law distributed with $\alpha = 0.4$. We use $\hat{\psi}_{\pm}(s) = 1/(1 + s^{0.4})$ for different times t increasing from the topmost to the lowermost curves. The dots on the left and on the right show $C(t, 0) = \langle I(t) \rangle$ and $C(t, \infty) = \langle I(t) \rangle / 2$ respectively. The process starts in the state *on*.

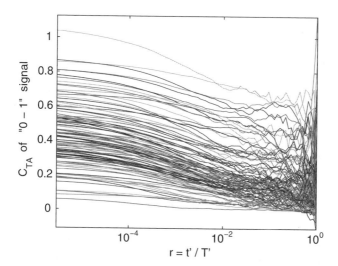

Figure 7. One hundred experimental time-averaged correlation functions (one of which is obtained from the signal shown in Fig. 1), after "renormalizing" the average on and off intensities to be 1 and 0, respectively. Note logarithmic abscissa.

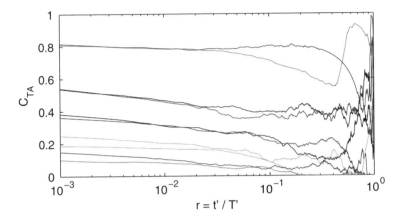

Figure 8.   Ten typical simulated realizations of C_{TA} for $\alpha = 0.8$.

otherwise). For each trajectory we calculate its own time average correlation. As we show in Fig. 8 the trajectories exhibit ergodicity breaking. The most striking feature of the figure is that even though the trajectories are statistically identical, the correlation function of the process is random, similar to experimental observations. In complete contrast, if we consider a two-state process with on and off times following exponential statistics, then all the correlation functions would be identical, and all of them would follow the same master curve: the ensemble average correlation function.

In this section we consider the nonergodic properties of blinking NCs using a stochastic approach. We assume that all the NCs are statistically identical in agreement with [5], and we restrict ourselves to case 3. For the simplicity we only consider the case when the distribution of *on* times is identical to distribution of *off* times, namely $\alpha_- = \alpha_+ = \alpha$ and $A_+ = A_- = A$. Generalization to $A_+ \neq A_-$ is straightforward [34]. The nonergodicity is found only for $\alpha < 1$, when the mean transition time is infinite, and should therefore disappear when exponential cutoffs of off and on times become relevant [3]—that is, when the mean transition times become of the order of, or less than, the experimental time. The model described, however, is valid in a wide time window spanning many orders of magnitude for the NCs, and is relevant to other systems, as mentioned in Section II.

A. Distribution of Time-Averaged Intensity

As mentioned in the introduction, blinking NCs exhibit a nonergodic behavior. In particular the ensemble average intensity $\langle I \rangle$ is not equal to the time average $\bar{I}$. Of course in the ergodic phase—that is, when both the mean *on* and *off* times are finite—we have $\langle I \rangle = \bar{I}$, in the limit of long measurement time. More generally we may conceive of $\bar{I}$ as a random function of time, which will vary from one measurement to another. In the ergodic phase and in the asymptotic limit the distribution of $\bar{I}$ approaches a delta function

$$P(\bar{I}) \rightarrow \delta(\bar{I} - \langle I \rangle) \tag{30}$$

The theory of nonergodic processes deals with the question what is the distribution of $P(\bar{I})$ in the nonergodic phase. For the two-state stochastic model

$$\bar{I} = \frac{T^+}{T} \tag{31}$$

where T^+ is the total time spent in state *on*.

A well-known example of similar ergodicity breaking is regular diffusion, or a binomial random walk on a line. The walker starts at the origin and can go left or right randomly, at each step. Let the measurement time be t, and let the position of the random walker be $x(t)$. The total time the walker remains on the right of the origin $x(t) > 0$ is T^+. The PDF of return time (or of the number of steps) τ to the origin decays as $\tau^{-3/2}$ for large τ, so that $\alpha = 1/2$. The two semiaxes at both sides of the origin can be thought of as the two states, on and off, of the random walker. The well-established result is that the fraction $\bar{I}$ of the total time spent by the walker on either side in the long time limit is given by the arcsine law [25,35]

$$P(\bar{I}) = \frac{1}{\pi\sqrt{\bar{I}(1 - \bar{I})}}$$

A main feature of this PDF is its divergence at $\bar{I} = 0$, 1, indicating that the random walker will most probably spend most of its time on one side (either left or right) of the origin. In particular, the naive expectation that the particle will spend half of its time on the right and half on the left, in the limit of long measurement time, is wrong. In fact the minimum of the arcsine PDF is on $\bar{I} = \langle \bar{I} \rangle = 1/2$. In other words the ensemble average $\langle \bar{I} \rangle = 1/2$ is the least likely event. This result might seem counterintuitive at first, but it is because the mean time for return to the origin is *infinite*, indicating that the particle gets randomly stuck in $x < 0$ or in $x > 0$ for a period which is of the order of the measurement time, no matter how long this measurement time is.

In the more general case $0 < \alpha < 1$ the distribution of $\bar{I}$ can be calculated based on the work of Lamperti [36] (see also Ref. 25), and one finds

$$l_\alpha(\bar{I}) = \frac{\sin(\pi\alpha)}{\pi} \frac{\bar{I}^{\alpha-1}(1 - \bar{I})^{\alpha-1}}{\bar{I}^{2\alpha} + (1 - \bar{I})^{2\alpha} + 2\cos(\pi\alpha)\bar{I}^\alpha(1 - \bar{I})^\alpha} \tag{32}$$

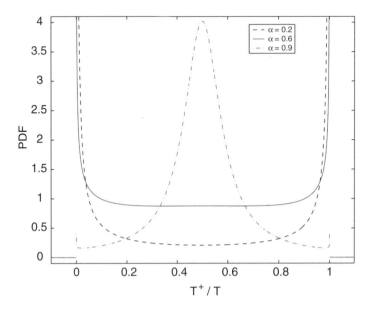

Figure 9. The probability density function of $\bar{I} = T^+/T$ for the case $\psi_+(t) = \psi_-(t) \propto t^{-(1+\alpha)}$. For the ergodic phase, $\alpha > 1$ $P(\bar{I})$ is a delta function on $\langle I \rangle = 1/2$. In the nonergodic phase, $\bar{I}$ is a random function; for small values of α the $P(\bar{I})$ is peaked on $\bar{I} = 0$ and $\bar{I} = 1$, indicating a trajectory which is in state *off* or *on* for a period which is of the order of measurement time T.

which is shown in Fig. 9. When $\alpha \to 0$ the PDF of $\bar{I}$ is peaked around $\bar{I} = 0$ and $\bar{I} = 1$, corresponding to blinking trajectories which for most of the observation time T are in state *off* or state *on*, respectively. When $\alpha \to 1$, we see that $l_\alpha(\bar{I})$ attains a maximum when $\bar{I} = \langle I \rangle = 1/2$, indeed in the ergodic phase $\alpha > 1$ we obtain as expected a delta peak centered on $\bar{I} = 1/2$, as we mentioned. There exists a critical $\alpha_c = 0.594611\ldots$ above (below) which $l_\alpha(\bar{I})$ has a maximum (minimum) on $\bar{I} = 1/2$. Note that the Lamperti PDF in Eq. (32) is not sensitive to the precise shapes of the *on* and *off* time distributions (besides α of course). For situations in which $A_- \neq A_+$ the symmetry of the Lamperti PDF will not hold. Note that line shapes with structures similar to those in Fig. 9, were obtained by Jung et al. [37] in a related problem. Similar expressions are also used in stochastic models of spin dynamics [29], and in general, the problem of occupation times, and a related persistence concept, are of interest in a wide variety of phenomena [34,38,39].

Next we extend our understanding of the *distribution* of time-averaged intensity to the time averaged correlation functions defined in Eq. (12).

B. Distribution of Time-Averaged Correlation Function

We first consider the nonergodic properties of the correlation function for the case $t' = 0$. It is useful to define

$$\mathcal{I}_{[a,b]} = \int_a^b I(t)dt/(b-a) \tag{33}$$

the time average intensity between time a and time $b > a$, and

$$T = T' - t', \qquad r = \frac{t'}{T'}$$

Using Eq. (12) and for $t' = 0$ the time-averaged correlation function is identical to the time average intensity

$$C_{TA}(T,0) = \mathcal{I}_{[0,T]} = \frac{T^+}{T} \tag{34}$$

and its PDF is given by Eq. (32). Figures 10, 11, and 12 for the case $r = 0$ show these distributions for $\alpha = 0.3$, 0.5 and 0.8, respectively, together with the numerical results.

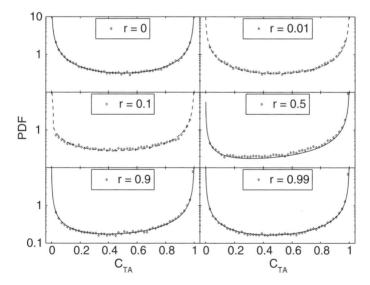

Figure 10.   PDF of $C_{TA}(T',t')$ for different $r = t'/T'$ and $\alpha = 0.3$. Abscissas are possible values of $C_{TA}(T',t')$. Diamonds are numerical simulations. Curves are analytical results without fitting: For $r = 0$, Eq. (32) is used (full line); for $r = 0.01$ and 0.1, Eq. (35) is used (dashed line); and for $r = 0.5$, 0.9, and 0.99, Eq. (38) is used (full line).

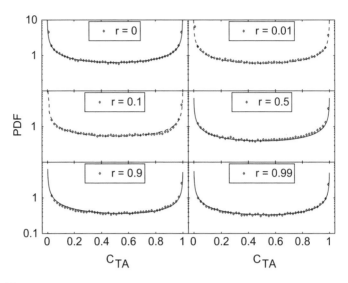

Figure 11. PDF of $C_{TA}(T',t')$ for different $r = t'/T'$ and $\alpha = 0.5$. Diamonds are numerical simulations. Curves are analytical results without fitting: For $r = 0$ Eq. (32) is used (full line); for $r = 0.01$ and 0.1, Eq. (35) is used (dashed line) and for $r = 0.5$, 0.9, and 0.99, Eq. (38) is used (full line).

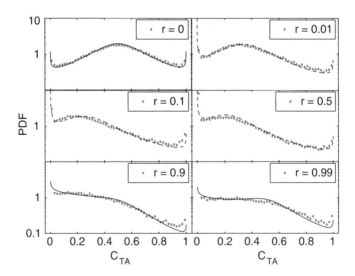

Figure 12. PDF of $C_{TA}(T',t')$ for different $r = t'/T'$ and $\alpha = 0.8$. Diamonds are numerical simulations. Curves are analytical results without fitting: For $r = 0$, Eq. (32) is used (full line); for $r = 0.01$, 0.1 and 0.5, Eq. (35) is used (dashed); and for $r = 0.9$ and 0.99, Eq. (38) is used (full lines). If compared with the cases $\alpha = 0.3$ and 0.5, the distribution function exhibits a weaker nonergodic behavior, namely for $r = 0$ the distribution function peaks on the ensemble average value of $1/2$.

An analytical approach for estimating the distributions $P_{C_{TA}(T',t')}(z)$ of $C_{TA}(T',t') = z$ for nonzero t' was developed in Margolin and Barkai [26, 40]. To treat the problem a nonergodic mean field approximation was used, in which various time averages were replaced by the time average intensity $\mathcal{I}_{[0,T]}$, *specific to a given realization*. For short $t' \ll T'$ the result is

$$
C_{TA}(t', T') \simeq
$$
$$
\begin{cases}
\mathcal{I}_{[0,T]}\left\{ 1 - (1 - \mathcal{I}_{[0,T]}) \left[\left(\frac{r}{(1-r)\mathcal{I}_{[0,T]}} \right)^{1-\alpha} \left(\frac{\sin \pi\alpha}{\pi\alpha} + 1 \right) - \frac{\sin \pi\alpha}{\pi\alpha} \frac{r}{(1-r)\mathcal{I}_{[0,T]}} \right] \right\}, \\
\hspace{8cm} t' < T^+ \\
\mathcal{I}^2_{[0,T]}, \qquad t' > T^+
\end{cases}
$$

$$(35)$$

Equation (35) yields the correlation function, however unlike standard ergodic theories the correlation function here is a random function since it depends on $\mathcal{I}_{[0,T]}$. The distribution of $C_{TA}(T',t')$ is now easy to find using the chain rule, and Eqs. (32), (34), and (35). In Figs. 10, 11, and 12 we plot the PDF of $C_{TA}(T',t')$ (dashed curves) together with numerical simulations (diamonds) and find excellent agreement between theory and simulation, for the cases where our approximations are expected to hold $r < 1/2$. We observe that unlike the $r = 0$ case the PDF of the correlation function exhibits a nonsymmetrical shape. To understand this, note that trajectories with short but finite total time in state *on* $(T^+ \ll T)$ will have finite correlation functions when $t' = 0$. However, when t' is increased, the corresponding correlation functions will typically decay very quickly to zero. On the other hand, correlation functions of trajectories with $T^+ \sim T$ alter very little when t' is increased (as long as $t' \ll T^+$). This leads to the gradual nonuniform shift to the left, and "absorption" into $C_{TA}(T',t') = 0$, of the Lamperti distribution shape, and thus to the nonsymmetrical shape of the PDFs of the correlation function whenever $r \neq 0$.

We now turn to the case $T \ll t'$. Then a decoupling approximation [26] yields

$$
C_{TA}(T',t') \simeq \mathcal{I}_{[0,T]}\mathcal{I}_{[t',T']} \tag{36}
$$

In the limit $t'/T' \to 1$ this yields

$$
P_{C_{TA}(T',t')}(z) \sim [\ell_\alpha(z) + \delta(z)]/2 \tag{37}
$$

which is easily understood if one realizes that in this limit $\mathcal{I}_{[t',T']}$ in Eq. (36) is either 0 or 1 with probabilities 1/2, and that the PDF of $\mathcal{I}_{[0,T]}$ is Lamperti's

PDF, Eq. (32). More generally, using the Lamperti distribution for $\mathcal{I}_{[0,T]}$, and probabilistic arguments [26], the PDF of $C_{TA}(T', t')$ is approximated by

$$
P_{C_{TA}(T',t')}(z) \simeq [1 - P_0(T, T')] \left\{ [1 - P_0(t', T')] \int_z^1 \frac{l_\alpha(x)}{x} dx \right.
$$
$$
\left. + \frac{P_0(t', T')}{2} [l_\alpha(z) + \delta(z)] \right\} + P_0(T, T') \left[z l_\alpha(z) + \frac{\delta(z)}{2} \right] \qquad (38)
$$

where $P_0(a, b)$ is the persistence probability, Eq. (28). Note that to derive Eq. (38) we used the fact that $\mathcal{I}_{[0,T]}$ and $\mathcal{I}_{[t',T']}$ are correlated. In Figs. 10, 11, and 12 we plot these PDFs of $C_{TA}(T', t')$ (solid curves) together with numerical simulations (diamonds) and find good agreement between theory and simulation where these approximations are expected to hold, that is, $r > 1/2$. In the limit $t'/T' \to 1$, Eq. (38) simplifies to Eq. (37).

VI.   EXPERIMENTAL EVIDENCE

In this section we analyze experimental data and make comparisons with theory. Data were obtained for 100 CdSe–ZnS nanocrystals at room temperature.[1] We first performed data analysis (similar to standard approach) based on the distribution of on and off times and found that $\alpha_+ = 0.735 \pm 0.167$ and $\alpha_- = 0.770 \pm 0.106$,[2] for the total duration time $T' = T = 3600$ s (bin size 10 ms, threshold was taken as $0.16 \max I(t)$ for each trajectory). Within error of measurement, $\alpha_+ \approx \alpha_- \approx 0.75$. The value of $\alpha \approx 0.75$ implies that the simple diffusion model with $\alpha = 0.5$ is not valid in this case. An important issue is whether the exponents vary from one NC to another. In Fig. 13 (top) we show the distribution of α obtained from data analysis of power spectra. The power spectrum method [26] yields a single exponent α_{psd} for each stochastic trajectory (which is in our case $\alpha_+ \approx \alpha_- \approx \alpha_{psd}$). Figure 13 illustrates that the spread of α in the interval $0 < \alpha < 1$ is not large. Numerical simulation of 100 trajectories switching between 1 and 0, with $\psi_+(\tau) = \psi_-(\tau)$ and $\alpha = 0.8$, and with the same number of bins as the experimental trajectories, was performed and the

---

[1] Core radius 2.7 nm with less than 10% dispersion, 3 monolayers of ZnS, covered by mixture of TOPO, TOP, and TDPA. Quantum dots were spin coated on a flamed fused silica substrate. CW excitation at 488 nm of Ar^+ laser was used, excitation intensity in the focus of oil immersion objective (NA = 1.45) was $\sim 600\,W/cm^2$.

[2] The standard deviation figures for $\alpha_\pm$ here and for α_{psd} in Fig. 13 represent the standard deviations of the distributions of the corresponding exponents, and not the errors in determination of their mean value. We also note that the on time distributions are less close to the power-law decays than the off times, partly due to the exponential cutoffs and partly due to varying intensities in the on state (cf. Figs. 1 and 2).

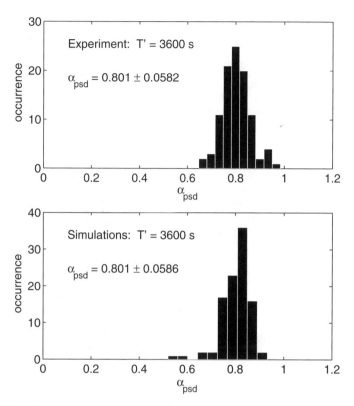

Figure 13.   Histograms of experimental (top) and simulated (bottom) fitted values of α for 100 trajectories. Fits are made to the power spectral densities of individual trajectories.

distribution of α values estimated from power spectra is also shown in Fig. 13 (bottom). We observe some spread in the measured values of α, which is similar to experimental behavior. This indicates that experimental data are compatible with the assumption that all dots are statistically identical (in our sample), in agreement with [5,6].

We also tested our nonergodic theory and calculated the distribution of the relative on times T^+/T—that is, of the ratios of the total time spent in the state on to the total measurement time. These relative on times are equivalent to the experimental time averaged intensities after their "renormalization" in a way such that the average intensity in state on/off is 1/0, respectively, by analogy with our stochastic model. Experimental and simulated distributions shown in Fig. 14 are, for the most part, in good agreement. Two important conclusions may be made from these distributions of relative on times. First the data clearly exhibits ergodicity breaking: Distribution of relative on times is not delta

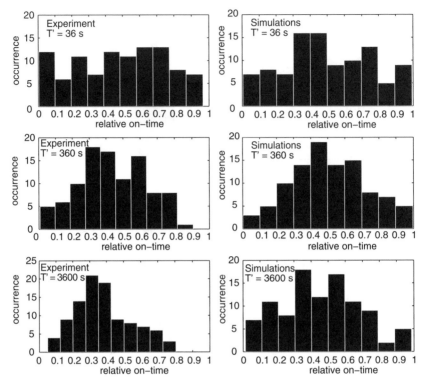

Figure 14. Histograms of relative on times T^+/T for 100 experimental (left) and 100 simulated (right) intensity trajectories, for different T'.

peaked, instead it is wide in the interval between 0 and 1, for different T'. The second important conclusion is that for a reasonably chosen threshold (cf. Fig. 1), the experimental data are compatible with the assumption

$$\psi_+(\tau) \approx \psi_-(\tau)$$

at least for a wide time window relevant to the experiments—in other words, not only $\alpha_+ \approx \alpha_-$ (ignoring the cutoffs) but also $A_+ \approx A_-$. This observation cannot be obtained directly from the on and off time histograms like Fig. 2 because if only power-law tails are seen, as in Fig. 2, these histograms cannot be normalized. To see that $A_+ \approx A_-$, note that the distributions of relative on times are roughly symmetric with respect to the median value of $1/2$ (cf. Fig. 14), and the ensemble average of relative on times is also close to $1/2$, while in general the ensemble average in our model process is given by $A_+/(A_+ + A_-)$. In addition, the variance of the experimental distributions for different T' is close to the variance of the Lamperti distribution $(1 - \alpha)/4$ [26] for $\alpha \approx 0.8$. There are a few comments to make. First, 100 trajectories are insufficient to produce accurate histograms, as can be seen from the right side of Fig. 14: ideally, these histograms should be identical for different T' and are given by the Lamperti distribution Eq. (32). Second, there is an effect due to the signal discretization, leading to a flatter and wider histogram at $T' = 36$ s. Third, there is a certain slow narrowing of the experimental histogram as T' increases, and the average relative on time slowly decreases. Both of these trends are probably due to cutoffs in the power-law distributions, especially for on times, as can be seen in Fig. 2. These trends slightly depend on the choice of the threshold separating on and off states.

As mentioned previously, the groups of Dahan and Bawendi [5,7] measure values of $\alpha_+ \approx \alpha_- \approx 0.5$ for hundreds of quantum dots (see Table I), while we report on a higher value of α. An important difference between our samples and Dahan/Bawendi groups is that in those works the dots are embedded in PMMA, while in our case they are not (see also Ref. 6).

VII. SUMMARY AND CONCLUSIONS

Our main points are the following:

1. A simple three-dimensional diffusion model can be used to explain the exponent $\alpha = 1/2$ observed in many experiments. In some cases deviations from $\alpha = 1/2$ are observed, and modifications of Onsager theory are needed. We cannot exclude other models.

2. A simple model of diffusion may lead to ergodicity breaking. Thus ergodicity breaking in single molecule spectroscopy should not be considered exotic or strange.

3. The time average correlation function is random. The ensemble average correlation function exhibits aging. Hence data analysis should be made with care.

4. Our data analysis shows $A_+ \approx A_-$, $\alpha_+ \approx \alpha_-$ (before the possible cutoffs) and that the distribution of α is narrow. It is important to check the validity of this result in other samples of nanocrystals, since so far the main focus of the experimentalist has been on values of α and not on the ratio of the amplitudes A_+/A_-.

How general are our results? From a stochastic point of view ergodicity breaking, Lévy statistics, anomalous diffusion, aging, and fractional calculus, are all related. In particular ergodicity breaking is found in other models with power-law distributions, related to the underlying stochastic model (the Lévy walk). For example, the well known continuous time random walk model also exhibits ergodicity breaking [34], and hence a natural conflict with standard Boltzmann statistics emerges. Since power-law distributions are very common in nature, we expect that single particle ergodicity breaking will be a common theme. Further, since we showed that a simple diffusion model can generate ergodicity breaking, for nano-crystals, we expect that ergodicity breaking may be observed in other single-molecule systems. One simple conclusion is that predictions cannot be made, based on ensemble averages. In fact the time averages of physical observables remain random even in the limit of long measurement time. The fact that the time-averaged correlation function is a random function means that some of the experimental published results, on time average correlation functions, are not reproducible.

Remark: After this work was completed, a theory of blinking manocrystals was proposed by Marcus and co-workers [49,50].

Acknowledgments

Acknowledgment is made to the National Science Foundation for support of this research with award CHE-0344930. EB thanks the Center of Complexity, Jerusalem and the Israel Science Foundation for financial support. MK thanks the ACS PRF for financial support.

References

1. G. Schlegel, J. Bohnenberger, I. Potapova, and A. Mews, Fluorescence decay time of single semiconductor nanocrystals. *Phys. Rev. Lett.* **88**(13):137401 (2002).
2. B. R. Fisher, H.-J. Eisler, N. E. Stott, and M. G. Bawendi, Emission intensity dependence and single-exponential behavior in single colloidal quantum dot fluorescence lifetimes. *J. Phys. Chem. B* **108**:143–148 (2004).
3. I. Chung and M. G. Bawendi, Relationship between single quantum-dot intermittency and fluorescence intensity decays from collections of dots. *Phys. Rev. B*, **70**:165304 (2004).
4. R. Verberk, A. M. van Oijen, and M. Orrit, Simple model for the power-law blinking of single semiconductor nanocrystals. *Phys. Rev. B* **66**:233202 (2002).
5. X. Brokmann, J. P. Hermier, G. Messin, P. Desbiolles, J.-P. Bouchaud, and M. Dahan, Statistical aging and nonergodicity in the fluorescence of single nanocrystals. *Phys. Rev. Lett.* **90**(12):120601 (2003).
6. A. Issac, C. von Borczyskowski, and F. Cichos, Correlation between photoluminescence intermittency of CdSe quantum dots and self-trapped states in dielectric media. *Phys. Rev. B* **71**:161302(R) (2005).
7. K. T. Shimizu, R. G. Neuhauser, C. A. Leatherdale, S. A. Empedocles, W. K. Woo, and M. G. Bawendi, Blinking statistics in single semiconductor nanocrystal quantum dots. *Phys. Rev. B* **63**(20):205316 (2001).
8. A. L. Efros and M. Rosen, Random telegraph signal in the photoluminescence intensity of a single quantum dot. *Phys. Rev. Lett.* **78**(6):1110–1113 (1997).
9. M. Kuno, D. P. Fromm, H. F. Hamann, A. Gallagher, and D. J. Nesbitt, "On"/"Off" fluorescence intermittency of single semiconductor quantum dots. *J. Chem. Phys.* **115**(2):1028–1040 (2001).

10. M. Kuno, D. P. Fromm, S. T. Johnson, A. Gallagher, and D. J. Nesbitt, Modeling distributed kinetics in isolated semiconductor quantum dots. *Phys. Rev. B* **67**:125304 (2003).

11. K. M. Hong and J. Noolandi, Solution of the Smoluchowski equation with a Coulomb potential. I. General results. *J. Chem. Phys.* **68**(11):5163–5171 (1978).

12. H. Sano and M. Tachiya, Partially diffusion-controlled recombination. *J. Chem. Phys.* **71**(3):1276–1282 (1979).

13. W. Nadler and D. L. Stein, Biological transport processes and space dimension. *Proc. Natl. Acad. Sci. USA* **88**:6750–6754 (1991).

14. W. Nadler and D. L. Stein, Reaction-diffusion description of biological transport processes in general dimension. *J. Chem. Phys.* **104**(5):1918–1936 (1996).

15. T. D. Krauss and L. E. Brus, Charge, polarizability and photoionization of single semiconductor nanocrystals. *Phys. Rev. Lett.* **83**(23):4840–4843 (1999).

16. K. M. Hong, J. Noolandi, and R. A. Street, Theory of radiative recombination by diffusion and tunneling in amorphous Si:H. *Phys. Rev. B* **23**(6):2967–2976 (1981).

17. E. Lifshitz, L. Fradkin, A. Glozman, and L. Langof, Optically detected magnetic resonance studies of colloidal semiconductor nanocrystals. *Annu. Rev. Phys. Chem.* **55**:509–557 (2004).

18. G. Zumofen, J. Hohlbein, and C. G. Hübner, Recurrence and photon statistics in fluorescence fluctuation spectroscopy. *Phys. Rev. Lett.* **93**:260601 (2004).

19. M. Haase, C. G. Hübner, E. Reuther, A. Herrmann, K. Müllen, and T. Basché, Exponential and power-law kinetics in single-molecule fluorescence intermittency. *J. Phys. Chem. B*, **108**:10445–10450 (2004).

20. A. R. Bizzarri and S. Cannistraro, Lèvy statistics of vibrational mode fluctuations of single molecules from surface-enhanced Raman scattering. *Phys. Rev. Lett.* **94**:068303 (2005).

21. I. Goychuk and P. Hänggi, Ion channel gating: A first-passage time analysis of the Kramers type. *Proc. Natl. Acad. Sci. USA* **99**(6):3552–3556 (2002).

22. I. Goychuk and P. Hänggi, The role of conformational diffusion in ion channel gating. *Physica A*, **325**:9–18 (2003).

23. E. Korobkova, T. Emonet, J. M. G. Vilar, T. S. Shimizu, and P. Cluzel, From molecular noise to behavioural variability in a single bacterium. *Nature* **428**:574–578 (2004).

24. G. Zumofen and J. Klafter, Scale-invariant motion in intermittent chaotic systems. *Phys. Rev. E* **47**(2):851–863 (1993).

25. C. Godrèche and J. M. Luck, Statistics of the occupation time of renewal processes. *J. Stat. Phys.* **104**(3–4):489–524 (2001).

26. G. Margolin and E. Barkai, Nonergodicity of a time series obeying Lévy statistics. *cond-mat/0504454* (2005) and *J. Stat. Phys.* DOI: 10.1007/s10955-005-8076-9.

27. Y. Jung, E. Barkai, and R. J. Silbey, A Stochastic theory of single molecule spectroscopy. *Adv. Chem. Phys.* **123**:199–266 (2002).

28. D. Novikov, M. Drndic, L. Levitov, M. Kastner, M. Jarosz, and M. Bawendi, Anomalous transport in quantum dot arrays. *cond-mat/0307031* (2003).

29. A. Baldassarri, J.-P. Bouchaud, I. Dornic, and C. Godrèche, Statistics of persistent events: An exactly soluble model. *Phys. Rev. E*, **59**(1):R20–R23 (1999).

30. G. Aquino, L. Palatella, and P. Grigolini, Absorption and emission in the non-poissonian case. *Phys. Rev. Lett.* **93**(5):050601 (2004).

31. G. Margolin and E. Barkai, Aging correlation functions for blinking nanocrystals, and other on-off stochastic processes. *J. Chem. Phys.* **121**(3):1566–1577 (2004).

32. R. Verberk and M. Orrit, Photon statistics in the fluorescence of single molecules and nanocrystals: Correlation functions versus distributions of on- and off-times. *J. Chem. Phys.* **119**(4):2214–2222 (2003).

33. G. Messin, J. P. Hermier, E. Giacobino, P. Desbiolles, and M. Dahan, Bunching and antibunching in the fluorescence of semiconductor nanorystals. *Opt. Lett.* **26**(23):1891–1893 (2001).

34. G. Bel and E. Barkai, Weak ergodicity breaking in the continous-time random walk, *Phys. Rev. Lett.* **94**:240602 (2005).

35. W. Feller, *An Introduction to Probability Theory and Its Applications*, Vol. 2, Wiley, New York, 1966.

36. J. Lamperti, An occupation time theorem for a class of stochastic processes. *Trans. Am. Math. Soc.* **88**(2):380–387 (1958).

37. Y. Jung, E. Barkai, and R. Silbey, Current status of single-molecule spectroscopy: Theoretical aspects. *J. Chem. Phys.* **117**(24):10980–10995 (2002).

38. A. Dhar and S. N. Majumdar, Residence time distribution for a class of gaussian Markov processes. *Phys. Rev. E* **59**: 6413–6418 (1999).

39. S. N. Majumdar, Persistence in nonequilibrium systems. *cond-mat/9907407* (2004).

40. G. Margolin and E. Barkai, Nonergodicity of blinking nanocrystals and other Lévy-walk processes. *Phys. Rev. Lett.* **94**:080601 (2005).

41. M. Kuno, D. P. Fromm, H. F. Hamann, A. Gallagher, and D. J. Nesbitt, Nonexponential "blinking" kinetics of single CdSe quantum dots: A universal power law behavior. *J. Chem. Phys.* **112**(7):3117–3120 (2000).

42. M. Kuno, D. P. Fromm, A. Gallagher, D. J. Nesbitt, O. I. Micic, and A. J. Nozik, Fluorescence intermittency in single InP quantum dots. *Nano Letters* **1**(10):557–564 (2001).

43. F. Cichos, J. Martin, and C. von Borczyskowski, Emission intermittency in silicon nanocrystals. *Phys. Rev. B* **70**:115314 (2004).

44. S. Hohng and T. Ha, Near-complete suppression of quantum dot blinking in ambient conditions. *J. Am. Chem. Soc.* **126**:1324–1325 (2004).

45. J. Müller, J. M. Lupton, A. L. Rogach, and J. Feldmann, Air-induced fluorescence bursts from single semiconductor nanocrystals. *Appl. Phys. Lett.* **85**(3):381–383 (2004).

46. W. G. J. H. M. van Sark, P. L. T. M. Frederix, A. A. Bol, H. C. Gerritsen, and A. Meijerink, Blueing, bleaching, and blinking of single CdSe/ZnS quantum dots. *Chem. Phys. Chem.* **3**:871–879 (2002).

47. A. Yu. Kobitski, C. D. Heyes, and G. U. Nienhaus, Total internal reflection fluorescence microscopy—a powerful tool to study single quantum dots. *Appl. Surface Sci.* **234**:86–92 (2004).

48. J. P. Bouchaud, Weak ergodicity breaking and aging in disordered systems. *J. Phys. I* **2**:1705–1713 (1992).

49. J. Tang and R. A. Marcus, Diffusion-controlled electron transfer processes and power-law statistics of fluoresence intermittency of nanoparticles. *Phys. Rev. Lett.* **95**, 107401 (2005).

50. P. A. Frantsuzov and R. A. Marcus, Explanation of quantum dot blinking without the long-lived trap hypothesis. *Phys. Rev. B.* **72**, 155321 (2005).

CHAPTER 5

THE CONTINUOUS-TIME RANDOM WALK VERSUS THE GENERALIZED MASTER EQUATION

PAOLO GRIGOLINI

Department of Physics, University of North Texas, Denton, Texas 76203, USA; and Department of Physics, University of Pisa, Pisa, Italy

CONTENTS

Fractals, Diffusion, and Relaxation in Disordered Complex Systems: A Special Volume of Advances in Chemical Physics, Volume 133, Part A, edited by William T. Coffey and Yuri P. Kalmykov. Series editor Stuart A Rice.

I. INTRODUCTION

The main purpose of this review is to show that the latest experimental results concerning the spectroscopy of Blinking Quantum Dots (BQD) and single molecules [1–5] require a deep theoretical change: The traditional methods of statistical mechanics, based on either the quantum or classical Liouville equation and consequently on the adoption of densities, must be replaced by the adoption of Continuous Time Random Walks (CTRW). This means, in practice, that we have to rely on the direct observation of stochastic trajectories, characterized by uncorrelated and unpredictable jumps. To prove the necessity for this change, we review the recent work done to derive anomalous diffusion of Lévy type from within a Liouville formalism. We show that this way of proceeding, which is satisfactory in the case of Poisson statistics, is not adequate to reproduce the numerical and experimental results in the non-Poisson case. Using the CTRW formalism we determine the structure of the Generalized Master Equations

(GME) that the Liouville method should produce. We also prove that these GME are characterized by aging.

We show that the GME, with aging, even if it is properly derived so as to yield results equivalent to the corresponding CTRW picture, might fail the purpose of predicting correctly the system's response to an external perturbation. We show that to derive a correct result we have to adopt the single trajectory perspective and to study the field action on the single trajectory rather than on a set of trajectories. This means that the non-Poisson renewal condition renders the GME a theoretical tool of limited validity.

Another issue that we discuss in this chapter, inspired again by the BQD physics, is the foundation of a dynamic model that might account for two distinct, and apparently conflicting, properties. The former is the renewal character of BQD materials, a property that implies the occurrence of rare jumps, resetting to zero the system's memory. The latter is the cooperation process that is expected to generate the deviation from the Poisson statistics, a property that implies the emergence of infinite memory. We shall see that these apparently conflicting properties refer to two distinct variables, the former being a diffusional variable, whose persistence in one state obeys renewal, and the latter the fluctuation collected by this diffusional variable.

The crisis of the GME method is closely related to the crisis in the density matrix approach to wave-function collapse. We shall see that in the Poisson case the processes making the statistical density matrix become diagonal in the basis set of the measured variable and can be safely interpreted as generators of wave function collapse, thereby justifying the widely accepted conviction that quantum mechanics does not need either correction or generalization. In the non-Poisson case, this equivalence is lost, and, while the CTRW perspective yields correct results, no theoretical tool, based on density, exists yet to make the time evolution of a contracted Liouville equation, classical or quantum, reproduce them.

The outline of this review is as follows. In Section II we discuss the Pauli master equation and we show how to derive it from a coherent quantum mechanical process through the addition of a dephasing process.

Section III is devoted to illustrating the first theoretical tool under discussion in this review, the GME derived from the Liouville equation, classical or quantum, through the contraction over the irrelevant degrees of freedom. In Section III.A we illustrate Zwanzig's projection method. Then, in Section III.B, we show how to use this method to derive a GME from Anderson's tight binding Hamiltonian: The second-order approximation yields the Pauli master equation. This proves that the adoption of GME derived from a Hamiltonian picture requires, in principle, an infinite-order treatment. The case of a vanishing diffusion coefficient must be considered as a case of anomalous diffusion, and the second-order treatment is compatible only with the condition of ordinary diffusion. In Section IV we illustrate the second theoretical tool under discussion in this review, the CTRW.

Note that the crisis of the GME method, based on the density approach, is generated by non-Poisson renewal processes. Therefore, we find it important to illustrate, by means of Section V, the model for intermittence, on which our theoretical arguments are based.

We devote Section VI to illustrating the attempts made in recent past to derive superdiffusion of Lévy kind, using a Liouville-like approach. We show that in the Poisson case, the use of CTRW (trajectory perspective) yields the same result as the GME derived from within a Liouville-like picture (density perspective). In the non-Poisson case, this equivalence is lost. It has to be pointed out that a careful use of the trajectory perspective and a statistical analysis of the resulting symbolic sequences prove that the non-Poisson condition yields multiscaling, or, equivalently, a very slow transition to mono-scaling, lasting forever. This important section is organized as follows. In Section VI.A we review first the Generalized Central Limit Theorem (GCLT), with heuristic arguments. This is a way of constructing a Lévy flight process. Section VI.B discusses the derivation of Lévy diffusion by means of a phenomenological GME that yields a result identical to the prescription of the GCLT. In Section VI.C we move to the more ambitious task of a Liouville approach to the GME that should generate Lévy diffusion, and we show that the resulting equation, fitting the CTRW prediction in the Poisson case, in the non-Poisson case generates a scaling departing from the observation of trajectories. The correct connection between trajectory and density perspective is recovered by enforcing the Markov approximation on the Liouville GME. As discussed in Section VI.D, the same scaling, departing from the Lévy scaling, is obtained from a Liouville approach based on the assumption that the diffusion generator fluctuation ξ is gaussian. In Section VI.E we show the CTRW in action to produce the proper derivation of the Lévy scaling. In Section VI.F we discuss the reasons why the approach to Lévy diffusion based on rare random fluctuations generates multiscaling, rather than the conventional scaling condition. In Section VI.G, finally, we draw preliminary conclusions on the first part of this review.

As a step toward the study of thermodynamic equilibrium in the case of anomalous statistical physics, in Section VII we study how the generators of anomalous diffusion respond to external perturbation. The ordinary linear response theory is violated and, in some conditions, is replaced by a different kind of linear response. In Section VIII we review the results of an ambitious attempt at deriving thermodynamics from dynamics for the main purpose of exploring a dynamic approach to the still unsettled issue of the thermodynamics of Lévy statistics. The Lévy walk perspective seems to be the only possible way to establish a satisfactory connection between dynamics and thermodynamics in this case. The interaction between system and bath, generating friction as well as diffusion, should convert the infinitely lasting regime of transition from dynamics into thermodynamics into a regime of transition occurring in a finite time. The motivations for this conclusion are reasonable even if the interaction between system and bath, generating this important condition, is not yet well-defined.

Section IX is devoted to the important task of revealing the physical reason of the breakdown of the equivalence between density and trajectory perspective in the non-Poisson case. First of all, in Section IX.A, we show that several theoretical approaches to the generalized fluctuation–dissipation process rest implicitly on the assumption that the higher-order correlation functions of a dichotomous noise are factorized. In Section IX.B we show that the non-Poisson condition violates this factorization property, thereby explaining the departure of the density from the trajectory approach in the non-Poisson case.

We now have to address the delicate issue of whether or not the density perspective can be properly adapted to studying this difficult non-Poisson condition, generating aging. In Section X we introduce aging by means of heuristic arguments. In Section XI we discuss aging with the help of a proper formalism. In Section XII we show that aging makes the evaluation of the fourth- and higher-order correlation functions extremely difficult, if not impossible, from within the density theoretical perspective.

In Section XIII we show that a GME of a given age, difficult if not impossible to build up using the density perspective, can be satisfactorily derived from the CTRW perspective. Yet, there are reasons to believe that this kind of GME is of limited utility, given the fact that the important problem of absorption and emission in the non-Poisson case requires the use of a genuinely CTRW approach. This important fact is discussed in detail in Section XIV.

At this stage, we are confident that a clear connection between Lévy statistics and critical random events is established. We have also seen that non-Poisson renewal yields a class of GME with infinite memory, from within a perspective resting on trajectories with jumps that act as memory erasers. The non-Poisson and renewal character of these processes has two major effects. The former will be discussed in detail in Section XV, and the latter will be discussed in Section XVI. The first problem has to do with decoherence theory. As we shall see, decoherence theory denotes an approach avoiding the use of wave function collapses, with the supposedly equivalent adoption of quantum densities becoming diagonal in the pointer basis set. In Section XV we shall see that the decoherence theory is inadequate to derive non-Poisson renewal processes from quantum mechanics. In Section XVI we shall show that the non-Poisson renewal properties, revealed by the BQD experiments, rule out modulation as a possible approach to complexity.

After ruling out slow modulation as a possible approach to complexity, we are left with the search for a more satisfactory approach to complexity that accounts for the renewal BQD properties. Is it possible to propose a more exhaustive approach to complexity, which explains both non-Poisson statistics and renewal at the same time? We attempt at realizing this ambitious task in Section XVII. In Section XVII.A we show that a non-Ohmic bath can regarded as a source of memory and cooperation. It can be used for a dynamic approach to Fractional Brownian Motion, which, is, however, a theory without critical events. In Section XVIII.B we show, however, that the recursion process is renewal and fits the requests emerging from the statistical analysis of real data afforded by the researchers in the BQD field. In Section XVII.C we explain why this model might afford an exhaustive approach to complexity.

Finally, in Section XVIII we discuss the main results. For the reader's convenience this section is properly divided into several subsections.

II.   MASTER EQUATION: A PHENOMENOLOGICAL APPROACH

The master equation approach has a long story tracing back to the pioneering work of Pauli [6] and van Hove [7], motivated by the need of reconciling quantum mechanical coherence with the evident randomness of statistical physics. In this section we illustrate this transition from coherent to incoherent behavior with a simple, but paradigmatic, model that will be repeatedly adopted in this review. Let us imagine a quantum system with two states, $|1\rangle$ and $|2\rangle$. The incoherent process described by the Pauli master equation is illustrated by the following set of two equations

$$\frac{d}{dt}p_1(t) = -\lambda(p_1(t) - p_2(t))$$
$$\frac{d}{dt}p_2(t) = -\lambda(p_2(t) - p_1(t)) \tag{1}$$

These two equations suggest the picture of a particle jumping from the state $|1\rangle$ to the state $|2\rangle$ and back, in such an erratic way as to force us to adopt a probabilistic picture, with $p_1(t)$ and $p_2(t)$ denoting the probability of finding at time t the particle in the state $|1\rangle$ and $|2\rangle$, respectively. The rate of transition per unit of time from one state to the other is denoted by the symbol λ. For simplicity, here and throughout the whole review we shall assume that the two states are equivalent, so that the particle spends the same amount of time in each of these two states.

On behalf of the discussion that we shall present in this review, it is convenient to rewrite the Pauli master equation of Eq. (1) in the following matrix form:

$$\frac{d}{dt}\mathbf{p}(t) = -\lambda \mathbf{K}\mathbf{p}(t) \qquad (2)$$

where

$$\mathbf{p}(t) \equiv \begin{pmatrix} p_1(t) \\ p_2(t) \end{pmatrix} \qquad (3)$$

and

$$\mathbf{K} \equiv \begin{pmatrix} 1 & -1 \\ -1 & 1 \end{pmatrix} \qquad (4)$$

How can we reconcile quantum mechanical coherence with this classical and random picture? To stress the conceptual difficulties, let us notice that a plausible quantum mechanical Hamiltonian is given by

$$H = V(|1\rangle\langle 2| + |2\rangle\langle 1|) \qquad (5)$$

The quantum mechanical wave function corresponding to this Hamiltonian, $|\psi(t)\rangle$, with the system prepared at $t = 0$ in $|1\rangle$, undergoes the following time evolution

$$|\psi(t)\rangle = |1\rangle \cos(Vt) - i|2\rangle \sin(Vt) \qquad (6)$$

thereby implying that the system is found for most of time in a superposition of the two distinct states, while according to our classical interpretation of Eq. (1) the particle is located either in $|1\rangle$ or in $|2\rangle$. Furthermore, the motion of Eq. (6) is the prototype of regular motion. As well known [8], according to the traditional wisdom, we adopt the quantum Liouville equation associated to the Hamiltonian of Eq. (5) and we assume that, due to the unavoidable existence of an environment, the off-diagonal elements of the statistical density matrix $\rho(t)$ in a very short time, τ_m, reach the equilibrium value. This process is adequately described by the following set of four equations:

$$\frac{d}{dt}\rho_{1,1}(t) = iV\left(\rho_{1,2}(t) - \rho_{2,1}(t)\right) \qquad (7)$$

$$\frac{d}{dt}\rho_{1,2}(t) = iV(\rho_{1,1}(t) - \rho_{2,2}(t)) - \frac{1}{\tau_m}\rho_{1,2}(t) \qquad (8)$$

$$\frac{d}{dt}\rho_{2,1}(t) = -iV(\rho_{1,1}(t) - \rho_{2,2}(t)) - \frac{1}{\tau_m}\rho_{2,1}(t) \qquad (9)$$

and

$$\frac{d}{d\tau}\rho_{2,2}(\tau) = -iV(\rho_{1,2}(\tau) - \rho_{2,1}(\tau)) \tag{10}$$

The assumption of fast attainment of equilibrium for the off-diagonal elements makes it possible for us to set $d\rho_{1,2}(t)/dt = 0$ in Eq. (8) and $d\rho_{2,1}(t)/dt = 0$ in Eq. (9). This is the famous Smoluchowski approximation [9]. The off-diagonal density elements play the same role as the velocity of the Brownian particle [9], and their time derivatives are assumed to vanish. This allows us to express $\rho_{1,2}$ and $\rho_{2,1}$ as a function of the diagonal density matrix elements,

$$\rho_{1,2}(t) = iV\tau_m(\rho_{1,1}(t) - \rho_{2,2}(t)) \tag{11}$$

and

$$\rho_{2,1}(t) = iV\tau_m(\rho_{2,2}(t) - \rho_{1,2}(t)) \tag{12}$$

By plugging Eq. (11) and Eq. (12) into both Eq. (7) and Eq. (10), we derive Eq. (1) with

$$\lambda = 2V^2\tau_m \tag{13}$$

Note that according to the decoherence theory [10–12] the dephasing process is equivalent to assessing, through a measurement process, if the system is in the state $|1\rangle$ or in the state $|2\rangle$, thereby forcing the wave function to collapse either in $|1\rangle$ or in $|2\rangle$, with probability proportional to the modulus of expansion coefficients. The inverse of de-phasing time τ_m is interpreted as an indicator of the frequency of this measurement process. We see from Eq. (13) that the shorter the τ_m, namely, the higher the measurement frequency, the larger the relaxation time $1/\lambda$. This effect is well known as the quantum Zeno effect [13]. For a recent review on the subject, we recommend the excellent article in Ref. 14.

Of course the diagonal density elements $\rho_{1,1}(t)$, $\rho_{2,2}(t)$ are identified with the probabilities $p_1(t)$ and $p_2(t)$, respectively. One of the major tasks of this review is to point out the difficulties that are currently met with the extension of this interpretation to the case where the Markov condition of the ordinary Pauli master equation does not apply.

III. THE GENERALIZED MASTER EQUATION, AND THE ZWANZIG PROJECTION METHOD

This section has the twofold purpose of illustrating a popular approach to GME by means of a contracted Liouville equation. This is the approach proposed by Zwanzig [15,16]. We plan also to outline some technical difficulties with this approach, when the Markov approximation is not allowed. This refers to the case when either an infinite or a vanishing diffusion coefficient would emerge from the treatment. The second case, of a vanishing diffusion coefficient, is discussed in the second subsection of this section.

A. The Projection Method

The approach proposed by Zwanzig [15,16] for a formal derivation of the master equation proceeds as follows. We assume the whole Universe, namely, the system of interest and its environment, to obey the Liouville equation

$$\frac{d}{dt}\rho = L\rho(t) \tag{14}$$

We define a convenient projection operator P, which projects the total density matrix $\rho(t)$ into the part of interest $\rho_1(t)$, defined by

$$\rho_1(t) \equiv P\rho(t) \tag{15}$$

The complementary projection operator $Q \equiv 1 - P$ serves the purpose of defining the irrelevant part of the total density matrix, $\rho_2(t)$, through

$$\rho_2 \equiv Q\rho(t) \tag{16}$$

The two projection operators are chosen to be time-independent. Thus they commute with the differential operator on the left-hand side of Eq. (14). As a consequence, by applying to Eq. (14) both the operator P and the operator Q, and by applying the property $\rho(t) = (P + Q)\rho(t) = \rho_1(t) + \rho_2(t)$ as well, we split this equation into the following coupled equations

$$\frac{d}{dt}\rho_1(t) = PL(\rho_1(t) + \rho_2(t)) \tag{17}$$

and

$$\frac{d}{dt}\rho_2(t) = QL(\rho_1(t) + \rho_2(t)) \tag{18}$$

The formal solution of Eq. (18) is

$$\rho_2(t) = \int_0^t dt' \exp[QL(t - t')]QL\rho_1(t') + \exp(QLt)\rho_2(0) \tag{19}$$

By plugging Eq. (19) into Eq. (17), we obtain

$$\frac{d}{dt}\rho_1(t) = PL\rho_1(t) + \int_0^t dt' PL\exp[QL(t-t')]QL\rho_1(t') + PL\exp(QLt)\rho_2(0)$$

(20)

Equation (20) is the central result of the Zwanzig projection method, and it is one of the two theoretical tools under scrutiny in this chapter, the first being the Generalized Master Equation (GME), of which Eq. (20) is a remarkable example, and the second being the Continuous Time Random Walk (CTRW) [17]. It must be pointed out that to make Eq. (20) look like a master equation, it is necessary to make the third term on the right-hand side of it vanish. To do so, the easiest way is to make the following two assumptions:

$$\rho_2(0) = 0$$

(21)

and

$$PLP = 0$$

(22)

With these two assumptions, and using the properties $P\rho_1 = \rho_1$, which is a natural consequence of $P^2 = P$, we turn Eq. (20) into

$$\frac{d}{dt}\rho_1(t) = \int_0^t dt' K(t-t')\rho_1(t')$$

(23)

where the operator $K(t)$ is defined by

$$K(t) = PL\exp(QLt)QLP$$

(24)

The assumption of Eq. (22), as we shall see, has the effect of canceling possible drifts, which would make our discussion less transparent with no special physical benefit for physical understanding. The assumption of Eq. (21) is much more significant for the main purpose of this review. To appreciate this important fact, let us rewrite Eq. (23) in the equivalent form

$$\frac{d}{dt}\rho_1(t) = \int_0^t dt' K(t')\rho_1(t-t')$$

(25)

which shows that the time evolution of $\rho_1(t)$ depends on the earlier time evolution of it. This is the non-Markov property of the GME, conflicting with the Markov character of the Pauli master equation [see Eq. (2)]. To recover the

Markov structure of the ordinary master equation, we replace $\rho_1(t - t')$ with $\rho_1(t)$ and set the upper time integration limit equal to infinity, thereby obtaining

$$\frac{d}{dt}\rho_1(t) = \Lambda\rho_1(t) \tag{26}$$

with the Markov operator Λ defined by

$$\Lambda \equiv \int_0^\infty dt' K(t') \tag{27}$$

We observe the emergence of a first problem here. To make the Markov approximation, we require that the decay of the operator $K(t)$ be fast enough. An inverse power law with index $\beta < 1$ would be incompatible with the Markov approximation. On the other hand, the direct use of Eq. (25) would imply a dependence of the time evolution of $\rho_1(t)$ on the initial condition. It is important to stress again that Eq. (23) implies the fundamental assumption of Eq. (21), which, in turn, rests on a special preparation of the system. In other words, the non-Markov equation of Eq. (23) is not a *bona fide* master equation, insofar as we have always to refer the corresponding physical process to the special preparation adopted at $t = 0$. It is worth quoting the remarkable work of Fox [18], who discussed this important issue to establish the proper Fokker–Planck equation corresponding to the generalized Langevin equation. We shall see that this sets a limit to the benefits afforded by the GME, even when we do succeed in overcoming the conceptual difficulties posed by their foundation, in the non-Markov case. It is important to stress that in the non-Poisson case this initial preparation is closely related to the system age, and it is done to study the relaxation of the system subsequent to that specific preparation. Let us suppose that this initial preparation corresponds to the brand new system. This is the GME of age $t_a = 0$. However, while relaxing, the system keeps aging. This has remarkable consequences. If at later time, $t_a > 0$, a further perturbation is applied to the system so as to observe the corresponding new response, a new GME must be built up, insofar as perturbing the GME of age $t_a = 0$ would lead to incorrect prediction.

A second problem with the GME derived from the contraction over a Liouville equation, either classical or quantum, has to do with the correct evaluation of the memory kernel. Within the density perspective this memory kernel can be expressed in terms of correlation functions. If the linear response assumption is made, the two-time correlation function affords an exhaustive representation of the statistical process under study. In Section III.B we shall see with a simple quantum mechanical example, based on the Anderson localization, that the second-order approximation might lead to results conflicting with quantum mechanical coherence.

B. Anderson Localization

In order to prepare the ground for the central discussion of this review, it is worth reviewing here the main results of the article in Ref. 19. Let us imagine that the system is driven by the Hamiltonian

$$H = H_0 + W \tag{28}$$

where

$$H_0 = \sum_m E_m |m\rangle \langle m| \tag{29}$$

and

$$W = V \sum_m (|m\rangle \langle m+1| + |m+1\rangle \langle m|) \tag{30}$$

In the special case where the site energies are random fluctuations, this is the Anderson model [20,21]. It is well known that Anderson used this model to prove that randomness makes a crystal become an insulating material. Anderson localization is subtly related to subdiffusion, and consequently this important phenomenon can be interpreted as a form of anomalous diffusion, in conflict with the Markov master equation that is frequently adopted as the generator of ordinary diffusion. It is therefore surprising that this is essentially the same Hamiltonian as that adopted by Zwanzig for his celebrated derivation of the van Hove and, hence, of the Pauli master equation.

We notice that in the case of only two sites, the Hamiltonian here under study becomes similar to that illustrated in Section II. This is essential to understand the relations between the Anderson and the Zwanzig point of view. A realistic treatment of the Anderson model would imply the use of an infinite chain of states, through which, according to the Zwanzig perspective, the electron would move as a random walker. Thus, to a first sight it might be surprising that, along the lines of Ref. 19, we reach our main conclusions with the two-site version of this model. Note that we are considering a Gibbs ensemble of identical systems, with a random distribution of site energies. Thus, with only two sites, the energies of the states $|1\rangle$ and $|2\rangle$, rather than being identical, as assumed in Section II, fluctuate around the same mean value. The fluctuations that we are studying in this case do not depend on time, as they do in the case of the model of Section II but on the system of the Gibbs ensemble considered. This means that we are considering a set of systems, some of which undergo the coherent process of Eq. (6). However, the majority of the systems are characterized by a different behavior. For instance, those where the energy of $|2\rangle$ is much higher than the energy of $|1\rangle$ are characterized by a time evolution where the wave function never significantly departs from $|1\rangle$, if $|\psi(0)\rangle = |1\rangle$. As a consequence, if the Anderson fluctuations are very intense, and all the systems are prepared in the same initial condition $|\psi(0)\rangle = |1\rangle$, the Gibbs system remains virtually localized in the state $|1\rangle$. This is not as trivial as it might seem at first sight. In fact, the statistical average over the infinitely many two-site systems of the Gibbs ensemble corresponds to the statistical average that we should carry out on a single realization with infinitely many sites. Our purpose is to adapt to this simple system the Zwanzig prescription; this, as we shall see, sheds light into the intriguing issue of why the same Hamiltonian can be used to produce either ordinary (Zwanzig) or anomalous (Anderson) statistical physics.

We assume that

$$E_m = \epsilon + \phi_m \tag{31}$$

As mentioned earlier, we are assuming that with changing site there is a fluctuation ϕ_m around a common value. We assume no correlation among different sites, namely

$$\langle \phi_m \phi_{m'} \rangle = A \delta_{m,m'} \tag{32}$$

Then we assume that the fluctuating quantity ϕ is drawn from the Cauchy prescription

$$p(\phi) = \frac{1}{\pi} \frac{\gamma}{\gamma^2 + \phi^2} \tag{33}$$

which yields

$$\langle \phi \rangle = 0 \tag{34}$$

At this stage we have to define a convenient form of projection operator. We denote this projection operator with the symbol Π rather than P to stress the fact that the choice we adopt is inspired to that made by Kenkre [22] to study, as we do here, a Gibbs system. We have to stress, in fact, that in the present case the statistical density matrix under study is

$$\rho_G = \int d\{\phi\} w(\{\phi\}) \rho(\{\phi\}, t) \tag{35}$$

The symbol $\rho(\{\phi\}, t)$ denotes the ordinary statistical density matrix, namely a set of Gibbs systems with the same random distribution of site energies. Moving from one statistical density matrix to another is equivalent to establishing a new random distribution of site energies and to considering an ensemble of systems with the same random distribution. The symbol $w(\{\phi\})$ is the weight of these distribution. To be more precise,

$$w(\{\phi\}) = \ldots p(\phi_{m-1})p(\phi_m)p(\phi_{m+1}) \ldots \tag{36}$$

Here we have to face two distinct ways of getting the same result. The first is based on assuming that the density matrix here under study is not $\rho(t)$, but it is $\rho_G(t)$ of Eq. (35). In this case we define the projection operator P as follows:

$$P\rho_G \equiv \sum_m |m\rangle\langle m|\rho_G|m\rangle\langle m| \tag{37}$$

This is identical to the projection operator adopted by Zwanzig [16]. The second possibility rests on the adoption of the ordinary statistical density matrix $\rho(t)$. In this case, to obtain the same result we must use a different projection operator, denoted by the symbol Π and defined by

$$\Pi\rho \equiv \int w(\{\phi\})d\{w\{\phi\}\} \sum_m |m\rangle\langle m|\rho\{\phi\}|m\rangle\langle m| \tag{38}$$

This operator is of the same kind as that adopted by Kenkre [22], namely, the Zwanzig projection operator applied to a coarse grained density matrix.

Note that in the present case the condition of Eq. (22) is ensured by Eq. (34). To fulfill the condition of Eq. (21) we assume that at the initial time the electron is localized in one of the sites. Thus, we obtain Eq. (23) that in this specific case reads:

$$\frac{d}{dt}p_n(t) = -\sum_{m \neq n} \int_0^t dt' \Xi_{m,n}(t - t')[p_n(t') - p_m(t')] \tag{39}$$

where

$$p_m(t) \equiv \langle m|\rho_G(t)|m\rangle \tag{40}$$

and

$$\Xi_{m,n}(t - t') \equiv \langle n|\{\Pi L \exp[(1 - \Pi)L(t - t')](1 - \Pi)L|m\rangle\langle m|\}|n\rangle \tag{41}$$

with

$$L\rho \equiv -\frac{i}{\hbar}[H, \rho] \tag{42}$$

H being the Hamiltonian of Eq. (28). From now on, for the sake of simplicity, we set $\hbar = 1$. Note that the contribution W of Eq. (30) can be judged small or large by comparison with the energy difference between two consecutive sites. This, in turn, is determined by the random distribution of energies, whose size is γ. We set the condition

$$V \ll \gamma \tag{43}$$

This condition means Anderson noise of large intensity, and, as we have seen, W is a weak perturbation. Note that on the extreme left and extreme right of the second term of Eq. (41) we have $\Pi L \ldots = -i\Pi[W, \ldots]$ and $(1 - \Pi)L \ldots = -i(1 - \Pi)[W, \ldots]$. This means that the second term of Eq. (39) is of second order. We aim at illustrating the consequence of making a second-order approximation. To keep our treatment at the second perturbation order, we neglect the perturbation appearing in the exponential of Eq. (41). This makes the calculation of the memory kernel very easy. Using the Cauchy distribution of Eq. (33), we obtain

$$\Pi L \exp[(1 - \Pi)L(t - t')](1 - \Pi)|m\rangle\langle m| = 2V^2 \exp[-2\gamma|t - t'|]$$
$$\times \left(|m + 1\rangle\langle m + 1| + |m - 1\rangle\langle m - 1| - 2|m\rangle\langle m|\right) \tag{44}$$

Let us plug this result in Eq. (39). The exponential decay of the kernel is compatible with the Markov approximation, which yields

$$\frac{d}{dt}p_n(t) = -\frac{2V^2}{\gamma}p_n(t) + \frac{V^2}{\gamma}p_{n+1}(t) + \frac{V^2}{\gamma}p_{n-1}(t) \tag{45}$$

In the two-site case we recover again Eq. (1) with

$$\lambda = V^2/\gamma \tag{46}$$

We would be tempted to conclude that the Anderson noise plays the same role as the dephasing producing environmental fluctuations of Section II. This conclusion would not be totally wrong. In fact, the work of Ref. 19 proves that the lack of Anderson localization emerging from Eq. (45) is due to the second-order approximation. A re-summation up to infinite order yields a memory kernel with a fast exponential decay identical to the one emerging from the second-order approximation followed by a weak but persistent negative tail that is responsible for subdiffusion, and localization as well. The authors of Ref. 19 prove that an external fluctuation has the effect of killing this negative tail. The stronger the Anderson noise, the weaker the intensity of the external fluctuation necessary to kill the negative tail. This result is of some interest insofar as it proves that, although the Markov approximation yields correct results, it might be inconsistent with the exact solution of the model under study. Thus the Markov approximation may correspond to the tacit assumption that ingredients foreign to the model under study enter into play. It is remarkable that the Markov approximation might turn out to be inconsistent with the claim that a rigorous quantum mechanical treatment is adopted, while a tacit violation of quantum mechanics is assumed instead [23]. These are aspects of the GME and of the derivation of the Pauli master equation from it that will be taken into account in the final discussion of this review.

On the other hand, the adoption of the Markov approximation, although yielding no mathematical inconsistencies, might correspond to annihilating important physical effects. Let us illustrate this important fact with a process related to the Anderson localization issue. Let us depict the Anderson localization as an ordinary fluctuation–dissipation process. Let us consider the Langevin equation

$$\frac{d}{dt}x = -\lambda x(t) + f(t) \tag{47}$$

The symbol $f(t)$ denotes a white gaussian noise, with the correlation function

$$\langle f(t_1)f(t_2)\rangle = 2D\delta(|t_1 - t_2|) \tag{48}$$

This Langevin equation fulfills the fluctuation–dissipation relation

$$\langle x^2\rangle_{eq} = \frac{D}{\lambda} \tag{49}$$

which sets a limit to the free diffusion of the variable x. The formal solution of Eq. (47) is

$$x(t) = \int_0^t \exp(-\lambda(t - t'))f(t')dt' + \exp(-\lambda t)x(0) \tag{50}$$

Let us explore the case of free diffusion from the equilibrium condition $x(0) = 0$. Let us rewrite Eq. (47) as follows:

$$\frac{d}{dt}x = \xi(t) \tag{51}$$

where $\xi(t)$ is a fluctuating variable defined by

$$\xi(t) = -\lambda \int_0^t dt' \exp(-\lambda(t - t')) + f(t) \tag{52}$$

It is evident that this is a way of rewriting the exact solution of Eq. (47). However, it is interesting to recover the fluctuation–dissipation prediction from a perspective that might lead to a free diffusion with no upper limit if an error is made that does not take into account the statistical properties of the fluctuation $\xi(t)$. Let us evaluate the correlation function of $\xi(t)$. Using the property of Eq. (48) and moving to the asymptotic time limit reflecting the microscopic equilibrium condition, we obtain

$$\langle \xi(t_1)\xi(t_2)\rangle = 2D\delta(|t_1 - t_2|) - D\lambda \exp(-\lambda|t_1 - t_2|) \qquad (53)$$

which is an idealization of the more realistic condition

$$\Phi_\xi(|t_1 - t_2|) \equiv \frac{\langle \xi(t_1)\xi(t_2)\rangle}{\langle \xi^2\rangle} = \frac{\Lambda \exp(-\Lambda|t_1 - t_2|) - \lambda \exp(-\lambda|t_1 - t_2|)}{(\Lambda - \lambda)} \qquad (54)$$

with $\Lambda \gg \lambda$. We see that the negative tail preventing us from doing the Markov approximation is an important physical property, responsible for the realization of the correct thermodynamic equilibrium pertaining to the ordinary fluctuation–dissipation condition.

This is an important fact that deserves some additional comments. The first is that the apparent conflict between Anderson and Zwanzig can be resolved by invoking, as usually done by the advocates of decoherence theory [24], the interaction with an external environment [25–29]. In fact, as pointed out in Ref. 19, the stronger the Anderson randomness, the weaker the intensity of environmental fluctuation necessary to inhibit localization. This argument is used to justify the correspondence principle [30], in the paradigmatic case of deterministic chaos. The quantum kicked rotor is known to yield localization due to the same cause as the one producing Anderson localization, namely, quantum coherence. Environmental fluctuations destroy quantum coherence and localization with it. These conditions are very reassuring insofar as they explain the transition from the quantum to the classical world [24]. However, we want to point out that the environmental fluctuations responsible for quantum decoherence are ordinary, and the deterministic chaos of the system of interest corresponds to a condition of total randomness. In this review we want to show that departing from the conditions of ordinary statistical physics has the effect of generating conditions that force us to question the arguments used by the advocates of decoherence theory. This is a reassuring perspective confirming the validity of the correspondence principle and, thus, of quantum mechanics itself, as a universal theory that requires neither changes nor extensions. We are convinced that this is limited to the case of ordinary statistical processes. For instance, if the kicked quantum rotor is studied in the so-called accelerating state [31], the quantum system generates a process of quantum-induced ordinary diffusion, and this property violates the correspondence principle. This is so because, as we shall see in Section XV, classical physics would produce superdiffusion and the uncorrelated environmental fluctuations produce normal diffusion. In other words, the medicine would produce the same effect as the disease. On the other hand, in this special condition, the quantum coherence provoking the breakdown of the correspondence principle becomes robust against environmental fluctuations, thereby making the medicine proposed by the advocates of decoherence theory ineffective to cure this quantum mechanical disease.

It is well known that the phenomenon of Anderson localization can be broken, or quenched, by internal correlation, namely correlation among the energy fluctuations ϵ_m of the tight-binding Hamiltonian of Eq. (28) [32–34]. To the best of our knowledge, it seems that this important research work establishes the onset of a localization enhancement, without assessing the nature of the ensuing diffusion process. We note also that the correlation produced by a DNA sequence [35] has the effect of significantly enhancing the localization length, and the kind of correlation and memory established is an effect of the non-Poisson but renewal conditions that are the main focus of this review. Therefore, it seems to be plausible that the inhibition of Anderson localization might be studied with the same perspective as that recently adopted to deal with the phenomenon of roughness saturation emerging from the random growth of surfaces. This anomalous phenomenon has been studied [36] adopting the new perspective of subordination to an ordinary fluctuation–dissipation process rather than the conventional generalized fluctuation–dissipation relations. On the basis of these arguments, we expect the DNA correlation to yield the same effect as the quantum kicked rotor, an effect robust against external fluctuations and thus requiring the adoption of the new perspective that we aim at illustrating in this review.

C. Conclusions

In conclusion, in this section we have proved that the Markov approximation requires some caution. The Markov approximation may be incompatible with the quantum mechanical nature of the system under study. It leads to the Pauli master equation, and thus it is compatible with the classical picture of a particle randomly jumping from one site to another, a property conflicting, however, with the rigorous quantum mechanical treatment, which yields Anderson localization.

Both the case where the Laplace transform of $K(t)$ of Eq. (24) diverge (superdiffusion) or vanish (subdiffusion) must be treated with caution. These conditions will be the main subject under study in this review. The existence of environment fluctuations makes it possible for us to interpret the electron transport as resulting from random jumps, without involving the notion of wave-function collapse, but this is limited to the case of Poisson statistics. Anderson randomness, with the ensuing localization, seems to conflict with this classical picture. However, in the case of uncorrelated and strong Anderson noise, in spite of the fact that it means quantum coherence, Anderson localization, equivalent to an ordinary fluctuation–dissipation process, can be easily broken by external fluctuations. We are convinced that in the case of anomalous statistical mechanics the quantum effects become more robust against external fluctuation, thereby making unavoidable the search for a more convenient perspective to explain why quantum mechanics does not emerge at a macroscopic level. We shall come back to this issue in Section XV.

In this review we show that there are two main sources of memory. One of them correspond to the memory responsible for Anderson localization, and it might become incompatible with a representation in terms of trajectories. The fluctuation–dissipation process used here to illustrate Anderson localization in the case of extremely large Anderson randomness is an idealized condition that might not work in the case of correlated Anderson noise. On the other hand, the non-Poisson renewal processes generate memory properties that may not be reproduced by the stationary correlation functions involved by the projection approach to the GME. Before ending this subsection, let us limit ourselves to anticipating the fundamental conclusion of this review: The CTRW is a correct theoretical tool to address the study of the non-Markov processes, if these correspond to trajectories undergoing unpredictable jumps.

IV. THE CONTINUOUS-TIME RANDOM WALK

It is interesting to notice that formally Eq. (39) can be written in the following matrix form:

$$\frac{d}{dt}\mathbf{p}(t) = -\int_0^t dt'\mathbf{K}(t-t')\cdot\mathbf{p}(t') \qquad (55)$$

The symbol $\mathbf{K}$ denotes the matrix

$$\mathbf{K}_{m,n}(t-t') \equiv \Xi_{m,n}(t-t') - \delta_{mn}\sum_{m'\neq n}\Xi_{m'n}(t-t') \qquad (56)$$

with $\Xi_{m,n}$ given by Eq. (41); the symbol $\mathbf{p}$ denotes a vector of infinite size, whose components are given by the probability of finding the particle in a given site, $\mathbf{p}_m = p_m$. The symbol $\mathbf{K}\cdot\mathbf{p}$ denotes the usual product row by column.

In the continuous limit, Eq. (39) reads

$$\frac{\partial}{\partial t}p(x,t) = \int_0^t dt' \int_{-\infty}^{\infty} dx'K(x-x',t-t')p(x',t') \qquad (57)$$

where

$$K(x,t) = \Pi(x,t) - \delta(x) \int_{-\infty}^{\infty} \Pi(x',t)dx' \qquad (58)$$

We shall discuss the GME, written in this form in Section VI. It is worth mentioning that the GME with this form has been introduced by Kenkre et al. [37].

In the special case where the time and space transitions are not correlated, Eq. (55) becomes

$$\frac{d}{dt}\mathbf{p}(t) = -\int_0^t dt' \Phi(t - t')\mathbf{K} \cdot \mathbf{p}(t') \qquad (59)$$

where the memory function $\Phi(t)$ takes care of the time delay and $\mathbf{K}$ is the matrix responsible for the transitions among sites.

Let us illustrate now the second of the two main theoretical tools under discussion in this review, this being the Continuous Time Random Walk (CTRW) proposed many years ago by Montroll and Weiss [17]. According to the CTRW we write

$$\mathbf{p}(t) = \sum_{n=0}^{\infty} \int_0^t dt' \psi^{(n)}(t') \Psi(t - t')\mathbf{M}^n \cdot \mathbf{p}(0) \qquad (60)$$

It is important to explain the meaning of this equation. First of all, let us point out that $\mathbf{p}(0)$ denotes the initial random walker distribution. The main purpose of this treatment is to determine the distribution at time t, $\mathbf{p}(t)$, as a function of the initial condition $\mathbf{p}(0)$. Equation (60) serves this important purpose as follows. It determines $\mathbf{p}(t)$ through the occurrence of random events. A random event corresponds to the occurrence of jumps from one site to other sites, with a probability described by the matrix $\mathbf{M}$. The times of occurrence of these jumps are described by the function $\psi^{(n)}(t)$. When we observe the system at time t, the distribution $\mathbf{p}$ is the result of an arbitrary number n of these random events. The function $\psi^{(n)}(t)$ denotes the probability that at time t, and exactly at time t, the nth of a sequence of n random events occurs. Of course, we have also to consider the case when no event occurs, this being expressed by $\psi^{(0)}(t) = \delta(t)$. It is evident that

$$\psi^{(n)}(t) = \int_0^t dt' \psi^{(n-1)}(t')\psi^{(1)}(t - t') \qquad (61)$$

The time convoluted nature of this connection allows us to express the Laplace transform of $\psi^{(n)}(t)$, $\hat{\psi}^{(n)}(u)$, in terms of the Laplace transform of $\psi^{(1)}(t)$. For simplicity the sake of we omit the superscript 1, and we adopt the convention $\psi(t) \equiv \psi^{(1)}(t)$, $\hat{\psi}(u) = \hat{\psi}^{(1)}(u)$. Thus, we obtain

$$\hat{\psi}^{(n)}(u) = [\hat{\psi}(u)]^n \tag{62}$$

Note that the last of a sequence of n random events does not necessarily occur at time t, at which we observe the system. It might occur at an earlier time $t' < t$. Thus, for our prescription to become reliable, we have to set the constraint that no event occurs between time t' and time t. In general, the probability that no event occurs up to a given time t is given by the function $\Psi(t)$ defined by

$$\Psi(t) \equiv \int_t^\infty \psi(t')dt' \tag{63}$$

The Laplace transform of $\Psi(t)$, $\hat{\Psi}(u)$, is related to $\hat{\psi}(u)$, through

$$\hat{\Psi}(u) = \frac{1 - \hat{\psi}(u)}{u} \tag{64}$$

Note that in between t', the time at which the nth of a chain of n events occurs, and the observation time t, the distribution keeps the value $\mathbf{M}^n$, the matrix $\mathbf{M}$ being the prescription establishing the jumps that the random walkers make when an event occurs. Since n events occurred, this matrix is applied n times to the initial condition $\mathbf{p}(0)$.

Using Eqs. (62) and (64) and the property $\sum_{n=0}^\infty \hat{\psi}^n(u) = 1/(1 - \hat{\psi}(u))$, we write the Laplace transform of $\mathbf{p}(t)$, $\hat{\mathbf{p}}(u)$, of Eq. (60) as follows:

$$\hat{\mathbf{p}}(u) = \frac{1}{u + \hat{\Phi}(u)(\mathbf{M} - 1)}\mathbf{p}(0) \tag{65}$$

where $\hat{\Phi}(u)$, the Laplace transform of $\Phi(t)$, is related to the Laplace transform of $\psi(t)$, through

$$\hat{\Phi}(u) = \frac{u\hat{\psi}(u)}{1 - \hat{\psi}(u)} \tag{66}$$

On the other hand, we see that the Laplace transform of Eq. (59) yields

$$\hat{\mathbf{p}}(u) = \frac{1}{u + \hat{\Phi}(u)\mathbf{K}}\mathbf{p}(0) \tag{67}$$

thereby proving that the CTRW is equivalent to the GME of Eq. (59), with the condition

$$\mathbf{M} = \mathbf{K} + 1 \tag{68}$$

It is important to discuss the consequence of assuming

$$\psi(t) = \lambda \exp(-\lambda t) \tag{69}$$

By Laplace transforming $\psi(t)$ and using Eq. (66), we obtain

$$\hat{\Phi}(u) = \lambda \tag{70}$$

which is the Laplace transform of

$$\Phi(t) = \lambda \delta(t) \tag{71}$$

Thus, Eq. (55) becomes identical to Eq. (2), thereby recovering an important result obtained many years ago [38]: A Poisson process, with $\psi(t)$ given by an exponential function of t, yields a Markov master equation.

Note that the formal equivalence between Eq. (59) and the GME stemming from the adoption of the Zwanzig method, for example, Eq. (23), suggests that $\Phi(t)$ is a function proportional to a correlation function. For instance, the memory kernel of Eq. (41), through the second-order approximation becomes the correlation function of the quantum coupling V, fluctuating as a consequence of the erratic value of the site energies. In the non-Poisson case, this correspondence is lost. In Section XVIII we shall discuss a non-Poisson case where $\Phi(t)$ is not proportional to an equilibrium correlation function, but it is related through a more complex expression, Eq. (323), to a non-stationary correlation function.

V. AN INTERMITTENT DYNAMIC MODEL

In the special case where there are only two sites, the CTRW procedure, supplemented by the Poisson assumption of Eq. (69), yields the Pauli master equation of Eq. (2). This means that the Pauli master equation is compatible with a random picture, where a particle with erratic motion jumps back and forth from the one to the other state, with a condition of persistence expressed by the exponential waiting time distribution of Eq. (69). Recent fast technological advances are allowing us to observe in mesoscopic systems analogous intermittent properties, with distinct nonexponential distribution of waiting times. This is the reason why in this section we focus our attention on how to derive a $\psi(t)$ with a non-Poisson character.

A. Experimental Versus Theoretical Laminar Regions

The dynamic model that we use was originally introduced to study turbulence, a phenomenon characterized by long times of sojourn in a condition of regular (laminar) motion interrupted by abrupt burst of randomness. Note that this idealized turbulence model will be adopted to generate a sequence of time lengths $\{\tau_i\}$, dividing the time axis into intervals that we refer to as *theoretical laminar regions*. This model is adopted as a dynamical prescription to generate a sequence of "light-on" and "light-off" states, with a non-Poisson statistics. We make the assumption that the experimental sequence of "light-on" and "light-off" states is obtained by assigning to each theoretical laminar region the sign $+$ ("light-on") or the sign $-$ ("light-off"), according to a coin tossing prescription. The time intervals with the same signs are denoted as *experimental laminar regions*. Of course a given experimental laminar region might contain more than one theoretical laminar regions, to which the coin tossing prescription has assigned the same sign. We shall adopt the same prescription for the dynamic generation of Lévy statistics. We shall generate a fluctuating velocity $\xi(t)$, with the value W, corresponding to the experimental "light-on" laminar region, and the value $-W$, corresponding to the experimental "light-off" laminar region.

We keep the symbol $\psi(t)$ to denote the distribution of time durations of the theoretical laminar regions, and we adopt the symbol $\psi_{\exp}(t)$ to denote the distribution of time duration of the experimental laminar regions. It is possible to relate $\psi(t)$ to $\psi_{\exp}(t)$. It is evident that the experimental is derived from the theoretical distribution by means of the expression

$$\psi_{\exp}(t) = \sum_{n=1}^{\infty} \psi_n(t) \frac{1}{2^n} \tag{72}$$

In fact, infinitely many random events might have occurred, but only with a probability $1/2$ of generating the same symbol, for instance W. Taking into account $\hat{\psi}_n(u) = \hat{\psi}(u)^n$, we get

$$\hat{\psi}_{\exp}(u) = \frac{\hat{\psi}(u)}{2 - \hat{\psi}(u)} \tag{73}$$

The connection between experimental and theoretical waiting time distribution, established by Eq. (73) allows us to discuss the master equation of Section II from within the CTRW perspective. In Section IV we have studied the motion of a random walker jumping from one site to another of an infinitely extended lattice. However, the formalism adopted can be applied also to the case of only two sites, $|1\rangle$ and $|2\rangle$, which correspond to the physical condition studied in Section II. In this case the matrix $\mathbf{K}$ of Eq. (59) becomes identical to the matrix defined by Eq. (4). Let us define the function

$$\Pi(t) = p_1(t) - p_2(t) \tag{74}$$

Using Eq. (59) we obtain

$$\frac{d\Pi(t)}{dt} = -\int_0^t d\tau \Phi(t-\tau)\Pi(\tau) \tag{75}$$

whose Laplace transform yields for $\hat{\Pi}(u)$, the Laplace transform of $\Pi(t)$, the following equation

$$\hat{\Pi}(u) = \frac{1}{(u + 2\hat{\Phi}(u))} \tag{76}$$

Note that according the notation established in this subsection, we have to write $\hat{\Phi}(u)$ of Eq. (66) under the following form

$$\hat{\Phi}(u) = \frac{u\hat{\psi}_{exp}(u)}{1 - \hat{\psi}_{exp}(u)} \tag{77}$$

In fact, the waiting function of sojourn times in one site must be identified, according to the definitions of this subsection, with the experimental waiting time distribution. On the other hand, using Eq. (73), we obtain

$$\frac{u\hat{\psi}_{exp}(u)}{1 - \hat{\psi}_{exp}(u)} = \frac{1}{2}\left[\frac{u\hat{\psi}(u)}{1 - \hat{\psi}(u)}\right] \tag{78}$$

Thus, Eq. (76) becomes

$$\hat{\Pi}(u) = \frac{1 - \hat{\psi}(u)}{u} = \Psi(u) \tag{79}$$

To obtain this interesting result we used also Eq (64). In conclusion the function $\Pi(t)$ coincides with the function $\Psi(t)$ of Eq. (63). It is natural to call this quantity *survival probability*.

It is important to stress that in the Poisson case, the experimental waiting distribution is an exponential function. Let us assign to it the form

$$\psi_{exp}(t) = \frac{r}{2}\exp\left(-\frac{r}{2}t\right) \tag{80}$$

From Eq. (73), we derive

$$\psi(t) = r\exp(-rt) \tag{81}$$

Thus the survival probability $\Psi(t)$ of Eq. (63) becomes

$$\Psi(t) = \exp(-rt) \tag{82}$$

In the case discussed in Section II, as we have seen, $\lambda = 2V^2\tau_m$. According to the notation of this subsection $r/2 = \lambda$. Thus, we find

$$\Psi(t) = \exp(-4V^2\tau_m t) \tag{83}$$

which is the survival probability of the Pauli master equation of Eq. (1). As explained in Section XV, according to the spirit of decoherence theory a statistical agreement between a quantum mechanical picture and a classical stochastic picture, has to be interpreted as a complete equivalence. Thus, the equivalence between CTRW and GME means that the treatment of Section II derives from quantum mechanics a stochastic trajectory jumping from one state to the other, with the exponential waiting time distribution of Eq. (81). This is a striking conclusion, given the fact that in quantum mechanics there are no trajectories. From an intuitive point of view, we can interpret this equivalence as corresponding to the occurrence of quantum mechanical collapses of the wave function, triggered by the measurement process mimicked by the dephasing process, acting in Eqs. (8) and (9).

B. Manneville Map and Its Idealized Version

We have now to illustrate the dynamic model used to derive $\psi(t)$. It rests on the physics of intermittent processes, namely, on a model adopted to account for turbulence [39]. This is a popular prototype of deterministic approach to turbulence given by the Manneville map [40]. This map reads

$$y_{n+1} = y_n + y_n^z, \quad \text{mod } 1 \tag{84}$$

with $z \geq 1$. This map consists of two regions separated by the value d, defined by $d + d^z = 1$. The region $1 \geq y \leq d$ and $y > d$ define the laminar and the chaotic state, respectively. If we run this map, we obtain a trajectory that spends a large amount of time in the laminar region, and only a few steps in the chaotic region, thereby mimicking a turbulent fluid that is characterized by regular motion interrupted by unpredictable bursts of disordered evolution. An idealized version of this celebrated map is given by the following model

$$\frac{dy}{d\tau} = \alpha y^z \tag{85}$$

with $z \geq 1, 0 < \alpha \ll 1$. The trajectory moves always from the left to the right within the interval $I = [0, 1]$. When the trajectory attains the boundary $y = 1$, it is injected back to a new initial condition y_0, selected randomly and with uniform probability among the positive real numbers smaller than the unity. Throughout this review we use this prescription to generate the sequence $\{\tau_i\}$ that will help us to illustrate in general the central issue of the comparison between GME and CTRW. This is the sequence of sojourn times within the interval I.

We record the time t_{i-1} of a back injection and the time t_i of the next back injection and we define

$$\tau_i \equiv t_i - t_{i-1} \tag{86}$$

The distribution density $\psi(\tau)$ satisfies the request of defining the probability dp_τ of finding a sojourn time located in the infinitesimally small interval $\tau, \tau + d\tau$, through the prescription

$$dp_\tau = \psi(\tau)d\tau \tag{87}$$

By integrating the dynamical prescription of Eq. (85), we obtain

$$y(t) = \frac{1}{\left(\frac{1}{y_0^{z-1}} - (z-1)\tau \right)^{\frac{1}{z-1}}} \tag{88}$$

Since the boundary has the coordinate $y = 1$, the general solution of Eq. (88) affords a way to establish the connection between the initial condition y_0 and the time duration of sojourn, τ, through the following expression:

$$\tau = \frac{1}{\alpha} \left[\frac{1}{1-z} - \frac{y_0^{1-z}}{1-z} \right] \tag{89}$$

With the prescription of Eq. (89) we turn the distribution of random numbers y_0, with probability density $p(y_0)$, into the probability density of the sojourn times τ. In fact,

$$\psi(\tau)d\tau = p(y_0)dy_0 \tag{90}$$

The assumption of uniform back injection yields $p(y_0) = 1$, thereby turning Eq. (90) into

$$\psi(\tau) = |dy_0/d\tau| \tag{91}$$

Thus, using Eq. (89), we get

$$\psi(t) = (\mu - 1)\frac{T^{\mu-1}}{(t+T)^\mu} \tag{92}$$

with

$$\mu \equiv \frac{z}{z-1} \tag{93}$$

and

$$T \equiv \frac{(\mu - 1)}{\alpha} \tag{94}$$

In practice, to generate the time sequence $\{\tau_i\}$, which plays a fundamental role for the dynamic approach to complexity discussed in this review, we do not run either the Manneville map or its idealized version. We draw the numbers y_0 from a uniform distribution on the interval I, and we convert each of them into the corresponding τ_i, using Eq. (89). This turns out to be a procedure much faster than running a map, either the Manneville map or its idealized version. However, these maps are important, since they are an attractive example of dynamic model behind the waiting time distribution of Eq. (92).

C. The Nutting Law

Let us notice that the analytical form emerging from the idealized version of the Manneville map of Eq. (85) fits two additional requests. The first is to reproduce the inverse power law properties revealed by experimental observation, the second being the normalization constraint $\int_0^\infty dt' \psi(t') = 1$. The analytical form adopted is a simple way to meet both requests. Consequently, the parameter T must be thought of as a time denoting the time duration of the regime of transition from microscopic dynamics, where the effect of cooperation is not yet perceived, to the long-time regime, where the inverse power law shows up. With Metzler and Nonnenmacher [41] we refer to this form as the *Nutting form*.

Note that for the dynamic approach to Lévy diffusion we set $\mu > 2$. In this case the mean waiting time, $\langle t \rangle$, is finite, yielding

$$\langle t \rangle = \frac{T}{(\mu - 2)} \tag{95}$$

This formula proves that in the case $\mu > 2$ the parameter T affords also an indication of the time scale beyond which the single trajectories lose any memory of their initial conditions. To study the BQD processes we have to consider the condition $\mu < 2$ [2]. This is a crucial inequality, incompatible with the stationary condition, which is still possible with $\mu > 2$.

D. Final Remarks

It is important to remark that in the literature there are very attractive examples of deterministic maps that produce waiting time distributions with the same asymptotic properties as the analytical function of Eq. (92), with the additional effect of assigning a sign, either positive or negative, to the theoretical laminar region of duration τ_i, totally equivalent to the coin tossing prescription, yet, derived from the deterministic dynamics of the map itself. We refer to Ref. 42 as a remarkable example of this kind. These authors use, in fact, a map that can be considered to be a generalization of the Manneville map, insofar as it consists of three regions, a left laminar region corresponding to the negative sign, a right laminar region corresponding to the positive sign, and a chaotic region in between. When the trajectory reaches the chaotic region, it can jump with equal probability either to the right or to the left region, a fact corresponding to the coin tossing procedure that we adopt in this review to establish the sign of the laminar region of time duration τ randomly selected from the distribution of Eq. (92).

Before ending this section, let us notice that the correlation function of the fluctuation $\xi(t)$ that we use for the dynamic derivation of Lévy diffusion, $\Phi_\xi(t)$, as shown in Section VI, is expressed in terms of the theoretical distribution function $\psi(t)$ [see Eq. (47)]. To get a simple analytical expression for this correlation function, therefore, it is convenient to assign to $\psi(t)$ the Nutting form, namely the form of Eq. (92). For theoretical discussions of experimental results it is convenient to assign the Nutting form to $\psi_{\exp}(t)$. The theoretical discussion of Section XIV rests, in fact, on assigning the Nutting form to the distribution $\psi_{\exp}(t)$. It is important to stress that in the inverse power-law case, although different, the distributions $\psi(t)$ and $\psi_{\exp}(t)$ share the same inverse power-law form, with the same index μ. In this review we assign the Nutting form to either the experimental or to the theoretical waiting time distribution following the criterion of using the simplest as possible analytical expressions.

VI. AN ATTEMPT AT EXPLORING THE SUPERDIFFUSION CONDITION

It is evident that the results of Section IV establishes a nice connection between the GME and the condition of subdiffusion explored with success in the last few years [43]. In fact, the more extended the sojourn time of a random walker on a given site, the slower the resulting diffusion process. Thus, if the exponential waiting time distribution of Eq. (69) is replaced by a slower decay, subdiffusion, namely a diffusion slower than ordinary diffusion, can emerge.

If we adopt the different walking rule of making the walker travel with constant velocity in between two unpredictable non-Poisson time events, subdiffusion is turned into superdiffusion. The CTRW method can be easily adapted to take care of this different walking rule [42,44]. However, in this case, as we shall see, there does not exist yet an exhaustive approach connecting the CTRW prescriptions to the GME structure discussed in Section III. This means that, not even in principle, it is yet known how to derive this kind of superdiffusion from the conventional Liouville prescriptions of nonequilibrium statistical physics. In this section we plan to make a preliminary illustration of this delicate issue.

A. The Generalized Central Limit Theorem (GCLT)

Let us imagine that the random walker, as in the ordinary random walk, makes jumps ahead or backward, at the regular times $n = 1, n = 2, n = 3, \ldots$. Let us imagine that the probability of making a jump is given by $\Pi(\xi)$, with $|\xi|$ denoting the jump intensity, and the sign of ξ indicating if the jump is in the positive, $\xi > 0$, or the negative, $\xi < 0$, direction. We shall be referring to $\Pi(\xi)$ as *propagator*. It is evident that the probability distribution function (pdf) $p(x, n)$ is defined as follows. We have

$$p(x, 1) = \Pi(x = \xi) \tag{96}$$

The distribution of random walkers after the first step is properly normalized since we assume

$$\int\limits_{-\infty}^{+\infty} d\xi \Pi(\xi) = 1 \tag{97}$$

To define the next step, we exploit the fact that the events of this sequence obey renewal theory, thereby producing no correlation among different waiting times. The distribution at the second step is

$$p(x, 2) = \int\limits_{-\infty}^{+\infty} d\xi p(\xi, 1)\Pi(x - \xi) \tag{98}$$

In fact, the probability of finding the random walker at x at time t is the result of two independent events. The first event is a jump by the distance ξ, occurring at time $n - 1$. The second event is a jump occurring at time $n = 2$ and allowing the walker to cover the remaining distance $x - \xi$, so as to arrive at the position x. The total probability is the product $p(\xi, 1)\Pi(x - \xi)$, thanks to the statistical independence of the two events. The integration from $-\infty$ to $+\infty$ takes into account the fact that there are infinitely many intermediate paths.

This prescription can be easily extended to find the pdf at a generic time n, this general prescription being

$$p(x, n) = \int_{-\infty}^{+\infty} d\xi\, p(x - \xi, n - 1)\Pi(\xi) \tag{99}$$

It is evident that the Fourier representation

$$\hat{p}(k, n) \equiv \int_{-\infty}^{+\infty} dk\, \exp(-ikx) p(x, n) \tag{100}$$

is especially convenient, due to the convolution theorem and the space convolution of Eq. (99). Thus we find

$$\hat{p}(k, n) = \hat{\Pi}(k)^n \tag{101}$$

The next issue to address is about the shape of $\hat{p}(k, n)$ for $n \to \infty$. Are there stable forms for this asymptotic condition? To proceed with our heuristic approach, let us consider the case where

$$\hat{\Pi}(k) = \exp(-b|k|^\alpha) \tag{102}$$

In this case, using Eq. (101), we get

$$\hat{p}(k, n) = \exp(-bn|k|^\alpha) \tag{103}$$

Thus the diffusion process is stable. In the asymptotic time limit, the pdf does not change if we set the constraint

$$|k|^\alpha n = \text{const} \tag{104}$$

Since k is proportional to the inverse of x, we reach immediately the conclusion that

$$x \propto n^\delta \tag{105}$$

where δ is the scaling parameter defined by

$$\delta \equiv \frac{1}{\alpha} \tag{106}$$

After assessing the stability of these processes, we have to address the more delicate question of whether or not they have also the attraction power. Thus a generic propagator yields, through Eq. (99), a pdf that falls in the basin of attraction Eq. (103). Before dealing with this delicate issue, let us make some preliminary remarks about the scaling prescription of Eq. (106). We see that for $\alpha > 2$ it would result in subdiffusion, a property that is satisfactorily studied by means of the CTRW (see Section IV). It is therefore difficult to imagine that a propagator exists yielding subdiffusion, a process slower than ordinary diffusion, through a sequence of uncorrelated jumps.

In the case $\alpha = 2$, we note that the anti-Fourier transform of Eq. (103) yields

$$p(x,n) = \frac{1}{\sqrt{4\pi nb}} \exp\left(-\frac{x^2}{4nb}\right) \tag{107}$$

This gaussian form is predicted by the central limit theorem to correspond to the basin of attraction of propagators $\Pi(\xi)$ of whatever form, provided that their second moment $\langle \xi^2 \rangle \equiv \int_{-\infty}^{+\infty} \xi^2 \Pi(\xi)$ is finite.

In conclusion, the ordinary central limit theorem (CLT) establishes that all the propagators sharing the same second moment $\langle \xi^2 \rangle$ yields in the time asymptotic limit the same gaussian pdf. What is the property that the propagators with a diverging second moment must have in common to produce the same pdf in the time asymptotic limit? The answer to this question is equivalent to establishing the GCLT [45]. The answer to this important question rests on the anti-Fourier transform of the expression of Eq. (102), which turns out [46] to yield for $|x| \rightarrow \infty$ a distribution proportional to $1/|\xi|^\mu$, with

$$\mu = \alpha + 1 \tag{108}$$

We note that α must be kept smaller than 2 to keep $\Pi(\xi)$ non-negative [46]. On top of that, the propagators with $\mu > 3$ would end up in the basin of attraction of the ordinary central limit theorem.

It is therefore natural to accept the GCLT, which establishes, in fact, that all the propagators with the asymptotic form $1/|\xi|^\mu$ fall in the same basin, given by Eq. (103). The anti-Fourier transform of $p(k,n)$ does not admit in general any analytical expression. However, the anti-Fourier transform method applied to Eq. (103) [46] shows that

$$\text{limit}_{n\rightarrow\infty} p(x,n) \propto \frac{n}{|x|^\mu} \tag{109}$$

Note that this allows us to express the scaling δ of Eq. (106) as

$$\delta = \frac{1}{(\mu - 1)} \tag{110}$$

In conclusion, the ordinary central limit theorem establishes that the stable condition of Eq. (103) is the attractor, with $\alpha = 2$, of the diffusion processes generated by the propagators $\Pi(\xi)$ with finite second moment. The GCLT establishes in a rigorous way a fact made natural by our earlier remarks. The stable condition of Eq. (103), with $\alpha < 2$, is the attractor of all those propagators that have the same asymptotic inverse power-law structure as the one of Eq. (109). All those propagators fitting the normalization condition of Eq. (97) with diverging second moment, which for large values of $|\xi|$ are proportional to $1/|\xi|^\mu$, fall in the Lévy basin of attraction, and for repeated iterations they generate a diffusion process, whose Fourier transform get the form of Eq. (103). It is evident that the region of action of the GCLT is given by

$$1 < \mu \leq 3 \tag{111}$$

In fact, in the case $\mu > 3$ the second moment of the propagator is finite, and the ordinary central limit theorem enters into play. The condition $\mu > 1$ is required by the fact that the propagator must be normalized, see Eq. (97). This is enough to rule out $\alpha > 2$, thereby making it plausible to accept the rigorous demonstration of Lévy [47], which establishes the positivity of $p(x, n)$ only for $\alpha \leq 2$.

We have tried to make plausible and physically acceptable a result of fundamental importance. In no way should the adoption of the earlier heuristic arguments diminish in the eyes of nonexpert readers the importance of these results for physics and mathematics. Lévy processes seem to be ubiquitous, and it is a challenge for theoretical physicists to find a way to establish if the traditional approaches of nonequilibrium statistical mechanics can satisfactorily account for them.

B. Toward a Dynamic Derivation of Lévy Processes

Physicists find it difficult to imagine physical conditions corresponding to the random walker producing Lévy diffusion. It is not clear how it is possible for a random walker to make jump of arbitrarily large intensities, implying infinitely large velocities, so as to realize Lévy diffusion. For this reason, some physicists [48] have introduced the concept of Lévy walk to contrast the mathematical picture of Lévy that is given the name of Lévy flight. The idea of the Lévy walk is that the random walker makes jumps whose intensities are selected from an inverse power-law distribution, while maintaining a finite velocity. Let us assume that the random walker moves with a velocity of fixed intensity, and that the signs of this velocity change randomly with time. This means that the random walker moves with fixed velocity in the positive direction for a given time, then the direction is abruptly changed, and the random walker begins moving in the opposite direction, keeping it for a while before abruptly changing direction again, and so on. For this walker to produce a diffusion process as close as possible to the Lévy flight, it is enough to assume that the random walker spends with a given velocity a time proportional to the jump intensity of the Lévy flight. Rather than assuming persistence for the sojourn in one spatial site, we assume persistence in one of two states of velocity, with the value W, for motion in positive direction, and $-W$, for motion in negative direction.

The Lévy walk is physically more plausible than the Lévy flight. How to derive the Lévy walk from a Liouville approach of the kind described in Section III? Here, we illustrate a path explored some years ago, to establish a connection between GME and this kind of superdiffusion [49,50]. We assume that there exists a waiting time distribution $\psi(\tau)$, prescribed, for instance, by the dynamic model illustrated in Section V. This function corresponds to a distribution of uncorrelated times. We can imagine the ideal experiment of creating the sequence $\{\tau_i\}$, by drawing in succession the numbers of this distribution. Then we create the fluctuating velocity $\xi(t)$, according to the procedure illustrated in Section V.

The first attempt at deriving the resulting diffusion is done using a phenomenological GME with the same structure as that of Eq. (57). According to Eq. (58), the memory kernel $K(x,t)$ of this equation is expressed in terms of the propagator $\Pi(x,t)$. We remind the reader that the sign of the fluctuating velocity, with the fixed intensity W is determined by a coin tossing procedure. Thus, we are naturally led to make the assumption

$$\Pi(x,t) = \frac{1}{2\langle t\rangle}\delta(|x| - Wt)\psi(t) \qquad (112)$$

It is convenient [49] to define

$$\psi(x,t) \equiv \frac{1}{2}\delta(|x| - Wt)\psi(t) \qquad (113)$$

The physical meaning of the assumption of Eq. (112) is evident. We are setting the condition that in this case persistence refers to velocity rather than position, while the division by $\langle t\rangle$, setting the correct physical dimensions, establishes the time scale of the process. The GME of Eq. (57) implying convolution in both space and time, makes it easy to evaluate the Fourier–Laplace transform of $p(x,t)$, $\hat{p}(k,u)$, which reads [49]

$$\hat{p}(k,u) = \frac{1}{u - \hat{K}(k,u)} \qquad (114)$$

where $\hat{K}(k,u)$ is the Fourier–Laplace transform of the memory kernel of Eq. (58), which involves the Fourier–Laplace transform of $\Pi(x,t)$. Thus, according to the assumption of Eq. (112) we obtain

$$\hat{K}(k,u) = \frac{\hat{\psi}(k,u) - \hat{\psi}(u)}{\langle t \rangle} \tag{115}$$

The details of the calculation are given in Ref. 49. Here we limit ourselves to remarking that to reach their results the authors of Ref. 49 made the assumption that the asymptotic regime is characterized by scaling, namely, $x \propto t^\delta = t^{1/\alpha}$. This implies that $u \propto k^\alpha$. Let us notice that $\delta > 0.5$. The scaling δ, on the other hand, cannot exceed the ballistic condition $\delta = 1$. In fact, according to $\mu = \alpha + 1$ [Eq. (108)] and $\delta = 1/\alpha$, we find that $\delta = 1$ yields $\mu = 2$. This, in turn, is the border between two regions with different physical properties. In the region $\mu > 2$ the mean waiting $\langle t \rangle$ is finite, whereas in the region $\mu < 2$, this mean waiting time is divergent. These mean values are obtained using the waiting time distribution of Eq. (92). The adoption of Eq. (115) indicates that we are referring ourselves to the finite mean time condition, and thus to the condition $\mu > 2$. Thus $\alpha > 1$ and $u = k^\alpha$ tends to zero faster than k. Using this property, the authors of Ref. 49 got

$$\hat{p}(k,u) = \frac{1}{u + bk^{\mu-1}} \tag{116}$$

with

$$b \equiv W(WT)^{\mu-2} \sin\left(\frac{(\mu-2)\pi}{2}\right)\Gamma(3-\mu) \tag{117}$$

This is simply the Fourier–Laplace transform of a Lévy diffusion. The explicit expression for b coincides with the result afforded by the direct use of the GCLT [51].

Lévy diffusion is a Markov process corresponding to the conditions established by the ordinary random walk approach with the random walker making jumps at regular time values. To explain why the GME, with the assumption of Eq. (112), yields Lévy diffusion, we notice [50] that the waiting time distribution is converted into a transition probability $\pi(x)$ through

$$\pi(x) = \frac{1}{2}\frac{1}{\langle t \rangle W}\psi\left(\frac{|x|}{W}\right) \tag{118}$$

with $\langle t \rangle$ given by Eq. (95).

This conversion of time into space distance yields

$$\pi(x) = \frac{1}{2T} \frac{(\mu - 1)(\mu - 2)T^{\mu-1}W^{\mu-1}}{(TW + |x|)^{\mu}} \tag{119}$$

Note that in Ref. 52 the choice of Eq. (119) was done using the prescriptions of nonextensive thermodynamics [53]. Thanks to the connection between $\pi(x)$ and $\psi(t)$, the anomalous character of $\pi(x)$ can be transmitted to $\psi(t)$ and viceversa.

We note that this way of proceeding explains the anomaly of the jump intensity distribution as an effect of the anomaly of the sojourn time distribution. Therefore, to have a satisfactory dynamic approach to Lévy diffusion, we need to explain the reasons for the non-Poisson distribution of waiting times. In the case of BQD intermittence, these waiting times must fit the renewal condition. We shall see that the model of Ref. 54 fits this constraint. In the conclusions of this review we shall argue that the origin of complexity might rest on the renormalization group approach to the non-Poisson form of $\psi(t)$ [54], which fits in fact the renewal condition. This, in turn, makes the sequences of "light on" and "light off" states statistically identical to the dichotomous fluctuation that is proved in this section to generate Lévy walk, and with it the condition that is denoted as the *living state of matter*. This will be discussed in Section F. Here we limit ourselves to quoting the work of Ignaccolo *et al.* [55] as an attempt at assessing the connection between dynamics and thermo-dynamics within the theoretical framework of nonextensive thermodynamics. Actually, this work established that the nonextensive thermodynamics approach does not yield a satisfactory connection between thermodynamics and microscopic dynamics. This is due to the special state of matter that characterizes the Lévy walk, as pointed out in Section VI.F.

Let us go back to the GME of Eq. (57). Let us adopt for $\Pi(x,t)$ the expression of Eq. (112) and let us make the time integration. This prescription yields [50] the Markov master equation

$$\frac{\partial}{\partial t}p(x,t) = \int_{-\infty}^{\infty} k(x - x')p(x',t)\,dx' \tag{120}$$

with

$$k(x) = \pi(x) - \delta(x)\int_{-\infty}^{\infty} \pi(x')\,dx' \tag{121}$$

which, with the choice of Eq. (119), becomes

$$\frac{\partial}{\partial t}p(x,t) = K\left[\int_{-\infty}^{\infty} \frac{dx'p(x',t)}{(TW + |x - x'|)^{\mu-1}} - \int_{-\infty}^{\infty} \frac{dx'.p(x,t)}{(TW + |x'|)^{\mu-1}}\right] \tag{122}$$

with $K \equiv (\mu - 1)(\mu - 2)(TW)^{\mu-1}/(2T)$. The authors of Ref. 50 prove [56] that in the limiting case $T \rightarrow 0$, the Fourier transform of Eq. (122) yields

$$\frac{\partial}{\partial t}\hat{p}(k.t) = -b|k|^{\mu-1}\hat{p}(k.t) \tag{123}$$

with b given by Eq. (117) and $\hat{p}(k,t)$ denoting the Fourier transform of $p(x,t)$.

At the same time these authors show that in the limiting case $T \rightarrow 0$, the right-hand side of Eq. (122) becomes the regularized form [57] of the integral I defined by

$$I \equiv \int_{-\infty}^{\infty} \frac{dx' p(x',t) \, dx'}{|x - x'|^{\mu}} \tag{124}$$

thereby coinciding with the earlier result of Seshadri and West [58].

It is important to point out that the satisfactory agreement between master equation and GCLT is due to the Markov nature of the master equation that corresponds to approximating a Lévy walk with a Lévy flight, thereby annihilating the infinitely extended regime of transition from dynamics to thermodynamics that is discussed in detail in Section VI. As we shall see, the CTRW can be adapted to fit the important request of reproducing a condition of transition from dynamics to thermodynamics lasting forever.

C.   GME Versus CTRW

In this subsection we illustrate the attempt made in Ref. 59 to derive a generalized diffusion equation from the Liouville approach described in Section III. We are addressing the apparently simple problem of establishing a density equation corresponding to the simple diffusion equation

$$\frac{d}{dt}x = \xi(t) \tag{125}$$

where $\xi(t)$ is the dichotomous variable whose fluctuations are described by the waiting time distribution $\psi(t)$. We move from the Liouville equation of the whole Universe:

$$\frac{\partial}{\partial t}\rho(x,\xi,\mathbf{R},t) = L_T \rho(x,\xi,\mathbf{R},t) \tag{126}$$

The operator L_T has the following form:

$$L_T = L_{\text{int}} + L_B \tag{127}$$

namely, it is the sum of two operators; the first operator, corresponding to Eq. (125), is the interaction between system and its bath, and it reads

$$L_{int} = -\xi \frac{d}{dx} \tag{128}$$

The second term is responsible for the fluctuations of ξ, and the symbol $\mathbf{R}$ denotes the set of variables necessary to assign to the variable ξ the proper intermittent properties. This is a crucial assumption. The model might rest, for instance, on a double-well potential, within which the variable ξ moves, virtually attaining only the values corresponding to the bottoms of the two wells. The crucial issue is to make the distributions of time of sojourn at the bottom of these two wells distinctly non-Poisson and renewal at the same time. Here we limit ourselves to assuming that the theoretical waiting time distribution $\psi(t)$ has the form of Eq. (92) and that $\psi_{exp}(t)$ is related to it via Eq. (73). In the specific case that we are here describing, a convenient form for the projection operator P is

$$P\rho(x, \xi, \mathbf{R}, t) = p(x, t)\rho_{eq}(\xi, \mathbf{R}) \tag{129}$$

where

$$p(x, t) = \int d\xi d\mathbf{R} \rho(x, \xi, \mathbf{R}, t) \tag{130}$$

We assume that the bath responsible for the fluctuations of ξ is characterized by thermodynamic equilibrium, described by $\rho_{eq}(\xi, \mathbf{R})$, thereby making the assumption

$$L_B \rho_{eq}(\xi, \mathbf{R}) = 0 \tag{131}$$

At this stage we are equipped to adopt the results of Section III. Let us make the assumption of Eqs. (21) and (22). This means that we locate all the random walkers at $x = 0$ at time $t = 0$. Furthermore, we assume that $\langle \xi \rangle = 0$. In the absence of any bias the condition $PLP = 0$ is automatically fulfilled. Note that $PLQ = PL_{int}P$ and $QLP = QL_{int}P$. Let us explain why it is so. Thanks to the definition of projection operator P of Eq. (29), $L_B P \ldots$ is equivalent to $L_B \rho_{eq}(\xi, \mathbf{r}) \ldots$, which turns out to be a vanishing quantity, as a consequence of Eq. (131). Furthermore, we assume that L_B is a differential operator beginning with differentiation with respect to the bath variable. The definition of projection opeator rests on making, first of all, a trace over the bath variables. Thus, using the method of integration by parts, we get a vanishing quantity again.

Notice that this leads us to the generalized diffusion equation

$$\frac{\partial}{\partial t}p(x,t) = \int_0^t K\left(\frac{\partial}{\partial x}, t - t'\right)\frac{\partial^2}{\partial x^2}p(x,t')\,dt' \qquad (132)$$

which is not very attractive, since the memory function K is a functional of the differential operator $\partial/\partial x$. To make this generalized diffusion equation tractable we are tempted to make the assumption that the operator L_{int} acting in the exponent of Eq. (24) can be neglected. This assumption yields

$$\frac{\partial}{\partial t}p(x,t) = W^2 \int_0^t \Phi_\xi(t - t')\frac{\partial^2}{\partial x^2}p(x,t')\,dt' \qquad (133)$$

where

$$\Phi_\xi(|t_1 - t_2|) = \frac{\langle \xi(t_1)\xi(t_2)\rangle_{\text{eq}}}{W^2} \qquad (134)$$

namely, the stationary correlation function of the fluctuating velocity $\xi(t)$. It can be proved [60] that this generalized diffusion equation is exact under the assumption that the correlation functions of the fluctuation ξ obey the following prescription:

$$\langle \xi(t)\rangle_{\text{eq}} = 0 \qquad (135)$$

$$\langle \xi(t_1)\xi(t_2)\rangle_{\text{eq}} = W^2\Phi_\xi(|t_1 - t_2|) \qquad (136)$$

$$\langle \xi(t_1)\xi(t_2)\xi(t_3)\rangle_{\text{eq}} = 0 \qquad (137)$$

$$\langle \xi(t_1)\xi(t_2)\xi(t_3)\xi(t_4)\rangle_{\text{eq}} = \langle \xi(t_1)\xi(t_2)\rangle_{\text{eq}}\langle \xi(t_3)\xi(t_4)\rangle_{\text{eq}} \qquad (138)$$

$$\langle \xi(t_1)\xi(t_2)\xi(t_3)\xi(t_4)\xi(t_5)\rangle_{\text{eq}} = 0 \qquad (139)$$

and similar expressions for the higher-order correlation functions. These are the ordinary prescriptions for dichotomous noise, whose rationale is apparently given by the same motivations as those adopted to justify the definition of gaussian noise as the fluctuation $\xi(t)$, whose fourth-order correlation function is [61]

$$\begin{aligned}
\langle \xi(t_1)&\xi(t_2)\xi(t_3)\xi(t_4)\rangle_{\text{eq}} \\
&= \langle \xi(t_1)\xi(t_2)\rangle_{\text{eq}}\langle \xi(t_3)\xi(t_4)\rangle_{\text{eq}} \\
&\quad + \langle \xi(t_1)\xi(t_3)\rangle_{\text{eq}}\langle \xi(t_2)\xi(t_4)\rangle_{\text{eq}} \\
&\quad + \langle \xi(t_1)\xi(t_4)\rangle_{\text{eq}}\langle \xi(t_2)\xi(t_3)\rangle_{\text{eq}}
\end{aligned} \qquad (140)$$

with analogous prescription for the higher-order correlation function. In this case we have

$$\langle \xi^{2n} \rangle_{eq} = (2n-1)!! \langle \xi^2 \rangle^n \tag{141}$$

a property ensuring that the distribution of ξ is a gaussian function. By the same token, it seems natural to assign to the dichotomous fluctuations the property of Eq. (138) and similar properties to the higher-order correlation functions, insofar as in this case we obtain

$$\langle \xi^{2n} \rangle_{eq} = \langle \xi^2 \rangle_{eq}^n \tag{142}$$

This equality is compatible with the equilibrium distribution

$$p_{eq}(\xi) = \frac{1}{2}[\delta(\xi - W) + \delta(\xi + W)] \tag{143}$$

We shall be referring to these prescriptions for dichotomous noise as the Dichotomous Factorization (DF) assumption. We notice that the factorization assumption is widely used regardless of whether the correlation function is an exponential or not. See, for instance, the work of Fuliński [62] for the use of this assumption in the non-Poisson case. We shall come back to these important issues in later sections. For the time being, let us limit ourselves to noticing that in the Poisson case,

$$\Phi_\xi(t) = \exp(-\gamma t) \tag{144}$$

the Fourier–Laplace transform of Eq. (133) yields the attractive result

$$\hat{p}(k, u) = \frac{1}{u + \frac{k^2 W^2}{(\gamma + u)}} \tag{145}$$

This result is very attractive because it describes the transition from a short-time region dominated by ballistic transport to a time asymptotic behavior indistinguishable from ordinary diffusion. In conclusion, all this suggests that the assumption that the operator L_{int} acting in the exponent of Eq. (24) can be neglected is not an approximation, insofar as it yields an equation of motion, Eq. (133), that is exact under the DF assumption on the higher-order correlation function.

Let us see if this is correct or not. In general, the Fourier–Laplace transform of Eq. (133) reads

$$\hat{p}(k, u) = \frac{1}{u + k^2 W^2 \hat{\Phi}_\xi(u)} \tag{146}$$

It is plausible that the deviation of $\psi(t)$ from the Poisson condition generates the nonexponential relaxation of $\Phi_\xi(t)$. We have to establish a precise connection between $\Phi_\xi(t)$ and $\psi(t)$, in the nonexponential case, under the renewal condition. According to renewal theory [63], we have

$$\Phi_\xi(t) = \frac{1}{\langle \tau \rangle} \int_t^\infty (t' - t)\psi(t') \, dt' \tag{147}$$

thereby getting the simple expression

$$\Phi_\xi = \left(\frac{T}{t+T}\right)^{\mu-2} \tag{148}$$

Thus, we can go back to Eq. (146) and using the asymptotic expression for the Laplace transform of Eq. (148), we get

$$\hat{p}(k, u) = \frac{1}{u + \chi k^2 u^{\mu-3}} \tag{149}$$

with χ being a constant whose value does not matter at this stage. To find the explicit value of the scaling parameter δ, we set $k = u^\delta$, we plug this equality into Eq. (149) and we look for the value of δ making the resulting expression proportional to $1/u$, namely the Laplace transform of a constant (scaling as an expression of equilibrium). We find

$$\delta = \frac{(4 - \mu)}{2} \tag{150}$$

which is different from the Lévy scaling of Eq. (110).

We shall come back to this important fact later. Here we limit ourselves to pointing out that the scaling emerging from the earlier simple arguments is correct, but it conflicts with the original task of yielding the Lévy scaling of Eq. (110). This means that the generalized diffusion equation of Eq. (133) corresponds to a diffusion process different from the superdiffusion process of Lévy kind that the authors of Ref. 59 aimed at deriving from their dynamic approach. The prediction that the scaling of Eq. (150) is the correct scaling of the diffusion process depicted by Eq. (133) is supported, in fact, by the exact solution to the generalized diffusion equation of Eq. (133) [60]. This conclusion agrees with the earlier work of Metzler and Nonenmacher [64], who have in fact the merit of being the first to discover that Eq. (133) does not afford an appropriate approach to Lévy diffusion.

We notice also that the Lévy scaling of Eq. (110) is obtained by turning Eq. (133) into a Markov equation, through a procedure adopted by the authors of Ref. 60. Let us refer to this procedure as "*delta trick.*" This name suggests that with this procedure we force an equation, with infinite memory, to become Markovian, so as to reproduce the Lévy flight condition. It is worth illustrating how the "delta trick" works, insofar as it sheds light into why the density approach is inappropriate to describe diffusion processes generated by non-Poisson statistics.

In the case where the correlation function $\Phi_\xi(t)$ has the form of Eq. (148), with μ fitting the condition $2 < \mu < 3$, the generalized diffusion equation is irreducibly non-Markovian, thereby precluding any procedure to establish a Markov condition, which would be foreign to its nature. The source of this fundamental difficulty is that the density method converts the infinite memory of a non-Poisson renewal process into a different type of memory. The former type of memory is compatible with the occurrence of critical events resetting to zero the systems' memory. The second type of memory, on the contrary, implies that the single trajectories, if they exist, are determined by their initial conditions.

Nevertheless, let us rewrite this equation, with a change of time integration variables, in the totally equivalent form

$$\frac{\partial}{\partial t}p(x,t) = W^2 \int_0^t \Phi_\xi(t') \frac{\partial^2}{\partial x^2} p(x, t - t')\, dt' \tag{151}$$

Then identify $p(x, t - t')$ with $p(x, t) + \Delta p(x, t; t')$. The second term denotes the change to the pdf $p(x, t)$ occurring when time changes from t to $t - t'$. We make the approximation of considering only the changes corresponding to merely ballistic motion. The contribution stemming from $p(x, t)$ would produce a gaussian diffusion process, meaning that the contribution of this diffusion to $p(x, t)$ would drop to zero much earlier than the fat Lévy tail. Thus, in accordance with the earlier mentioned approximation criterion and with the authors of Ref. 60, we neglect this contribution. As to $\Delta p(x, t; t')$, this change has to fit the following merely ballistic relation:

$$\Delta p(x, t; t') = \frac{1}{2W} \int_{-\infty}^{\infty} \delta\left[t' - \frac{|x - x'|}{W}\right] p(x', t)\, dx'$$

$$- \frac{1}{2W} \int_{-\infty}^{\infty} \delta\left[t' - \frac{|x - x'|}{W}\right] p(x, t)\, dx' \tag{152}$$

The total change is zero, because it can be easily assessed by integrating $\Delta p(x, t; t')$ from $-\infty$ to $+\infty$. This is a consequence of the ballistic assumption. We impose the condition (152) on the form of the generalized diffusion equation of (133). At this stage, as done in Ref. 60, we commute the order of integration. We do first the integration on x' and then the integration on t'. Then, using the properties of the Dirac delta function, we express the second-order derivative with respect to x as a second-order derivative with respect to t'. We apply the method of integration by parts, which has the effect of applying the second-order time derivative to the function $\Phi_\xi(t')$. Finally, we make the integration over t', using again the property of the Dirac delta function. We thus obtain the master equation [50]

$$\frac{\partial p(x,t)}{\partial t} = \frac{1}{2W} \int\limits_{-\infty}^{\infty} W^2 \left[\frac{\partial^2}{\partial x'^2} \Phi_\xi \left(\frac{|x-x'|}{W} \right) \right] p(x',t) \, dx'$$

$$- \frac{1}{2W} \int\limits_{-\infty}^{\infty} W^2 \left[\frac{\partial^2}{\partial x'^2} \Phi_\xi \left(\frac{|x-x'|}{W} \right) \right] p(x,t) \, dx' \qquad (153)$$

Using Eq. (147), we turn Eq. (249) into

$$\frac{\partial}{\partial t} p(x,t) = \int\limits_{-\infty}^{\infty} dx' \pi(x') p(x',t) - \left(\int\limits_{-\infty}^{\infty} dx' \pi(x') \right) p(x,t) \qquad (154)$$

Thus, we recover the Markov master equation (122), and with it [50], taking the limit $TW \ll 1$, we obtain the important equation

$$\frac{\partial p(x,t)}{\partial t} = b \int\limits_{-\infty}^{+\infty} \frac{p(x',t) \, dx'}{|x-x'|^\mu} \qquad (155)$$

which is the Reisz fractional derivative discovered by Seshadri and West [58]. The Fourier transform of this equation coincides with the expression of Eq. (123).

It is now well understood that this is not an approximation; rather it is a way to force an equation with infinite memory to become compatible with Lévy diffusion. The assumption (152) makes it possible for us to get rid of the time convolution nature of the generalized diffusion equation (133). At the same time, this key relation replaces the correlation function $\Phi_\xi(t)$ with its second-order derivative and, as a consequence of Eq. (147), with the waiting time distribution $\psi(t)$. This fact is very important. In fact, any Liouville-like approach makes the correlation function $\Phi_\xi(t)$ enter into play. The CTRW is a perspective resting on trajectories and consequently on $\psi_t(t)$. Establishing a connection between the two pictures implies the conversion of $\Phi_\xi(t)$ into $\psi(t)$, or vice versa. Here, this conversion has been realized paying the price of altering the physics of the generalized diffusion equation (133).

In Ref. 65 the conjecture was made that the emergence at macroscopic level of processes requiring fractional calculus is the consequence of infinitely extended memory breaking the separation between microscopic dynamics, supposed to be not continuous and not differentiable. The same conjecture has been adopted by Stanislavsky [66]. Thus, in a sense, the failure of Eq. (133) to reproduce Lévy diffusion can be interpreted as follows. The adoption of the density method converts the infinite memory of non-Poisson critical events, implying nondifferentiable trajectories, into a second type of memory. This second kind of memory has macroscopic manifestations different from the macroscopic manifestation of the non-Poisson renewal memory. As we shall see hereby, the adoption of trajectory perspective yields the correct Lévy scaling. Thus, Lévy statistics and the corresponding fractional derivative are a macroscopic manifestation of a microscopic nondifferentiable dynamics. With a wide time-scale separation between dynamics and thermodynamics, the density method works well, since the central limit theorem is insensitive to the details of the microscopic dynamics. When, on the contrary, the time scale separation between microscopic dynamics and macroscopic diffusion is abolished, the existence of critical events occurring at the microscopic level forces the diffusion process to adopt the Lévy scaling. On the other hand, according to the results illustrated in Section VI.F, one could conclude that the terms microscopic and macroscopic themselves may be misleading, insofar as a mesoscopic condition may exist in a perennial state of change.

In conclusion, the "delta trick" forces the diffusion process make a transition from the scaling of Eq. (150) to the Lévy scaling of Eq. (110). The dynamic approach to Lévy diffusion by means of the Lévy walk perspective generates memory in the fluctuation $\xi(t)$ and consequently a breakdown of the CLT due to the time correlation of the fluctuation ξ rather than to the diverging second moment of this fluctuation. This generates an extremely extended, virtually infinite, regime of transition from dynamics to thermodynamics that is revealed by the careful statistical analysis illustrated in Section VI.F. The CTRW is expected to reproduce this regime of extremely slow transition from dynamics to thermodynamics. The adoption of the density method behind Eq. (133) generates, correctly, a diffusion process with infinite memory. However, this infinite memory would correspond to a microscopic dynamics with no abrupt jumps, thereby generating a macroscopic diffusion different from Lévy diffusion. The "delta trick" enforces on this diffusion process a Markov condition corresponding to the abrupt interruption of large time regions of ballistic motion, a condition mimicking very well the physical properties of this process, as they are described by numerical simulation. Of course, the "delta trick" cannot reproduce the condition of the infinitely lasting transition from dynamics to thermodynamics that we describe in Section VI.F; it reproduces only the final result of this transition process.

D. Gaussian Case

Before addressing the problem using the CTRW, let us illustrate a case of anomalous diffusion that can be easily expressed in terms of the correlation function $\Phi_\xi(t)$. This is the case where the fluctuation $\xi(t)$ is gaussian. Let us consider the equation of motion

$$\frac{d}{dt}x = \xi(t) \tag{156}$$

Let us integrate it:

$$x(t) = \int_0^t dt'\xi(t') \tag{157}$$

For simplicity, we assume that this and all the other trajectories of the Gibbs ensemble have the same initial condition, $x = 0$. We assume that the fluctuation does not have any bias, $\langle\xi(t)\rangle = 0$, and that is stationary and gaussian, according to the definition of Eq. (140). Let us evaluate the second moment of the diffusion process. This means making first the square of Eq. (157) and then averaging over the Gibbs ensemble

$$\langle x^2(t)\rangle = 2\int_0^t dt_1\int_0^t dt_2\langle\xi(t_1)\xi(t_2)\rangle \tag{158}$$

The characteristic function of this diffusion process is by definition the Fourier transform of $p(x, t)$, namely

$$\hat{p}(k, t) = \langle\exp(ikx)\rangle_t \tag{159}$$

the mean value of $\exp(ikx)$ being done using $p(x, t)$ as a statistical weight. In the gaussian case [61]

$$\langle\exp(ikx)\rangle_t = \exp\left(-\frac{k^2\langle x^2(t)\rangle}{2}\right) \tag{160}$$

The anti-Fourier transform of the characteristic function is the pdf we are looking for, and it reads

$$p(x, t) = \frac{1}{\sqrt{2\pi\langle x^2(t)\rangle}}\exp\left(-\frac{x^2}{2\langle x^2(t)\rangle}\right) \tag{161}$$

It is straightforward to prove that this gaussian pdf is the solution of the following equation of motion:

$$\frac{\partial}{\partial t}p(x, t) = \langle\xi^2\rangle\left(\int_0^t dt'\Phi_\xi(t')\,dt'\right)\frac{\partial^2}{\partial x^2}p(x, t) \tag{162}$$

This is a plausible way to prove that Eq. (162) is the diffusion equation that applies to the gaussian condition. It is important to point out that a more satisfactory derivation of this exact result can be obtained by using the Zwanzig projection approach of Section III [67,68]. Thus, Eq. (162) as well as Eq. (133) must be considered as generalized diffusion equations compatible with a Liouville origin.

It is also important to observe that the asymptotic time evolution of the second moment of Eq. (158) is

$$\langle x^2(t) \rangle \propto \text{const} \cdot t^\delta \tag{163}$$

with

$$\delta = 1 - \frac{(\mu - 2)}{2}, \quad 2 \leq \mu \leq 3 \tag{164}$$

because it can be easily proved by differentiating twice with respect to time $\langle \xi^2(t) \rangle$ of Eq. (158). In the case where the correlation function has a negative tail with index β fitting the condition $2 \leq \beta \leq 3$, this same kind of calculation leads to

$$\delta = 1 - \frac{\beta}{2} \tag{165}$$

therefore producing subdiffusion. In conclusion, in the asymptotic time limit Eq. (161) yields

$$p(x, t) = \frac{1}{\sqrt{4\pi D t^\delta}} \exp\left(-\frac{x^2}{4D t^\delta}\right) \tag{166}$$

with $0 \leq \delta \leq 1$. This is just the Fractional Brownian Motion (FBM) proposed by Mandelbrot [69]. We conclude therefore that FBM is also compatible with a Liouville derivation.

E. Approach Based on Trajectories

Let us now show the CTRW in action. The picture of Section IV has to be properly modified so as to fit the physical condition of the Lévy walk. Hence following Ref. 70 we have to define the probability of moving by a quantity $|x|$ in either the positive, $x > 0$, or negative direction, $x < 0$, in a time t. This quantity is given by

$$\psi(x, t) \equiv \frac{1}{2} (\delta(x - Wt) + \delta(x + Wt))\psi(t) \tag{167}$$

The probability for the walker to travel for a given quantity $|x|$, in either the positive, $x > 0$, or negative direction, $x < 0$, in a time t, without the occurrence of any event is described by the function $\Psi(x,t)$, defined by

$$\Psi(x,t) \equiv \frac{1}{2}(\delta(x - Wt) + \delta(x + Wt))\Psi(t) \tag{168}$$

Note that $\psi(t)$ and $\Psi(t)$ have the same meaning as the corresponding functions appearing in Eq. (63). Thus, we can express the pdf $p(x,t)$ in the form

$$p(x,t) = \sum_{n=0}^{\infty} \int_0^t dt' \int_{-\infty}^{\infty} dx' \psi^{(n)}(x',t')\Psi(x - x', t - t') \tag{169}$$

with $\psi^{(n)}(x,t)$ indicating, analogously to $\psi^{(n)}$ of Eq. (61), the probability that the event corresponding to moving by the quantity $|x|$, in the positive or negative direction according to whether $x > 0$ or $x < 0$, is the last of a sequence of n events. Of course, now the adoption of statistical independence of different events yields

$$\psi^{(n)}(x,t) = \int_0^t dt' \int_{-\infty}^{+\infty} dx' \psi^{(n-1)}(x',t')\psi^{(1)}(x - x', t - t') \tag{170}$$

Thanks to the space and time convoluted nature of Eq. (170) and to the Laplace–Fourier convolution theorem, we get the counterpart of Eq. (62), which reads

$$\hat{\psi}^{(n)}(k,u) = [\hat{\psi}(k,u)]^n \tag{171}$$

Even in this case we adopt the convention

$$\psi^{(1)}(x,t) \equiv \psi(x,t) \tag{172}$$

Taking the Fourier–Laplace transform of Eq. (169), using Eq. (171) and summing, we get the following result:

$$\hat{p}(k,u) = \frac{\hat{\Psi}(k,u)}{1 - \hat{\psi}(k,u)} \tag{173}$$

In the Poisson case of Eq. (81), with a straightforward calculation, we find that Eq. (173) becomes identical to Eq. (145). This is a very important result. In fact, it means that in the Poisson case the two theoretical tools, under scrutiny in this review, yield the same result.

In the non-Poisson case, this equivalence is broken. This can be proved by using Eq. (173) with the choice of Eq. (92) for the function $\psi(t)$ to plug into Eq. (167). The authors of Ref. 70 use this prescription to derive Eq. (116), indicating that CTRW agrees with the GCLT, yielding the same scaling, namely the Lévy scaling of Eq. (110).

F. Multiscaling

In the earlier subsection we have seen that the CTRW yields the correct Lévy scaling. This is a time asymptotic analysis that does not reveal how rich, and long-lasting as well, the regime of transition to this Lévy scaling regime is. This interesting regime of transition was found in the work of Ref. 71.

Let us illustrate, in short, the results found by the authors of Ref. 71. Let us study the dynamic approach to Lévy diffusion using the Lévy walk perspective. Let us use the condition $\mu > 2$, which is compatible with the existence of the stationary correlation function $\Phi_\xi(t)$. As we have shown earlier, the Lévy scaling can be derived using the arguments behind Eq. (118). This equation implies that the number of events is proportional to time. This is not quite correct. The exact formula was found by Feller [72], and it reads

$$\langle N \rangle = \frac{t}{\langle \tau \rangle} \left[1 + \frac{T^{\mu-2}}{(3-\mu)} \frac{1}{t^{\mu-2}} \right] \qquad (174)$$

We notice that the correction to the regime of constant rate of event production decays with the same inverse power law as the correlation function of Eq. (148). Note that the correction is produced by the random walkers that throughout the time interval t did not make any transition from one velocity to the other velocity state. As a consequence, the pdf of the corresponding diffusion process is truncated by two ballistic peaks, in agreement with the theoretical predictions of earlier work [59,73].

Thus, we have two ballistic peaks, moving in opposite directions with constant velocity, thereby yielding the scaling $\delta = 1$. The central part of the pdf is proven [71] to yield the Lévy scaling of Eq. (110). Formally this means that the fractional moment $\langle |x|^q(t) \rangle$ obeys the prescription given by the formula

$$\langle |x|^q(t) \rangle \equiv \int\limits_{-\infty}^{+\infty} p(x,t)|x|^q dx \propto t^{\xi_q} \qquad (175)$$

with the index ξ_q playing a crucial role as an indicator of the scaling properties of the process. In the case where the diffusion is mono-scaling, namely, the pdf obeys the property

$$p(x,t) = \frac{1}{t^\delta} F\left(\frac{x}{t^\delta}\right) \qquad (176)$$

the indicator ξ_q is a linear function of q, namely a straight line with the slope given by δ:

$$\xi_q = \delta q \tag{177}$$

In the Lévy walk case, this is true for $q < \mu - 1$. For $q > \mu - 1$, the slope becomes equal to 1, ballistic motion, according to

$$\xi_q = q - \mu + 2 \tag{178}$$

These properties have been proved theoretically with a very good agreement with numerical results [71]. In conclusion, Lévy walk is richer than Lévy flight, and it can be used to argue the existence of a new physical condition. If we denote as dynamic the short-time region, preceding the emergence of anomalous scaling, and as thermodynamic the asymptotic time region exhibiting scaling, we have to conclude that an intermediate condition exists, which can be perceived as a state of transition from dynamics to thermodynamics. This is the strange condition under discussion in this review. Notice that this interesting physical condition has been denoted as the *living state of matter* [71] because it implies a form of aging, in spite of the fact that $\mu > 2$ ensures the existence of a stationary condition. This is the condition that we are using to study diffusion. However, the diffusion process makes a form of aging appear. As discussed in Ref. 71, aging in this case means that the observation of the diffusion process makes it possible to assess when the initial condition with all the walkers located at $x = 0$ was realized. In fact the ratio of the population at the propagation front to the population contributing the central part of the distribution is a unique deterministic function of time, which allows us to establish the initial condition. In the ordinary case the mere observation of the diffusion width does not allow us to establish the time of preparation, insofar as the same width can be produced by a weak fluctuation working for a large time, or by a strong fluctuation working only for a short time.

G. Conclusions

Here we make a preliminary conclusion. The transition from the Poisson to the non-Poisson case corresponds to the breakdown of the agreement between trajectories (CTRW) and densities (generalized diffusion equation). Why does the generalized diffusion equation (133) depart from the CTRW in the non-Poisson case? In Section IX we shall explain the reasons for this disagreement. At the moment of writing this review, a satisfactory treatment of non-Poisson intermittent process, resting on a Liouville approach, is not yet known. This might the sign that the Liouville approach is inadequate to study non-Poisson statistics and that the CTRW perspective must be adopted instead.

VII.   DYNAMIC APPROACH TO ANOMALOUS DIFFUSION: RESPONSE TO PERTURBATION

This review is devoted to illustrating the problems that the processes of anomalous diffusion, of renewal nature, are raising in the adaptation of the traditional prescriptions of nonequilibrium statistical physics. In this section we explore the delicate issue of the response to external perturbations.

Let us imagine a set of particles whose spatial coordinates x obey the diffusion prescription of Eq. (125). Let us imagine that a time $t = 0$ an external perturbation is suddenly applied to these particles. Using the standard Green–Kubo method [61], we find [49] for the response of the system the following expression:

$$\frac{d}{dt}\langle x(t)\rangle_{\text{pert}} = K \int\limits_0^t dt' \langle \xi(0)\xi(t')\rangle_{\text{eq}} \qquad (179)$$

where the parameter K depends on the intensity and on the physical details of the system considered. Using the definition of normalized correlation function of Eq. (134) (note that here we use the more general notation $\langle \xi^2 \rangle$ rather than W^2, which refers to the case where the variable ξ is dichotomous), we obtain

$$\frac{d}{dt}\langle x(t)\rangle_{\text{pert}} = K\langle \xi^2 \rangle_{\text{eq}} \int\limits_0^t dt' \Phi_\xi(t') \qquad (180)$$

On the other hand, by integrating the equation of motion [Eq. (125)], squaring $x(t)$, and making the Gibbs average, under the stationarity assumption, we obtain also [74]

$$\langle x^2(t)\rangle_{\text{unp}} = 2\langle \xi^2 \rangle_{\text{eq}} \int\limits_0^t dt' \int\limits_0^{t'} dt'' \Phi_\xi(t'') \qquad (181)$$

which yields the unperturbed second moment of x in the case where all the particles are located on the origin $x = 0$ at $t = 0$.

Note that by integrating Eq. (180), by taking into account that the perturbation is abruptly applied at time $t = 0$, when the system is at equilibrium, and no drift exists, $\langle x(0)\rangle = 0$, and using Eq. (181), we get [75]

$$\langle x(t)\rangle_{\text{pert}} = \frac{K}{2\langle \xi^2 \rangle_{\text{eq}}} \langle x^2(t)\rangle_{\text{unp}} \qquad (182)$$

This is the celebrated Einstein relation, which gives the false impression that perturbing a process of anomalous diffusion does not produce anomalous effects.

Actually, if we focus our attention on Eq. (180) and we consider the case where the correlation function $\Phi_\xi(t)$ has the analytical form of Eq. (148), with $2 < \mu < 3$, we reach the conclusion that the conductivity of the system would become infinite in this case, given the fact that the form of Eq. (148) makes non integrable the correlation function Φ_ξ. The current $j(t)$ is the time derivative of $\langle x(t) \rangle_{eq}$. Thus, time differentiating Eq. (182) and using Eq. (163), we see that $j(t) \propto t^{\delta-1}$. Hence the generalized Einstein relation of Eq. (182) is not a problem for subdiffusion [76], since the current tends to vanish in the time asymptotic limit. It becomes a problem for superdiffusion, since in this case the current tends to diverge for $t \to \infty$.

Thus, we find that a dilemma is created. We have to make a choice between two possible ways out. The first is that the Green–Kubo relation of Eq. (180) does not apply, and the second is that the Green–Kubo relation of Eq. (180) does apply, but the common sense expectation that a weak perturbation creates a finite current is violated.

The authors of Ref. 49 and of Refs. 77 and 78 reached similar conclusions. After a transient, where the system obeys the Green–Kubo prediction, a transition to a regime of finite conductivity is made. Thus, we find that the finite conductivity condition is possible as a consequence of the violation of the Green–Kubo prediction. Hence after the abrupt application of a perturbation the system follows the ordinary prescription that in this specific case may force it to adopt an asymmetric form of response [79]. Eventually, the system, departing from the Green–Kubo behavior, makes a transition to a regime of drift linear in time [49].

Let us make a final comment, concerning the violation of the Green–Kubo relation. There is a close connection between the breakdown of this fundamental prescription of nonequilibrium statistical physics and the breakdown of the agreement between the density and trajectory approach. We have seen that the CTRW theory, which rests on trajectories undergoing abrupt and unpredictable jumps, establishes the pdf time evolution on the basis of $\psi(t)$, whereas the density approach to GME, resting on the Liouville equation, either classical or quantum, and on the convenient contraction over the irrelevant degrees of freedom, eventually establishes the pdf time evolution on the basis of a correlation function, the correlation function $\Phi_\xi(t)$ in the dynamical case discussed in Section VI. In conclusion the CTRW rests on $\psi(t)$ and the GME, derived from the Liouville equation, rests on $\Phi_\xi(t)$. The fundamental equation (147) establishes the connection between $\psi(t)$ and $\Phi_\xi(t)$, thereby making it possible, in principle, to establish a connection between the CTRW and the Liouville approach to the GME. This approach, however, does not seem to be compatible with the occurrence of crucial events. The same remark applies to the Green–Kubo relation, whose foundation [61] rests, in fact, on the Liouville equation. It seems that the correct response of a system to an external perturbation is determined by the perturbed $\psi(t)$. This is equivalent to determining the influence of external perturbation on the crucial events responsible for the anomalous statistical properties under study.

Let us see now how to study the effect of an external perturbation, when the unperturbed dynamics of a given system is driven by $\psi(t)$. It is evident that in this case we are led to study the effect that this external perturbation might have on $\psi(t)$. This leads us to express the response on the basis of the perturbed $\psi(t)$ and, if the perturbation is very weak, on the basis of the unperturbed $\psi(t)$, thereby making us move in a direction different from the path adopted by the conventional approach to the response to external perturbation. If the function $\psi(t)$ has an inverse power-law form, the external perturbation may have the effect of truncating this inverse power-law form. We notice that a weak perturbation affects the low modes of the system of interest, which are responsible for the long-time property of the function $\psi(t)$, if it has an inverse power-law form. Thus, a power-law truncation may well be realized, with a consequent significant departure from the prediction of the Green–Kubo theory.

VIII.   DYNAMIC VERSUS THERMODYNAMIC APPROACH TO NONCANONICAL EQUILIBRIUM

In this section we plan to discuss the dynamic versus the thermodynamic approach to noncanonical equilibrium. We plan to show that the thermodynamic method, in the form of an information approach, does not yield satisfactory result. The dynamic approach is more promising than the information approach, even if it involves difficult conceptual problems that have not yet been solved. We argue, however, that the strange response of fluctuations with inverse power law correlation to perturbation, illustrated in Section VII, might turn out to be useful in settling these difficult conceptual problems.

A.   Information Approach to Noncanonical Equilibrium

In this section we use the arguments of Ref. 80 to show how to derive from an information approach a form of thermodynamic equilibrium departing from the ordinary Boltzmann prescription. Let us discuss the derivation of gaussian equilibrium first. Let us imagine that the problem to be solved is the determination of the most probable form of square summable function $f(x)$ belonging to the real axis $[-\infty, \infty]$. We know that this function is symmetric around $x = 0$ and we know the first two nonvanishing terms of its Taylor series expansion about $x = 0$,

$$f(x) = \frac{c_0}{2\pi} - \frac{c_2}{2\pi}x^2 + \cdots \tag{183}$$

Keeping this Taylor series expansion truncated at the second order, is equivalent to assuming that the only information available to us is expressed by

$$f(0) = \frac{c_0}{2\pi} \tag{184}$$

and

$$\frac{1}{2}\frac{d^2}{dx^2}f(x) = -\frac{c_2}{2\pi} \tag{185}$$

If we adopt the maximum entropy method to reconstruct a given function [81] $g(\eta)$, with η being a real variable ranging from $-\infty$ to ∞, the physical meaning of η does not matter, and our attention has to focus only on the maximization of the Shannon entropy

$$H(g(\eta)) = -\int_{-\infty}^{\infty} g(\eta) \log g(\eta)\, d\eta \tag{186}$$

under the condition of taking the proper constraints into account through the Lagrange multiplier method. Thus, there are no reasons why we cannot identify $g(\eta)$ with $\hat{f}(k)$, which is the Fourier transform of the function $f(x)$. Here the constraints of Eqs. (184) and (185) become

$$\int dk \hat{f}(k) dk = c_0 \tag{187}$$

and

$$\int dk \hat{f}(k) k^2 dk = c_2 \tag{188}$$

respectively. The principle of entropy maximization yields, in this case,

$$\hat{f}(k) = A \exp(-\tau k^2) \tag{189}$$

this resulting form being the Fourier transform of an ordinary gaussian distribution.

The derivation of the Lévy distribution from the proper extension of these arguments is straightforward. We assume that the second piece of information available to us, rather than being expressed in terms of the second-order derivative of Eq. (185), is given by

$$\frac{d^\alpha}{d|x|^\alpha}f(x)\bigg|_{x=0} = -\frac{c_\alpha}{2\pi} \tag{190}$$

while the form of the first piece of information of Eq. (184) is left unchanged. Note that the symbol $d^{\alpha}/d|x|^{\alpha}$ denotes the symmetric fractional derivative defined by its action on the Fourier space [82], which reads

$$F\left(\frac{d^{\alpha}}{d|x|^{\alpha}}f(x)\right) = -|k|^{\alpha}\hat{f}(k) \qquad (191)$$

Hence in the Fourier space the constraint on the second moment of Eq. (188) is now replaced by

$$\int dk\hat{f}(k)|k|^{\alpha}dk = c_{\alpha} \qquad (192)$$

Here the method of entropy maximization yields

$$\hat{f}(k) = \exp(-\tau|k|^{\alpha}) \qquad (193)$$

which is well known to be the Fourier transform of an α-stable Lévy process [83].

Notice that this information approach to Lévy statistics is even more direct than the nonextensive thermodynamic approach. As shown in Ref. 52, the adoption of the method of entropy maximization, with the Shannon entropy replaced by the Tsallis entropy [53], does not yield directly the Lévy distribution, but a probability density function $\Pi(x)$ whereby reiterated application of the convolution generates the stable Lévy distribution.

B. Dynamic Approach to Canonical Equilibrium

However, this derivation of Lévy statistics does not seem to reproduce satisfactorily all the properties that emerged from the illustration of superdiffusion processes of Section VI. We think that this formal way of proceeding does not address directly what seems to us a still unsolved issue of modern physics, this being the connection between dynamics and thermodynamics. As pointed out by the authors of Ref. 80, a satisfactory connection between a noncanonical form of equilibrium and thermodynamics should satisfactorily address the problem of thermal contact between a system at equilibrium in a canonical state and a system at equilibrium in a noncanonical condition. In other words, we have to establish which is the temperature of Lévy system, if we believe that this form of noncanonical equilibrium is compatible with thermodynamic equilibrium.

To see how challenging and difficult this issue is, we can shortly summarize the results of an attempt made some years ago at dealing with it in the ordinary case of canonical equilibrium [84]. An oscillator with coordinate x and velocity

v is coupled to a dynamic system, called a *booster* to emphasize the fact that no thermodynamic property is already used, as occurs with ordinary thermal baths. The Hamiltonian coupling is given by $kx\xi$, where ξ is the coordinate of one of the particles of the booster, referred to as a *doorway* variable. The booster is a dynamic system in a condition of strong chaos. We do not assign to it any thermodynamic property, but a given energy E, and we assume that the condition of strong chaos makes it reach the microcanonical state. We write the Liouville equation of the whole system, oscillator plus bath, and we derive from this equation, with the projection method of Section III. the equation of motion of the oscillator. We prove that, under the condition of time scale separation, this reduced Liouville equation becomes equivalent to the Fokker–Planck equation, thereby leading to a canonical form of equilibrium. No use of thermodynamics has been done to reach this important conclusion, but only dynamic properties have been invoked [84]. At this stage, we assume that the width of the resulting oscillator equilibrium distribution can be identified with the temperature T of the booster, and we obtain [84]

$$k_B T = \left[\frac{\partial}{\partial E} \ln W(E) + \frac{\partial}{\partial E} \ln \langle \xi^2 \rangle_{eq} \mathrm{Re} \hat{\Phi}_\xi(\omega) \right] \qquad (194)$$

where $W(E)$ denotes the volume of the booster multidimensional surface with energy E and $\hat{\Phi}_\xi(\omega)$ is the Laplace transform of the correlation function of the doorway variable ξ, evaluated at ω, the oscillator frequency. We know that $W(E)$ is proportional to E^N where N is the number of degrees of freedom of the booster. Thus, in the thermodynamic limit, only the first term within the square bracket of Eq. (194) survives, thereby recovering the ordinary form of Boltzmann principle. More recently, Adib [85] has argued that the correction to the ordinary Boltzmann expression, considered by the authors of Ref. 84 to be of dynamic origin, is on the contrary a consequence of the fact that for systems of finite size the volume rather than the surface version of the Boltzmann principle must be adopted.

This observation reinforces, rather than weakens, the importance of the result of Ref. 84 insofar as it shows that with merely dynamic arguments the authors of this paper did derive the appropriate form of Boltzmann principle. This result sets the challenge for the derivation of the thermodynamic properties of Lévy statistics from the same dynamic approach as that used in Ref. 84 to derive canonical equilibrium.

C. Dynamic Approach to Noncanonical Equilibrium

In Section VI we have seen that the Lévy walk is physically more attractive than the Lévy flight. It is therefore reasonable to address the issue of the derivation of noncanonical equilibrium by using the Lévy walk perspective. On the other hand, in Section VII we have found that the dynamic generators of the Lévy walk depart from the Green–Kubo recipe, insofar as they respond linearly to perturbations according to a prescription involving the waiting time distribution $\psi(t)$ rather than the correlation function $\Phi_\xi(t)$.

For these reasons, it becomes natural to imagine an equation of motion as

$$\frac{dx}{dt} = -\gamma x(t) + \xi(t) \tag{195}$$

The variable $\xi(t)$ is the correlated variable studied in Section VII, with the nonintegrable correlation function of Eq. (148) and $2 < \mu < 3$. According to the important prescriptions of Ref. 84, dynamic equilibrium, to be compared to thermodynamic equilibrium, stems from the joint action of two processes, the bath action on the system, here described by the fluctuation $\xi(t)$, and the action of the system on the bath, here described by the friction term $-\gamma x(t)$. Note that the discussion we are making in this section refers to variables, $\xi(t)$ and $x(t)$, with a generic physical meaning. For the purpose of this discussion, we can imagine them to be the force and the velocity, respectively, of a particle with mass $m = 1$. The action of the stochastic but correlated force would produce a steady increase of kinetic energy, and this would ultimately violate the second principle of thermodynamics without the balancing effect of the action exerted by the system on the bath. The action of the system on the bath has to be compared to the effect of an external perturbation on the booster of the dynamic model adopted by the authors of Ref. 84. We have seen that the authors of Ref. 84 have used as a booster a dynamic system characterized by strong chaos. What about replacing the booster with a dynamic system characterized by intermittent dynamics and weak chaos? The investigation illustrated in Section VII suggests that a booster with a doorway variable $\xi(t)$, characterized by the correlation function $\Phi_\xi(t)$ of Eq. (148), cannot obey the conventional linear response theory that would produce infinite conductivity, but a nonconventional form of linear response making the mean value respond linearly to external perturbation, although with susceptibilities that are related to the waiting time distribution rather than to to the correlation function. This makes it legitimate to adopt the simple friction term $-\gamma x(t)$ of Eq. (195).

Equation (195) has been studied numerically by Annunziato and Grigolini [51]. At first sight Eq. (195) could be confused with

$$\frac{dx}{dt} = -\gamma x(t) + \eta_L(t) \tag{196}$$

where $\eta_L(t)$ indicates the stochastic source of Lévy flight. This equation was originally studied by West and Seshadri [86] through the associated equation for the pdf $p(x,t)$. It is evident that the function $p(x,t)$ is driven by the following equation of motion:

$$\frac{d}{dt}p(x,t) = \left[\gamma\frac{d}{dx}x + b\frac{d^\alpha}{d|x|^\alpha}\right]p(x,t) \qquad (197)$$

where

$$\frac{d^\alpha}{d|x|^\alpha}p(x,t) = \int_{-\infty}^{\infty}\frac{p(x',t)}{|x-x'|^\mu} \qquad (198)$$

This is the fractional derivative that according to the prescription of Ref. 58 [see also Eq. (190)] describes the diffusion process generated by the fluctuation $\eta_L(t)$.

The Langevin equation (196) is a Langevin equation, with the ordinary friction term, $-\gamma x(t)$, and with the Lévy noise $\eta_L(t)$ replacing the conventional gaussian white noise of the usual Langevin equation. The first term on the right-hand side of Eq. (197) corresponds to the usual friction term $-\gamma x(t)$, and the second term, as earlier pointed out, describes the diffusion effect of $\eta_L(t)$. This is a nice example of a nonconventional fluctuation–dissipation process, sharing the Markov nature of the conventional fluctuation–dissipation process, which is derived in fact from Eq. (197) by setting $\alpha = 2$. The Fourier transform of Eq. (197), with $\hat{p}(k,t)$ denoting the Fourier transform of $p(x,t)$, obeys the time evolution equation

$$\frac{\partial}{\partial t}\hat{p}(k,t) = -b|k|^\alpha\hat{p}(k,t) - \gamma k\frac{\partial}{\partial k}\hat{p}(k,t) \qquad (199)$$

This equation yields the equilibrium distribution

$$\hat{p}(k,\infty) = \exp\left(-\frac{b}{\alpha\gamma}\right) \qquad (200)$$

which we refer to as West–Seshadri (WS) noncanonical equilibrium. It is important to point out that this form of equilibrium distribution has slow tails inversely proportional to $|x|^\mu$. In other words, in the case $2 < \mu < 3$ the equilibrium distribution keeps unchanged the power-law nature of the original fluctuation, as well as the corresponding index.

The adoption of Eq. (195) yields an equilibrium distribution that is similar to the WS statistics, with the main difference, however, that the inverse power law is truncated by two peaks, at $\xi = W/\gamma$ and $\xi = -W/\gamma$. Note that the Lévy walk noise $\xi(t)$ is generated according to the renewal prescriptions of Section VI; that is, we use the waiting time distribution $\psi(t)$ of Eq. (92) and, according to the renewal prescription of Eq. (147), the correlation function $\Phi_\xi(t)$ with the analytical form of Eq. (148). The intensity of friction γ can be so large that the variable x reaches its equilibrium before the completion of a laminar region. This means that the free trajectory, with no friction, moves with either velocity W or $-W$, for a time τ, decided by the experimental distribution $\psi_{\exp}(t)$, which has the same asymptotic properties as the distribution $\psi(t)$ of Eq. (92). When we perturb this free motion with the addition of the friction term $-\gamma x(t)$, this trajectory with uniform motion is transformed into a trajectory that at a time of the order of $1/\gamma$ stops this free motion. If the friction is large enough, this happens before the completion of a given laminar region. If, on the contrary, the time $1/\gamma$ is so large as to leave the motion almost identical to the free motion for the whole time duration of a laminar region, a completely different physical condition emerges. The work of Ref. 51 shows that in the former case the equilibrium distribution has a U-shape form. Decreasing the intensity of γ has the effect of filling the empty space between the two peaks $x = W/\gamma$ and $x = -W/\gamma$. Using both the central limit theorem, when $\mu > 3$, and the generalized central limit theorem, when $2 < \mu < 3$, these authors prove that the central portion of the distribution yields the gaussian and the Lévy form, respectively. Actually, the Lévy form is virtually indistinguishable from the WS statistics, even if the WS statistics yield a divergent second moment and the dynamic approach of Ref. 51 yields a finite second moment, because the distribution vanishes at distances larger than W/γ from $x = 0$.

This is a property of remarkable interest for the settlement of the still open problem of defining the temperature associated to a given Lévy distribution [80]. In fact, thanks to this truncation, the dynamic process of Eq. (195) is shown [87] to reach an equilibrium distribution characterized by a finite second moment. The explicit expression of this second moment is

$$\langle x^2(\infty) \rangle = \langle \xi^2 \rangle_{\text{eq}} T^{\mu-2} \exp(\gamma(\mu-2)T) \frac{\Gamma(3-\mu, \gamma T)}{\gamma^{4-\mu}} \qquad (201)$$

where $\Gamma(\alpha, z)$ denotes the incomplete Gamma function. According to the authors of Ref. 80 it should be possible to exploit this condition of finite second moment to establish a condition of equilibrium between a Gauss and a Lévy system, along the lines of Ref. 88. These authors studied the process of heat transfer from a Gauss system at higher temperature to a Gauss system at lower temperature. Using this approach, it should be possible to replace the system at higher temperature with a Lévy system, and it should be possible to establish which temperature the Gauss system has to be assigned to thermally equilibrate with a given Lévy system, whose temperature has to be established. Since the connection between the width of a Gauss distribution and a temperature is a well-assessed fact, it should be possible to settle through this approach the unresolved problem of defining the temperature of a Lévy system. To the best of our knowledge, this attempt has never been made so far, and we think that there are good reasons to try.

D. Conclusions

Before ending this section, we want to notice that a completely dynamical approach to the noncanonical equilibrium, of the WS form, is still missing. The adoption of the same dynamical models as those used in Ref. 49 to establish the existence of a form of linear response violating the Green–Kubo theory seems to be impractical for numerical reasons. A more accessible dynamic model was studied in Ref. 89, resulting, however, in an unstable form of noncanonical equilibrium, although characterized by an extremely extended lifetime. With all these warnings in mind, we reach the conclusion that a noncanonical form of equilibrium of the WS type cannot be ruled out.

The physical realization of this form of noncanonical equilibrium implies the existence of a system–bath interaction establishing an upper time limit to the living state matter condition illustrated in Section VI.F. In other words, the feedback of the system on its bath, necessary to produce friction, would also have the effect of assigning a finite lifetime to the long-lasting regime of transition from dynamics to thermodynamics that we called the living state of matter. This is another way to conclude that the deviation from canonical distribution could be an out of equilibrium property, in a sense different from the more popular idea of out of equilibrium thermodynamics. In fact, as pointed out in Section VI.F, the *living state of matter* is not a thermodynamic condition.

IX. NON-POISSON DICHOTOMOUS NOISE AND HIGHER-ORDER CORRELATION FUNCTIONS

This section is devoted to discussing the higher-order correlation functions of a non-Poisson dichotomous process. In Section VI we have assigned to a dichotomous fluctuation the properties of Eqs. (135)–(139). In the Poisson case, we have found an exact agreement between the generalized diffusion equation (133) and the CTRW expression (173), both yielding the same prediction of Eq. (145) for the Fourier–Laplace transform of the pdf. One would be tempted to conclude that this agreement is not accidental, insofar as it reflects the expected equivalence between the picture based on stochastic trajectories and the picture based on the time evolution of densities. Note that, in principle, the picture based on densities does not even require that trajectories really exist. Thus the agreement between the density and the trajectory perspective has deep consequences in quantum mechanics, where localized trajectories do not exist. It is well known that to explain the existence of localized trajectories, the hypothesis has been made of the existence of wave-function collapses. The statistical agreement between a density and a trajectory treatment suggests that this hypothesis is not necessary if we adopt the philosophy of considering statistical equivalence as synonymous of identity. As we shall show in Section XV, with this kind of equivalence in mind, the advocates of decoherence theory consider the quantum wave-function collapse hypothesis to be obsolete. It has to be stressed that the failure of decoherence theory would force us to reconsider the hypothesis of wave-function collapses, and this would imply an extension or generalization of quantum mechanics, insofar as the wave-function collapses are not dynamic processes described by the ordinary Schrödinger equation.

This equivalence between jumping trajectories and coherence relaxation is confined to the Poisson case. In this section we plan to study higher-order correlation functions, and we plan to prove the emergence from non-Poisson statistics of unexpected properties violating the condition for the trajectory-density equivalence. These properties, as we shall see in Section XV, weaken the conviction that the wave-function issue is settled.

A. On a Wide Consensus on the General Validity of the DF Assumption

We note that in Ref. 87 the projection method was used to study the effect of friction on a process that, in the free case, is known to produce a Lévy walk. As already discussed in Section VIII, the key point has been that the violation of the Green–Kubo relation must imply a different form of linear response, used later to justify the WS form of noncanonical equilibrium. It is worth mentioning that an interesting result of that article has been the following equation of motion:

$$\frac{\partial}{\partial t}p(x,t) = \langle \xi^2 \rangle_{\text{eq}} \int_0^t \Phi_\xi(t-t') \frac{\partial^2}{\partial x^2}p(x,t')dt' + \gamma\frac{\partial}{\partial x}xp(x,t) \qquad (202)$$

This same equation has been independently derived by Cáceres [90] using van Kampen's lemma [91] and the Bourret–Frisch–Pouquet theorem [92], while the theory adopted by Annunziato et al. [87] rests essentially on the Zwanzig approach of Section III, namely, a Liouville-like perspective.

Note that the non-Markov equation (202), in the absence of friction, becomes identical to the generalized diffusion equation of Eq. (133), thereby implying, with Eq. (133), the validity of the prescriptions of Eqs. (135)–(139). We remind the reader that we have denoted the fulfillment of this condition as DF assumption. We note that Eq. (202) yields Eq. (201), the finite second moment that seems to be so important in ensuring the condition of thermal equilibrium between a Gauss and a Lévy system, as pointed out in Section VIII. Note that the numerical work of Ref. 87 confirms the validity of this expression for the second moment. In conclusion, on the basis of theoretical and numerical arguments, we would be tempted to conclude that the DF property extends from the Poisson to the non-Poisson case. Using the arguments of Allegrini et al. [93], we plan to prove that the non-Poisson condition yields the breakdown of the DF property.

B. Four-Time Correlation Function

It has to be pointed out that our arguments are based on examining a single sequence ξ, and thus on time averages, rather than on ensemble averages. This is possible because we are considering the condition $\mu > 2$. In this case the use of a mobile window moving along a very large sequence, assumed to be infinitely large, has the effect of realizing the stationary condition that makes the correlation function $\Phi_\xi(t_1, t_2)$ depend only on the time difference, so as to become $\Phi_\xi(|t_1 - t_2|)$. We assume that the theoretical sequence $\{\tau_i\}$ is derived, as described in Section VI, from the idealized Manneville map of Eq. (85). We refer to this sequence, which is not observable, as a theoretical sequence. We have already pointed out that the random sign assignment to the time intervals of this sequence, has the effect of establishing a fluctuating velocity $\xi(t)$, with the values W or $-W$. As in Section VI, we denote the waiting time distribution of the observable sequence with the symbol $\psi_{\exp}(\tau)$. The two waiting time distributions are related to each other via their Laplace transforms, through Eq. (73). Note that both distributions have the same time asymptotic properties as Eq. (92). We select this analytical form for the theoretical function, $\psi(t)$, thereby implying that $\psi_{\exp}(t)$ departs from it, although the time asymptotic property with a given power index μ is unchanged. In the special case of BQD systems the experimental waiting time distribution is found to be an inverse power law with $\mu < 2$ [2]. Here we consider, as mentioned earlier, the case $\mu > 2$, so as to realize, on the basis of the indications afforded by the idealized Manneville map, a condition compatible with the existence of a stationary correlation function for $\xi(t)$. For the sake of simplicity, we assign to this correlation function the simple analytical form of Eq. (148), and consequently we assign to $\psi(t)$ the Nutting form.

Due to the theoretical prescription that we adopt to realize the dichotomic sequence under study, a given experimental laminar region, namely, a time interval where $\xi(t)$ keeps the same sign, might correspond to an arbitrarily large number of theoretical time intervals, to which the coin tossing procedure assigns the same sign. Let us rewrite Eq. (147) under the form

$$\Phi_\xi(t_2 - t_1) = \frac{\int_{t_2 - t_1}^\infty d\tau[\tau - (t_2 - t_1)]\psi(\tau)}{\int_0^\infty d\tau\tau\psi(\tau)} \tag{203}$$

where we assume $t_2 > t_1$. This equation for the correlation function implies that, with a window of size $t_2 - t_1$, we move along the entire (infinite) sequence of theoretical laminar regions and count how many window positions are compatible with the window being located within a theoretical laminar region, which must have a length larger than the window size. In addition, we have to count the total number of window positions. In other words, the stationary correlation function $\Phi_\xi(t_2 - t_1)$ is just the probability that the two times t_1 and t_2 are located within the same theoretical laminar region.

We evaluate the four-time correlation function using the same arguments. Consider four times, ordered as $t_1 < t_2 < t_3 < t_4$. The corresponding correlation function exists under given conditions. The first is that all four times are located in the same theoretical laminar region. The second condition is that the pairs (t_1, t_2) and (t_3, t_4) are located in distinct theoretical laminar regions, each of them being, however, in the same theoretical laminar region. This means that no random event occurs between the first two and the last two times, while at least one random event occurs between t_2 and t_3.

We use the notation $p(ij)$ to denote the probability that t_i and t_j belong to the same theoretical laminar region. Thus the prescription for the correlation function given by Eq. (203) can be expressed as the probability function $p(ij)$,

$$\Phi_\xi(t_2 - t_1) = p(1, 2) \tag{204}$$

We also use the notation

$$p(\overline{\overline{ij}}) \equiv 1 - p(ij) \tag{205}$$

to denote that at least one transition occurs between times t_i and t_j.

Let us introduce the concept of conditional probability. We denote the joint probability of events A and B by $p(A, B)$ and the conditional probability of occurrence of event A given event B by $p(A|B)$. Thus, we have

$$p(A|B) = \frac{p(A, B)}{p(B)} \tag{206}$$

We denote the conditional probability that event A occurs, given that event B does not, by $p(A|\overline{B})$. Using the same prescription as that of Eq. (206), we should write

$$p(A|\overline{B}) = \frac{p(A, \overline{B})}{p(\overline{B})} \tag{207}$$

On the other hand,

$$p(A) = p(A, B) + p(A, \overline{B}) \tag{208}$$

and

$$p(\bar{B}) = 1 - p(B) \tag{209}$$

Using Eq. (208) and Eq. (209), we turn Eq. (207) into

$$p(A|\bar{B}) = \frac{p(A) - p(A, B)}{1 - p(B)} \tag{210}$$

This is an example of the way to proceed to evaluate the four-time correlation function.

Using this procedure, we find the following exact expressions for the four-time correlation function

$$\frac{\langle \xi(t_1)\xi(t_2)\xi(t_3)\xi(t_4) \rangle}{\langle \xi^2 \rangle} = p(12, 34) \tag{211}$$

and

$$p(12, 34) = p(14) + p(34|12, 2\bar{3})p(12|2\bar{3})p(2\bar{3}) \tag{212}$$

In fact, as earlier pointed out, the four-time correlation function is the sum of two probabilities, the first being that t_1 and t_4 are in the same theoretical laminar region, and the second being that one or more random events occur between t_2 and t_3, but no event occurs between t_1 and t_2 and between t_3 and t_4, either. The second term of Eq. (212) takes this condition into account. In fact, $p(2\bar{3})$ ensures that a random event takes place between t_2 and t_3, $p(12|2\bar{3})$ establishes that no event occurs between t_1 and t_2, given the fact one or more events occur between t_2 and t_3, while, finally, $p(34|12, 2\bar{3})$ establishes that no event occurs between t_3 and t_4 under the condition that no event occurs between t_1 and t_2 and one or more events occur between t_2 and t_3. It is evident that the product of these three probabilities is the probability that t_1 and t_2 belong to the same theoretical laminar region, so do t_3 and t_4, while one or more events occur between t_2 and t_3. In conclusion, Eq. (212) affords an exact expression for the four-time correlation function.

To make our discussion more transparent, it is useful to express the four-time correlation function in terms of the two-time correlation function. The exact expression of Eq. (212) is not suitable for this important purpose. For this reason we make recourse to the approximation

$$p(34|12, 2\bar{3}) = p(34|2\bar{3}) \tag{213}$$

This kind of approximation will be repeatedly made in the second part of this review so as to allow a straightforward numerical assessment of renewal aging. Let us see what the rationale for this assumption is all about. If one or more events occur between t_3 and t_2, due to the renewal character of the process, any memory is lost of the past history of the trajectory, including the fact that t_2 was in the same theoretical laminar region as t_2. Thus, the property of Eq. (213) at first sight seems to be exact, being a natural consequence of the fact that the trajectories do not have any memory of their past. The occurrence of a random event resets to zero the memory of the system. Yet, the adoption of a probabilistic approach establishes a total independence of the past only in the Poisson case, as clearly shown by Eq. (70). A deviation from the Poisson condition makes the system non-Markovian. Consequently, Eq. (213) can be accepted only as an approximation. We shall return to this issue later on in this review. We shall see that an important problem to address when dealing with real processes is to establish if the memory experimentally detected stems from the single trajectories, retaining dependence on their past history or it is due to the non-Poisson and renewal nature of the process under study. In Section 1.17 we shall illustrate an attempt at building a model accounting for both properties. For the time being let us limit ourselves to remarking that the authors of Ref. 93 estimated the error of this approximation and found it to yield accurate results. Thus, we shall proceed with our discussion on the basis of this approximation. Hence, we write

$$p(12, 34) = p(14) + p(12|2\bar{3})p(34|2\bar{3})p(2\bar{3}) \tag{214}$$

Using Eq. (206) we write

$$p(12|2\bar{3}) = \frac{p(12, 2\bar{3})}{p(2\bar{3})} \tag{215}$$

and

$$p(34|2\bar{3}) = \frac{p(34, 2\bar{3})}{p(2\bar{3})} \tag{216}$$

On the other hand, we have

$$p(12) = p(13) + p(12, 2\bar{3}) \tag{217}$$

In fact, it is evident that that, given the fact that $t_3 > t_2$, the absence of events between t_1 and t_2 is realized in two possible ways. In the former no event occurs between t_3 and t_2, either. In the latter, one or more events can occur between t_2 and t_3. By the same token we write

$$p(34) = p(24) + p(34, 2\overline{3}) \tag{218}$$

Using Eqs. (217), (218), (215), and (216), we obtain

$$p(12|2\overline{3})p(34|2\overline{3})p(2\overline{3}) = \frac{(p(12) - p(13))(p(34) - p(42))}{p(2\overline{3})} \tag{219}$$

Thanks to the fundamental property of Eq. (204), we are now in a position to turn Eq. (214) and Eq. (211) into

$$\frac{\langle \xi(t_1)\xi(t_2)\xi(t_3)\xi(t_4) \rangle}{\langle \xi^2 \rangle^2} = \Phi_\xi(t_4 - t_1)$$

$$+ \frac{[\Phi_\xi(t_2 - t_1) - \Phi_\xi(t_3 - t_1)][\Phi_\xi(t_4 - t_3) - \Phi_\xi(t_4 - t_2)]}{1 - \Phi_\xi(t_3 - t_2)} \tag{220}$$

It is straightforward to prove that when the Poisson condition applies, $\Phi_\xi(t) = \exp(-\gamma t)$, Eq. (220) yields

$$\frac{\langle \xi(t_1)\xi(t_2)\xi(t_3)\xi(t_4) \rangle}{\langle \xi^2 \rangle} = \Phi_\xi(t_2 - t_1)\Phi_\xi(t_4 - t_3) \tag{221}$$

Hence, in the Poisson case, the prescriptions of Eqs. (135)–(139) are correct, and so is the generalized diffusion equation (133).

On the basis of this important result one might be tempted to conclude that the claim made by the authors of Ref. 60 is incorrect. In fact, Bologna et al. noticed that in the non-Poisson case the generalized diffusion equation of Eq. (133) yields a scaling different from the Lévy walk process that this equation is supposed to describe, thereby suggesting that in the non-Poisson case the equivalence between the trajectories and density description is lost. This section shows that this breakdown is due to the failure of the DF assumption, described by the prescriptions of Eqs. (135)–(139), on which this equation is based. Thus, there is no density-trajectory conflict after all. However, this conclusion is not unquestionable. In fact, if we advocate the equivalence between density and trajectory picture, we should prove our conviction by building up a density equation that turns out to be exactly equivalent to the trajectories picture, and this should be done with no recourse whatsoever to trajectory arguments. To the best of our knowledge, a satisfactory treatment of this kind does not yet exist, and the lack of a density-based derivation of Lévy superdiffusion could be due to the fact that the Liouville approach to Lévy statistics meets internal inconsistencies that cannot be resolved.

We are convinced that these internal inconsistencies are generated by aging, a property of non-Poisson statistics that the density method should properly address. Aging is a property of non-Poisson statistics subtly related to the emergence of memory when a density (probabilistic) treatment is adopted. However, we must point out that usually the GME time-convoluted structure, of Liouville origin, is a manifestation of trajectory memory, if trajectories exist, and the memory kernel is a stationary correlation function. The generalized diffusion equation (133) is a notable example of this kind. The GME emerging from the CTRW, on the contrary, is directly related to a waiting time distribution rather than to a correlation function, and it is compatible with trajectories losing memory of their past after the occurrence of a jump. In the non-Poisson case, this condition is responsible for the GME time-convoluted structure; however, at the same time it generates a renewal aging condition, which is difficult, if not impossible, to reproduce using the density perspective. After the illustration of renewal aging, made in Sections X and XI, in Section XII we shall show why using the density perspective it is difficult to describe correctly the breakdown of the DF assumption, and thus to establish a correct connection with the trajectory approach. This is so because this breakdown is subtly related to the existence of renewal aging.

X. NON-POISSON PROCESSES AND AGING: AN INTUITIVE APPROACH

Let us imagine an ideal experiment made with electric bulbs. Imagine that we have available a virtually infinite number of electric bulbs of the same brand. This means that the electric bulbs are totally identical. Let us imagine that a time $t = 0$ the first electric bulb is switched on and is left in this state up to the moment of its failure, occurring at time $t = \tau_1$. At this time, another electric bulb is immediately switched on. At time $t = \tau_1 + \tau_2$, also the second lamp fails and it is immediately replaced by a third lamp, and so on. Let us imagine that we have at our disposal an infinite number of lamps, and also an infinite observation time, so as to create the infinite sequence $\{\tau_i\}$. For the purpose of our experiment, it might be convenient to have at our disposal a Gibbs system, meaning that simultaneously at $t = 0$ infinitely many sequences of the earlier described type should be set up. In practice, as we shall do with the BQD experimental case, we can derive infinitely many sequences from the same sequence. For instance, a second sequence might be derived from the earlier sequence, by assuming that a time $t = 0$ the second lamp is switched on. This means that in this second sequence the first failure occurs at time $t = \tau_1' = \tau_2$, the second failure at time $t = \tau_2' = \tau_3$, and so on.

Now, let us imagine that having created these infinitely many sequences, we want to assess the efficiency of these electric lamps, and we want to establish the probability that a generic lamp fails at a time located in the infinitely small interval $[\tau, \tau + d\tau]$. Thus we need to establish the waiting distribution $\psi(\tau)$ which would allow us to state that this probability is indeed $\psi(\tau) \, d\tau$. In practice, we should consider a generic sequence, and establish which is the value τ_1 of it. Then, we should observe another sequence to determine the corresponding τ_1, so as to plot an histogram which becomes the function $\psi(\tau)$.

Let us imagine now that the same experiment is done, rather than at time $t = 0$, at a later time $t = t_a > 0$, so that for any sequence we have to record the time at which the first failure occurs. The probability that the lamp under observation was turned on exactly at time $t = t_a$ is virtually zero. The lamp was turned on earlier, and, if t_a is large, it is probable that many lamps have been switched on and have failed, prior to the unknown time at which the lamp under observation has been switched on. Thus we realize our histogram recording duration times that are shorter than the real ones. Nevertheless, because we are using the convenient normalization procedure, it is not impossible that the resulting waiting time distribution that we should denote as $\psi_{t_a}(t)$, may obey

$$\psi_{t_a}(t) = \psi(t) \tag{222}$$

Let us imagine that as a consequence of the delayed observation we observe a portion τ' of the total duration given by $\tau' = k\tau$, with $k < 1$. It is evident that this does not have any effect on the resulting distribution and that the property of Eq. (222) is fulfilled. This is what happens in the Poisson case. If, on the contrary, the cutting coefficient k for short times is larger than the cutting coefficient for large times, the resulting waiting time will be deformed. The statistical weight of the short times will be lower and the statistical weight of the long times will become higher. Thus the waiting time distribution $\psi_{t_a}(t)$ will show a decay as a function of τ slower than the decay of $\psi(\tau)$. This is what happens when we adopt the non-Poisson waiting time distribution of Eq. (92). This is the aging phenomenon.

As simple and heuristic these arguments are, they serve the purpose of clarifying an issue that is causing some confusion in literature. This has to do with the fact that a renewal process, where the occurrence of a failure is totally unpredictable, has the effect of resetting to zero the memory of the system; however, a form of infinite memory may exist. Let us see why it is so. We have seen that in the Poisson case there is no way to establish, through experimental observation, when the Gibbs system was prepared. In the non-Poisson case, it is not so. In that case, the decay of $\psi_{t_a}(t)$ is slower than the decay of $\psi(t) = \psi_{t_a=0}(t)$, and the experimental observation of the former, once the latter is theoretically known, allows us to establish with absolute precision when the material was originally prepared.

XI. AGING: A MORE FORMAL APPROACH

The most natural way to introduce aging is through the *age-specific failure rate* illustrated in the nice book of Cox [94]. To apply Cox's arguments to the BQD physics, we have to identify the failure of a component, for instance an electric light bulb, with the occurrence of an event, as discussed in Section X. According to Cox, let us consider a sequence of the Gibbs ensemble of Section X known not to have produced an event at time t and let $r(t)$ be the limit of the ratio to Δt of the probability of event occurrence in $[t, t + \Delta]$. In the usual notation for conditional probability,

$$r(t) = \lim_{\Delta t \to 0+} \frac{p(t < T \leq t + \Delta t | t < T)}{\Delta t} \tag{223}$$

According to Cox [94], $r(t)$ gives the probability of almost immediate event occurrence of a sequence known to be of age t.

Using the definition of conditional probability, Cox relates the property of Eq. (223) to $\psi(t)$ and to $\Psi(t)$, the survival probability defined by

$$\Psi(t) = \int_t^\infty dt' \psi(t') \tag{224}$$

yielding the very attractive formula

$$r(t) = -\frac{\frac{d}{dt}\Psi(t)}{\Psi(t)} \tag{225}$$

By straightforward algebra, it may be shown that the choice of Eq. (92) for $\psi(t)$ yields for the survival probability $\Psi(t)$ the following analytical expression:

$$\Psi(t) = \left[\frac{T}{T+t}\right]^{\mu-1} \tag{226}$$

Then, using the definition of $r(t)$ given by Eq. (225), we obtain

$$r(t) = \frac{r_0}{1 + r_1 t} \tag{227}$$

where

$$r_0 \equiv \frac{\mu - 1}{T} \tag{228}$$

and

$$r_1 \equiv \frac{1}{T} \qquad (229)$$

We refer to r_0 and r_1 as the *rate of the brand new system* and *dynamical rate*, respectively. It is remarkable that the power law index μ is determined by the ratio of the rate of the brand new system to the dynamic rate,

$$\mu = 1 + \frac{r_0}{r_1} \qquad (230)$$

Is is also remarkable that, according to Eq. (227), aging is an apparently unavoidable consequence of any finite value of μ. Only, in the case $\mu = \infty$, does the rate $r(t)$ become time independent, thereby producing no aging. It is worth comparing Eq. (230) to Eq. (93), and to the renewal model of Eq. (85). This shows that the parameter α is the rate of the brand new system and $1/T$ is the dynamic rate, thereby clarifying the renewal nature of the model proposed in Section VI.

The definition of aging as a process yielding the time dependence of $r(t)$, defined by Eq. (225), might generate the wrong impression that any process with a nonexponential $\Psi(t)$ has aging. It is important to stress that this is correct only in the case where the renewal condition applies. For recent work on renewal theory and aging, see, for instance, Refs. 95 and 96. The time-dependent rate of Eq. (227) can be interpreted as a manifestation of aging only when the waiting time distribution density $\psi(t)$ of Eq. (92) is derived from a histogram of times that have no correlation whatsoever among themselves. In Section XVI we shall see that modulation can generate for the sojourn times histograms identical to those of the renewal theory. Yet, the times of these histograms are subtly correlated. Thus, to establish whether a process is derived from renewal or from modulation we have to assess the existence of aging using arguments different from the mere deviation of $\psi(t)$ and $\Psi(t)$ from exponential decay.

To establish whether the deviation from exponential decay is of renewal kind or not, we proceed as follows. With the same prescription as that illustrated in Section X, we create a sufficiently large number of sequences $\{\tau_i\}$, to produce a reliable Gibbs ensemble. If we set the observation time equal to $t = 0$, and we count how many of the theoretical laminar region beginning at $t = 0$ ends at a later time, we create a histogram that, obviously, yields the waiting time distribution $\psi(\tau)$ of Eq. (92). Then, we begin observing the sequences at a given time $t_a > 0$. For any sequence we do not know if the first sojourn begins at the moment when we start observing, or not. Using renewal theory, it is shown [97] that the resulting distribution is accurately described by the formula

$$\psi_{t_a}(\tau) \approx \frac{\int_0^{t_a} dy \psi(y + \tau)}{K_{t_a}} \tag{231}$$

where K_{t_a} is the normalization constant. This formula is approximate, but, in the case of an inverse power law, it has been shown [97] to be very accurate. In the exponential case this formula does not yield any deviation from the original waiting time distribution, showing therefore that in the exponential case no aging is possible, even if the process is renewal.

In the BQD case, we have at our disposal only one sequence. Yet, it is possible to create a very large set of distinct sequences as follows. We shift in time the original sequence by the quantity τ_1, thus creating the second sequence. Thus with the second sequence the first event occurs at $t = \tau_2$, the second at time $t = \tau_2 + \tau_3$, and so on. The third sequence is created by shifting the second by the quantity τ_2, and so on. With this method it is possible to assess not only the existence of aging, a fact already assessed by the authors of [98]. It is also shown [99] that the predictions of Eq. (231) are fulfilled with surprising accuracy. This seems to be a compelling proof that the BQD systems obey renewal theory.

XII.   CORRELATION FUNCTIONS AND QUANTUM-LIKE FORMALISM

In this section we address the issue of evaluating correlation function of any order, adopting the density rather than the trajectory perspective. The reader can find the mathematical details of this discussion in a recent article [100].

We refer ourselves to the the idealized version of the Manneville map of Eq. (85). The density time evolution corresponding to this equation reads

$$\frac{\partial}{\partial t} p(y, t) = \alpha \left[-\frac{\partial}{\partial y} y^z p(y, t) + p(1, t) \right] \tag{232}$$

It is evident that the second term on the right-hand side of this equation corresponds to the process of back injection to any point of the interval I, with equal probability. In fact, we see that this term affords a positive contribution to the time derivative on the left-hand side of Eq. (232), with the same rate of back injection, regardless the value of y. It is straightforward to show that if $z < 2$, and so $\mu > 2$, the equilibrium distribution $p_{eq}(y)$ exists and it is given by the properly normalized form

$$p_{eq}(y) = \frac{(2 - z)}{y^{z-1}} \tag{233}$$

We note that in the case $z > 2$, a distribution with the form $1/y^{z-1}$ would yield a divergent normalization factor because for $z > 2$ ($\mu < 2$) an invariant measure does not exist. If we set the system in an initial flat distribution, the pdf $p(y, t)$ will keep changing forever, becoming sharper and sharper in the vicinity of $y = 0$, without ever reaching any equilibrium distribution.

Using Eq. (232), we can address in a quantitative way the aging issue, discussed in Section X. We invite the reader to go back to Eq. (90). We notice that this important equation, with a uniform distribution $p(y)$ was adopted to define $\psi(t)$, the brand new distribution of sojourn times. As time goes on this initial flat distribution changes shape and the probability of smaller y's become larger than that of larger y's. Hence the statistical weight of short times decreases and the statistical weight of large times increases, in accordance with the heuristic arguments of Section X. The nonequilibrium flat distribution corresponds to preparing all the Gibbs sequences of Section X with the electric bulb switched on at $t = 0$. If the observation begins at some later time $t_a > 0$, some lamps have already failed and have been already replaced. Thus the distribution $p(y, t)$ has been evolving from the initial flat distribution to a distribution that is becoming denser and denser as we move to positions closer and closer to $y = 0$. Hence the t_a-old waiting time distribution is obtained by replacing Eq. (90) with

$$\psi_{t_a}(\tau)d\tau = p(y_0, t_a)dy_0 \qquad (234)$$

It is evident that always the resulting waiting time distribution changes with changing t_a, but in the case $z = 1$. In fact, in this case, as proved by Eq. (233), the initial flat distribution coincides with the invariant distribution. The authors of Ref. 96 used this approach to recover the same expression for the aging waiting time distribution as Barkai [95]. Actually, this approach yields exact expressions for the aging waiting time distribution. These expressions are not easy to use, and for practical purposes it turns out to be more convenient to use the approximate expression of Eq. (231), which has been proved [97] to be a very accurate approximation to the exact expression for the aged waiting time distribution. The nice property of this approximate expression is that it shows that the exponential form for $\psi(t)$ is independent of t_a.

It is easy to extend this model so as to create a physical condition closer to BQD physics. This extended dynamic model has two rather than only one laminar spatial region, divided by a border [100]. The particle in the left and right spatial laminar regions move toward the right and the left, respectively. Thus in both cases they move toward the border. When the particle reaches the border, it is injected back either to the right or to the left, with uniform probability. Let us write the corresponding density equation in the generic form

$$\frac{\partial}{\partial t}p(y, t) = Lp(y, t) \qquad (235)$$

The operator L has a form similar to that of Eq. (232). However, its explicit expression is slightly more complicated because it describes two rather than only one spatial laminar region. Its explicit form is of no interest for the discussion of this section, and we refer the interested reader to the work of Ref. 100.

We limit ourselves here to pointing out the difficulties that we meet in attempting to derive from the Liouville-like formalism the correlation function of higher order discussed in Section IX. Let us split the pdf $p(y, t)$ into the symmetric and antisymmetric component,

$$p_S(y, t) \equiv \frac{1}{2}(p(y, t) + p(-y, t)) \tag{236}$$

and

$$p_S(y, t) \equiv \frac{1}{2}(p(y, t) - p(-y, t)) \tag{237}$$

respectively. As a consequence, we split the Liouville-like equation of Eq. (235) into the two corresponding equations

$$\frac{\partial}{\partial t}p_S(y, t) = L_S p_S(y, t) \tag{238}$$

and

$$\frac{\partial}{\partial t}p_A(y, t) = L_A p_A(y, t) \tag{239}$$

respectively. It is not necessary to write the explicit expression of the two operators L_S and L_A. It is enough to remark that the term corresponding to the second term of Eq. (232), namely the contribution corresponding to the injection back process, appears only in the symmetric equation. This is responsible for the distribution $p(y, t)$ to reach the equilibrium state $p_{eq}(y)$, corresponding to $L_S p_{eq}(y) = 0$. The antisymmetric equation omits this injection back term. Thus, if we create an out-of-equilibrium distribution that is not symmetric, namely an initial condition with a nonvanishing p_A, this is annihilated with time, and at equilibrium the correct symmetric form is recovered.

The evaluation of the equilibrium correlation function can be made without a problem. In, fact, adopting a quantum-like formalism we can write, for $t_2 > t_1$,

$$\langle \xi(t_2)\xi(t_1)\rangle = \langle p_{eq}|\hat{\xi}\exp(L_A(t_2 - t_1))|\hat{\xi}|p_{eq}\rangle \tag{240}$$

On the right-hand side of this equation, we adopt the symbol $\hat{\xi}$ to stress that, with this quantum-like formalism the variable ξ becomes an operator that changes the distribution on its right side, which is now the symmetric equilibrium distribution, into the corresponding antisymmetric term. The time evolution of this antisymmetric form is driven by the antisymmetric operator L_A. Let us denote by $p_A^{eq}(y, 0)$ the equilibrium distribution turned into an asymmetric form by applying to it the operator $\hat{\xi}$. We have

$$\exp(L_A(t_2 - t_1))p_{eq}(y) = A(t_2 - t_1)p_A^{eq}(y, 0) + p_A^{noneq}(y, t_2 - t_1) \qquad (241)$$

In other words, we split the antisymmetric time evolution into two terms, one remaining proportional to the original equilibrium distribution, turned into an antisymmetric form, and the other corresponding to an antisymmetric form whose symmetric counterpart is supposed to be different from the equilibrium distribution. In fact, it were not so, this latter term would contribute to the former. In conclusion, we proceed as follows. First, we apply to the equilibrium distribution the operator $\hat{\xi}$. This has the effect of turning the equilibrium distribution into an unstable antisymmetric form that has to be driven by the antisymmetric operator L_A. At time $t = t_1$ we apply to this unstable antisymmetric distribution the evolution operator $\exp(L_A t)$ and we let this distribution evolve up to the time $t_2 > t_1$. At time $t = t_2$, we apply to this antisymmetric distribution the operator $\hat{\xi}$ again, so that $A(t_2 - t_1)p_A^{eq}(y, 0) + p_A^{noneq}(y, t_2 - t_1)$ becomes $A(t_2 - t_1)p_S^{eq}(y, 0) + p_S^{noneq}(y, t_2 - t_1)$. The authors of Ref. 100 proved that $A(t_2 - t_1) = \Phi_\xi(t_2 - t_1)$, with $\Phi_\xi(t_2 - t_1)$ given by Eq. (148). They also proved that

$$\langle p_{eq} | p_S^{noneq}(t_2 - t_1) \rangle = 0 \qquad (242)$$

This equality implies that $p_S^{noneq}(y, t_2 - t_1)$ is not positive definite, a price that we have to pay to ensure the equivalence between the density and trajectory picture in the non-Poisson case. Thus, the two-time correlation function is evaluated using only density prescriptions, and the result turns out to be identical to Eq. (148), which is known to correspond to the prescription of renewal theory [see Eq. (147)]. In the Poisson case the equilibrium distribution is flat. Thus, the contribution $p_S^{noneq}(y, t_2 - t_1)$ vanishes.

The DF breakdown is caused by the existence of a nonvanishing $p_S^{noneq}(y, t_2 - t_1)$. If this term vanishes, the fourth-time correlation function obeys the DF prescription. To evaluate the correct expression of the four-time correlation function, we have to focus our attention on the effects produced by this nonvanishing term. At time $t = t_2$, rather than making a trace over the equilibrium condition, we must study the time evolution of $p_S^{noneq}(y, t_2 - t_1)$ from t_2 to t_3, yielding $p_S^{noneq}(y, t_3, t_2, t_2 - t_1)$. This is a contribution generating some concern, since it activates again the back injection process, which is intimately related to aging and deviation from equilibrium. It is expected that the compatibility with equilibrium should rest on the property $\langle p_{eq} | p_S^{noneq}(y, t_3, t_2, t_2 - t_1) \rangle = 0$.

At time t_3 we have to apply the operator $\hat{\xi}$ again, and this allows us to use the antisymmetric representation, with the time evolution given by the operator $\exp(L_A(t_4 - t_3))$. At time t_4 we apply the operator $\hat{\xi}$ again, and we go back to the symmetric representation and we conclude the calculation by multiplying the resulting distribution by $\langle p_{eq}|$. At the moment of writing this review, it is not clear if a four-time correlation function equivalent to that found in Section IX can be found. It is evident, however, that the adoption of densities forces us to break, in some way, the condition of keeping the second term on the right-hand side of Eq. (232) inactive. The calculation of the two-time correlation function, as we have seen, does not involve the back injection term, and so it does not involve the aging process. The four-time correlation function, on the contrary, does it, thus violating the condition that an equilibrium calculation should keep the second term on the right-hand side of Eq. (232) inactive. This is a consequence of the non-Poisson nature of the process, and it is one of the reasons why we are convinced that non-Poisson statistics leads us to quit the traditional Liouville or Liouville-like approach. In other words, the study of non-Poisson processes seems to require the use of the CTRW perspective, compatible with the existence of trajectories with abrupt and unpredictable jumps. We are therefore led to abandon the use of the density method.

XIII. GENERALIZED MASTER EQUATION OF A GIVEN AGE

In the earlier sections we have seen the GME of Eq. (59) in action with the memory kernel defined by Eq. (66). In Section X we have seen that non-Poisson renewal processes are characterized by aging. Thus, this is the right moment to establish the age of the GME of Eq. (59). In this section we show that this is the GME corresponding to a brand new condition. Thus, we have also to establish the form of the aged GME, if this ever exists.

We notice that the correlation function defined by Eq. (147) is stationary. Thus, it fits the Onsager principle [101], which establishes that the regression to equilibrium of an infinitely aged system is described by the unperturbed correlation function. The authors of Ref. 102 have successfully addressed this issue, using the following arguments. According to an earlier work [96] the GME of infinite age has the same time convoluted structure as Eq. (59), with the memory kernel $\Phi(t)$ replaced by $\Phi_\infty(t)$. They proved that the Laplace transform of Φ_∞ is

$$\hat{\Phi}_\infty(u) = \frac{u\hat{\psi}_\infty(u)}{1 + \hat{\psi}(u) - 2\hat{\psi}_\infty(u)} \tag{243}$$

We do not go through the details of the demonstration done in Ref. 96 to get this result. We limit ourselves to noticing that if $\hat{\psi}_\infty(u)$ is replaced by the brand new property $\hat{\psi}(u)$, then the memory kernel of Eq. (243) becomes identical to the memory kernel of Eq. (66). Then, there exists an even more compelling reason to accept the result of Ref. 96. This is done through the realization of the Onsager principle via the GME of infinite age. This GME reads

$$\frac{d}{dt}\mathbf{p}(t) = -\int_0^t dt' \Phi_\infty(t-t')\mathbf{K} \cdot \mathbf{p}(t') \tag{244}$$

This GME, being of infinite age, fits the Onsager principle. Therefore, we can identify the stationary correlation function $\Phi_\xi(t)$ with the out of equilibrium distribution, namely,

$$\Phi_\xi(t) = \frac{p_1(t) - p_2(t)}{p_1(0) - p_2(0)} \tag{245}$$

The time evolution of the term on the right-hand side of this equation is determined by the GME of Eq. (244). Thus, using the definition of Eq. (4) for $\mathbf{K}$, from Eq. (244) we get

$$\frac{d}{dt}\Phi_\xi(t) = -2\int_0^t dt' \Phi_\infty(t-t')\Phi_\xi(t') \tag{246}$$

which yields for the Laplace transform of $\Phi_\xi(t)$ the following expression

$$\hat{\Phi}_\xi(u) = \frac{1}{u + 2\hat{\Phi}_\infty(u)} \tag{247}$$

To convince the reader that Eq. (243) is a proper expression for the memory kernel of infinite age, it is enough to show that this approach to the stationary correlation function $\Phi_\xi(t)$ yields the same result as the prediction of the renewal arguments, namely, Eq. (147). To prove this important fact, we use the definition of the infinitely aged memory kernel given by Eq. (243), we plug it in Eq. (247) and we compare the resulting expression with the Laplace transform of Eq. (147) so as to assess whether we get the same analytical expression.

These two expressions are found to coincide one with the other. This encouraging result can also be used to move in the same direction to find an analytical expression for the t_a-old correlation function $\Phi_\xi^{(t_a)}(t)$. It is plausible that the memory kernel $\Phi_{t_a}(t)$, of a generic age, intermediate between $\Phi_{t_a(t)=0} = \Phi(t)$ and $\Phi_{t_a(t)=\infty} = \Phi_\infty(t)$, is given by

$$\hat{\Phi}_{t_a}(u) = \frac{u\hat{\psi}_{t_a}(u)}{1 + \hat{\psi}(u) - 2\hat{\psi}_{t_a}(u)} \tag{248}$$

To derive the expression for the correlation function $\Phi_\xi^{(t_a)}(t)$ we adapt Eq. (247) to the intermediate age defined by t_a, thereby obtaining

$$\hat{\Phi}_\xi^{(t_a)}(u) = \frac{1}{u + 2\hat{\Phi}_{t_a}(u)} \tag{249}$$

Of course, we have to use the exact expression for $\hat{\psi}_{t_a}(u)$, of which Eq. (231) is only an accurate approximation. The exact expression for $\hat{\psi}_{t_a}(u)$ can be found, for instance, in Ref. 96. Then we plug Eq. (248) into Eq. (249) thus obtaining the Laplace transform of the correlation function of $\xi(t)$, of age t_a. The authors of Ref. 102 prove that the result of this procedure is an exact expression for $\hat{\Phi}_\xi^{(t_a)}(u)$, which coincides indeed with the recent exact result of Godréche and Luck [103].

XIV. LIMITS OF THE GENERALIZED MASTER EQUATIONS

The GME of a given age illustrated in Section XIII allowed us to find a path to the aging correlation function based on physical rather than mathematical arguments. This is a fact of some interest. However, the resulting GME is not as useful as one would hope it is. The reason is that the GME has a given age, and it can be used only by making reference to that age. This property requires some explanation. Let us discuss the case where $\mu > 2$, so as to make it possible for the condition of infinite age to exist. We prepare the material at time $t = 0$. Hence all the sequences begin at time $t = 0$, with the beginning of an experimental laminar region. Let us assume that all these experimental laminar regions correspond to "light-on" states. Upon time change some of the Gibbs systems will jump to the "light-off" state. Thus, for $t \to \infty$, we shall reach a condition where half of the systems are in the "light-on" state and half are in the "light-off" state. However, the age of the system depends on trajectory properties that are not established by the equal distribution in the "light-on" and "light-off" states. In fact, the practical realization of the Onsager principle rests on selecting only the trajectories that at a given time $t = t_a = \infty$ are in the "light-on" states. Then, from this moment on, we record the time evolution of only these trajectories. The authors of Ref. 96 have proved that the decay of $p(1,t) - p(2,t)$ yields the stationary correlation function. The authors of Ref. 102 have proved that the same procedure has to be applied at a generic time $t = t_a > 0$, to define the correlation function with that specific age. However, once the GME of that specific age has been built up, its form remains fixed forever, in spite of the fact that the individual trajectories keep aging.

This property generates an effect that makes the use of GME of a given age questionable. If we perturb the GME of a given age with an external field at times $t > t_a$, the GME response might not have anything to do with the trajectory response. In fact, if we make a field perturb a trajectory at any time t subsequent to the preparation at time $t = 0$, regardless of whether it is smaller or larger than time t_a, the interaction between field and trajectory corresponds to the physics of the process. The interaction of the field with the GME of a given age, on the contrary, corresponds to a condition that does not properly reflect reality. This is the reason why Sokolov et al., [104] found that the two procedures generate different results. The reason for this disconcerting result is that the GME keeps forever a dependence on the procedure adopted at $t = t_a$ to extend the Onsager principle to the case of trajectories with a given age. After this preparation the trajectories keep aging, while the GME maintains forever the dependence on the condition established at $t = t_a$. As disconcerting as this fact is, it is another manifestation of the fact that the later jumps of a given trajectory do not retain a memory of the earlier jumps, while the GME has an infinite memory and never forgets the initial condition. This nonintuitive property is a consequence of the non-Poisson nature of the process. No matter how far we are from the preparation made at $t = 0$, there is still a significant probability that some trajectories did not depart from the initial state. The GME is characterized by a memory kernel determined by this initial condition, but a field perturbing it at later times corresponds in practice to perturbing single trajectories of changing age. This makes the trajectory perspective adequate to describe the effects of external perturbation, while the adoption of the GME would force the perturbing field to interfere with density memory properties, with results that do not necessarily correspond to the perturbation of the single trajectories.

In this section we show the trajectory perspective in action for the evaluation of absorption and emission spectra [105]. In the last few years, as a consequence of an increasingly faster technological advance, it has become clear that the conditions of ordinary statistical mechanics assumed by the line-shape theory of Kubo and Anderson [106] are violated by some of the new materials. For instance, the experimental research work of Neuhauser et al. [4] has established that the fluorescence emission of single nanocrystals exhibits interesting intermittent behavior, namely, a sequence of "light-on" and "light-off" states, departing from the Poisson statistics. In fact, the waiting time distribution in both states is not exponential, and it shows a universal power law [2]. In this section, as done in Ref. 105, we assign to both states the same waiting time distribution with the inverse power-law structure of Eq. (92). Actually, we assign this form to the experimental waiting time distribution of Eq. (72). The two distributions are different but share, in the time asymptotic regime, the same inverse power-law index. Note that Brokmann et al. [98] have applied to real data a statistical analysis aiming at assessing the existence of aging. We shall come back to this issue. Here we limit ourselves to stressing that there is compelling evidence that these materials yield renewal aging.

Let us use the following stochastic equation:

$$\frac{d}{dt}d(t) = i[\omega_0 + \xi(t)]d(t) \tag{250}$$

The quantity $d(t)$ is a complex number, corresponding to the operator $|e\rangle\langle g|$ of the more rigorous quantum mechanical treatment [107], where $|e\rangle$ and $|g\rangle$ are the excited and the ground state, respectively, ω_0 is the energy difference between the excited and the ground state, and $\xi(t)$ denotes the energy fluctuations caused by the cooperative environment of this system. In the presence of the coherent excitation, Eq. (250) becomes

$$\frac{d}{dt}d(t) = i[\omega_0 + \xi(t)]d(t) + k\exp(i\omega t) \tag{251}$$

where ω denotes the radiation field frequency. It is convenient to adopt the rotating-wave approximation. Let us express Eq. (251) by means of the transformation $\tilde{d}(t) = \exp(i\omega t)d(t)$. After some algebra, we get a simple equation of motion for $\tilde{d}(t)$. For simplicity, we denote $\tilde{d}(t)$ again with the symbol $d(t)$, thereby making the resulting equation read

$$\frac{d}{dt}d(t) = i[\delta + \xi(t)]d(t) + k \tag{252}$$

where $\delta \equiv \omega_0 - \omega$. The reader can easily establish the connection between this picture and the stochastic Bloch equation of Ref. 107 by setting $d = v + iu$. Note that the three components of the Bloch vector in Ref. 107, (u, v, w), are related to the rotating-wave representation of the density matrix ρ, v, and u being the imaginary and the real part of $\exp(-i\omega t)\rho_{ge}$, and w being defined by $w \equiv (\rho_{ee} - \rho_{gg})/2$. Note that the equivalence with the picture of Ref. 107 is established by assuming the radiative lifetime of the excited state to be infinitely large and the Rabi frequency $\Omega \equiv k$ vanishingly small.

It is straightforward to integrate Eq. (252), thereby getting

$$d(t) = k\int_0^t dt' \exp\left(i\int_{t'}^t \xi(t'')dt''\right)\exp(i\delta(t-t')) \tag{253}$$

with the dipole $d = 0$, when the exciting radiation is turned on. Now we have to address the intriguing problem of averaging Eq. (253) over a set of identical systems, in such a way as to take aging effects into account [95,96,98]. In fact, the averaging process turns Eq. (253) into

$$\langle d(t)\rangle = k\int_0^t dt' \left\langle \exp\left(i\int_{t'}^t \xi(t'')dt''\right)\right\rangle_{t'}\exp(i\delta(t-t')) \tag{254}$$

with the subscript t' denoting that the system, brand new at $t = 0$, is t'-old when we evaluate the corresponding characteristic function, thereby implying that the distribution of waiting time before the first jumps, is not $\psi(t)$, and $\psi_{t'}(t)$ has to be used instead. Thus the radiation field at time t' has to induce transitions on trajectories that are reminiscent of the electric bulbs of the example of Section X.

It is evident that the adoption of CTRW method yields results that are incompatible with the adoption of a GME, corresponding to the preparation done at $t = 0$. This master equation would retain forever memory of the observation beginning at the moment of preparation itself. However, the corresponding trajectories would keep aging. Some of them would lose memory of this initial condition, others would remain in the initial state for a very extended time. The radiation field at time t' promotes transitions on t'-old trajectories, whereas the GME would retain a memory forever of the brand new initial condition at $t = 0$.

We point out that the number of photons emitted at time t is determined by $N(t) = \langle d(t)d^*(t) \rangle$. It is straightforward to prove that the rate of photon emission, namely $R(t) \equiv dN/dt$, obeys the relation

$$R(t) = 2k\mathrm{Re}\langle d(t) \rangle \tag{255}$$

Thus, we conclude that the real part of $\langle d(t) \rangle$ can be used to denote either emission or absorption at time t. We note that the approximation ensuring the equivalence between this picture and Ref. 107, for large photon count, has also the effect of making the absorption identical to the emission spectrum.

The authors of Ref. 105 used both a numerical and an analytical approach to the evaluation of $R(t)$ and also to the evaluation of $\langle d(t) \rangle$. They adopted a numerical procedure, and an analytical approach as well, which has the remarkable property of yielding a very good agreement with the numerical results. Here we limit ourselves to explaining the numerical procedure, since this is a way to afford a further illustration of the concept of aging trajectories, in spite of their renewal character. Using the non-Poisson distribution of Eq. (92), the authors of Ref. 105 run N distinct sequences $\{\tau_i\}$, with $N \gg 1$. For any of these N sequences the sojourn in one of the two states begins exactly at time $t = 0$ and ends at time $t = \tau_1$. For $N/2$ of these sequences they use the "light-on" state as initial condition, and for $N/2$ they use the "light-off" state. Let us consider the former kind of trajectories for illustration purposes. The "light-on" state begins at $t = 0$ and ends at $t = \tau_1$, at which time the "light-off" of state begins, ending at time $t = \tau_1 + \tau_2$, and so on. With this numerical method we produce a set of fluctuations $\xi(t)$. Then, we create a set of diffusion trajectories $\int_{t'}^{t} \xi(t'')\, dt''$, and hence the set of exponentials $\exp\left(i\int_{t'}^{t} \xi(t'')\right)$. Since all trajectories of this set are t'-old, the resulting numerical average is automatically equivalent to evaluating $\left\langle \exp\left(i \int_{t'}^{t} \xi(t'') dt''\right)\right\rangle_{t'}$ with the t'-old probability distribution of first sojourn times, $\psi_{t'}(t)$.

XV.   NON-POISSON AND RENEWAL PROCESSES: A PROBLEM FOR DECOHERENCE THEORY

This section is devoted to illustrating the problems caused to decoherence theory by the occurrence of crucial events obeying non-Poisson renewal. To make the illustration of our arguments clearer and more convincing, we divide this section into three subsections. Section XV.A is a concise review of decoherence theory. After reviewing this theory, we explain why non-Poisson renewal processes yield the breakdown of decoherence theory. In Section XV.B we consider the case where the source of anomalous statistical mechanics is external. In Section XV.C we show that decoherence fails in establishing the correspondence principle in the case when the system of interest itself yields non-Poisson renewal processes. Is this another manifestation of the density–trajectory conflict? To give more support to this conjecture, we show that the density–trajectory conflict manifests itself also at the level of the connection between the Kolmogorov–Sinai entropy and Clausius entropy. This is discussed in Section XV.D. In Section XV.E we argue that all these difficulties have the same origin, namely, the occurrence of random events according to a non-Poisson rather than a Poisson prescription.

A.   Decoherence Theory

As discussed in Ref. 108, decoherence theory implies that the quantum superposition of two macroscopically distinct states, $|A\rangle$ and $|B\rangle$, is forbidden by the entanglement process

$$(|A\rangle + |B\rangle)|E\rangle \rightarrow |A\rangle |E\rangle_A + |B\rangle|E\rangle_B \qquad (256)$$

where $|E\rangle$ is the initial environment state and $|E\rangle_A$ and $|E\rangle_B$ are the states expressing how the environment adapts itself to the system's state $|A\rangle$ and $|B\rangle$, respectively. As strange as this property might seem, that a big system adapts itself to a much smaller one, it is a genuine quantum mechanical property that occurs with no exchange of energy, if the energy of $|A\rangle$ is identical to that of $|B\rangle$, thus ensuring the condition for the second principle of thermodynamics to apply. This is correct and takes place naturally as an effect of a unitary transformation. It is interesting to examine this property from the perspective of the reduced density matrix. The reduced density matrix ρ corresponding to the left-hand side of Eq. (256) is that of a pure state, fitting the condition $\rho^2 = \rho$, thereby leading to a vanishing von Neumann entropy. When the entanglement process, indicated by the arrow of Eq. (256), is completed, the reduced density matrix becomes mixed, thereby producing a finite von Neumann entropy. Thus, if we look at this entanglement process from the perspective of the reduced density matrix, we see the density matrix becoming diagonal with respect to the basis set of the states $|A\rangle$ and $|B\rangle$. These two states are the eigenstates of an observable, for instance position, corresponding to two distinct eigenvalues of this observable, and are consequently orthogonal. We also find that the von Neumann entropy of the reduced density matrix reaches its maximum value. For this important condition to apply, we need the condition

$$\langle E_A|E_B\rangle = 0 \qquad (257)$$

In a recent article [109], Adler questioned the validity of this picture, on the basis of the fact that in his opinion Eq. (257) conflicts with the claim of unitary transformation made by the advocates of decoherence theory. Actually, the work of Bonci et al. [110] seems to confirm the validity of Eq. (257). These authors studied the special case where the environment of the the the two-state systems, with eigenstates $|A\rangle$ and $|B\rangle$, is an oscillator. The two states $|A\rangle$ and $|B\rangle$ are the two states of a $1/2$-spin system. The two states are coupled to each other in the same way as the state $|1\rangle$ is coupled to the state $|2\rangle$ through the Hamilton operator of Eq. (5). The parameter V is denoted by the symbol ω_0 in Ref. 110. This quantum mechanical coupling forces the system to establish a linear superposition of the two states $|A\rangle$ and $|B\rangle$. For this reason, following Ref. 111 we call this term *inphasing coupling* and ω_0 *inphasing strength*. On the same token, we refer to the phenomenological terms appearing in Eqs. (8) and (9) as *dephasing term*.

It is enough to assign to the quantum oscillator a coherent physical condition corresponding to exciting infinitely many oscillator eigenstates so as to make the reduced density matrix diagonal, with a corresponding entropy increase. The process implies that after reaching the condition of maximum entanglement, corresponding to Eq. (257), the system returns to the initial condition. However, if we go closer and closer to the condition $\omega_0 = 0$, the recursion times become infinite, and the condition of Eq. (257) becomes irreversible. Actually, the time evolution of the states $|E_A\rangle$ and $|E_B\rangle$ is not independent of the system of interest, and this is probably the reason why the remarks of Adler on the unitary transformation do not apply. It is worth showing explicitly how to establish the connection between decoherence theory and ordinary diffusion processes, following the approach of Ref. 108. Let us assume that the Hamiltonian under study is

$$H = E(|A\rangle\langle A| + |B\rangle\langle B|) + g((|A\rangle\langle A| - |B\rangle\langle B|)\xi + H_E \qquad (258)$$

The time evolution of the variable ξ depends on the Hamiltonian H_E, referring to the environment, whose explicit expression is not important for our discussion.

We limit ourselves to imagining that the environment consists of virtually infinitely many degrees of freedom. The two states $|A\rangle$ and $|B\rangle$ are degenerate. The results of our discussion are independent of the value assigned to E, which, for simplicity, is assumed to vanish, $E = 0$. Note that the inphasing term is now missing and only the dephasing process is present, being represented by the second term on the right-hand side of Eq. (258). This term is inactive when the wave function contains only either the state $|A\rangle$ or the state $|B\rangle$. This corresponds to the equilibrium condition that would be reached as a result of a measurement process aiming at establishing if the system is in the state $|A\rangle$ or in the state $|B\rangle$. From a formal point of view, if the initial condition is given by $(|A\rangle + |B\rangle)/\sqrt{2}$, the wave-function collapses would produce the diagonal density matrix $\rho_\infty = (|A\rangle\langle A| + |B\rangle\langle B|)/2$. According to the spirit of decoherence theory, a statistical picture yielding the same final condition has to be considered to be totally equivalent to the result of a process of wave-function collapse.

It is convenient to use the interaction picture to represent the time evolution of the wave function $|\psi(t)\rangle$. By writing the formal solution of the Schrödinger equation in the interaction picture and returning to the laboratory reference frame, we obtain

$$|\psi(t)\rangle = \exp\left(-\frac{iH_Bt}{\hbar}\right) \exp\left[-ig(|A\rangle\langle A| - |B\rangle\langle B|)\int_0^t dt'\xi(t')\right]|\psi(0)\rangle \quad (259)$$

where

$$\xi(t) \equiv \exp(iH_Et)\xi\exp(-iH_Et) \quad (260)$$

Let as assume that the initial condition is

$$|\psi(0)\rangle = \frac{1}{\sqrt{2}}(|A\rangle + |B\rangle)|E\rangle \quad (261)$$

where the vector $|E\rangle$ denotes the initial environment state. At the initial time, the system and the environment are not entangled and are statistically independent. For the time evolution of Eq. (259), we obtain immediately the form

$$|\psi(t)\rangle = \frac{(|A\rangle|E(t)\rangle_A + |B\rangle|E(t)\rangle_B)}{\sqrt{2}} \quad (262)$$

where

$$|E(t)\rangle_A = \exp\left(-\frac{iH_Bt}{\hbar}\right) \exp\left[-i\frac{g}{\hbar}\int_0^t dt'\xi(t')\right]|E\rangle \quad (263)$$

and

$$|E(t)\rangle_B = \exp\left(-\frac{iH_Bt}{\hbar}\right)\exp\left[i\frac{g}{\hbar}\int_0^t dt'\xi(t')\right]|E\rangle \qquad (264)$$

According to decoherence theory the human observer perceives only the system of interest. Thus, we have to evaluate first the total density matrix corresponding to the wave function of Eq. (259) and to make a contraction over the environmental degrees of freedom, to derive out of it the contracted density matrix out, $\rho(t)$. As a result of this procedure we get

$$\rho(t) = \frac{(|A\rangle\langle A| + |B\rangle\langle B|)}{2} + |A\rangle\langle B|\Phi_{AB}(t) + |B\rangle\langle A|\Phi_{AB}^*(t) \qquad (265)$$

where

$$\Phi_{AB}(t) = \langle E|\exp\left[-2i\frac{g}{\hbar}\int_0^t dt'\xi(t')\right]|E\rangle \qquad (266)$$

The function $\Phi_{AB}(t)$ should be evaluated using a rigorous quantum mechanical calculation. In earlier work [108] we noted that decoherence theory simplifies this task by assuming that the time-dependent quantum operator $\xi(t)$ is a stochastic classical variable, so that the function $\Phi_{AB}(t)$ becomes a characteristic function. By assuming that $\xi(t)$ is an ordinary gaussian and uncorrelated fluctuation, it is shown that

$$\Phi_{AB}(t) = \exp(-2Dt) \qquad (267)$$

where

$$D = g^2\langle\xi^2\rangle/\hbar^2 \qquad (268)$$

In the specific case where $|A\rangle$ and $|B\rangle$ are two coherent states of the same macroscopic oscillator, located at a distance Δx apart from one another, it is shown [108] that the picture based on the prescription of the Hamiltonian of Eq. (258) can be maintained provided that we set the condition $g^2\langle\xi^2\rangle = (\Delta x)^2\Sigma$, where Σ is a parameter proportional to the product of temperature and friction. The details of this calculation are not important for this review. Therefore let us limit ourselves to observing that this procedure yields for the decoherence time the expression

$$t_D = \frac{\hbar^2}{2\Sigma(\Delta x)^2} \qquad (269)$$

The important fact is that the decoherence time may become so small as to ensure that there is no room for the superposition condition $|A\rangle + |B\rangle$ in classical physics, if $|A\rangle$ and $|B\rangle$ correspond to two distinct positions, macroscopically distant from one another.

One may question the claims of decoherence theory by adopting an individual system perspective. The states $|A\rangle$ and $|B\rangle$ are usually assumed to be the eigenstates of the operator to measure. Which is the wave-function structure corresponding to the reduced density matrix becoming diagonal, in the basis set of the states $|A\rangle$ and $|B\rangle$? The wave function could become widely extended in space so as to generate blurring [112] rather than a wave-function collapse. The authors of Refs. 113 and 114 made the conjecture that a real wave-function collapse might occur, as a result of the enhancement of spontaneous collapses, meant as true correction to ordinary quantum mechanics. An attractive example of theory of this kind is given by the work of Ghirardi et al. [115]. These authors made a wise choice of space and time scale of the process of spontaneous collapse so as to produce quite negligible corrections to ordinary quantum mechanics if the elementary microscopic constituents are individually studied. The macroscopic wave-function collapse is the result of an enhancement of these spontaneous collapses, triggered by the constraints among the microscopic constituents of a given macroscopic body. This enhancement can also be provoked by the interaction between a system and the environment. In fact, the conventional fluctuation–dissipation processes involve the interaction between the system of interest and a very large number of environmental degrees of freedom. Thus, the system–environment interaction necessary to make decoherence theory work can also activate the enhancement of spontaneous collapses [113,114]. However, these arguments do not represent a real failure of decoherence. Rather, from a practical point of view, this would be a success of decoherence theory, and the practical role of spontaneous collapses might even not be properly appreciated, because it would rest on complicated calculations leading to results identical to those predicted by decoherence theory with a much easier algebra. This is a consequence of the fact that recourse to ordinary statistical mechanics makes it impossible to defeat decoherence theory.

B. Nonordinary Environment

We would like to attract the attention of the reader to the case when the environment is a source of anomalous diffusion. Paz et al. [116] studied the decoherence process generated by a supra-ohmic bath, but they did not find any problem with the adoption of the decoherence theory. It is convenient to devote some attention to the case when the fluctuation ξ is a source of Lévy diffusion [59]. If the fluctuation ξ is an uncorrelated Lévy process, the characteristic function again decays exponentially, and the only significant change is that the parameter Σ appearing in Eq. (269) would have a sublinear dependence on temperature. However, it is more interesting to consider the case when ξ yields a Lévy walk rather than a Lévy flight [71]. In Section VI.F we have seen that the Lévy walk produces a condition of transition from dynamics to thermodynamics lasting forever. For this reason the Lévy walk is a paradigmatic dynamic model fitting the definition of complexity proposed by the authors of Ref. 117. The diffusion process departs from the condition of mono-scaling predicted by the generalized central limit theorem, and it lives forever in a state intermediate between multi-scaling (dynamics) and mono-scaling (thermodynamics). In the multiscaling case, the probability density of the diffusing variable keeps changing its form, thereby making it impossible for us to define a condition equivalent to thermodynamic equilibrium. Here the characteristic function $\Phi_{AB}(t)$, after a fast exponential decay, moves to an oscillatory regime, with a slowly decreasing intensity, which, in practice, takes an infinite time to vanish. In fact, as shown by Eq. (266), the decoherence function is nothing but than the characteristic function of a diffusion process. If this diffusion process is the Lévy walk diffusion process of Section VI.F, it is straightforward to prove [59] that the central part of the diffusion process, corresponding to Lévy diffusion, yields a fast exponential decay, while the ballistic peaks produce an oscillatory decay. The decay of this residual coherence is as slow as the decay of the number of random walkers contributing the ballistic peaks of the anomalous diffusion process described in Subsection VI.F.

It is important to point out that the non-Poisson renewal case discussed in Section XIV is incompatible with the assumptions made to justify the basic tenets of decoherence theory. As in Section XIV the relaxation function of Eq. (267) is a characteristic function. However, this is the characteristic function of an ordinary diffusion process. This implies that the bath producing coherence annihilation is an ordinary *Ohmic* bath. There is therefore an intimate relation between (a) decoherence theory working with no difficulty and (b) the assumption of ordinary statistical physics. If we refer ourselves to nonordinary rather than ordinary baths, this lucky condition is broken and with it, quite probably, the validity of decoherence theory as well. Decoherence theory works very well if the scaling condition applies, including the case of Lévy flight. The Lévy walk condition, studied in Section XIV and fitting the requests raised by the BQD experimental observation, implies the breakdown of the conditions behind the successful application of decoherence theory. In conclusion the *living state of matter* condition of Section XIV requiring an approach entirely resting on trajectories provokes at the same time the breakdown of the decoherence theory. In fact, the quantum decoherence does not follow a fast exponential relaxation, but it adopts an inverse power-law behavior, implying that coherence never vanishes completely. Of course, this aspect might also be used for the opposite purpose of increasing the significance of decoherence theory. In fact, if baths of the same kind as that discussed in Ref. 59 exist, then one might be tempted to have recourse to them for the main purpose of protecting quantum mechanical information. The process of decoherence that we are describing, in fact, can be thought of as an environment-induced waste of quantum information, and the adoption of the kind of environment conjectured in Ref. 59 might help the search for special procedures aiming at protecting quantum information from degrading, thereby helping the research work currently done for the realization of quantum computers.

C. Further Problems Caused by Non-Poisson Physics: Transition from the Quantum to the Classical Domain

In the BQD case the non-Poisson renewal condition is a property emerging from the interaction with a cooperative bath. We want to illustrate here another important case, although this is not of central importance for the main aim of this review. Research work done in the last 15 years proves also that the non-Poisson character of the system of interest, generated by internal dynamics, rather than by the interaction with an anomalous bath, creates problems in decoherence theory. Let us see why.

Quantum chaos is a field of research whose aim is to study the dynamics of those nonlinear systems that would be characterized by deterministic chaos in the classical limit. The interested reader can consult excellent review articles, such as Ref. 118, where the definition of quantum chaos is given through the spectral properties, thereby applying also to systems with no classical counterparts. However, here we limit ourselves to discussing a quantum process, with a classical limit: the kicked rotor [119]. This is a rotor kicked at regular intervals of time by an impulsive torque with a strength proportional to $K \sin \theta$, where θ is an angle denoting the rotor orientation. At any impulsive kick the rotation momentum x changes by a a given quantity ξ. Thus after many kicks the system dynamics become indistinguishable from

$$\frac{dx}{dt} = \xi(t) \tag{270}$$

In fact, if the control parameter K is large enough, the variable ξ is an uncorrelated fluctuation, and, in the long-time limit, it can be thought of as a noisy function of the continuous time t. The solution of Eq. (270) yields results agreeing with ordinary statistical mechanics, namely, a diffusion process making the second moment of x increase as a linear function of time. However, in the quantum case this linear increase has an upper bound in time. At times of the order of

$$t_L \propto \frac{K^2}{\hbar^2} \tag{271}$$

the second moment of x stops increasing [118,119]. In the case of a macroscopic rotor the momentum localization might be considered a macroscopic manifestation of quantum mechanics, consequently generating another quantum mechanical mystery. In fact, a macroscopic kicked rotor is expected to obey classical physics, and so its behavior should be identical to that of a classical kicked rotor, without localization. An obvious way to settle this paradox is by observing that for an actual macroscopic kicked rotor the action to $\hbar$ ratio is so large as to make the localization time of Eq. (271) infinitely much larger than the time duration of any realistic experiment. This explanation is not quite satisfactory because the localization process is the consequence of latent quantum memory that survives for a virtually infinite time, even if it does not yield physical effects at times shorter than t_L. The weak tail of Eq. (54), in fact, has been derived to account for the memory effects of quantum origin that the Zwanzig projection method with a re-summation at infinite order is expected to yield.

In the Poisson case, the decoherence theory affords a more satisfactory justification for the correspondence principle [20]. Adopting the Wigner formalism, it is possible to express quantum mechanical problems in terms of the classical phase space, and the Wigner quasi-probability is expected to remain positive definite until the instant at which a quantum transition occurs, according to the estimate of Ref. 120, at the time

$$t_\chi = \frac{1}{\lambda} \ln(\chi/\hbar) \qquad (272)$$

where χ is a scale parameter proportional to the nonlinear interaction. However, Zurek and Paz [120] show that the environmental fluctuations, as weak as they are, can erase the quantum mechanical coherence, thereby preventing the birth of quantum interference effects that should take place on the time scale of Eq. (272). In the specific case of the kicked rotor the time necessary to kill the quantum mechanical coherence is given by [121]

$$t_D = \frac{\hbar^2}{2\sigma(\Delta x)^2} \qquad (273)$$

where $\Delta x = K^2/\hbar$ and σ is the intensity of environmental fluctuation. It is possible to prove [120] that in realistic cases $t_D < t_\chi$, thereby preventing the localization process from occurring. This interpretation is attractive: At times $t > t_D$ there is no more latent quantum memory, and the system becomes genuinely classical. The decoherence theory prevents the emergence of macroscopic quantum effects, regardless of how far this goes into the future. If $\hbar$ is much smaller than the classical action, the system is classical, and no localization is permitted. Only ordinary diffusion is possible, as a consequence of the fact that in the classical limit, as we have seen, the variable ξ of Eq. (270) is an uncorrelated noise, with no memory whatsoever.

As pointed out in Section I, we want to examine the case of anomalous statistical mechanics. Anomalous diffusion is generated by Eq. (270) when ξ departs from the condition of uncorrelated noise. For the kicked rotor, this anomalous condition is realized by assigning to the control parameter K special values. These special values make two accelerator islands appear in the phase space of the kicked rotor [54], embedded in a chaotic sea. The surface of separation between the accelerator island and the chaotic sea is sticky. Thus the trajectory undergoes an erratic motion in the chaotic sea and, then, from time to time, as an effect of erratic diffusion, sticks to the surface of one of these two islands. Throughout the extended time of sojourn at the surface with one of these two islands, the momentum keeps changing by the same quantity, either W or $-W$, depending on the island. The important fact is that the waiting time distribution in one of the two states, $\psi(t)$, has the following time asymptotic property:

$$\lim_{t \to \infty} \psi(t) \propto \frac{1}{t^{\mu}} \tag{274}$$

where $2 < \mu < 3$. The condition of ordinary statistical mechanics, the Poisson condition, corresponds to an exponential waiting time distribution, namely, to $\mu = \infty$. Thus, we are in the presence of a physical condition where anomalous rather than normal statistical mechanics is involved. In the case studied in Ref. 31, $K = 6.9115$ and $\mu = 2.667$. The reasons why this separation surface yields the anomalous waiting distribution are well known, and the reader can find a detailed discussion in Ref. 54, for instance. The separation surface actually is not a curve with vanishing width. It is rather a layer of finite size, with channels that make it possible for the surrounding sea to penetrate the layer. The surrounding sea generates these channels through ramification that generates the branch channels of decreasing size as moving toward the interior of the layer. As a consequence of this geometric structure, in the classical case a trajectory sticks to the layer, and thus to the border with one of the two accelerator islands, for long times, with a distribution density given by Eq. (274). In the laboratory frame the second moment of x is proportional to $t^{4-\mu} = t^{1.333}$, which is significantly faster than the linear in time increase, reflecting the prescription of ordinary statistical mechanics. It is important to stress that this model generates a non-Poisson distribution of waiting times that fit the condition of being renewal. In this sense, this model would pass the aging test necessary to be considered a proper vehicle for describing BQD physics. Of course, this model should yield $\mu < 2$, which is in fact the crucial condition required by BQD physics. Here, on the contrary, to realize the Lévy walk condition we have to impose the inequality $\mu > 2$, as explained in Section VI.

Now, let us see why this condition of anomalous diffusion, resting on a non-Poissonian renewal process, yields the breakdown of the correspondence principle. Let us refer ourselves to the work of Ref. 31. The authors of Ref. 31 confirmed the new effect discovered by the authors of an earlier article [122]. This is as follows. The second moment of the momentum of quantum kicked rotor grows faster than in the ordinary case for a while, then, at a given time it makes a transition to a condition of linear in time growth. This effect can be accounted for as follows. Let us denote by $\Xi(0)$ the area of the portion of the separation surface, directly arrived by trajectories starting from the chaotic sea. In accordance with the assumption that the kicked rotor is virtually classical, we assume that $\hbar$ is much smaller than $\Xi(0)$. Notice that, according to the theory accounting for the origin of the inverse power law of Eq. (274), the trajectory penetrates within the boundary regions, through channels of decreasing size. Thus, the area $\Xi(t)$ is a decreasing function of time. On the other hand, quantum mechanics turns the bunch of classical trajectories into a wave function with the quantum uncertainty, $U(t)$, increasing as $\exp(\lambda t)$. The correspondence with classical physics is lost when

$$\Xi(t) = U(t) \tag{275}$$

The reason why the condition of Eq. (275) yields the breakdown of the correspondence between quantum and classical mechanics is evident. The quantum wave function can be identified with a trajectory if it is sharp enough, namely if $U(t) < \Xi(t)$. When the width of the wave function becomes as large as the width of the channel, within which the wave function moves, the wave function motion starts to depend on the structure of the surrounding phase space, and the correspondence is broken. Using a geometrical model making $\Xi(t)$ decrease exponentially with time [122], the condition of Eq. (275) is shown to occur at time $t = t_B$, where

$$t_B = \frac{1}{\lambda} \ln(1/\hbar) \tag{276}$$

This prediction has been confirmed by the results of Refs. 123 and 124. In fact, the numerical result of Ref. 123 indicates that the waiting time distribution of Eq. (274) has an exponential truncation, this being an effect of the tunneling from the boundary between chaotic sea and accelerator island, back to the chaotic sea. The authors of Refs. 31 and 122 argue that the quantum induced recovery of ordinary diffusion is followed by a corresponding localization process.

How does one use the decoherence theory to annihilate these quantum effects? There are problems. In fact, in classical physics the adoption of environmental noise produces a departure from anomalous diffusion [125,126].

Thus, the assumption of the decoherence theory that there are no isolated systems, and that we have always to consider the influence of environmental fluctuations, would kill anomalous diffusion. Furthermore, the numerical results of Ref. 31 show that the quantum-induced transition from anomalous to ordinary diffusion is a quantum effect more robust than the localization phenomenon itself. This indicates that in the presence of a weak environmental fluctuation is now insufficient to reestablish the correspondence principle.

We are now in the right position to reach a preliminary conclusion. Although the decoherence theory is an attractive and efficient way of defeating the emergence of quantum effects at a macroscopic level, the authors of Ref. 112 did not feel comfortable with it. The reason is that when the observer has the impression that a wave-function collapse occurs, actually the quantum mechanical coherence is becoming even more extended and macroscopic, since it spreads from the system to the environment, Eq. (256).

In the example discussed with Eq. (256), $|A\rangle$ and $|B\rangle$ are already supposed to be a macroscopic state, whose linear superposition has to be killed by the environmental fluctuations. It can be shown [127] that the measurement process itself has the effect of creating the linear superposition of two macroscopic states. In the measurement process $|\alpha\rangle$ and $|\beta\rangle$ are two eigenstates, with eigenvalues α and β, respectively, of a given observable O pertaining to a microscopic state $|m\rangle$. The environment of the microscopic system a is the experimental apparatus itself, with a pointer up, $|A\rangle$, and down, $|B\rangle$, indicating to the macroscopic observer that the observable measured has the value α and β, respectively. If the microscopic system is in a linear superposition of $|\alpha\rangle$ and $|\beta\rangle$, for instance, $|m\rangle = (|\alpha> +|\beta >)/\sqrt{2}$, the experimental apparatus would end up in the corresponding linear superposition of $|A\rangle$ and $|B\rangle$, a fact that would correspond to the emergence of quantum mechanics at the macroscopic level. Decoherence theory has to be invoked at this level, by introducing the environment E of the measurement apparatus, with two orthogonal states $|E\rangle_A$ and $|E\rangle_B$. The final effect of this process would be to create the condition

$$|\psi(t)\rangle = \frac{1}{\sqrt{2}}(|\alpha\rangle|A\rangle|E\rangle_A + |\beta\rangle|B\rangle|E\rangle_B) \qquad (277)$$

This equation is derived using the same procedure as that earlier used to derive Eq. (262). More specifically, we have to apply twice the entanglement prescription used to derive Eq. (262): the first time to describe the entanglement between the microscopic state and the macroscopic measurement apparatus, and the second time to entangle the macroscopic pointer with its own environment.

Thus, to make sense of all this, we have to assume that the observer cannot see the linear superposition $|m\rangle$, because this is a microscopic state. The observer cannot see the linear superposition of Eq. (277) either, because the observation focuses only the experimental apparatus. This latter assumption is subtly related to considering the second principle, and the transition from quantum to classical physics as well, as a consequence of human limitation rather than as an objective fact of nature.

For these reasons, the authors of Ref. 112 argue that decoherence theory does not produce genuine wave-function collapses, but rather an effect determined by the limited information available to the observer, a property that they define *subjective collapse*. They notice [112] that the theory of Ref. 115 generates genuine wave-function collapses. This theory is a generalization or extension of quantum mechanics, turning the wave-function collapse assumption of the founding fathers of quantum mechanics into an essential dynamical ingredient. There should be no concern for losing the wonderful achievements of quantum mechanics, for more than 100 years [128]. In fact, thanks to a wise choice of two new constants [115], the new Schrödinger equation, with a stochastic correction, is almost always equivalent to the ordinary Schrödinger equation. The corrections to quantum mechanics become active as a consequence of the measurement-induced cooperation process [112]. This has to do with the fact that the propagation of microscopic coherence to layers of increasing size, described by Eq. (277), involves an increasing number of microscopic constituents, which are forced to follow the same prescriptions. Thus, as pointed out in Refs. 113 and 114, the physical processes necessary to produce decoherence might promote the cooperative action necessary to activate genuine wave-function collapses. As a consequence, the main problem here is not so much the risk of losing the benefits of quantum mechanics, but rather it is how to make the theory of objective wave-function collapses distinguishable from decoherence theory. In fact, from a statistical point of view, the picture adopted by the authors of Ref. 112 is essentially equivalent to the decoherence theory.

On the same token, the increasing interest in the stochastic Schrödinger's equation in studying many-body problems [129–131] does not imply a conversion of these authors to the theory of genuine wave-function collapses. Some of these authors [131] did succeed in realizing the important condition of norm conservation, which would be an appealing property from the perspective of wave-function collapses (for a discussion of the wave-function collapse with no norm conservation, see, for instance, Ref. 132). However, the main aim of these authors is to create an efficient computational tool with a stochastic picture that yields results that are statistically equivalent to the Lindblad master equation [133], a well-known Markov master equation. The norm-preserving stochastic Schrödinger equation, statistically equivalent to the Lindbland master equation, was pioneered by Gisin and Percival [134] and is formally equivalent to the stochastic Schrödinger equation of Ref. 115, a fact reinforcing our conviction that it is very difficult, if not impossible, to distinguish subjective from objective wave-function collapses.

In conclusion, the condition of ordinary statistical physics makes the decoherence theory a valuable perspective, as well as an attractive way of deriving classical from quantum physics. The argument that the Markov approximation itself is subtly related to introducing ingredients that are foreign to quantum mechanics [23] cannot convince the advocates of decoherence theory to abandon the certainties of quantum theory for the uncertainties for a search for a new physics. The only possible way of converting a philosophical debate into a scientific issue, as suggested by the results that we have concisely reviewed in this section, is to study the conditions of anomalous statistical mechanics. In the next sections we shall explore with more attention these conditions.

What about the quantum breakdown of anomalous diffusion [31,122]? It seems that later theoretical work [124] confirmed the results of Refs. 31 and 122. The discovery that the new effect is robust against environmental fluctuations [31], if taken into account by the researchers in this field, might have the effect of triggering some experimental research work in this direction. However, it is expected that the experimental observation of this effect would be thought of as the discovery of an interesting way to produce macroscopic quantum effects: a strange condition that would not be perceived as a failure of quantum mechanics. In other words, decoherence theory would leave open some special channels for the emergence of macroscopic quantum effects. In the conclusion of this review we shall make conjectures on the possibility that the mechanism of intermittent fluorescence might yield instead the crisis of quantum mechanics itself as well as of decoherence theory.

D. Trajectory and Density Entropies

Before ending this section, we want to illustrate another manifestation of the conceptual problems caused by non-Poisson intermittence, whose connection with the quantum perspective adopted in this section will be explained at the end of this subsection.

The so-called Kolmogorov–Sinai (KS) entropy [135] is a property of a time sequence of symbols and can be interpreted as the mean entropy increase per unit of time. In the case of a dynamical system the sequence of symbols is generated by a trajectory running through a phase space divided into many cells of finite size and labeled with given symbols. In this case this form of entropy can be related to the Lyapunov coefficient [136]. If the density approach is used, we make the conjecture that the spreading of the density distribution is proportional to the Lyapunov coefficient. Thus, the ordinary Gibbs entropy is expected to increase linearly in time, with a rate that turns out to be proportional to the KS entropy. This vision, advocated by Latora and Baranger [137], has attracted the attention of many researchers. We note that the perspective advocated by these authors is based on the implicit assumption that trajectories are more fundamental than densities.

This is a crucial aspect subtly related to the main goal of this review. In Section VIII we have seen that the non-Poisson condition violates the DF property. This is a prescription for constructing the correlation functions of the dichotomous fluctuation $\xi(t)$, serving the purpose of defining the corresponding diffusion process. In Section XII we have illustrated the problems associated to the derivation of these higher-order correlation functions from a density perspective. It is worth while to address this issue again.

At first sight, questioning the equivalence between trajectories and densities does not make sense, given the fact that the density equations (Liouville, Liouville-like, and Frobenius–Perron equations alike) are built up for the specific purpose of reproducing the time evolution of a bunch of trajectories. In the literature the existence of a possible conflict between densities and trajectories can be found in the work of Petrosky and Prigogine [138]. Although disconcerting, their claim makes sense. In fact, according to these authors the adoption of densities, and consequently of the Liouville approach, implies the diagonalization of matrices. These authors pointed out that there exists a sort of equivalence between Hamiltonian systems with infinitely many degrees of freedom and low-dimensional chaotic systems. In both cases the dynamic operator driving the pdf, expanded on a suitable basis set, becomes a matrix of infinite size. In both cases recourse to the method of analytical continuation is done, this being the source of irreversibility within the context of physical laws which are invariant by time reversal. Thus they establish a distinction between trajectories and densities and consider the latter more fundamental than the former. In this review, we reach the opposite conclusion that trajectories are more fundamental than densities. However, the view of Petrosky and Prigogine serves here the very useful purpose of this subsection. This is that once the choice of a density is made, then the Gibbs entropy should be evaluated using only the information provided by the corresponding density equation.

There exists another deep reason in favor of studying the Gibbs entropy, or its quantum mechanical counterpart, the von Neumann entropy [139], by means of the corresponding density equation. In quantum mechanics there are no trajectories. In quantum statistical mechanics we use the quantum Liouville equation, which drives the time evolution of the statistical density matrix. The advocates of decoherence theory interpret a contracted quantum Liouville equation in terms of stochastic trajectories. As pointed out in Section XV, any claim about the individual system observation is not an argument convincing enough for the advocates of decoherence to abandon their view. This is so because, even if the observation of quantum jumps [140] implies the occurrence of events, and can be judged to be equivalent to the experimental observation of wave-function collapses, then the advocates of decoherence adopt a statistical perspective, which is equivalent to making averages on the Gibbs ensemble, even if the individual systems of this ensemble are directly observable. The only possible way out of this ambiguous condition seems to be offered by the breakdown of the equivalence between the trajectory and the density picture [138]. It is worthwhile to explore this issue, even if this discussion will lead us to the opposite conclusion of considering trajectories more fundamental than densities. To shed light on this intriguing issue, we are convinced that if a density perspective is adopted, then use of the corresponding equation of motion, and only of that, has to be made. In agreement with the earlier results of this review, we find that the condition of normal diffusion does not create conflicts between the density, Gibbs entropy, and trajectory, KS entropy, perspective. The condition of anomalous diffusion, on the contrary, does it.

The first reason that led Latora and Baranger to evaluate the time evolution of the Gibbs entropy by means of a bunch of trajectories moving in a phase space divided into many small cells is the following: In the Hamiltonian case the density equation must obey the Liouville theorem, namely it is a unitary transformation, which maintains the Gibbs entropy constant. However, this difficulty can be bypassed without abandoning the density picture. In line with the advocates of decoherence theory, we modify the density equation in such a way as to mimic the influence of external, extremely weak fluctuations [141]. It has to be pointed out that from this point of view, there is no essential difference with the case where these fluctuations correspond to a modified form of quantum mechanics [115].

There exists a second reason why Latora and Baranger have been forced to depart from the adoption of a density equation, thereby rather adopting the supposedly equivalent time evolution of a bunch of trajectories. This is due to the fact that the Lyapunov coefficients are local and might change with moving from one point of the phase space to another. It is important to stress this second reason because it is closely related to the directions which need to be followed to reveal by means of experiments the breakdown of the density perspective, and with it of quantum mechanics, in spite of the fact that so far the predictions of quantum mechanics have been found to fit very satisfactorily the experimental observation.

This form of disagreement between densities and trajectories has been discussed in detail by Bologna et al. [141]. They studied the Manneville map [40] of Eq. (84), of which Eq. (85) is an idealization. This important map, when $z = 1$ becomes identical to the Bernoulli map [142], which is characterized by the same Lyapunov coefficient, ln 2, throughout the whole definition interval $I = [0, 1]$. For any $z > 1$ the interval splits into two parts, one containing $x = 0$, laminar region, and one containing $x = 1$, chaotic region. The chaotic region is characterized by very large Lyapunov coefficients, while the laminar region is filled with Lyapunov coefficients becoming increasing smaller as x comes closer to $x = 0$. Notice that in this condition, it is impossible to make the rate of the Gibbs entropy increase become identical to the KS entropy. Notice that when

$z > 2$ the Manneville map does not admit an invariant measure anymore, nor does it admit any ergodic property. However, at $z < 2$, where the KS entropy h_{KS} is known to be finite ($h_{KS} > 0$) [143] and the system is ergodic, we are in the presence of this interesting scenario. A single trajectory attains in time the whole interval I, chaotic and laminar regions alike. This corresponds to a condition of thermodynamic equilibrium, which, in the density framework is given by the equilibrium distribution. If we adopt the density prescription, we must establish at the initial time an out-of-equilibrium condition, namely a delta function distribution located in some point of the interval I corresponding to a well-defined Lyapunov coefficient. However, as the distribution starts spreading, new regions with different Lyapunov coefficient are involved, thereby making it impossible for the initial growth of the Gibbs entropy to increase with a rate corresponding to a single Lyapunov coefficient.

E. Conclusions

The simple model of Eq. (85) is an idealization of the Manneville map. Both models have in common the property of involving an inhomogeneous distribution of Lyapunov coefficients. This makes it impossible to establish a proper connection between the KS entropy and the rate of increase of Gibbs entropy. This property has a quantum mechanical counterpart discussed in detail in the recent work of Ref. 139. This has to do with a quantum mechanical two-state system identical to the simple model studied in Section II. This means a system under the joint action of (a) an inphasing process that is the quantum coupling of intensity V between the state $|1\rangle$ and the state $|2\rangle$ and (b) the dephasing process, annihilating the off-diagonal density elements in a time τ_m. As we have seen in Section II, this produces a relaxation time $1/\lambda$ which tends to diverge for $\tau_m \to 0$, if the quantum coupling V is kept constant. We have seen that this model leads to the Pauli master equation. In Section 1.4 we have seen that the two-state version of the CTRW model yields the same statistical result, with the additional picture of a random walker jumping from one state to the other, with an exponential distribution of sojourn times. The work of Ref. 139 proves that now the rate of von Neumann entropy increase coincides with the KS entropy of the symbolic sequence that may be recorded by means of experimental observations of the quantum Zeno experiment.

This is the reason why there exists an universal conviction that quantum mechanics needs no extension. A well-known case of intermittent radiation-induced fluorescence is given by the Dehmelt experiment [44]. The experiment is done on a single ion with three energy levels, $|g\rangle$, $|e\rangle$, and $|s\rangle$. A strong laser excitation transfers the electron from the ground state $|g\rangle$ to the excited state $|e\rangle$, which emits light making the electron fall back to the ground state. With the help of a photodetector, this experiment would produce a continuous-time signal. However, there exists a second possible electron cycle, this being produced by a weaker laser excitation than that would make the electron move from the ground state $|g\rangle$ to the third state $|s\rangle$, called the shelving state by Dehmelt. The state $|s\rangle$ does not emit light. Thus, when only the weak laser excitation is present, there is no light signal. When both laser fields are on, the resulting effect is an intermittent fluorescent light. Hence the system lives in a "light-on" state for a given time τ_1, then it jumps to the "light-off" state, where it resides for a time interval τ_2, at the end of which it jumps back to the "light-on" state, and so on. This experiment has been studied by Cook and Kimble [145] with the help of a Markov master equation yielding, in agreement with the experimental results, Poisson distributions for the times of sojourn in the "light-on" and "light-off" states.

What about an hypothetical experiment yielding a non-Poisson distribution of "light-on" and "light-off" states? On the basis of the results of Section XV.E, we are led to believe that it would be impossible to make the rate of Gibbs entropy increase to coincide with the KS entropy. We are convinced that for the same reasons the results of the work of Ref. 139 could not be extended to this case, and it would be impossible, in conclusion, to establish a connection between the rate of von Neumann entropy increase and the KS of the experimental sequence of "light-on" and "ligh-off" states.

At this point, we have to make the reader aware that an anomalous condition of this kind exists, and it is afforded by the BQD phenomenon, the blinking phenomenon in semiconductor nanocrystallytes [2], an example of fluorescence intermittency to be termed complex, according to the definition of complexity given in Section I. In fact, in this case the waiting time distributions are found to fit an inverse power law for some time decades. If we turn the BQD experimental sequence into a symbolic sequence of $+$'s and $-$'s, we obtain a time series whose complexity is identical to that of the Manneville map, with $z > 2$. This implies a vanishing Lyapunov coefficient, a condition even more dramatic than the one we have earlier imagined, with a finite KS entropy corresponding to an inhomogeneous distribution of Lyapunov coefficients. It is interesting to notice that techniques of statistical analysis exist [146] to establish this condition on the basis of experimental data.

Is there a way to extend to this new condition the theoretical picture that Cook and Kimble have adopted to account for the Dehmelt experiment? A natural way of proceeding might rest on the modulation approach that we illustrate in Section XVI. We shall see that it does not work, and that recourse has to be made to the CTRW perspective.

XVI. NON-POISSON RENEWAL PROCESSES: A PROPERTY CONFLICTING WITH MODULATION THEORIES

The new field of complexity is attracting the attention of an increasing number of researchers, and it is triggering heated debates about its true meaning [147–149].

Here we adopt the simple-minded definition of complexity science, as the field of investigation of multicomponent systems characterized by noncanonical distributions. On intuitive grounds, this means that we trace back the deviation from the canonical form of equilibrium and relaxation to the breakdown of the conditions on which Boltzmann's view is based: short-range interaction, no memory, and no cooperation. Thus, the deviation from the canonical form, which implies total randomness, is a measure of the system complexity. However, this definition of complexity does not touch the delicate problem of the origin of the departure from Poisson statistics. Here we plan to illustrate a proposed approach to complexity, which we shall refer to as *modulation*. We shall show that this approach can generate a waiting time distribution identical to that of Eq. (92). This proposal is very attractive, since it shows where the physical roots to complexity may be. Thus, at first sight, one might be tempted to conclude that modulation is the final step in our illustration. We shall show that it is not so, given the fact that modulation does not produce the aging effects that have been proved to be a typical BQD feature. Thus, at least for this specific form of complex process, modulation cannot be used.

A. Modulation

We define as modulation theory an approach to a noncanonical distribution based on the modulation of processes that with no modulation would yield canonical distributions. For instance, a double-well potential under the influence of white noise yields a Poisson distribution of the time of sojourn in the two wells [150]. In the case of a symmetric double-well potential we have

$$\psi(t) = \lambda \exp(-\lambda t) \tag{278}$$

The parameter λ is determined by the Arrhenius formula

$$\lambda = k \exp\left(-\frac{Q}{k_B T}\right) \tag{279}$$

In the case when either the barrier intensity Q [150] or temperature T [151] are slowly modulated, the resulting waiting time distribution becomes a super-position of infinitely many exponentials. At least since the important work of Shlesinger and Hughes [152], and probably earlier, it is known that a superposition of infinitely many exponentially decaying functions can generate an inverse power law.

In recent times, the term superstatistics has been coined [153] to denote an approach to non-Poisson statistics, of any form, not only the Nutting (Tsallis) form, as in the original work of Beck [154]. We note that Cohen points out explicitly [153] that the time scale to change from a Poisson distribution to another must be much larger than the time scale of each Poisson process. Thus, we can qualify superstatistics as a form of modulation. Therefore, from now on we shall refer to this approach to complexity indifferently either as modulation or superstatistics.

In conclusion, according to the modulation theory we write the waiting time distribution $\psi(t)$ under the following form:

$$\psi(\tau) = \int d\lambda \Pi(\lambda)\lambda \exp(-\lambda t) \tag{280}$$

where $\Pi(\lambda)$ is the Γ distribution of order $\mu - 1$ given by

$$\Pi(\lambda) = \frac{T^{\mu-1}}{\Gamma(\mu-1)}\lambda^{\mu-2}\exp(-\lambda T) \tag{281}$$

This formula was proposed by Beck [154] and used in a later work [155].

To understand the physical consequences of modulation, we make the assumption of being able to generate time series with no computer time and computer memory limitation. Of course, this is an ideal condition, and in practice we shall have to deal with the numerical limits of the mathematical recipe that we adopt here to understand modulation. The reader might imagine that we have a box with infinitely many labelled balls. The label of any ball is a given number λ. There are many balls with the same λ, so as to fit the probability density of Eq. (281). We randomly draw the balls from the box, and after reading the label we place the ball back in the box. Of course, this procedure implies that we are working with discrete rather than continuous numbers. However, we make the assumption that it is possible to freely increase the ball number so as to come arbitrarily close to the continuous prescription of Eq. (281).

After creating the sequence $\{\lambda_j\}$, we create the sequence $\{\tau_i\}$ with the following protocol. For any number λ_j, let us imagine that we have available a box with another set of infinitely many balls. Each ball is labeled with a number τ, and in this case the distribution density is given by $\psi(\tau) = \lambda \exp(-\lambda\tau)$. We create a subsequence $\{\tau_i^{(j)}\}$ by making N_d drawing from this box, with N_d being a very large, virtually infinite, number. The sequence $\{\tau_i\}$ is defined by $\{\tau_i\} = \tau_1^{(1)}, \tau_2^{(1)}, \ldots, \tau_{N_d-1}^{(1)}, \tau_{N_d}^{(1)}, \tau_1^{(2)}, \tau_2^{(2)}, \ldots, \tau_{N_d-1}^{(2)}, \tau_{N_d}^{2}, \ldots$. Notice that for any subsequence $\{\tau_i^{(j)}\}$, the correlation function of the fluctuation ξ, $\Phi_\xi(\tau)$, is equal to $\exp(-\lambda_j\tau)$. We note also that the smaller λ the larger is the time interval corresponding to it. Consequently, for a proper definition of the effect of modulation on $\Phi_\xi(\tau)$, we have to use the statistical weight $\Pi(\lambda)/\lambda$ [155], which yields

$$\Phi_\xi(\tau) = \frac{\int d\lambda \frac{\Pi(\lambda)}{\lambda}\exp(-\lambda\tau)}{\int d\lambda \frac{\Pi(\lambda)}{\lambda}} \tag{282}$$

Thus the waiting time distribution $\psi(\tau)$, of Eq. (280), is proportional to the second time derivative of Eq. (282), this being a known consequence of renewal theory [63], as the reader can easily establish by means of Eq. (147). Hence, for $\mu > 2$, modulation and renewal yield not only the same $\psi(\tau)$, but also the same correlation function. In the next subsection, however, we shall see that modulation does not yield the renewal aging of Section XI, which was found by the authors of Ref. 98 to be a genuine property of BQD intermittent fluorescence.

B. Modulation: No Aging

Here we show that slow modulation does not yield any aging and that consequently superstatistics is not the proper approach to to the BQD complexity. Let us assume that the renewal condition applies and that Eq. (234) can be used. Let us assume that the initial condition is the flat distribution, $p(y, 0) = 1$ for any value of y from $y = 0$ to $y = 1$. We decide to start the observation process at $t = t_a > 0$. The waiting time distribution of the first sojourn times is given by

$$p(y, t_a) dy = \psi_{t_a}(t) \, dt \tag{283}$$

yielding

$$\psi_{t_a}(t) = p(y, t_a) |dt/dy| \tag{284}$$

which defines $\psi_{t_a}(t)$; on the right side, y is expressed as a function of the time necessary to get the border moving from y, this function being the inverse of Eq. (89).

The authors of Ref. 96 have proved that this approach can be used to derive the same analytical expressions as those derived by Barkai [95] in an earlier publication. According to Ref. 97, these analytical formulas can be proven to be equivalent to Eq. (231). In conclusion, Eq. (283) is a reliable way to express aging. We use this approach to prove that superstatistics yields no aging. In fact, the time evolution of $p(y, t)$ using modulation theory reads

$$\frac{d}{dt} p(y, t) = \lambda(t) \left[-\frac{d}{dy} y p(y, t) + p(1, t) \right] \tag{285}$$

This equation can be studied using the lines suggested by Kubo et al. [61], through the enlarged equation on motion

$$\frac{d}{dt} P(y, \lambda, t) = \left[-\lambda \frac{d}{dy} y P(y, \lambda, t) + P(1, \lambda, t) + R P(y, \lambda, t) \right] \tag{286}$$

The rationale for this equation has been widely discussed in the past [156]. According to the spirit of the Reduced Model Theory (RMT) [156], the system of interest is assumed to interact with a set of auxiliary variables whose pdf obey a conventional Markov equation of motion. Thus, in the case here under discussion, the auxiliary variable is one-dimensional, and it coincides with λ itself. We assume that its pdf, $\rho(\lambda, t)$, obeys the equation of motion $d\rho(\lambda, t)/dt = R\rho(\lambda, t)$. Thus, Eq. (286) corresponds to the Liouville-like equation of motion of the whole Universe, the variable of interest, y, and the variable λ. The first two terms on the right-hand side of Eq. (286) describe the motion of y under the influence of a given λ; that is, they have to do with the interaction between y and λ. The third term is responsible for the fluctuations of the variable λ.

The general solution of this equation is not straightforward except in (a) fast modulation and (b) very slow modulation. In case (a), it is easy to derive from Eq. (286) an equation of motion for the reduced distribution

$$p(y, t) = \int_0^\infty P(y, \lambda, t) \qquad (287)$$

The equation of motion for this reduced picture reads

$$\frac{d}{dt} p(y, t) = \langle \lambda \rangle_{eq} \left[-\frac{d}{dy} y p(y, t) + p(1, t) \right] \qquad (288)$$

with

$$\langle \lambda \rangle_{eq} = \int_0^\infty d\lambda \, \lambda \Pi(\lambda) \qquad (289)$$

In conclusion, due to the fast fluctuations of λ, the system of interest is driven by the mean value of this variable.

In case (b), the time evolution of

$$p(y, t) = \int_0^\infty d\lambda \, \Pi(\lambda) p_\lambda(y, t) \qquad (290)$$

is determined by the time evolution of $p_\lambda(y, t)$, which, in turn, is driven by

$$\frac{d}{dt} p_\lambda(y, t) = \lambda \left[-\frac{d}{dy} y p(y, t) + p(1, t) \right] \qquad (291)$$

Hence, because the fluctuations of λ are infinitely slow, the time evolution of the system of interest depends on the superposition of infinitely many processes, for each of which no λ-fluctuation occurs.

We see that in both case (a), Eq. (288), and case (b), Eq. (291), the flat distribution coincides with the equilibrium distribution. Consequently,

$$p(y, t_a) = p(y, 0) \tag{292}$$

Thus, using Eq. (283), we get

$$\psi_{t_a}(t) = \psi(t) \tag{293}$$

Hence, both very fast and very slow modulation yield no aging. As we have seen in Section XVI, superstatistics is equivalent to infinitely slow modulation. Thus, we can conclude that superstatistics yields no aging. On the other hand, since the BQD intermittent fluorescence is characterized by aging [98], we rule out superstatistics as a proper approach to complexity in the BQD case.

It is important to stress that this conclusion leaves open the case where the modulation time scale is in between condition (a) and condition (b). We want also to mention that there are interesting issues to solve concerning the statistical analysis of a single sequence. In the case where the sequence $\{\lambda_i\}$ is realized with an ordering prescription, for instance, with $\lambda_{i+1} \geq \lambda_i$, a form of aging might appear that should depart, however, from the renewal prediction. Here we limit ourselves to pointing out that a numerical experiment done [157] with N_d very large yields no aging, in accordance with the conclusion of this section.

XVII. AN EXHAUSTIVE PROPOSAL FOR COMPLEXITY

This section is devoted to illustrating a proposal to the BQD complexity that accounts for two fundamental requests: (a) This proposal must yield aging (b) This proposal must account for the non-Poisson nature of this renewal process and must relate it to a form of cooperation.

A.   Non-Ohmic Bath

Let us imagine that a bath of infinitely many oscillators exists with all possible frequencies. Let us imagine that there exists a prescription to generate a fluctuation $\xi(t)$, changing the uncorrelated dynamics of these oscillators into a cooperative process. The prescription we adopt is the following

$$\xi(t) = \sum_{i=1}^{N} c_i \left[x_i(0) \cos \omega_i t + \frac{v_i(0)}{\omega_i} \sin \omega_i t \right] \tag{294}$$

with $N \gg 1$. This means that we consider a large number of oscillators, with different frequencies ω_i and with random initial conditions. Thus the correlation function $\langle \xi(t_1)\xi(t_2) \rangle$ reads

$$\langle \xi(t_1)\xi(t_2) \rangle = \sum_{i=0}^{N} c_i^2 \left[\langle x_i^2(0) \rangle \sin \omega_i t_1 \sin \omega_i t_2 + \frac{\langle v_i^2(0) \rangle}{\omega_i^2} \cos \omega_i t_1 \cos \omega_i t_2 \right]$$

(295)

Let us assume that the initial conditions stem from a canonical equilibrium corresponding to temperature T and let us set the mass of the oscillators equal to 1. Then

$$\langle v_i^2(0) \rangle = \omega_i^2 \langle x_i^2 \rangle = k_B T$$

(296)

As a consequence, Eq. (295) becomes

$$\langle \xi(t_1)\xi(t_2) \rangle = \sum_{i=0}^{N} \frac{c_i^2}{\omega_i^2} \langle x_i^2(0) \rangle \cos \omega_i(t_1 - t_2)$$

(297)

Let us assume

$$c_i = k\omega_i^{\frac{\delta+1}{2}}$$

(298)

where

$$0 \le \delta \le 2$$

(299)

With this assumption the normalized correlation function $\Phi_\xi(t_1 - t_2)$ becomes

$$\Phi_\xi(t_1 - t_2) = \frac{\sum_i \omega_i^{\delta-1} \cos \omega_i(t_1 - t_2)}{\sum_i \omega_i^{\delta-1}}$$

(300)

Let us discuss the physical meaning of this formal result. The form of Eq. (300) suggests that δ is a control parameter generating cooperative and noncooperative properties. We see that the choice $\delta = 1$ corresponds to creating a collective motion that is totally insensitive to the frequencies of the bath oscillators and, consequently, yields no cooperation. The collective variable ξ becomes equivalent to white noise.

The asymptotic properties are determined by the low-frequency oscillators, with $\omega_i \to 0$. However, when we set $\delta > 1$ we make the weight of these frequencies smaller and smaller as their values become smaller and smaller. The choice $\delta < 1$, on the contrary, makes the slower oscillators more important than the faster. This assigns to the correlation function $\Phi_\xi(t)$ different time asymptotic properties. We have

$$\lim_{t \to \infty} \Phi_\xi(t) = \frac{\text{const}}{t^\delta} \qquad (301)$$

with $0 < \delta < 1$ and

$$\lim_{t \to \infty} \Phi_\xi(t) = -\frac{\text{const}}{t^\delta} \qquad (302)$$

with $1 < \delta < 2$. In accordance with the earlier remarks, with $\delta = 1$ the correlation function is the Dirac delta function: This condition yields no organization and no memory. The choice $\delta < 1$ has the effect of establishing an infinite memory. The correlation function $\Phi_\xi(t)$ in this case is not integrable. In the case $\delta > 1$, the correlation function is integrable. However, also the case $\delta > 1$, as well as $\delta < 1$, generates anomalous diffusion.

To appreciate this important fact, let us study the diffusion process generated by the fluctuation $\xi(t)$. This is a gaussian fluctuation that brings us back to gaussian case of Section VII and especially to Eq. (158). Using this equation, we write the time derivative of $\langle x^2(t) \rangle$ as follows:

$$\frac{d}{dt} \langle x^2(t) \rangle = 2D(t) \qquad (303)$$

where $D(t)$ denotes the time-dependent diffusion coefficient defined by

$$D(t) \equiv \langle \xi^2 \rangle \int_0^t dt' \Phi_\xi(t') \qquad (304)$$

We see that for $t \to \infty$, the time evolution of the second moment $\langle x^2(t) \rangle$ is determined by the properties of the Laplace transform of $\Phi_\xi(t)$,

$$\hat{\Phi}_\xi(u) \equiv \int_0^\infty \exp(-ut)\Phi_\xi(t)\, dt \qquad (305)$$

for $u \to 0$.

For this reason, it is convenient to study the Laplace transform of the correlation function of the fluctuation $\xi(t)$. The Laplace transform of $\Phi_\xi(t)$ has the following asymptotic expression for $u \to 0$:

$$\lim_{u \to 0} \hat{\Phi}_\xi(u) = c_\delta u^{\delta-1} \qquad (306)$$

Thus, for $\delta < 1$, the diffusion coefficient $D(t)$ tends to diverge for $t \to \infty$, thereby implying the superdiffusion condition, an anomalous condition. In the time asymptotic limit the case $\delta > 1$ yields a vanishing diffusion condition, this being another anomalous condition, corresponding to the anomalous case discussed in Section VI. There is in fact a connection between the correlation function corresponding to $\delta > 1$ and the correlation function of Eq. (54). In a sense, the non-Ohmic condition with $2 > \delta > 1$ is a generalization of the anomalous condition described by Eq. (54), which yields in fact a vanishing diffusion coefficient generating the Anderson localization. As discussed in Section VI, the Anderson localization is generated by the violation of the ordinary condition that we termed the Zwanzig condition. The absolute value of the negative area spanned by the negative tail of the correlation function is identical to the positive area. The equality of the two areas applies also to the case here under discussion, with the only difference that now the negative tail, rather than being exponential is an inverse power law, with index δ. In conclusion, the fact that the negative tail is integrable does not imply a normal condition. The subdiffusion case is as anomalous as the superdiffusion case.

In conclusion, the non-Ohmic picture is an attractive way to generate cooperative behavior. In Section VI we have found that this approach is a dynamic generation of FBM, yielding in the time asymptotic limit Eq. (166). Here we are recovering the same result with a negative tail given by an inverse power law rather than by an exponential function. However, the key point is the same. In both cases we get localization as a result of the balance between the negative slow tail and the fast drop at short times. We want to point out that both superdiffusion and subdiffusion are a manifestation of memory. Notice that the FBM scaling, usually denoted by the symbol H [158] , is related to the non-Ohmic parameter δ by the relation

$$H = 1 - \delta/2 \tag{307}$$

Thus, $\delta > 1$ yields subdiffusion and $\delta < 1$ superdiffusion. Both cases imply memory. The condition $\delta < 1$ implies that a strong fluctuation in the positive direction is most probably followed by a fluctuation in the same direction. The condition $\delta > 1$, on the contrary, implies that after a positive fluctuation, a fluctuation in the opposite direction ensues. Both conditions are a form of memory. In conclusion, the dynamic approach to FBM is a nice way to create memory, with the memory of the variable ξ being the signature of a coordinated and cooperative process, provided that $\delta \neq 1$.

B. Recurrences to $x = 0$

Let us now discuss a property that according to many authors relates to the recursion of a diffusing variable x to the origin. Some recent work on the random growth of surfaces [159] have raised the attention of the researchers in the field of persistence on a property discussed in the earlier work of many authors. This has to do with the recursion to the origin of a diffusional trajectory $x(t)$. If the diffusion process generated by this trajectory is characterized by the scaling H, and the distribution of recursion times is an inverse power law with index μ, then H and μ are related by the following constraint:

$$H = 2 - \mu \tag{308}$$

According to Ding and Yang [160] this important property is derived from the fractal dimension of a FBM trajectory. A more recent derivation of Eq. (308) was made by Failla et al. [36]. It is worth devoting some room to this demonstration. Let us assume that a set of trajectories $x(t)$ generates diffusion with scaling. This means [see also Eq. (166)]

$$p(x,t) = \frac{1}{t^H} F\left(\frac{x}{t^H}\right) \tag{309}$$

We notice that in the scaling regime, a trajectory $x(t)$, which at $t = 0$ is located at the origin $x = 0$, must move out of this initial condition, and it can contribute to the population at origin only through its successive re-crossings. Let us define

$$R(t) = p(0,t)\Delta_x \tag{310}$$

This is the population contained in the sharp interval $[-\Delta_x/2, \Delta_x/2]$, with Δ_x being so small as to make it legitimate to approximate $p(x,t)$ with $p(0,t)$. We can write

$$R(t) = k \sum_{i=1}^{\infty} \psi_n(t) \tag{311}$$

In the Laplace transform representation we have

$$\hat{R}(u) = k \frac{1 - \hat{\psi}(u)}{\hat{\psi}(u)} \tag{312}$$

Let us now make the assumption that $\psi(t)$ has the inverse power law form of Eq. (92), with $\mu < 2$. We know that in this case

$$\hat{\psi}(u) = 1 - cu^{\mu-1} \tag{313}$$

We thus get the following result:

$$u^{H-1} = ku^{1-\mu} \tag{314}$$

which yields Eq. (308).

We notice that probably the reason why the proposal made by some authors to derive the renewal sequence $\{\tau_i\}$ from the diffusion process of $x(t)$ was not further pursued is because the BQD phenomenon does not lead exactly to $\mu = 1.5$. We have also seen that a small deviation from $\delta = 1$ yields a jump into either the sub-Ohmic or super-Ohmic condition. This is not a problem. In fact, recent numerical work [161] proves that the recursion times obey renewal theory for any value of H.

C. Conclusion

In conclusion, we think that in this section an attractive model for BQD complexity emerges. The non-Ohmic bath creates a trajectory $x(t)$ with infinite memory. When this trajectory crosses the point $x = 0$ and for a given time remains in the positive semiplane $(x > 0)$, it activates the photon emission. Then, as a result of this diffusion process the trajectory can re-cross the point $x = 0$ again, so as to enter the negative semiplane, where it can sojourn for another amount of time. In this region the photon emission process is turned off.

One might argue that ordinary diffusion is enough to generate an anomalous coefficient μ. In fact, on the basis of Eq. (308) we know that $H = 0.5$ yields $\mu = 1.5$. However, we are inclined to reverse the statement of the authors of Ref. 162. We agree with them, on the fact that *anomalous is normal*. On the basis of the earlier remarks, showing that $H = 0.5$ is a singularity, we conclude that *normal is anomalous*.

XVIII. CONCLUDING REMARKS

The aim of this section is to discuss the main results of this review. For the sake of reader's convenience we divide this section in several subsections, one subsection for each significant result.

A. In Search of a Theory of Complexity

One of the most important results of this review has to do with the origin of complexity, which is here meant to denote statistical processes significantly departing from the canonical distributions. The generalized central limit theorem serves the purpose of establishing a stable form of diffusion, with a well defined scaling. This stable form of diffusion is thought to be a thermodynamic condition, since it emerges as a distribution of Gauss type and thus as a form of canonical equilibrium. Since the central limit theorem applies to the superposition of a very large number of uncorrelated fluctuations [163], we conclude that thermodynamic equilibrium is the macroscopic expression of microscopic randomness. Thus, the emergence of a noncanonical distribution seems to imply by necessity some form of cooperation and organization determined by the microscopic interactions.

Lévy diffusion is an attractive example of violation of the canonical prescription. From a physical point of view a Lévy walk is more attractive than a Lévy flight, thereby forcing us to focus our attention on a waiting time distribution rather than on a jump-intensity distribution. As proved by the authors of Ref. 71, the Lévy walk condition, deriving Lévy diffusion from the waiting time distribution $\psi(t)$, with $\mu > 2$, generates a multiscaling process, or, equivalently, a dynamic process that becomes mono-scaling after an infinitely long transient time. This infinitely lasting transient suggests that complexity might not have a thermodynamic significance, and might rather correspond to a condition intermediate between dynamics and thermodynamics. The authors of Ref. 117 borrowed from Buiatti and Buiatti [164] the evocative term *living state of matter* to denote this condition. Of course, actual living systems are characterized by additional important properties discussed in detail by the authors of Ref. 164. We refer the interested reader Ref. 164 for an excellent illustration of the key properties of biological systems. Here we limit ourselves to noticing that the problems met in the microscopic derivation of a fluctuation–dissipation process yielding a Lévy form of equilibrium, may be a sign that thermodynamic equilibrium is incompatible with the survival of the complexity condition, and that thermodynamic equilibrium is compatible only with canonical distributions.

The Lévy walk diffusion emerges from a preparation condition, corresponding to assigning to all the walkers the position $x = 0$ at $t = 0$. The diffusion distribution keeps changing its shape forever, thereby making it possible for us to determine the preparation time from experimental observation. This is a form of aging, closely related to the renewal aging discussed in this review. However, it is realized in the ideal case of a stationary preparation, which is possible in this case, because the Lévy walk condition requires $\mu > 2$. This form of aging requires the adoption of the renewal condition, namely, the condition that makes each waiting time totally independent of the other waiting times of the same sequence.

The renormalization group model illustrated by Zaslavsky [54] fits this renewal condition and can be safely adopted to generate a Lévy walk, and, through it, a Lévy flight, as a time asymptotic condition never realized in practice. However, the QBD physics requires $\mu < 2$. This requests sets a condition where no invariant distribution exists. In this condition, the dynamic approach to Lévy diffusion, through a Lévy walk, does not work, since the maximum possible scaling compatible with a constant velocity, is $\delta = 1$, whereas the Lévy scaling, $\delta = 1/(\mu - 1)$, would exceed the ballistic scaling. In this case, a well-defined diffusion scaling emerges if we impose the condition that the random walker does not move between two critical events [165]. This corresponds to a dynamic derivation of subdiffusion, which is now a well-established result [43]. In this condition renewal aging is unavoidable, since the stationary condition cannot be realized, and consequently the Onsager principle is not possible, not even as an ideal property.

There is no reason to believe that the renormalization group approach of Ref. 54 cannot be used in this case, even if Zaslavsky [54] focused on $\mu > 2$ and on the emerging Lévy walk. However, a correct model to account for the BQD physics is given by the model of Section XVII. This model is confined to the condition $\mu < 2$, but it offers the attractive property of explaining both renewal and memory properties. It is so because this model rests on the joint use of the diffusional variable $x(t)$ and of the fluctuating "velocity" $\xi(t) = dx/dt$. The diffusional variable $x(t)$ fluctuates between two states, $x > 0$, "light-on" state, and $x < 0$, "light-off" state. The transition from one state to the other obeys renewal theory [161], while the fluctuation $\xi(t)$ exhibits memory and is a signature of the cooperative process used to generate its memory properties [161].

The result of Ref. 161, proved by means of numerical calculations, may have an even more general significance. In fact, in the current literature on complex processes there exists some confusion between the infinite memory generated by renewal and non-Poisson processes and the memory of single trajectories. The former is a form of memory emerging from the observation of a set of walker spreading in time after an initial preparation. This form of renewal memory is discussed in Section X. The latter kind of memory is the infinite memory of the single fluctuation $\xi(t)$ created by means of the non-Ohmic model of Section XVII. The $\xi(t)$ trajectories have an infinite memory, as a consequence of the fact that the corresponding correlation function $\Phi_{\xi}(t)$ generates anomalous diffusion, without involving the action of renewal jumps. The diffusion trajectory generates "light-on", $x(t) > 0$, and "light-off", $x(t) < 0$, states with no time correlation. The signature on $x(t)$ of the original memory of $\xi(t)$ appears as a deviation of the corresponding distribution of sojourn times from the Poisson form. It is probable that whatever repartition of the x-axis is made to define different states, the transition from one to another state is characterized by renewal properties.

B. Consequences on the Search for Invisible Crucial Events

Another important result is that, in spite of yielding the same waiting time distribution and the same correlation function, renewal and modulation may produce diffusion with different scaling properties. In fact, if the modulation of Section XVI is done with the request that a given λ is used for a large but fixed time T_d rather than for a fixed number N_d of time drawings, then the correlation function of Eq. (282) must be replaced by

$$\Phi_\xi(\tau) = \frac{\int d\lambda \Pi(\lambda)\lambda \exp(-\lambda\tau)}{\int d\lambda \Pi(\lambda)} \tag{315}$$

The waiting time distribution $\psi(t)$ is obtained from this correlation function by differentiating it with respect to time only once. Thus the case $2 < \mu < 3$ would render this correlation function integrable, thereby departing from the condition of anomalous diffusion. We conclude that the experimental observation of $\psi(t)$ is not enough to establish the statistical properties of a system. The recent work of Allegrini et al. [66] shows that, quite surprisingly, critical renewal events are compatible with modulation. This depends on the way modulation is realized. If the time distance between the drawing of a given λ and the next is not Poisson, then the scaling of the resulting diffusion process is anomalous. This sets a challenge for the detection of crucial and invisible events. The statistical properties of these events can be detected with the adoption of a proper analysis technique—for instance, the method of diffusion entropy [165,167].

C. CTRW Versus GME

In Section IV we have shown that the CTRW can be thought of as being the solution of an equation of motion with the same time convoluted structure as the GME derived from a Liouville or Liouville-like approach by means of a contraction on the irrelevant degrees of freedom. However, this formal equivalence may not imply an equivalence of physical meanings. This review affords the information necessary to establish this important point. Let us consider, for instance, the diffusion process $dx/dt = \xi$, widely discussed in this chapter. From within the density method treatment, this equation has been proven to generate two possible generalized diffusion equations, which are written here again for the reader's convenience.

The first generalized diffusion equation is described by Eq. (133), which reads

$$\frac{\partial}{\partial t}p(x,t) = \langle \xi^2 \rangle \int_0^t \Phi_\xi(t-t')\frac{\partial^2}{\partial x^2}p(x,t')\,dt' \tag{316}$$

with W^2 replaced by $\langle \xi^2 \rangle$ to render in more easy fashion the comparison with the second equation. As discussed in Section VI, this equation was originally derived for the purpose of describing the diffusion process generated by the variable $\xi(t)$, when this is a dichotomous fluctuation.

The second generalized diffusion equation, Eq. (162), is

$$\frac{\partial}{\partial t}p(x,t) = \langle \xi^2 \rangle \left(\int_0^t dt' \Phi_\xi(t')\,dt' \right) \frac{\partial^2}{\partial x^2}p(x,t) \tag{317}$$

This equation rests on the assumption that $\xi(t)$ is a gaussian process.

The CTRW approach generates a third generalized diffusion equation, whose form is

$$\frac{\partial}{\partial t} p(x,t) = D \int_0^t dt' \Phi(t - t') \, dt' \frac{\partial^2}{\partial x^2} p(x,t') \qquad (318)$$

This equation is derived from Eq. (60) by setting

$$\mathbf{M} = \frac{1}{2} \sum_i (|i\rangle\langle i + 1| + |i + 1\rangle\langle i|) \qquad (319)$$

Thus we are considering the case of a random walker moving in an infinitely extended lattice, and making jumps from one site of this lattice to one of the two nearest neighbor sites with equal probability. Thanks to Eq. (68), the matrix $\mathbf{K}$ of Eq. (59) becomes

$$\mathbf{K} = -\sum_i |i\rangle\langle i| + \frac{1}{2} \sum_i (|i\rangle\langle i + 1| + |i + 1\rangle\langle i|) \qquad (320)$$

Moving from the discrete site to the continuous space representation, and replacing the distance between two nearest-neighbor sites and two consecutive jumps times, which are assumed to be equal to 1, with the earlier expressions for $\mathbf{M}$ and $\mathbf{K}$, with a and τ_u, respectively, we may prove that Eq. (59) yields Eq. (318), with

$$D = \frac{a^2}{2\tau_u} \qquad (321)$$

We can easily prove that these three equations yield the same time evolution for the second moment $\langle x^2(t) \rangle$ if the first and the second refer to the same correlation function $\Phi_\xi(t)$, and if the memory kernel $\Phi(t)$ of the third equation fulfills the condition

$$D\Phi(t) = \langle \xi^2 \rangle \Phi_\xi(t) \qquad (322)$$

This equality seems to suggest a total equivalence between CTRW, from which the third equation is derived, and the Liouville approach, based on density. We note that the first equation was originally derived from the Zwanzig GME. Therefore, all this seems to be in line with the claimed equivalence between CTRW and GME.

In the Poisson case, as discussed in this review, this equivalence is unquestionable. If we move from the Poisson case to the non-Poisson case, however, the comparison between the two perspectives becomes controversial. In the superdiffusion case, corresponding to Eq. (92) with $\mu > 2$, the first equation conflicts with the CTRW, for reasons that are explained in detail in Section IX. The search for the corresponding GME is still open [168].

Let us consider now the case $\mu < 2$. In this case the third generalized diffusion equation, derived from a CTRW, is well established [43] as a paradigmatic case of subdiffusion. The correlation function $\langle \xi(t)\xi(t')\rangle$ of the diffusion generating fluctuation $\xi(t)$ is related to the memory kernel $\Phi(t)$ by means of the following relation:

$$D\Phi(t) = \langle \xi^2(t)\rangle + \int_0^t dt' \frac{d}{dt}\langle \xi(t)\xi(t')\rangle \tag{323}$$

Note that in this case the correlation function $\langle \xi(t)\xi(t')\rangle$ is not stationary, and the mean value $\langle \xi^2(t)\rangle$ depends on time. Consequently, the memory kernel $\Phi(t)$, although formally identical to the stationary correlation function $\Phi_\xi(t)$, cannot even be interpreted as a correlation function, and its connection with a correlation function is given in fact by Eq. (323), which reveals the nonstationary nature of this property.

D. Consequences on the Physics of Blinking Quantum Dots

The results of this review show that the emergence of a non-Poisson distribution of "light on" and "light off" states cannot be derived from the modulation of a Poisson model, which, without modulation, would account for the experimental results of Dehmelt [140]. This issue is still unsettled. However, it seems to be evident that the model must be of the renewal type, as proved by Brokmann et al. [98]. More recently, other results have been found [99] confirming with surprising accuracy the renewal character of these non-Poisson processes.

An acceptable BQD model could be obtained through subordination. This model would be subordinated to the quantum Zeno model of Section II in the same way as the subdiffusion process of Eq. (318) is subordinated to the ordinary random walk model. This would produce non-Poisson distributions of sojourn times in the "light-on" and "light-off" states, and these distributions would be of renewal type [69].

Another crucial result was illustrated in Section XIV: The authors of [105] in order to properly address the evaluation of absorption and emission spectra had to go beyond the restrictions of the conventional approaches, and to rest on the notion of trajectory (random trajectory) rather than that of density, and of the corresponding Liouville-like equation. Is it possible to derive the same result by using a proper GME? This GME should take aging into account and should be derived from a Liouville or Liouville-like approach. To the best of our knowledge, there does not exist yet a derivation of this kind. We do not know yet whether this is possible. However, we are already in a position to prove that if this derivation is possible, a new perspective must be adopted anyway. In fact,

Sokolov et al. [104] have recently shown something that in our opinion is very disconcerting, namely, the response of the GME to an external perturbation does not coincide with the response to external perturbation of the corresponding CTRW.

These results mean that, once the GME coinciding with the CTRW has been built up, we cannot look at it as a fundamental law of nature. If this GME were the expression of a law of nature, it would be possible to use it to study the response to external perturbations. The linear response theory is based on this fundamental assumption and its impressive success is an indirect confirmation that ordinary quantum and statistical mechanics are indeed a fair representation of the laws of nature. But, as proved by the authors of Ref. 104, this is no longer true in the non-Poisson case discussed in this review.

E. Consequences on Quantum Measurement Processes

It seems that the conventional approach to the quantum mechanical master equation relies on the equilibrium correlation function. Thus the CTRW method used by the authors of Ref. 105, yielding time-convoluted forms of GME [96], can be made compatible with the GME derived from the adoption of the projection approach of Section III only when $\mu > 2$. The derivation of this form of GME, within the context of measurement processes, was discussed in Ref. 155. The authors of Ref. 155 studied the relaxation process of the measurement pointer itself, described by the $1/2$-spin operator Σ_z. The pointer interacts with another $1/2$-spin operator, called σ_x, through the interaction Hamiltonian

$$H_{int} = G\Sigma_z\sigma_x \qquad (324)$$

The variable σ_x has the eigenstates $|+\rangle$ and $|-\rangle$ and, as a result of the interaction with its own bath, is assumed to dwell in those two states with a nonexponential waiting time distribution. This is similar to the model of Ref. 105, where the role of the pointer is played by the dipole $d(t)$. The adoption of the ordinary approach to GME yields

$$\frac{d}{dt}\rho_\Sigma = 2\left(\frac{G^2}{\hbar^2}\right)\int_0^t dt'\Phi_\sigma(t-t')[\Sigma_z\rho_\sigma(t')\Sigma_z - \rho_\sigma(t')] \qquad (325)$$

We apply to this equation the same remarks as those adopted for the comparison among Eqs. (316)–(318). We note, first of all, that the structure of Eq. (325) is very attractive, because it implies a time convolution with a Lindblad form, thereby yielding the condition of positivity that many quantum GME violate. However, if we identify the memory kernel with the correlation function of the $1/2$-spin operator σ_x, assumed to be identical to the dichotomous fluctuation ξ studied in Section XIV, we get a reliable result only if this correlation function is exponential. In the non-Poisson case, this equation has the same weakness as the generalized diffusion equation (133). This structure is probably compatible with the modulation approach to the anomalous waiting time distribution, if modulation can ever be realized in practice without producing critical events, and with the adoption of the ordinary Liouville-like approach [155]. It is shown [60] that this structure implies that the higher-order correlation functions factorize according to the prescription $\langle 4321 \rangle = \langle 43 \rangle \langle 21 \rangle$. This factorization condition, valid in the Poisson case, is broken when the statistics of the system departs from the Poisson condition [93]. In the Poisson case the adoption of the density perspective yields results that are indistinguishable from the statistical properties generated by real jumps. In the non-Poisson case the two pictures yield totally different results. Whether or not a GME can be derived from the density approach without using implicitly or explicitly the factorization condition is still an open issue [100].

It is interesting to notice that Eq. (325) can also be derived from the Lindblad master equation using the same subordination approach as that adopted to derive Eq. (318). Here, however, the memory kernel of this master equation does not have the meaning of a correlation function.

When the correlation function of $\sigma_x(t)$ is identified with the dichotomous fluctuation $\xi(t)$ and this has the form of a nonintegrable inverse power law, the relaxation of the pointer can be evaluated with the trajectory method, and it is proven to be an exponential decay followed by oscillations whose intensity decays with an inverse power law [59].

After more than one 100 years of unquestionable successes [128], there is a general agreement that quantum mechanics affords a reliable description of the physical world. The phenomenon of quantum jumps, which can be experimentally detected, should force the physicists to extend this theory so as to turn the wave-function collapse assumption, made by the founding fathers of quantum mechanics, into a dynamical process, probably corresponding to an extremely weak random fluctuation. This dynamical process can be neglected in the absence of the enhancement effects, triggered either by the deliberate measurement act or by the fluctuation–dissipation phenomena such as Brownian motion. This enhancement process must remain within the limits of ordinary statistical physics. In this limiting case, the new theory must become identical to quantum mechanics.

We have seen that decoherence theory, according to its advocates [128], makes the wave-function collapse assumption obsolete: The environmental fluctuations are enough to destroy quantum mechanical coherence and generate statistical properties indistinguishable from those produced by genuine wave-function collapses. All this is unquestionable, and if a disagreement exists, it rests more on philosophy than on physical facts. Thus, there is apparently no need for a new theory. However, we have seen that all this implies the assumption that the environment produces white noise and that the system of interest, in the classical limit, produces ordinary diffusion. As we move from normal to anomalous diffusion, the environmental fluctuations do not have the effect of forcing the system to recover the classical properties, under the form of the corresponding anomalous diffusion. This is the symptom of a deficiency that seems to be generated by the departure from the condition of ordinary statistical mechanics. The experimental observation of quantum jumps is compatible with the Lindblad structure for a master equation [133], which, in turn, is generally regarded as being compatible with quantum mechanics, in spite of the fact that Markov approximation might conflict with the condition that the bath of the system of interest is really quantum [23]. The conflict between Anderson localization and Zwanzig free diffusion discussed in Section III is another example of a case where the conventional Markov approximation, which would be incompatible with a rigorous quantum mechanical calculation, yields an ordinary diffusion, indicating also that the assumption made to study ordinary statistical processes may imply both a departure from quantum mechanics and the tacit assumption that new rules are acting. The transition from Poisson to non-Poisson physics has the effect of making these aspects manifest.

F.   Conclusion Summary

In conclusion, the experimental observation of non-Poisson intermittent fluorescence is revealing the existence of renewal aging. This crucial effect is accounted for by using the CTRW perspective. Since the search for an exhaustive theoretical model is still open, it is not possible to make conclusive statements. However, the content of this review reveals that there are significant signs of a turning point: New theoretical tools must be adopted and the trajectory perspective must be preferred to the density one. In a recent article, Laughlin and Pines [170] argue that the main task of theoretical physics of the twenty-first century will be "to catalogue and understand emergent behavior in its many guises, including potentially life itself." We think that the problems highlighted by this review, concerning the traditional density method, are a reflection of emergening properties that must be studied with new tools. Our conclusion is that the proper theoretical tools could be offered by the CTRW perspective.

Acknowledgment

We gratefully acknowledge the Welch foundation, for financial support of this research work through grant # 70525.

References

1. U. Banin, M. Bruchez, A. P. Alivisatos, T. Ha, S. Weiss, and D.S. Chemla, *J. Chem. Phys.* **110**, 1195 (1999).

2. M. Kuno, D. P. Fromm, H. F. Hamann, A. Gallagher, and D. J. Nesbitt, *J. Chem. Phys.* **115**, 1028 (2001).

3. K. T. Shimizu, R. G. Neuhaser, C. A. Leatherdale, S. A. Empedocles, W. K. Woo, and M. G. Bawendi, *Phys. Rev. B* **63**, 205316 (2001).

4. R. G. Neuhauser, K. T. Shimizu, W. K. Woo, S. A. Empedocles, and M. G. Bawendi, *Phys. Rev. Lett.* **85**, 3301(2000).

5. M. Kuno, D. P. Fromm, H. F. Hamann, and A. Gallagher, *J. Chem. Phys.* **112**, 3117 (2000).

6. W. Pauli, in *Probleme der Modernen Physik*, Festschrift zum 60, Geburstag A. Sommerfeld, P. Debye, ed., Hirzel, Leipzig, 1928.

7. L. van Hove, *Physica* **21**, 517 (1955); **23**, 441 (1957).

8. P. Grigolini, *Quantum Irreversibility and Measurement*, World Scientific, Singapore, 1993.

9. M. Smoluchowski, *Ann. Phys.* **48**, 1103 (1915).

10. A. Joos, in *Decoherence and the Appearance of a Classical World in Quantum Theory*, D. Giulini et al., Springer, Berlin, 1999.

11. J. P. Paz and W. H. Zurek, *Phys. Rev. Lett.* **82**, 5181 (1999).

12. W. H. Zurek, *Prog. Theor. Phys.* **89**, 281 (1993).

13. B. Misra and E. C. G. Sudarshan, *J. Math. Phys.* **18**, 756 (1977).

14. K. Koshino and A. Shimizu, quant-ph/0411145.

15. R. Zwanzig, *J. Chem. Phys.* **33**, 1338 (1960).

16. R. Zwanzig, in *Lectures in Theoretical Physics*, Vol. 3, W. E. Brittin et al., eds., Interscience, New York, 1961, pp. 106–141.

17. E. W. Montroll and G. H. Weiss, *J. Math. Phys.* **6**, 167 (1965).

18. R. F. Fox, *J. Math. Phys.* **18**, 2331 (1977).

19. A. Mazza and P. Grigolini, *Phys. Lett. A* **238**, 169 (1998).

20. P. W. Anderson, *Phys. Rev.* **109**, 1492 (1958).

21. P. W. Anderson, *Rev. Mod. Phys.* **50**, 191 (1978).

22. V. M. Kenkre, in *Excitation Dynamics in Molecular Crystal and Aggregates*, Springer Tracts in Modern Physics. Vol. 94, Springer, Berlin, 1982, p. 1.

23. A. Rocco and P. Grigolini, *Phys. Lett. A* **252**, 115 (1999).

24. W. H. Zurek, *Phys. Today* **44** (10), 36 (1991).

25. T. Dittrich and R. Graham, *Europhys. Lett.* **11**, 589 (1990).

26. T. Dittrich and R. Graham, *Phys. Rev. A* **42**, 4647 (1990).

27. J. Flores, *Phys. Rev. B* **60**, 30 (1999).

28. P. Facchi, S. Pascazio, and A. Scardicchio, *Phys. Rev. Lett.* **83**, 61 (1999).

29. J. C. Flores, *Phys. Rev. B* **69**, 012201 (2004).

30. J. Gong and P. Brumer, *Phys. Rev. E* **60**, 1643 (1999).

31. L. Bonci, P. Grigolini, A. Laux, and R. Roncaglia, *Phys. Rev. A* **54**, 112 (1996).

32. J. C. Flores, *J. Phys. Condens. Matter* **1**, 8471 (1989).

33. D. H. Dunlap, H.-L. Wu, and P. Philips, *Phys. Rev. Lett.* **65**, 88 (1990).

34. M. Hilke and J. C. Flores, *Phys. Rev. B* **55**, 10625 (1996).

35. P. Allegrini, L. Bonci, P. Grigolini, and B. J. West, *Phys. Rev. B* **54**, 11899 (1996).

36. R. Failla, M. Ignaccolo, P. Grigolini, and A. Schwettmann, *Phys. Rev. E* **70** (R), 010101(2004).

37. V. M. Kenkre, E. W. Montroll, and M. F. Shlesinger, *J. Stat. Phys.* **9**, 45 (1973).

38. D, Bedeaux, K. Lakatos-Lindenberg, and K. E. Shuler, *J. Math. Phys.* **12**, 2116 (1971).

39. P. Bergé. Y. Pomeau, and C. Vidal, *Order within Chaos: Towards a Deterministic Approach to Turbulence*, John Wiley & Sons, New York, 1986.

40. P. Manneville, *J. Phys. (Paris)* **41**, 1235 (1980).

41. R. Metzler and T. F. Nonnenmacher, *Int. J. Plasticity* **19**, 941 (2003).

42. G. Zumofen and J. Klafter, *Phys. Rev. E* **47**, 851 (1993).

43. R. Metzler and J. Klafter, *Phys. Rep.* **339**, 1 (2000).

44. G. Zumofen and J. Klafter, *Physica A* **196**, 102 (1993).

45. B. V. Gnedenko and A. N. Kolmogorov, *Limit Distributions for Sums of Independent Random Variables*, Addison-Wesley, Reading, MA, 1954.

46. E. W. Montroll and M. F. Shlesinger, in *Nonequilibrium Phenomema II: From Stocastics to Hydrodynamics*, Studies in Statistical Mechanics, Vol. XI, E. W. Montroll and J. L. Lebowitz, eds., North-Holland, Amsterdam, 1984 pp. 1–121.

47. P. Lévy, *Théory de l'Addition de Variables Alétories*, Gauthier-Villars, Paris, 1937.

48. M. F. Shlesinger, G. M. Zaslavsky, and J. Klafter, *Nature* **363**, 31 (1993).

49. G. Trefán, E. Floriani, B. J. West, and P. Grigolini, *Phys. Rev. E* **50**, 2564 (1994).

50. M. Bologna, P. Grigolini, and J. Riccardi, *Phys. Rev. E* **60**, 6435 (1999).

51. M. Annunziato and P. Grigolini, *Phys. Lett. A* **269**, 31 (2000).

52. M. Buiatti, P. Grigolini, and A. Montagnini, *Phys. Rev. Lett.* **82**, 3383 (1999).

53. C. Tsallis, *J. Stat. Phys.* **52**, 479 (1988).

54. G. M. Zaslavsky, *Physics of Chaos in Hamiltonian Systems*, Imperial College Press, London, 1998.

55. M. Ignaccolo, P. Grigolini, and A. Rosa, *Phys. Rev. E* **64**, 026210 (2001).

56. It is worth remarking that the parameter b of Ref. 50 differs from the correct value of Eq. (117) due to the missing factor of $1/2$ in the expression (112) and to the adoption of T rather than $< t > = T/(\mu - 1)$.

57. L. M. Gelfand and G. E. Saletan, *Generalized Functions: Properties and Operations*, Vol. I, Academic Press, New York, 1964.

58. V. Seshadri and B. J. West, *Proc. Natl. Acad. Sci. USA* **79**, 45-1 (1982).

59. P. Allegrini, P. Grigolini, and B. J. West, *Phys. Rev. E* **54**, 4760 (1996).

60. M. Bologna, P. Grigolini, and B. J. West, *Chem. Phys.* **284**, 115 (2002).

61. R. Kubo, M. Toda, and H. Hashitsume, *Statistical Physics II. Nonequilibrium Statistical Mechanics*, Springer-Verlag, Berlin, 1985.

62. A. Fuliński, *Phys. Rev. E* **52**, 4523 (1995).

63. T. Geisel, J. Nierwetberg, and A. Zacherl, *Phys. Rev. Lett.* **54**, 616 (1985).

64. R. Metzler and T. F. Nonnenmacher, *Phys. Rev. E* **57**, 6409 (1998).

65. P. Grigolini, A. Rocco, and B. J. West, *Phys. Rev.* **59**, 2603 (1999).

66. A. A. Stanislavsky, *Phys. Rev. E* **61**, 4752 (2000).

67. P. Grigolini, *Phys. Lett. A* **119**, 157 (1986).

68. M. Rahman, *Phys. Rev. E* **52**, 2486 (1995).

69. B. B. Mandelbrot, *The Fractal Geometry of Nature*, W. H. Freeman, New York, 1982.

70. G. Zumofen and J. Klafter, *Phys. Rev. E* **47**, 851 (1993).

71. P. Allegrini, J. Bellazzini, G. Bramanti, M. Ignaccolo, P. Grigolini, and J. Yang, *Phys. Rev.* **66**, 015101 (R) (2002).

72. W. Feller, *Trans. Am. Math. Soc.* **67**, 98, 1949.

73. J. Klafter and G. Zumofen, *Physica A* **196**, 102 (1993).

74. R. Mannella, B. J. West, and P. Grigolini, *Fractals* **2**, 81 (1994).

75. J. P. Bouchaud and A. Georges, *Phys. Rep.* **195**, 127 (1990).

76. E. Barkai and V. N. Fleurov, *Phys. Rev. E* **58**, 1296 (1998).

77. E. Barkai and J. Klafter, *Phys. Rev. Lett.* **79**, 2245 (1997).

78. E. Barkai and J. Klafter, *Phys. Rev. E* **57**, 5237 (1998).

79. E. Floriani, Gy. Trefán, P. Grigolini, and B. J. West, *Phys. Lett. A* **218**, 35 (1996).

80. M. Bologna, M. Campisi. and P. Grigolini, *Physica A* **305**, 89 (2002).

81. N. Wu, *The Maximum Entropy Method*, Springer, Heidelberg,1997.

82. B. J. West, M. Bologna, and P. Grigolini, *Physics of Fractal Operators*, Springer-Verlag, New York, 2003.

83. E. W. Montroll and F. Shlesinger, in *Nonequilibrium Phenomena II: From Stochastics to Hydrodynamics*, J. L. Lebowitz, E. W. Montroll, eds., North-Holland, Amsterdam, 1984.

84. M. Bianucci, R. Mannella, B. J. West, and P. Grigolini, *Phys. Rev.* **51**, 3002 (1995).

85. A. B. Adib, *J. Stat. Phys.* **117**, 581 (2004).

86. B. J. West and V. Seshadri, *Physica A* **196**, 203 (1982).

87. M. Annunziato, P. Grigolini, and J. Riccardi, *Phys. Rev. E* **61**, 4801 (2000).

88. S. Faetti, L. Fronzoni, and P. Grigolini, *Phys. Rev. A* **32**, 1150 (1985).

89. M. Annunziato, P. Grigolini, and B. J. West, *Phys. Rev. E* **64**, 011107 (2001).

90. M. O. Cáceres, *Phys. Rev. E* **67**, 016102 (2003).

91. N. G. van Kampen, *Stochastic Processes in Physics and Chemistry*, 2nd ed., North-Holland, Amsterdam, 1992.

92. R. C. Bourret, U. Frisch, and A. Puquet, *Physica (Amsterdam)* **65**, 303 (1973).

93. P. Allegrini, P. Grigolini, L. Palatella, and B. J. West, *Phys. Rev. E* **70**, 046118 (2004).

94. D. R. Cox, *Renewal Theory*, Chapman and Hall, London, 1962.

95. E. Barkai, *Phys. Rev. Lett.* **90**, 104101(2003).

96. P. Allegrini, G. Aquino, P. Grigolini, L. Palatella, and A. Rosa, *Phys. Rev. E* **68**, 056123 (2003).

97. G. Aquino, M. Bologna, P. Grigolini, and B. J. West, *Phys. Rev. E* **70**, 036105 (2004).

98. X. Brokmann, J-P. Hermier, G. Messin, P. Desbiolles, J. -P. Bouchaud, and M. Dahan, *Phys. Rev. Lett.* **90**, 120601 (2003).

99. S. Bianco, P. Grigolini, and P. Paradisi, in preparation. *J. Chem. Phys.* **123**, 174704 (2005).

100. P. Allegrini, P. Grigolini, L. Palatella, A. Rosa, and B. J. West, *Physica A* **347**, 268 (2005).

101. L. Onsager, *Phys. Rev.* **38**, 2265 (1931); **37**, 405 (1931).

102. P. Allegrini, G. Aquino, P. Grigolini, L. Palatella, A. Rosa, and B. J. West, *Phys. Rev. E* **71** 066109 (2005).

103. G. Godréche and J. M. Luck, *J. Stat. Phys.* **104**, 489 (2001).

104. I. M. Sokolov, A. Blumen, and J. Klafter, *Europhys. Lett.* **56**, 175 (2001).

105. G. Aquino, L. Palatella, and P. Grigolini, *Phys. Rev. Lett.* **93**, 050601 (2004).

106. R. Kubo, *J. Phys. Soc. Jpn.* **9**, 935 (1954); *Adv. Chem. Phys.* **15**, 101 (1969); P. W. Anderson, *J. Phys. Soc. Jpn* **9**, 316 (1954).

107. Y. J. Jung, E. Barkai, and R. J. Silbey, *Chem. Phys.* **284**, 181 (2002).

108. P. Grigolini, *Anomalous Diffusion, Spontaneous Localization and the Correspondence Principle*, Lecture Notes in Physics, Vol. 457, Springer-Verlag, Berlin, 1995, p. 101.

109. S. L. Adler, *Stud. Hist. Philos. Mod. Phys.*, **34**, 135 (2003).

110. L. Bonci, R. Roncaglia, B. J. West, and P. Grigolini, *Phys. Rev. Lett.* **67**, 2593 (1991).

111. V. Capek and I. Barvík, *Physica A* **294**, 388 (2001).

112. L. Bonci, P. Grigolini, G. Morabito, L. Tessieri, and D. Vitali, *Phys. Lett. A* **209**, 129 (1995).

113. D. Vitali, L. Tessieri, and P. Grigolini, *Phys. Rev. A* **50**, 967 (1994).

114. L. Tessieri, D. Vitali, and P. Grigolini, *Phys. Rev. A* **51**, 4404 (1995).

115. G. C. Ghirardi, P. Pearle, and A. Rimini, *Phys. Rev. A* **42**, 78 (1990).

116. J. P. Paz, S. Habib, and W. H. Zurek, *Phys. Rev. D* **47**, 488 (1993).

117. P. Allegrini, M. Giuntoli, P. Grigolini, and B. J. West, Chaos, *Solitons and Fractals* **20**, 11 (2004).

118. F. M. Izrailev, *Phys. Rep.* **196**, 299 (1990).

119. F. G. Casati, B. V. Chirikov, F. M. Izrailev, and J. Ford, in *Stochastic Behavior in Classical and Quantum Hamiltonian Systems*, Lecture Notes in Physics, Vol. 93, F. G. Casati and J. Ford, eds., Berlin, Springer-Verlag, 1979.

120. W. H. Zurek and J. P. Paz, *Phys. Rev. Lett.* **72**, 2508 (1994).

121. K. Shiokawa and B. L. Hu, *Phys. Rev. E* **52**, 2497 (1995).

122. R. Roncaglia, L. Bonci, B. J. West, and P. Grigolini, *Phys. Rev. E* **51**, 5524 (1995).

123. M. Stefancich, P. Allegrini, L. Bonci, P. Grigolini, and B. J. West, *Phys. Rev. E* **58**, 6625 (1998).

124. A. Iomin, S. Fishman, and G. M. Zaslavsky, *Phys. Rev. E* **65**, 036215 (2002).

125. R. Bettin, R. Mannella, B. J. West, and P. Grigolini, *Phys. Rev. E* **51**, 212 (1995).

126. E. Floriani, R. Mannella, and P. Grigolini, *Phys. Rev. E* **52**, 5910 (1995).

127. P. Grigolini, *Quantum Mechanical Irreversibility and Measurement*, World Scientific, Singapore, 1994.

128. M. Tegmark and J. A. Wheeler, *Sci. Am.* **284**, 68 (2001).

129. H. P. Breuer, *Phys. Rev. A* **68**, 032105 (2003).

130. J. Carusotto and Y. Castin, *Phys. Rev. Lett.* **90**, 030401 (2003).

131. L. Tessieri, J. Wilkie, and M. Cetinbas, arXiv: quant-ph/0406069.

132. V. Giovannetti, P. Grigolini, G. Tesi, and D. Vitali, *Phys. Lett. A* **224**, 31 (1996).

133. G. Lindblad, *Commun. Math. Phys.* **48**, 119 (1976).

134. N. Gisin and I. C. Percival, *J. Phys. A: Math. Gen.* **25**, 5677 (1992).

135. A. N. Kolmogorov, *Doklady Academii Nauk SSSR* **119**, 861 (1958); Y. G. Sinai, *Doklady Akademii Nauk SSSR* **124**, 768 (1959).

136. Y. B. Pesin, *Akademiya Nauk SSSR i Moskovskoe Matematicheskoe Obshchestvo. Uspeckhi Matematicheskikh Nauk* **32**, 55 (1977).

137. V. Latora and M. Baranger, *Phys. Rev. Lett.* **82**, 520 (1999).

138. T. Petrosky and I. Prigogine, Chaos, Solitons and Fractals **11**, 373 (2000).

139. P. Grigolini, M. G. Pala, and L. Palatella, *Phys. Lett. A* **285**, 49 (2001).

140. H. Dehmelt, *Rev. Mod. Phys.* **62**, 525 (1990).

141. M. Bologna, P. Grigolini, M. Karagiorgis, and A. Rosa, *Phys. Rev. E* **64**, 016223 (2001).

142. D. J. Driebe, *Fully Chaotic Maps and Broken Time Symmetry*, Kluwer Academic Publishers, Dordrecht, 1999.

143. P. Gaspard and X. J. Wang, *Proc. Natl. Acad. Sci. USA* **85**, 4591 (1988).

144. H. Dehmelt, *Bull. Am. Phys. Soc.* **20**, 60 (1975).

145. R. J. Cook and H. J. Kimble, *Phys. Rev. Lett.* **54**, 1023 (1985).

146. P. Allegrini, V. Benci, P. Grigolini, P. Hamilton, M. Ignaccolo, G. Menconi, L. Palatella, G. Raffaelli, N. Scafetta, M. Virgilio, and J. Yang, *Chaos, Solitons and Fractals* **15**, 517 (2003).

147. S. M. Mason, *Geoforum* **32**, 405 (2001).

148. F. Reitsma, *Geoforum* **34**, 13 (2003).

149. D. C. Mikulecky, *Compu. Chem.* **25**, 341 (2001).

150. P. Allegrini, P. Grigolini, and A. Rocco, *Phys. Lett. A* **233**, 309 (1997).

151. M. Compiani, T. Fonseca, P. Grigolini, and R. Serra, *Chem. Phys. Lett.* **114**, 503 (1985).

152. M. F. Shlesinger and B. D. Hughes, *Physica A* **109**, 597 (1981).

153. E. G. D. Cohen, *Physica D* **193**, 35 (2004).

154. C. Beck, *Phys. Rev. Lett.* **87** (2001).

155. M. Bologna, P. Grigolini, M. Pala, and L. Palatella, Chaos, *Solitons and Fractals* **17**, 601 (2003).

156. P. Grigolini, *Chem. Phys.* **38**, 389 (1979).

157. F. Barbi, P. Grigolini, and P. Paradisi, work in progress.

158. J. Feder, *Fractals*, Plenum Press, New York, 1988.

159. J. Merikoski, J. Maunuksela, M. Myllis, and J. Timoken, *Phys. Rev. Lett.* **90**, 024501 (2003).

160. M. Ding and W. Yang, *Phys. Rev. E* **52**, 207 (1995).

161. R. Cakir, P. Grigolini, and A. Krochin, in preparation.

162. A. M. Lacasta, J. M. Sancho, A. H. Romero, I. M. Sokolov, and K. Lindenberg, *Phys. Rev. E* **70**, 051104 (2004).

163. A. I. Khinchin, *Mathematical Foundations of Statistical Mechanics*, Dover Publications, New York, 1949.

164. M. Buiatti and M. Buiatti, *Chaos, Solitons and Fractals* **20**, 55 (2004).

165. P. Grigolini, L. Palatella, and G. Raffaelli, *Fractals* **9**, 439 (2001).

166. P. Allegrini, F. Barbi, and P. Grigolini, *Phys. Rev. E.*, submitted.

167. N. Scafetta and P. Grigolini, *Phys. Rev. E* **66**, 036130 (2002).

168. I. M. Sokolov and R. Metzler, *Phys. Rev. E* **67**, 010101 (2003).

169. F. Girardi and P. Grigolini, *Phys. Rev. A.*, submitted.

170. R. B. Laughlin and D. Pines, *Proc. Natl. Acad. Sci. USA* **97**, 28 (2000).

AUTHOR INDEX

Numbers in parentheses are reference numbers and indicate that the author's work is referred to although his name is not mentioned in the text. Numbers in *italic* show the pages on which the complete references are listed. Letter in **boldface** indicates the volume.

Abbes, K., **B:**556(286), *590*
Abel, J., **B:**174–175(94), *280*
Abetz, V., **A:**11(42), *118*, **B:**569(331), *591*
Abou, B., **A:**259–260(12), 303(12,56), 307(12,56), 315(12), 317(12), 319(12), *321,323*
Abragam, A., **A:**150(81), 154(81), *246*
Abraham, R. J., **B:**54(69), *90*
Abramowitz, M., **B:**316(51), 319(51), 323(51), 332(51), 334(51), 341(51), 367–369(51), 381–382(51), 385(51), 389–390(51), 403(51), 421(51), 414–426(51), *436*
Adachi, K., **B:**572(341), *592*
Adachi, K. B., **B:**601(15), *631*
Adam, G., **B:**499(8), *582*
Adam, M., **B:**518(158), 547(158), *586*
Adams, G., **A:**13(63), 94(63), 103(63), *118*, 156(100), *246*
Adib, A. B., **A:**410(85), *472*
Adichtchev, S., **A:**52(203), 62(203), *122*, 164(136), 170(181,183), 175(136,183), 177(136), 179(136,183,230), 181–182(183), 183(136), 188–189(136,230), 191(230), 194(230), 202–203(230), 205(230), 209(230), 216(183), 222(183), 223(136,230), 230(183), 231(410), 232(183), 235(410), *247–248, 250, 255*
Adler, J., **B:**135(48), *278*
Adler, S. L., **A:**436(109), *473*
Adolf, D., **A:**11(39), *118*
Afanasyev, V., **B:**445(39), *493*
Affouard, F., **A:**173(210), *249*, **B:**548(277), *590*
Agmon, N., **A:**52(206), 112(206), *123*
Agranat, A. J., **A:**24(126), 40(177–180), 45(179), 46(257), 47–48(179), 93–94(179),

96(179), 101(179,257), 102(257), *120, 122, 124*
Aharony, A., **A:**38(152), 62(152), 65–66(152), 72(152), 80(152), *121*, **B:**96–97(1), 132–133(1), 135(1,48), 138(1), 147(1), 156(1), 160(1), *277–278*
Akagi, Y., **A:**234(414), *255*
Alba-Simionesco, C., **A:**170(184), 179(229), 182(229), 210(343), 223(229), 224(378–379), *249–250, 253–254*, **B:**518(149), 532(231–232), 553(231–232), 579(149), *586,589*
Alcoutlabi, M., **B:**566–567(317), *591*
Alder, B. J., **A:**142(57), *245*
Alegría, A., **A:**11(41), 106(263), *118, 124*, 180(234), 202(318), *250, 252*, **B:**502(50), 503(69), 518(142), 558(294), 567(322), 572(340), *584, 586, 590–592*
Alencar, A. M., **B:**4(7), 13(24), *87–88*
Alexander, S., **A:**10(24), *117*
Alexandrov, Y., **A:**6(9), 12(47), 33(149), 35(47), 56(47), 68(225), 72(47), *117–118, 121, 123*, **B:**235(204), *283*
Alexandrowicz, Z., **B:**137(56), *279*
Alig, I., **B:**575(353–354), *592*
Alivisatos, A. P., **A:**358(1), *469*
Allain, M., **B:**132–133(31), *278*
Allegrini, P., **A:**374(35), 392(59), 396(59), 401(71), 403(59,71), 404(71), 415(93), 419(93), 424(96), 425–426(100), 429(96,102), 430(96), 431(96,102), 433(96), 439(59), 440(59,71,117), 441(59), 444(123), 451(146), 452(150), 454(96), 462(117), 464(166), 466(100), 467(96), 468(59,93,100), *470–473, 476*, **B:**35(49), *89*

Fractals, Diffusion, and Relaxation in Disordered Complex Systems: A Special Volume of Advances in Chemical Physics, Volume 133, Part A, edited by William T. Coffey and Yuri P. Kalmykov. Series editor Stuart A Rice.

SUBJECT INDEX

Letter in **boldface** indicates the volume.

Accelerating state, Zwanzig projection method,
    Anderson localization, **A**:373–374
Ac conductivity, porous glasses relaxation
    response, **A**:40–41
Action potentials, fractal-based time series
    analysis, **B**:23–26
Adam-Gibbs theory
    confined system relaxation kinetics, **A**:103–104
    entropy model, vitrification of liquids,
        B:499–502
    ferroelectric crystals, liquid-like behavior,
        A:94–95
    temperature/pressure superpositioning,
        B:510–511
Admittance
    broadband dielectric spectroscopy, **A**:16–18
    time-domain spectroscopy, **A**:20–21
Aftereffect function
    dielectric relaxation
        Cole-Davidson and Havriliak-Negami
            behavior, **B**:314–316
        fractional Smoluchowski equation solution,
            B:320–325
        inverse Fourier transform calculations, **B**:424
Agglomerates of particles, fractal viscoelastic
    properties, **B**:224–235
Aging effects
    anomalous diffusion, non-Ohmic models,
        A:302–303
    noise and friction, **A**:296–297
    one- and two-time dynamics, **A**:297–303
        Mittag-Leffler relaxation, **A**:298–299
        particle coordinate and displacement,
            A:299–300
        time-dependent quantum diffusion
            coefficient, **A**:300–302

velocity correlation function, **A**:299
    quantum aging, **A**:303
blinking quantum dots, **A**:337–342
    on-off correlation function, **A**:338
    on-off mean intensity, **A**:337–338
Brownian motion
    Langevin model, **A**:279–284
        displacement response and correlation
            function, **A**:280
        fluctuation-dissipation ratio, **A**:280–281
        temperature effects, fluctuation-
            dissipation ratio, **A**:283–284
        time-dependent diffusion coefficient,
            A:281–283
        velocity correlation function, **A**:279–280
    overdamped classical motion, **A**:277–279
        displacement response and correlation
            functions, **A**:277–278
        fluctuation-dissipation ratio, **A**:278–279
    overview, **A**:276–277
    quantum motion, **A**:284–296
        displacement response
            correlation function, **A**:288–289
            time-dependent diffusion coefficient,
                A:289–291
        effective temperature determination, **A**:292
        modified fluctuation-dissipation
            theorem, **A**:291–292
        Ohmic model temperature, **A**:292–295
        time-dependent diffusion coefficient,
            A:286–291
        velocity correlation function, **A**:285–286
continuous time random walk
    applications, **A**:429–431
    formal approach, **A**:423–425
    non-Poisson processes, **A**:421–423

Fractals, Diffusion, and Relaxation in Disordered Complex Systems: A Special Volume of Advances in Chemical Physics, Volume 133, Part A, edited by William T. Coffey and Yuri P. Kalmykov. Series editor Stuart A Rice.
Copyright © 2006 John Wiley & Sons, Inc.